OLIGOMERIZATION OF CHEMICAL AND BIOLOGICAL COMPOUNDS

Edited by **Claire Lesieur**

Oligomerization of Chemical and Biological Compounds
http://dx.doi.org/10.5772/57075
Edited by Claire Lesieur

Contributors

Bozidarka Zaric, Claire Lesieur, Laurent Vuillon, Calin Jianu, Ke-Qin Zhang, Xinfang Liu, Darija Jurasin, Maja Dutour Sikiric, Giovanni Gotte, massimo libonati, Kunio Kawamura, Ajda Taler-Verčič, Eva Zerovnik, Mira Polajnar, Thomas Chenal, Marc Visseaux, Dai Mizuno, Masahiro Kawahara, Csaba Fehér, Jeno Hancsok, Eszter Krivan, Rita Skoda-Foldes, Zoltán Eller, Giovanni Feverati, Guy Cousineau

Published by InTech

Janeza Trdine 9, 51000 Rijeka, Croatia

Publishing Process Manager Iva Lipovic
Technical Editor InTech DTP team
Cover InTech Design team

First published June, 2014
Printed in Croatia

Additional hard copies can be obtained from orders@intechopen.com

Oligomerization of Chemical and Biological Compounds, Edited by Claire Lesieur
p. cm.
ISBN 978-953-51-1617-2

For EU product safety concerns:
IN TECH d.o.o., Prolaz Marije Krucifikse Kozulić 3, 51000 Rijeka, Croatia,
info@intechopen.com or visit our website at intechopen.com.

Meet the editor

After getting a master's degree in physics and chemistry, Dr. Claire Lesieur switched to biochemistry and biophysics for her PhD. She worked on the pore-forming toxin aerolysin, her favorite example of protein fold plasticity: it starts as a soluble monomer and ends as a heptameric pore. Dr. Lesieur did a post doc on the oligomerization of the cholera toxin B into pentamers aimed at isolating assembly intermediates. The difficulty of producing such intermediates experimentally drove her to explore computational approaches. The idea was to identify the amino acids determinant for the oligomerization and to investigate how they conduct subunit association. Ultimately, one may hope that the molecular detail of the process will help in designing better inhibitors against pathological oligomers.

Contents

Preface

The book is on the oligomerization of chemical and biological compounds. Oligomers are built on the association of several copies of a unit. The units that compose a chemical oligomer have a broad range of chemical compositions. The units that compose biological oligomers are more restricted. There are polymers of amino acid units, called proteins and polymers of nucleotides called DNA and RNA. They are also amphipathic polymers called lipids which are made of a polar head and polymers of carbon atoms, commonly C14 to C24. Those are covalent biological polymers, the units are covalently bound to one another. Then there are protein oligomers whose unit is a protein chain. In both chemical and biological oligomers, there are covalent and non-covalent oligomers as described in several chapters (e.g. chapters 5 and 8). The book is divided in three sections.

The section 1 includes six chapters to highlight the grounds shared by the two types of oligomers. The chapters 1 to 3 explore the use of chemical and biological oligomers as innovative materials. The first and the second chapters deal with the synthesis of olefin oligomers. They expose the challenges of synthetizing such chemical molecules with particular properties and their application in the industrial world, in particular in the oil industry. Thus, the chapters present technical challenges as well as applications. The first chapter by Thomas Chenal and Marc Visseaux describes Coordinative Chain Transfer Polymerization (CCTP) and how this method allows the production of functionalized end-capped olefin oligomers. The second chapter by Csaba Fehér and co-authors is on the use of ionic liquids as catalyst in the oligomerization of alkenes/olefins. The authors recall how olefin oligomers are necessary in the oil industry, present an overview of the different methods of synthesis, the different oligomeric products and finally focus on the promising use of ionic liquid as an efficient catalyst for producing olefin compounds. The use of ionic liquid, also called molten salts, is an alternative to two-phase catalysis. The detail understanding of the mechanism of the olefins oligomerization illustrates the gap of knowledge between chemical and biological oligomerization. In chemical oligomerization, it is already possible, although still very challenging, to create tailor-made molecules with expected/directed properties. In contrast, in biological oligomerization, the research still mainly focuses on studying the biological oligomers. The level of understanding required to produce biological compounds in a design manner is not yet reached. Nevertheless, synthetic biology is a booming field and the gap is certainly closing, the future is opened to nano-biomaterial. The third chapter by Xinfang Liu and Ke-Qin Zhang on silkworm and spider fibers is a perfect example of the recent progress towards that direction. The authors have produced a thorough review, giving information from the basic structural elements of the fibers to their incredible mechanical properties. Finally, the authors discuss their use as biomaterial in the textile industry.

The chapters 4 to 6 address more fundamental mechanistic problem and illustrate that chemical and biological oligomerization share basic prerequisites. The definition of an oligomer is the association of several copies of a "unit" to achieve complex and diverse (novel) conformations. The "unit" may change enormously in composition, the formation of an oligomer always entails a remarkable cut in the chemical/building cost because a unique chemical "piece" is used in a combinatory manner to produce distinct tridimensional local arrangements and yield a broad variety of global conformations. To build equivalent conformations with variable "pieces" would be chemically more expensive, or more costly in terms of coding for biological oligomers. How little local change produces a large variety of conformations and functions is well illustrated in the fourth chapter by Jianu Calin. The author describes oligomeric derivatives of the ethylene oxide molecule R1(OCH2CH2O)R2 achieved by simply changing R1 and R2 and the environment. In the chapter five by D. Jurašin and M. Dutour Sikirić, they introduce conventional surfactants, which are amphipathic molecules made of polar and hydrophobic group. They describe how surfactants are able to transit from a population of disperse monomers to a bilayer (elongated oligomer), then to a micelle (cyclic oligomer) and up to supramolecular assembly such as liquid crystal (phase diagram). In other words, surfactant makes oligomers which grow in one direction (micelle and bilayer) and oligomers which grow in more than one direction (supramolecular assembly). Interestingly protein oligomers which grow in one direction (cyclic oligomer) and protein fibers which grow in two directions are their biological equivalent. Thus the phase diagram also stands for the transition from protein oligomer to protein fiber. The importance of the growth direction in the formation of protein fibers is explored in the chapter 13 by Laurent Vuillon and myself. Moreover, D. Jurašin and M. Dutour Sikirić discuss the novelty of introducing a covalent linkage, referred to as a spacer, between monomers and the effects of the spacer (nature, length, rigidity) on the phase diagram and the oligomeric state of the surfactant. Again, there is a proteic equivalent to such spacer, named a hinge loop which introduces sufficient fold plasticity to allow a protein to change its oligomeric state. This mechanism called domain swapping is discussed in the chapter 8 by Gotte and Libonati as well as its consequences when the function of the protein is lost upon the conformational change, leading to diseases, called conformational diseases (e.g. Alzheimer's disease). Conformational diseases and fibers are also discussed in details in the chapters 9 and 10. These analogies are interesting and hopefully the reading of the chapters on chemical oligomers would open new perspectives for investigating biological oligomers and vice-versa.

Of course, it is tempting to make another analogy between a soup of amino acids interacting via weak bonds and conventional surfactant and between covalently bound amino acids, namely proteins and covalent surfactant. The length/rigidity/nature of the "spacer" could be considered as the backbone of the protein (loop, beta strand, alpha helice) which would introduce the flexibility to produce tridimensional structures. Loops would introduce some "laxity" while beta-strands and alpha-helices some rigidity. The authors show that the covalent bond reduces the CMC (Critical Micellization Concentration) and allows the formation of supramolecular structures at lower "unit" concentration. In that sense, introducing a peptide bond between amino acids may have promoted the formation of proteins at lower amino acid concentration acting like a catalyst. Likewise the "hinge loop" in the domain swapping may allow oligomerization at lower protein concentration.

The presence of cyclic and elongated oligomers takes a different meaning when reading the chapter 6 by Kunio Kawamura on the oligomerization of nucleic acids and peptides under the primitive earth conditions. Escaping cyclic oligomerization seems a decisive step towards the formation of more complex molecules of life. This chapter makes an appropriate bridge between the chemical and the biological world and on the key stages necessary for life. It retraces the state of the art of the origin of life from chemical molecules to DNA/RNA and proteins. The author brings us on a fascinating journey through time and space.

The section 2 includes the chapters 7 to 10 and covers biological oligomerization. It starts with the chapter 7 by Bozidarka L. Zaric which describes the RNA world and Sm/LSm proteins, protein oligomers that participate to the maturation of RNA. It makes a good bridge between the chapter by Kunio Kawamura and the chapters 8 to 10 which focus on protein oligomerization. The chapter 7 evokes RNA maturation and splicing mechanism, a nice example of the complexity of biological oligomers and the combinatory aspect of such complexity. The splicesome, the cellular machinery which controls the splicing and is made of protein and RNA entities, is dissected for the readers to understand this huge biological assembly and how all the partners are orchestrated to yield one of the most important cellular activities in higher organisms. Chapter 8 by Gotte and Libonati is already mentioned. The authors provide a thorough review on protein oligomerization, illustrate the broad diversity of protein oligomers, the different technical methods to study such compounds and discuss the plasticity of certain protein to undergo a fold transition between distinct oligomeric states. The chapters 9 to 10 explain how such fold plasticity may unfortunately lead to diseases, so called conformational diseases. The chapter 9 by Dai and Kawahara describes more exclusively the proteins involved in pathological oligomerization and neurodegenerative diseases. The authors present biophysical and biochemical results as well as the molecular and cellular context. Finally, the chapter 10 by Ajda Taler-Verčič and co-authors discusses the particular case of Stefin B Oligomers/cystatin and how it is a good model of conformational diseases. Interestingly, they discuss several familial mutations known to promote the transition from a healthy cystatin B to pathological cystatin B oligomers and to lead to myoclonus epilepsy. Another example of how a local change may have great impact on the conformation of the molecule. The authors also describe some mechanisms of pathological assemblies and the role of cellular factors such as lipids in the mechanism.

The section 3 includes the chapters 11 to 14 and is intended to present results from computational and theoretical approaches to stress their value as alternative and complementary methods to experimental approaches. The chapter 11 by myself makes a bridge, discussing experimental and computational results. It also emphasizes novel advances in the understanding of protein oligomerization using graph and network theories. The chapter 12 by Giovanni Feverati focuses on the area of contact between two adjacent chains in a protein oligomer, so called the protein interface. The author proposes an algorithm to identify hot spots, amino acids determinant for the formation of a protein interface and analyzes some properties of the interfaces over a dataset of 40 protein oligomers. This chapter is a good illustration of the result obtained by a computational approach. The chapter 13 by Laurent Vuillon and myself, already mentioned, proposes a mathematical framework to fiber formation considering 2D-tiling and symmetry operations. A particular effort has been made to escort the reader through the mathematics that grasp the structural determinants required for fiber formation. The result may have implications beyond protein fibers in biological quasi-crystal/viral assembly, nevertheless these aspects are not treated. The main breakthrough is the mean of identifying on the 3D-structure, the basic local properties that make a protein oligomer more susceptible to fiber formation. Hopefully, this would open new venue to design therapeutic strategies to protect from pathological oligomerization. The book ends on the chapter 14 by Guy Cousineau where the notion of 2D-tiling is extended to other symmetries. This chapter is pure mathematics but comes as an extension of chapter 13, and we hope it will become handy to understand the construction of some biological oligomers.

The large conformational plasticity of oligomers due to global consequences arising from local features is the common notion that transpires throughout the chapters. It is the main message of the book.

Claire Lesieur
AGIM, UGA-CNRS, France

Section 1

What Chemical and Biological Oligomers Have in Common

Chapter 1

End-capped Oligomers of Ethylene, Olefins and Dienes, by means of Coordinative Chain Transfer Polymerization using Rare Earth Catalysts

Thomas Chenal and Marc Visseaux

Additional information is available at the end of the chapter

http://dx.doi.org/10.5772/58217

1. Introduction

Polymerization catalysis has seen a huge development with the progress of organometallic chemistry. Metallocenes, post-metallocenes, and constrained geometry complexes (CGC), have been used as single-site catalysts by polymerists, with the aim to elaborate polymeric materials with improved properties that can't be attained by other synthetic strategies [1]. Since the beginning of the 21st century and besides the search for new organometallic structures that could be exploited as potential catalysts, new methods and concepts have been developed, aiming at better controlling polymerization catalysis. Living (up to Immortal) Polymerization [2], Chain Walking [3] Chain Shuttling [4] have emerged. Mastering transfer reactions in polymerization catalysis, as Coordinative Chain Transfer Polymerization (CCTP) [5], has reappeared recently as a tool that would allow a better control of the whole process, and would also open the way to unprecedented macromolecular architectures. On the other hand, mastering of transfer reactions in polymerization catalysis is also of high interest because it allows the preparation of functionalized oligomers.

We will describe in this chapter some representative examples of our recent results that illustrate the possibilities offered by controlling transfer reactions in rare earth mediated oligomerization catalysis. In particular, the association of dialkylmagnesium with rare earth (RE) precatalysts afforded active combinations for the oligomerization of a series of olefins including ethylene, octene, styrene, butadiene, isoprene and β-myrcene. Chain transfer on the magnesium atom afforded long chain di(polyolefin)magnesium derivatives (see scheme 1). Characterizations and applications of these end-capped oligomers will then be largely developed.

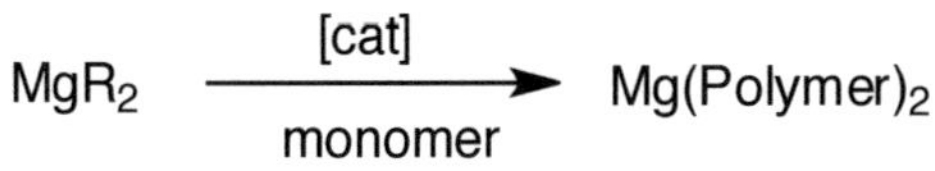

Scheme 1. End-capped oligomers starting from dialkyl magnesium

Particular emphasis will be given to the ease of implementation of the process demonstrating its usefulness as a versatile tool for the elaboration of high-added value macromolecular objects. This facility is illustrated by the possibility that we disclosed, to assess a large set of RE/MgR_2 combinations, including *in situ* synthesized catalysts as depicted on scheme 2. In particular, homoleptic, mono or bis-substituted complexes can straightforwardly be prepared in the catalytic mixture, simply starting with RE salts [6].

[cat] = Rare Earth Salt + H-Ligand + MgR_2

Scheme 2. *In situ* synthesis of catalytic species as a convenient process.

These results are rooted in the pioneering breakthrough of Pr Mortreux [7-9] and their impact is well-illustrated by subsequent exploitation by other groups [10-19] and a booming development in chain transfer agents. Moreover, in the case where the alkyl-Mg group contains a functional moiety (ie aromatic ether, amino or alkenyl), this may afford telechelic oligomers [12] (see scheme 3)

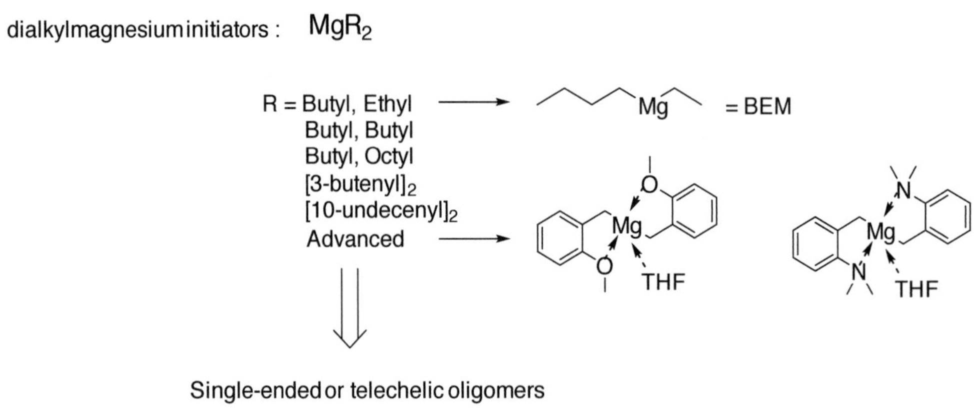

Scheme 3. Examples of dialkylmagnesium initiators

The starting compounds of the process are based on dialkyl magnesium compounds MgR_2 (see scheme 3). Solutions in hydrocarbons are commercially available, including at the industrial scale, for example: dibutylmagnesium (Aldrich, 1 M in hexane), butylethylmagnesium (Texas Alkyls, 20% in heptane) or butyloctylmagnesium (Akzo Nobel, 20% in heptane). Alternatively, more sophisticated dialkylmagnesium compounds were synthesized as sake of functionalized initiators, for example bis(ortho-dimethylaminobenzyl)magnesium(THF), bis(ortho-methoxybenzyl)magnesium(THF) [20], bis(3-butenyl)magnesium(dibutylether)$_2$ and bis(10-undecenyl)magnesium(dibutylether)$_2$ [12]. Shorter compounds i. e. the diethyl or dimethyl and even the dihydrido magnesium compounds were excluded of the field of this study due to their insolubility. Some difficulties were encountered when using the well-known RMgX Grignard reagents for two distinct reasons: first, the halide atoms clearly compete with alkyl groups on the active centers and disturb equilibria resulting to the formation of the active Cat-R species; second, ethers are the usual solvents of Grignard species while they are inhibitors of the polymerization due to high oxophilicity of the catalyst metal. Hence, specially designed initiators were conveniently obtained by conversion (Schlenk equilibrium, see scheme 4) of Grignard reagents into dialkylmagnesium compounds by means of MgX_2(dioxane) precipitation and concentration of the filtrated solution to a minimum level of ethers.

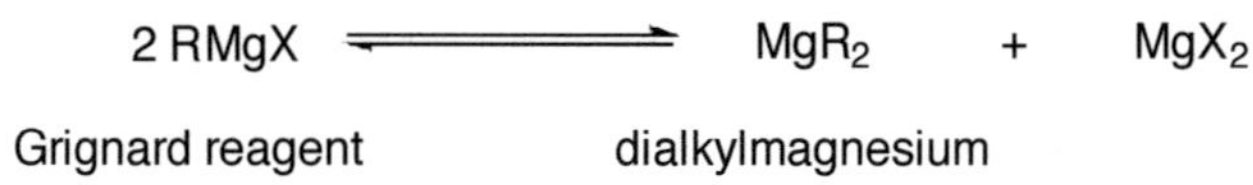

Scheme 4. Schlenk equilibrium

2. Oligomers of Ethylene

2.1. Ease of implementation

Unlike for liquid monomers, which can be handled with classical Schlenk techniques, gaseous ethylene (at standard conditions) requires particular equipment as, for example, with solution process. The convenient solvents tested were hydrocarbons, and among them linear or cyclic alkanes, aromatics, possibly halogenated. Rigorous exclusion of air, moisture and protic impurities was reached by working with a sealed reactor equipped with a vacuum line (see scheme 5). After a purge with clean gas, a simple wash with a sacrificial batch of dialkylmagnesium solution enabled the required purity level. Since ethylene polymerization is highly exothermic, the reactor was correctly thermostated. In the same way, kinetic limitations due to dissolution rates of ethylene were overtaken by powerful stirring. Ethylene inlet was regulated at a constant pressure, at atmospheric or higher pressure. A mass flowmeter and a data acquisition system recorded the exact amount of ethylene consumed all along the reaction, hence allowing deducing the average molar mass of the chains at any time.

The catalytic species for ethylene insertions in alkyl chains were obtained *in situ* by adding to the dialkylmagnesium reagent a solution of a rare earth salt containing also the necessary amount of ligand under its protio form, to obtain the pre-catalyst in solution (see next paragraph). As a typical example, two equiv of pentamethylcyclopentadiene ligand will afford the *in situ* formation of a decamethylmetallocene. Several metals were tested, for example yttrium, lanthanum, neodymium, samarium and lutetium [21-22]. Although the catalytic

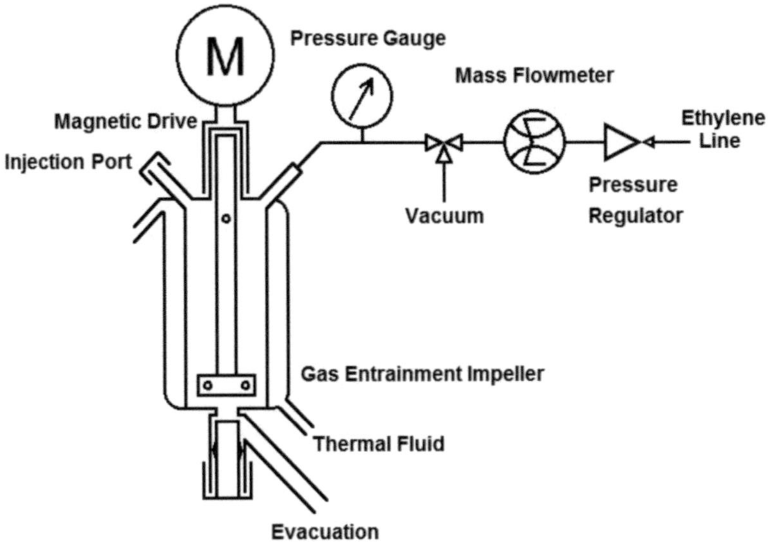

Scheme 5. Reactor equipment

activity and selectivity were clearly different depending on the rare earth nature, these differences were of low importance, as far as ethylene polymerization is concerned [9]. The neodymium species were however much more studied in the literature, probably in relation to their interesting properties for polymerization of other monomers, especially diene monomers. Different anions of the rare earth species were used, for example chloride, borohydride, alkoxides (tertiobutylates, phenates), carboxylates (versatate, 2-ethylhexanoate), phosphates and phosphonates. Once more, differences in reactivity were noticed, but the most important feature concerning the nature of the anionic ligand was the solubility of the corresponding salts. A comparative study between chloride and versatate salts of neodymium gave exactly the same results in terms of oligomers distribution for the same ethylene conversion, but the reaction started immediately with the versatate which was soluble in toluene, while it was sluggish with the insoluble chloride salt. The neodymium versatate has the other advantage to be available at an industrial scale (Solvay-Rhodia Rare Earth Systems), while at a laboratory scale, the neodymium borohydride [23] was preferred for its well-defined structure and its purity, allowing easiest spectroscopic characterizations, hence better control of the generation of catalytic centers.

Well-defined organometallic compounds bearing two pentamethylcyclopentadienyl and an alkyl group in the coordination sphere of the rare earth metal were initially established as best-suited [24]. But these catalytic species, while being very active in polymerization conditions, were found excessively prone to decomposition in the absence of monomer and impossible to store. Hence, several synthetic routes were proposed to generate the catalyst *in situ*, starting from a series of precatalysts (see scheme 6). The earliest route (A) reported in the literature was involving hydride species [21], which can be formed by hydrogenolysis of a crowded alkyl: the bis(trimethylsilyl)methyl (TMS_2CH-). A next route was developed at the turn of the

http://dx.doi.org/10.5772/58217

21st century [25-26] and consisted in the use of a crystalline bis(pentamethylcyclopentadienyl) rare earth chloride (B). The latest route, since a few years [6,27-28], proposed the direct use of a "naked" rare earth salts, as catalyst precursor, which was mixed with two equivalents of ligand precursor (pentamethylcyclopentadiene, C_5Me_5H, noted as Cp*H). When this mixture was added to a solution of dialkylmagnesium in excess, under atmospheric pressure of ethylene, it took only a few seconds to see the activity rising up at the same level as with the use of well-defined catalysts. This was interpreted in terms of alkyl/X ionic metathesis occurring between the rare earth and the magnesium atom affording the formation of rare earth alkyls. In a subsequent step, the Ln-alkyl species was assumed to react with the ligand precursor Cp*H with alkane elimination as irreversible step, to lead to the formation of a {Cp^*_2LnR} compound. Clues for this mechanism arose from an NMR study that excluded the other possible pathway: a mixture of dialkyl magnesium and two equivalents of Cp*H gave the expected complex Cp^*_2Mg very slowly compared to the rate of catalyst formation.

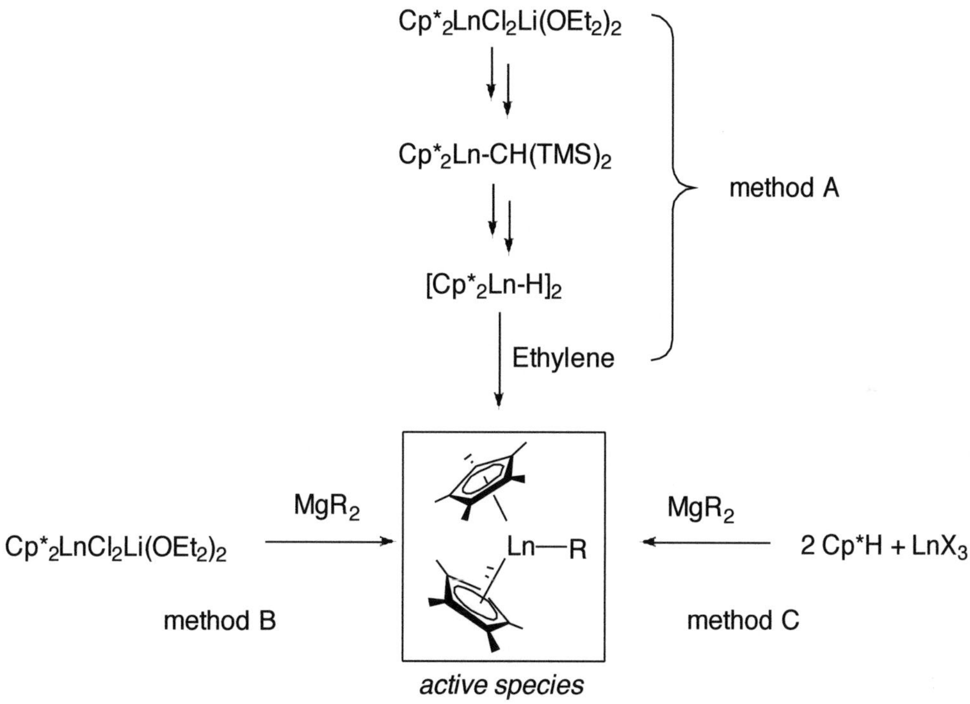

Scheme 6. Various strategies leading to the formation of the active species.

Implementation of the reaction was tested at a laboratory scale with the following details given here as an example. In a glove box, the butylethylmagnesium (BEM, 200 µmol) was weighed in a syringe and diluted in 20 mL of toluene (which was degassed by argon bubbling and stored over molecular sieves). The polymerization reactor was purged under vacuum and filled with ethylene. Injection of the BEM solution (200 µmol, 20 mL toluene) in the reactor and stirring at 90°C led to passivation of the inner walls of the reactor. Meanwhile, a new set of BEM solution was prepared in the glove box. In a 2 mL syringe, neodymium versatate ($NdVs_3$, 13.6 mg, 20.7 μmol) and pentamethylcyclopentadiene (Cp*H, 7.0 mg, 51 μmol) were weighed and diluted with 1 mL of the above BEM solution. After evacuation of the passivation solution, the remaining 19 mL BEM solution was injected in the reactor and stirred until pressure and temperature were stable. Injection of the small syringe solution gave the start of the polymerization. The monitoring of ethylene consumption allowed to measure the amount of ethylene consumed over time, and consequently the theoretical chain length of the polyethylene produced. Reaction can be stopped at a desired degree of polymerization or, if allowed to go free, it will stop spontaneously after a short burst of activity, the average molar mass being around 1000-2000 g/mol. Polymerization quench was possible with any known reactive of Grignard species, typically an electrophile, for example protic compounds produced alkanes, aldehydes or ketones gave secondary or tertiary alcohols, carbon dioxide gave carboxylic acids, iodine gave iodinated alkanes, etc. (see further functionalization section).

2.2. Mechanism of control

Preferred paradigm for understanding the present reaction came from the concept of "Aufbaureaktion" early developed by Karl Ziegler in the fifties [29]. The process under study at this time aimed at growing the linear chains of alkyl aluminum by insertion of ethylene. When implementing potential catalysts, the investigations were particular fruitful using titanium chloride and were oriented in a new field nowadays known as "Ziegler-Natta polymerization". While being originally restricted to alkyl aluminum compounds, the concept of Aufbaureaktion can be transposed to alkyl magnesium compounds at the condition that a suitable catalyst would be found. Fortunately, in our laboratory, the studies of rare earth complexes as model compounds of Ziegler-Natta catalysts revealed that dialkylmagnesium compounds behaved as efficient transfer agents during the ethylene polymerization with these rare earth complexes. The striking fact was that the transfer process was fast and reversible between magnesium and rare earth. That means that, in the presence of a large Mg/RE ratio, the alkyl chains were essentially observed on magnesium, nevertheless insertion of ethylene seemed to happen homogeneously on all these chains with the activity and selectivity of the rare earth catalyst. Each chain can be considered as continuously transferred from one to another metallic atom, with insertion of ethylene only when it was found on a lanthanide unit. Alkyl chains were observed inactive for ethylene insertion when located on magnesium, so they were named "dormant species" in this case, since they were only waiting for a transfer to lanthanide (see scheme 7).

The mechanism proposed for the observed chain growth on dialkyl magnesium fundamentally contrasts with an anionic mechanism, for example in the chain growth of alkyl lithium as observed in the relevant work of Bergbreiter [30]. In these studies, the alkyl lithium species were modified by the use of tetramethyl ethylene diamine (TMEDA) as a ligand of the lithium atom, rendering the alkyl anion sufficiently reactive to attack the ethylene monomer. Regardless the cost problem arising from the need of up to three equivalent of TMEDA ligand per produced oligomer, the high reactivity of the species and the resulting secondary reactions required to work at very low temperature for further uses.

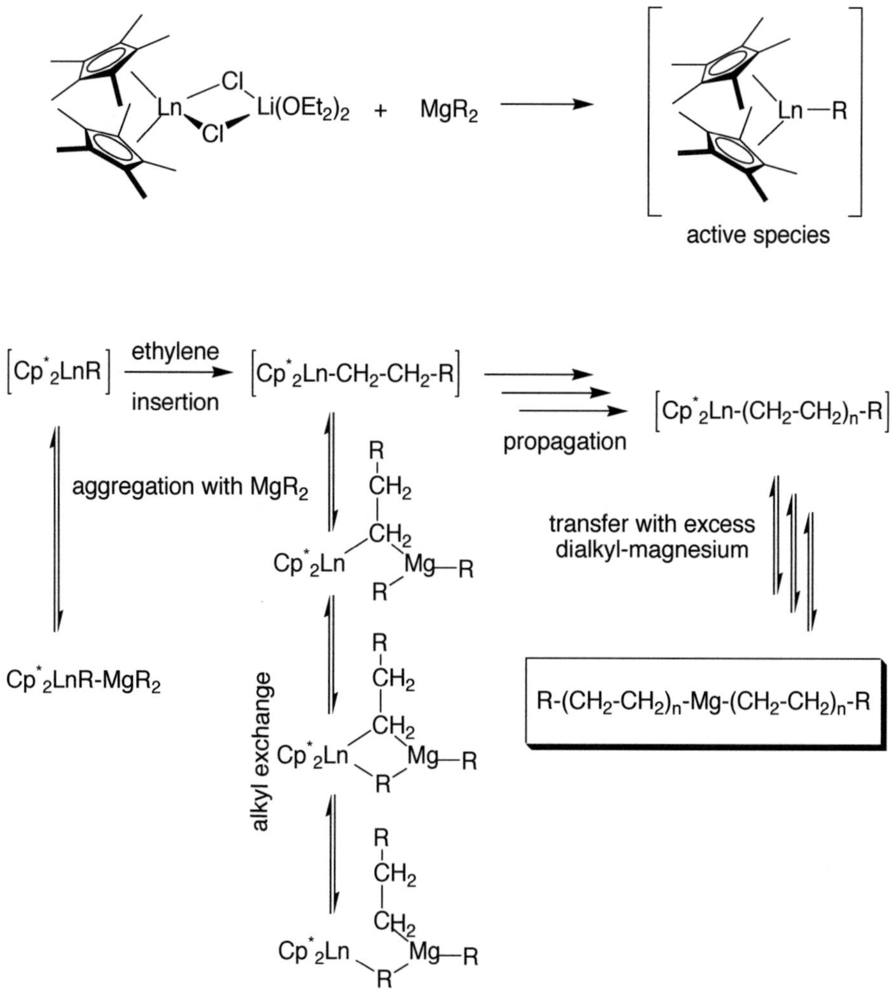

Scheme 7. Postulated mechanism for chain growth on dialkylmagnesium.

The mechanism described here clearly belongs to the "catalyzed chain transfer polymerization" concept (CCTP otherwise named coordinative chain transfer polymerization) as recently reviewed by our group [31]. CCTP is part of the new concepts developed in coordination polymerization catalysis in the recent years, also including Living Degenerative Group-Transfer Coordination Polymerization [32], Chain Walking Polymerization (CWP) [33] and Chain Shuttling Polymerization (CSP) [4]. In the area of radical polymerization, a parallel can be seen between Controlled Radical Polymerization mechanisms (Atom Transfer Radical Polymerization, ATRP, Nitroxide Mediated Polymerization, NMP or Radical Addition Fragmentation Transfer RAFT, for example [34]) and the CCTP concept, which is restricted to mechanisms by coordination polymerization. By streamlining, CCTP can be considered as a generalized concept of the original Aufbaureaktion, describing the fast and reversible transfer of chains between a large amount of inexpensive species and a suitable catalyst. Other couples of such combinations with different metals have been employed, beside the herein described dialkyl magnesium / rare earth metallocene association. For example with diethyl zinc, chain growth was observed using iron complexes with pyridinediimine ligand [5] and with zirconium complexes with salicylaldimine ligand [35]. With trialkyl aluminum compounds, yttrium complexes with pyridinamine ligand [36] and chromium complexes with Cp* ligand were also found successful [37-38]. However, the resulting oligomers in these cases were end caped with zinc and aluminum, respectively, consequently having a different reactivity than the magnesium end capped oligomers.

2.3. Chain length distribution and other characterizations

At the end of the ethylene uptake, a simple way to characterize the oligomers produced was to hydrolyze the C-Mg chains end by any protic reagent; for example by pouring the homogeneous solution of the reactor at 90°C in cold methanol. The slurry had to be well stirred to obtain a fine precipitate, while bubbling argon was necessary in order to avoid oxygen side reactions. Meanwhile, the presence of trace amounts of hydrochloric acid favored the dissolution of all metallic residues. When oligomers with a carbon number under 20 could be neglected, the precipitate was straightforwardly filtered and dried, otherwise a low amount of methanol was used in order to recover the entire mixture for further analyses.

NMR ^{1}H and ^{13}C, as well as IR analyses established the linearity of the chains; absolutely no branching was observed. The chain ends consisted of normal alkyl $-(CH_2)_n-CH_3$ with only trace amounts of vinyl ends $-CH=CH_2$. Some olefins could indeed be produced by a secondary reaction with mechanisms named either β-H elimination, or transfer to ethylene monomer. Low temperature and high magnesium concentration favored the very low amount of such defects. Typically, a temperature of 90°C and toluene solutions between 10^{-1} and 10^{-2} M represented a good compromise.

Several analytical methods were used to characterize the distribution of the chain sizes. For mixtures having rather high molar mass, the size exclusion chromatography (SEC or gel permeation chromatography, GPC) at 135°C in trichlorobenzene were found to be the best choice, an example is given in figure 1.

Gas chromatography (GC) was found best suited for oligomers having less than 100 carbons, if the apparatus was equipped with a Simdist metal column, or 60 carbons when using a usual fused silica column. Advantages of this analytical method lie in the identification and quantification of each kind of molecule with a high precision, including the separation of impurities like olefins up to 22 carbons. Use of alkane standards, commercially available for 36, 50 and even 60 carbons, ensured reliable calibration of the flame ionization detector, bypassing an obvious segregation during the evaporation step in the injection port. An example is given in figure 2 [39].

Ultimate characterization of oligomers was reached with mass spectrometry using a time of flight detector and a matrix assisted laser desorption ionization process (MALDI-TOF). However, considering that alkanes represent ones of the most difficult molecules to be ionized,

http://dx.doi.org/10.5772/58217

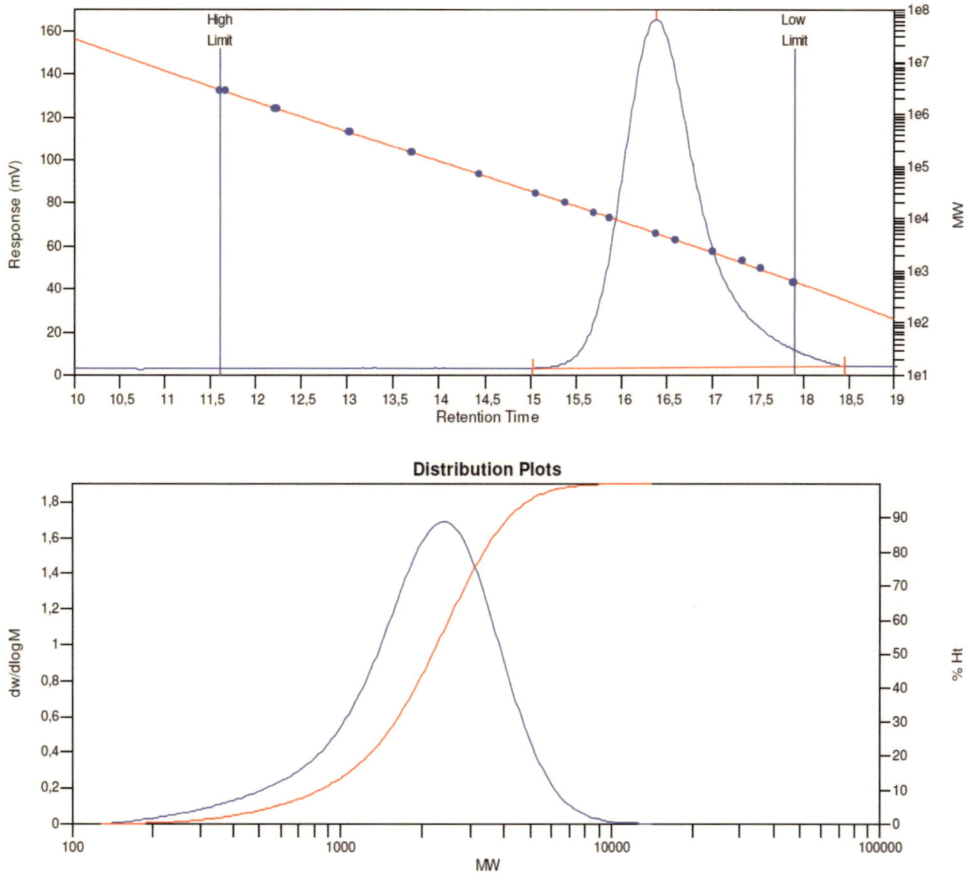

Figure 1. SEC analysis of an oligomer sample with $\bar{M}_n$=1530 g/mol and dispersity Đ=1.55

it is not surprising that we failed to analyze the sample after hydrolysis. Hence we tried to functionalize the crude oligomer mixture, when it was still constituted of alkyl chains end capped with magnesium, with a tag that was easily ionisable [40]. As tag precursor, we looked for a carbonyl compound with fused aromatic cycles and electrodonor substituents. Carbonyl compounds are known to undergo addition of organomagnesium compounds, while fused aromatic cycles with electrodonor substituents would give stable adducts with H^+. With these criteria, the cheap and nicely colored Rhodamine B base was selected. Advantage of the MALDI-TOF analytical method once again is the recognition of individual molecules, complementary to the GC method. However, quantification was distorted by segregation at the evaporation step and we are aware that only the living fraction of the oligomers mixture, once functionalized, can be detected (dead chains, for example olefins, can't be observed by this technique). The spectrum is given in figure 3, with an enlargement showing the good fit between data and a simulated distribution based on Poisson statistics and ^{13}C isotopic abundance.

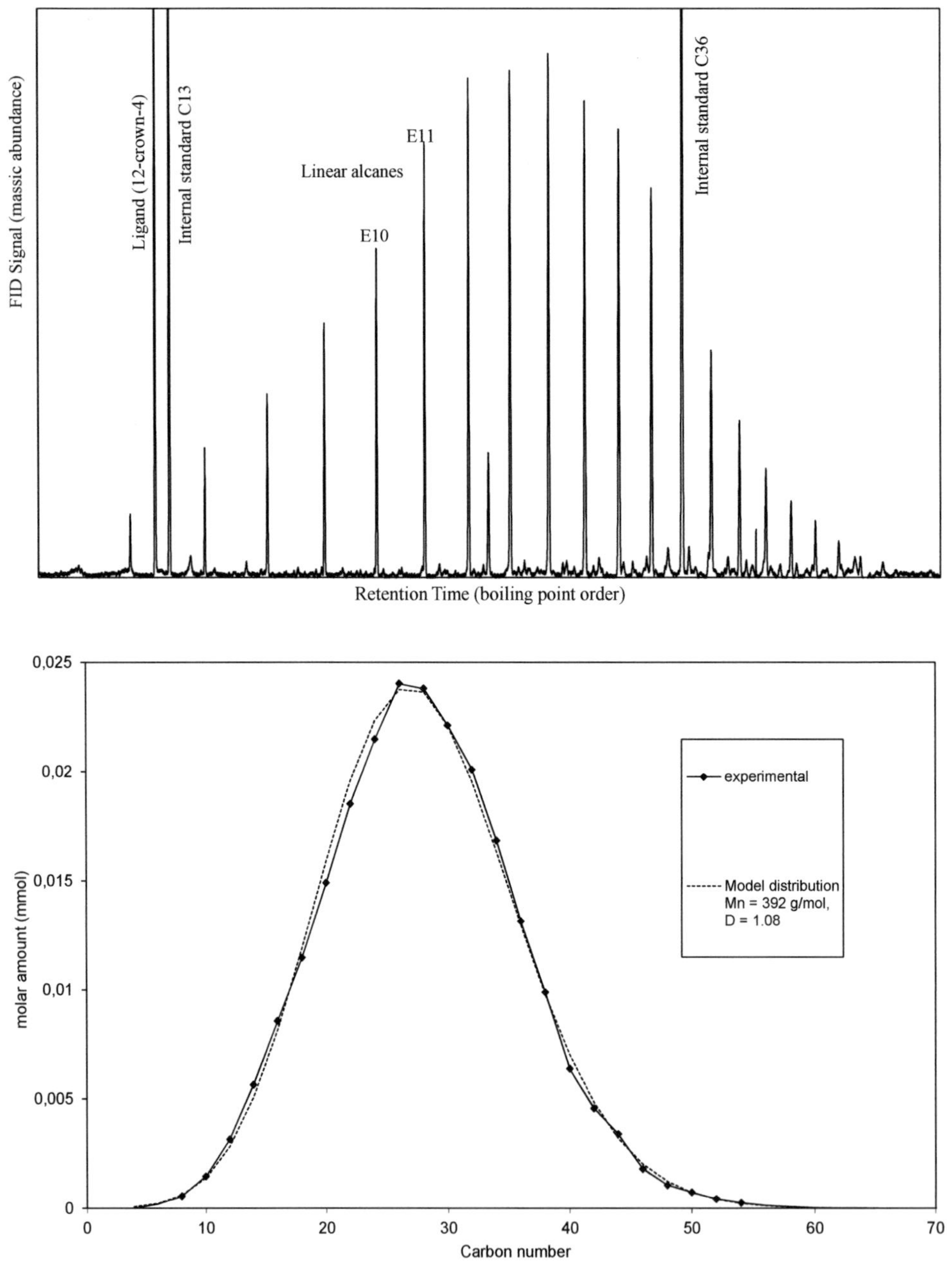

Figure 2. GC analysis of an oligomer sample with $\bar{M}$n=392 g/mol and dispersity Đ=1.08

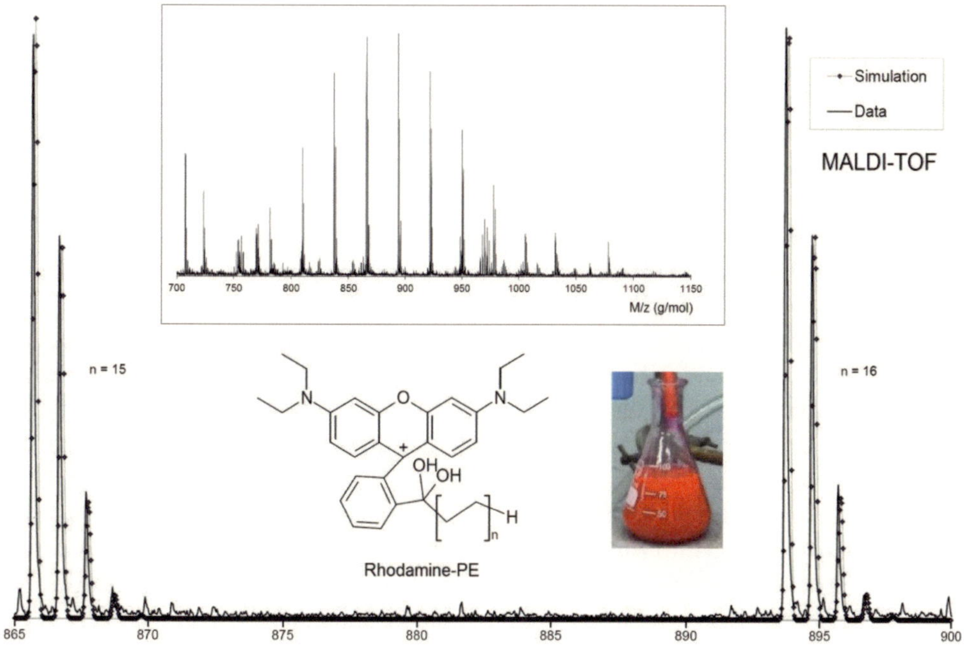

Figure 3. Mass spectrum by MALDI-TOF of an oligomer sample functionalized by Rhodamine B base.

2.4. Functionalizations

The functionalization of ethylene oligomers was demonstrated in the above paragraph by the use of Rhodamine B base. This reaction was an addition of the long chain alkyl magnesium to the carbonyl group of an aromatic lactone. As seen on figure 3, the resulting molecule was an ethylene oligomer covalently bonded to a polar dye. The red color of the dye was retained in the final material, but other physical properties, like solubility for example, were severely modified. As a matter of fact, it was easy to clean by washing with methanol all Rhodamine residues, but it was impossible to decolorize by any way a plastic container (LDPE) in which had been stored the oligomers capped with Rhodamine. A possible explanation would be that the long chains were integrated in the bulk of the container walls. The dye, which is covalently bonded to these long chains, had consequently negligible leaching rates. Applications of this material as permanent ink, non-leaching and readily dispersible in the bulk of non-polar polymers, without any treatment or additives, appeared promising.

Long chain primary alcohols were obtained by a straightforward functionalization of the oligomers with oxygen and then hydrolysis. Indeed, one had simply to follow the next step of the Aufbaureaktion in its application as an industrial process. An analogy can be seen between aluminum and magnesium as far as the reactivity of alkyl metal with oxygen is concerned.

Kinetics of the oxidation of alkyl magnesium were found fast, perhaps faster than hydrolysis, since we detected long chain alcohols as impurities of alkanes if hydrolysis of alkyl magnesium was realized in non-degassed methanol. On the contrary, when long chain alcohols were the desired product, alkane impurities were observed with a peculiar specificity: the distribution of these alkanes was higher in mean molar mass, sometimes up to twice the mean molar mass relative to the parent distribution of alcohols. A possible explanation would be to consider a coupling mechanism of two alkyls resulting either from a reductive elimination or from a Würtz-like reaction. The coupling mechanism would be induced by oxygen on unsolvated dialkyl magnesium compounds. We bypassed this difficulty by using ether additives. Best results (93% functionalization efficiency, see figure 4) were obtained using crown ether of small size (12-crown-4). An advantage of the latter compound was its inertness relative to the rare earth catalyst, allowing its addition at the beginning of the ethylene oligomerization. A this stage, the crown ether had a beneficial effect on the CCTP mechanism by impeding the aggregation of dialkyl magnesium compounds: first, the ethylene insertion kept a high rate irrespective of the magnesium concentration; second, exchange of alkyl groups were found faster, as seen with an improvement of the dispersity. For example figure 2 (Đ=1.08) was obtained with crown ether while figure 1 (Đ=1.55) was obtained without it. Noteworthy, a large range of oxygenated impurities would have more or less the same character (n donor or Lewis base), provided the stoichiometry with respect to magnesium would be between 1 and 2. This explained why it is possible to run the oligomerization process with relatively unclean materials, with good activity and dispersity, but with doubtful and irreproducible results. Figure 4 is the GC trace of a sample of long chain linear alcohols obtained in our best conditions, *i. e.* with 1 equivalent of 12-crown-4. An analysis by ^{13}C NMR spectroscopy is presented in figure 5 for a similar sample but with slightly shorter chains and after precipitation and filtration [6].

Many other functionalizations have been achieved with magnesium end-capped oligomers of ethylene which were synthesized by rare earth/magnesium based CCTP. For example carboxylic acids were obtained in our laboratory by addition of carbon dioxide [9]. Long-chain alkyl tin or alkyl silicium in a star shape were successfully tested [9,41]. An important work has also been realized by Boisson and D'Agosto who focused on the synthesis of thiol and iodine end-capped oligomers by addition of sulphur and iodine on the long-chain dialkyl magnesium, respectively [11]. The iodo compounds were then transformed into the corresponding long chain azides, which were used in click chemistry. Alternatively, the latter azides were hydrogenated into the corresponding amino end-capped oligomers. These examples are not exhaustive and the field of application should be widen since this represents only a part of the larger theme of Catalytic Chain Transfer as recently reviewed by Marks [42].

2.5. Uses as macro-initiators

A special application of magnesium end-capped oligomers was to consider them as macro-initiators and to synthesize block copolymers. A polyethylenyl block can then be grafted to a second block of different nature giving rise to a specialty polymer. Since the linear alkyl group has a lipophilic character, with good crystallization ability, the interest was to associate a

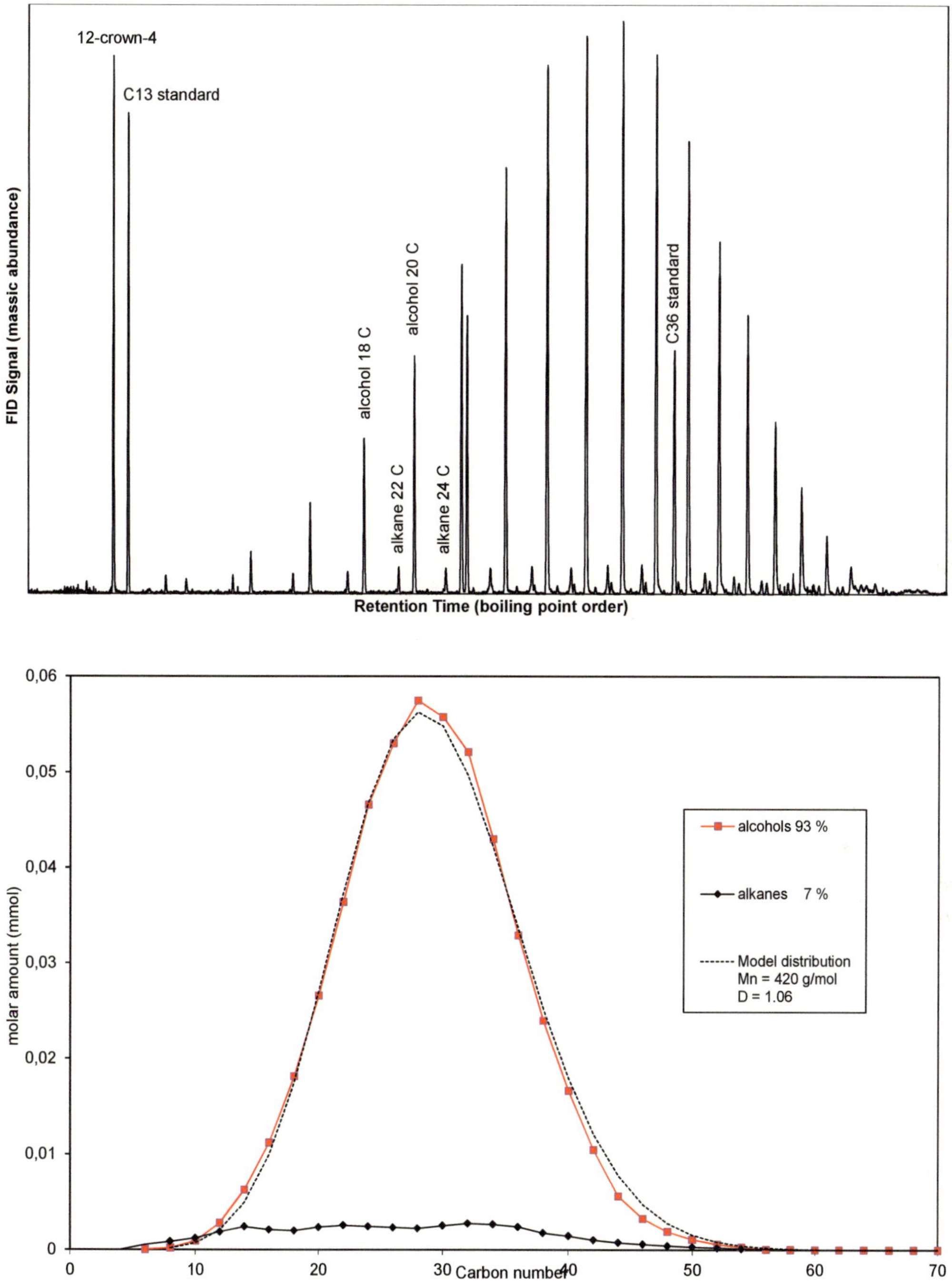

Figure 4. GC analysis of an oligomer sample functionalized to primary alcohols by oxygen.

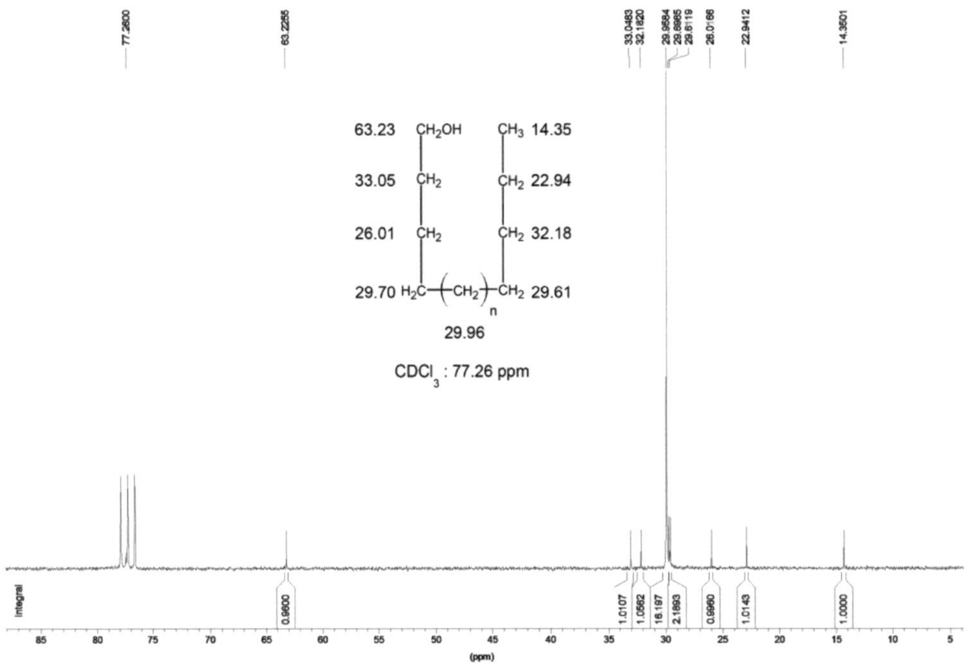

Figure 5. NMR ($CDCl_3$, 75 MHz) of an oligomer sample functionalized to primary alcohols by oxygen

second block more polar or more amorphous. From a catalytic point of view, the second step generally needs its own catalyst, which was added simultaneously to the second monomer, since different monomers require usually different catalysts. But a particularly elegant process, in terms of atom economy, should be highlighted when the magnesium atom itself was the catalytic centre of the second step. As an example of the latter case, we tested the polymerization of methyl methacrylate with magnesium end-capped macro-initiators and we obtained the corresponding PE-PMMA biblock [9]. A different situation was encountered when ε-caprolactone monomer was chosen for the second step. Indeed the traces of rare earth compounds (1 to 2 % of the magnesium amount) were found very active and compete with the magnesium activity, resulting in a polymodal distribution of PE-PCL block copolymers. A third example of block copolymerization, described recently [39], used isoprene as second monomer and needed the addition at the second step of one additional equivalent of rare earth salt per magnesium macro-initiator. The polyethylene-polyisoprene copolymers were obtained in a one-pot procedure (see scheme 8).

After several steps of functionalization as described above, the end-capped oligomers of ethylene can be designed as sophisticated macro-initiators, which were adapted to some particular polymerization mechanism. For example, RAFT or NMP mechanisms were successfully initiated with special macro-initiators based on long-chain alkyls issued from CCTP [11]. As for functionalization, the field of applications of block copolymers is largely open and new products can be synthesized purposely, depending on the precise properties required (see scheme 9).[9,11-19]

http://dx.doi.org/10.5772/58217

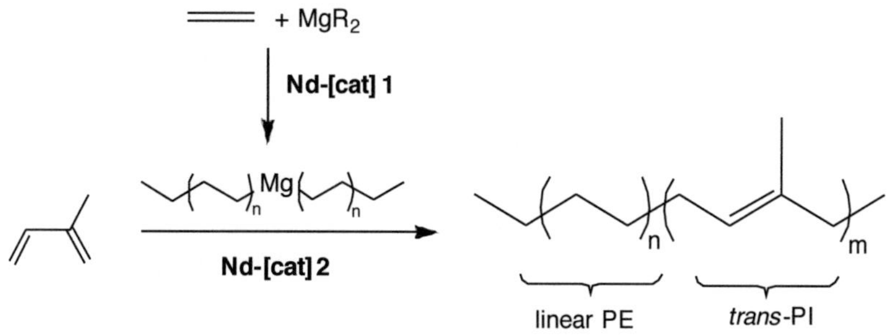

Scheme 8. Block copolymer of ethylene and isoprene with *trans* stereospecificity.

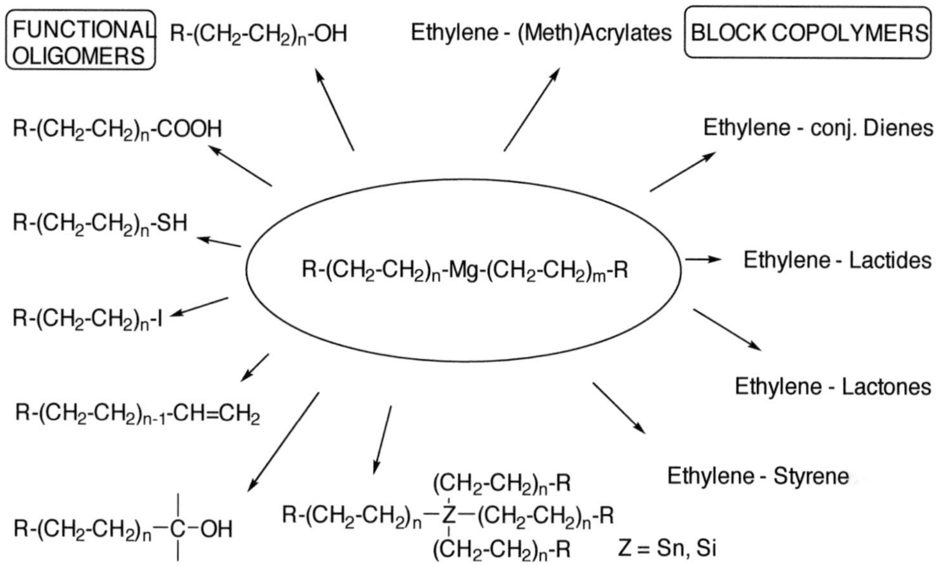

Scheme 9. Functionalization of ethylene oligomers after CCTP process.

3. Oligomers and co-oligomers of olefins and conjugated dienes

3.1. Octene

In contrast to the high activity of the rare earth / magnesium bimetallic system for ethylene polymerization, when using pentamethylcyclopentadiene Cp* ligands, the activity for other olefins such as octene was found very low. Moreover, poor selectivity hampered the propagation step by a secondary transfer reaction. This is an irreversible hydride transfer reaction, named π-allyl formation, between a living alkyl-metal catalyst and an olefin, giving rise to a dead alkane and a π-allyl-metallic complex, the latter having low, if any, catalytic ability to insert a new olefin monomer (higher activation energy of the insertion step in the catalytic cycle). This problem was overcome by changing the ligand in the coordination sphere of the rare earth. In particular, replacement of the methyl substituents on the cyclopentadiene ring of Cp* by several crowded silyl groups fulfilled the electronic and steric requirements for a better activity and selectivity for the octene polymerization. The synthesis of a peculiar ligand (see scheme 10) and isolation of the corresponding rare earth complex represented a prerequisite to test the ability of the CCTP concept for the synthesis of magnesium end-capped oligomers of octene. Successful results were obtained, ($\overline{M}$n=400-1300 g/mol, Đ=1.11-1.65) with longer reaction times than for ethylene (24 h). [43-44]

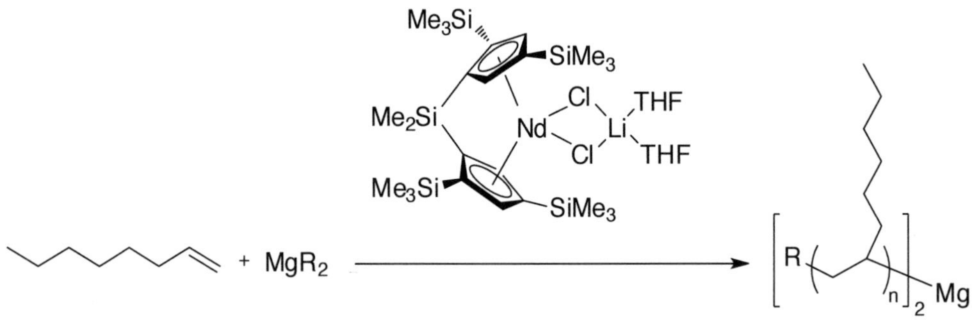

Scheme 10. Oligomerization of octene by CCTP with neodymium / magnesium system.

3.2. Styrene

As already mentioned for α-olefins, the activity of the decamethylmetallocene RE / Mg system was observed as negligible for the insertion step of styrene in a polymerization catalytic cycle. Higher temperature gave effective chain transfer of oligostyrene on dialkylmagnesium, but with competitive radicalar and/or anionic mechanisms besides the coordination one [45-48]. By switching from Cp* to other ligands, a stereoselective CCTP process yielded near-perfect syndiospecific and isospecific oligostyrenes end-capped with magnesium [49-51]. The use of half-metallocene systems (with only one Cp* ligand per rare earth) and even of inorganic rare earth salts like chloride, borohydride, alkoxide or phenoxide [52], gave successful results in oligomerization of styrene in a chain transfer mechanism to dialkyl magnesium. In a study centered to the structure/reactivity relationships of the pre-catalyst, it was shown that $Ln(BH_4)_3(THF)_x$ (x=3, Ln=Nd, La) as well as the mixed $La(BH_4)_2Cl(THF)_{2.6}$ led to an efficient transmetallation of the growing polystyrene chain with the Mg chain transfer agent. However, ^1H NMR and MALDI-TOF studies established the simultaneous occurrence of some β-H abstraction. Such uncontrolled termination reactions were absent with $LaCl_3(THF)_3$, $Cp^*Nd(BH_4)_2(THF)_2$ and $Cp^*La(BH_4)_2(THF)_2$. The quantitative transfer efficiency observed led us to conclude to a Catalyzed Chain Growth on magnesium. Moreover, the reaction remained significantly syndioselective (85 %) with the two latter ones, as observed previously when combined with only 1 equiv BEM (see scheme 11) [53-54].

http://dx.doi.org/10.5772/58217

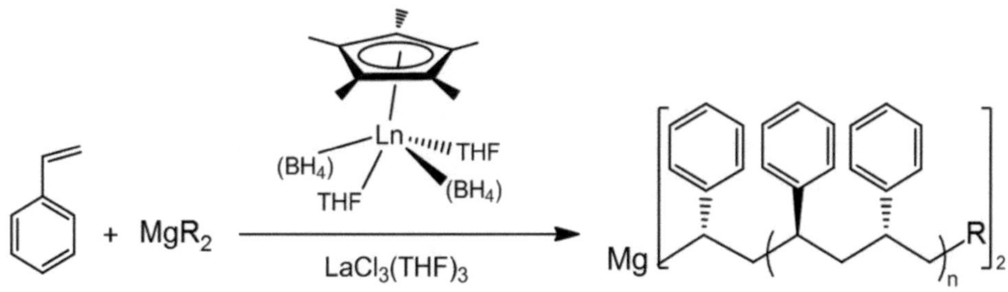

Scheme 11. Oligomerization of styrene by CCTP with rare earth / magnesium system. 85% syndiotactic selectivity for $Cp^*Ln(BH_4)_2(THF)_2$, Ln=Nd, La, and atactic polystyrene for $LaCl_3(THF)_3$.

Statistical copolymerization was assessed with 1-hexene and styrene monomers in the presence of the $Cp^*La(BH_4)_2(THF)_2$/n-butylethylmagnesium (BEM) catalytic system. Poly(styrene-*co*-hexene) statistical copolymers were obtained with up to 46% yield, and 23% 1-hexene content (see scheme 12). The occurrence of chain transfer reactions in the presence of excess BEM was established in the course of the statistical co-polymerization, through significant molecular weights decrease *vs.* 1 equiv BEM, along with narrowing of the dispersities. Thanks to this transfer process, the quantity of 1-hexene in the copolymer was increased substantially, from 8.6 to 23.2 %, for 80/20 1-hexene/styrene composition in the feed and in the presence of 10 equiv. BEM versus 1 equiv. [55]. These results extend the range of CCTP concept to a chain transfer induced control of the composition of statistical copolymers to poly(styrene-co-hexene) copolymers.

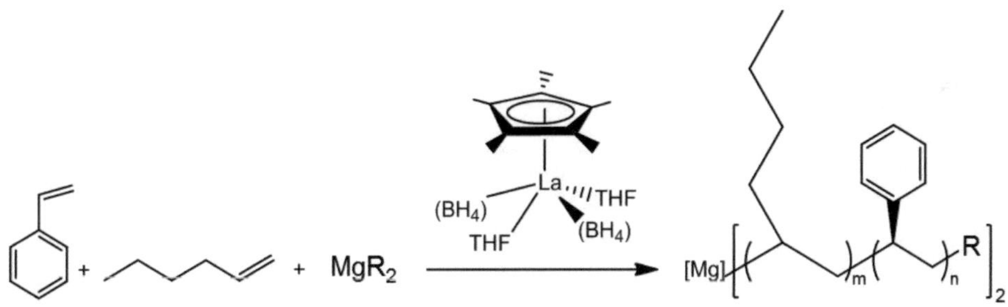

Scheme 12. Half-lanthanocene/BEM-mediated styrene-hexene Coordinative Chain Transfer co-Polymerization.

3.3. Butadiene

Polymerization of butadiene was known for a long time to be realized with catalytic systems based on rare earth salts and dialkyl magnesium [56-58]. Chain transfer efficiency to magnesium was of prime importance for example for the synthesis of functionalized polybutadienes with applications in the tyre industry [20]. Another striking point of this polymerization was the regioselectivity of the reaction which, in principle, can afford three kinds of sequences named 1,2-(or vinyl), 1,4-*trans* and 1,4-*cis* (see scheme 13). Using dialkyl magnesium in combination with rare earth salts was most generally effective for 1,4-*trans* selectivity [59], but

1,4-*cis* sequences have been obtained in some cases [60]. We recently established that the selectivity shifted slightly from high 1,4-*trans* (up to 97%) to more 1,2-regular (17%) simply by increasing from 1 to 10 the ratio of magnesium transfer agent relative to rare earth amounts. [61]. A chain transfer induced control of regioselectivity was hence evidenced (see table 1). Decrease of molar masses with increase of magnesium amounts was observed, delineating the transfer efficiency, although with a moderately controlled character (moderate dispersity). The magnesium presence at the tips of oligomer chains was attested by further functionalization, as mentioned below.

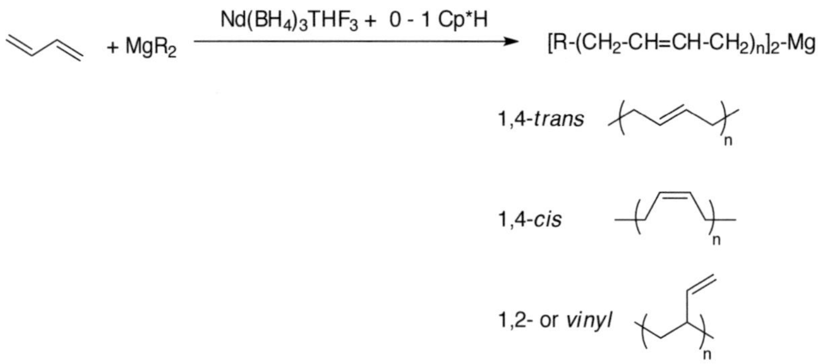

Scheme 13. Oligomerization of butadiene by CCTP with rare earth / magnesium catalytic system.

Run [a]	[Mg]/[Nd]	Time (h)	Isolated Yield (%)	Selectivity [b] (%)			$\bar{M}n$ [c] (kg.mol^{-1})	Đ [d]
				1,4-trans	*1,4-cis*	*1,2*		
1	1	2	96	95.0	3.5	1.5	49	1.29
2[e]	1	2	43	96.7	1.9	1.4	47	1.38
3	2	2	54	95.5	2.4	2.1	19	1.28
4	3	2	16	93.0	2.4	4.6	5.5	1.20
5	5	2	8	88.0	3.1	8.9	3.9	1.24
6	5	8	20	89.9	2.1	8.0	3.9	1.43
7	5	20	37	90.4	1.3	8.3	5.5	3.00
8	10	20	18	80.7	3.5	15.8	3.8	2.90

a: Reactions at 50 °C, in 10 mL of toluene; pre-catalyst: $Nd(BH_4)_3(THF)_3$ 20 µmol; co-catalyst: BEM; [butadiene]/[Nd]=1000

b: Determined by ^1H and ^{13}C NMR spectroscopy in $CDCl_3$

c: Number-average molecular weight measured by SEC with reference to PS standards. No correction factor applied.

d: Dispersity measured by SEC: Đ=$\bar{M}$ w / $\bar{M}$n

e: With addition of Cp*H (20 µmol) and supplementary BEM (10 µmol) in order to have an in situ prepared half-metallocene Cp^*NdX_2.

Table 1. Butadiene polymerization with Nd $(BH_4)_3(THF)_3$ and chain transfer with dialkyl magnesium.

Statistical copolymerizations were achieved with styrene and butadiene monomers under conditions similar to previous experiments in table 1. Corresponding oligomers were obtained with $\overline{M}$n ranging from 4.4 to 240 kg/mol, for Mg / Nd from 10 to 1, respectively, and styrene incorporation up to 16.9%. Regioselectivity was 1,4-*trans* sequences for butadiene and isolated styrene units inserted statistically in the chains, as identified by ^{13}C NMR. Insertion rates for butadiene were found higher than for styrene. Among these copolymerization experiments, one was short stopped by quenching with a ketone (benzhydrylidene anthrone, Sigma-Aldrich, see scheme 14) in order to have light oligomers, end-capped with a functional group. Analysis by MALDI-TOF mass spectrometry gave a detailed snapshot of the growing chains, evidencing the C-Mg reactivity of the magnesium end-capped oligomers (see figure 6). Statistical distributions of styrene and butadiene monomers were observed, that follow exactly the Poisson probability. The lower styrene insertion rate, as compared to the butadiene one, was clear: the average degree of polymerization, after a 15-minute reaction, was between 17 and 18 for butadiene monomer whereas it was between 1 and 2 for styrene monomer.

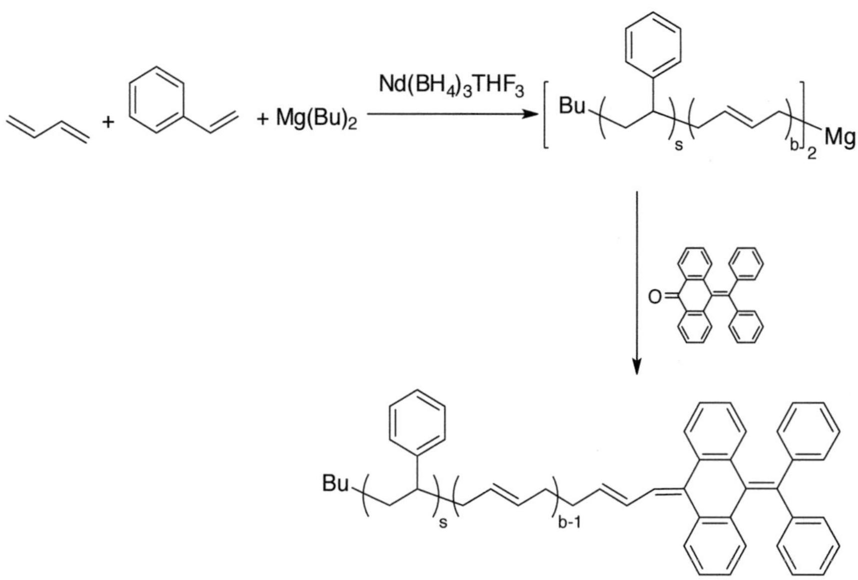

Scheme 14. Statistical copolymerization of butadiene and styrene under CCTP conditions and further functionalization with benzhydrylidene anthrone.

3.4. Isoprene

In the presence of 1 to 10 equiv alkyl magnesium as chain transfer agent, combined to $Cp^{*}Ln(BH_4)_2(THF)_n$ (Ln=La, n=2.5; Ln=Nd, n=2), the observed molecular weight distributions are monomodal, and the number average molecular weight is close to the calculated one considering two growing chains per magnesium atom. This, along with reasonable dispersities (Đ 1.3–1.9), highlights a rare earth catalyzed polyisoprene chain growth on magnesium (see scheme 15). With $Ln(BH_4)_3(THF)_3$ (Ln=La, Nd) in the same conditions transfer efficiency is

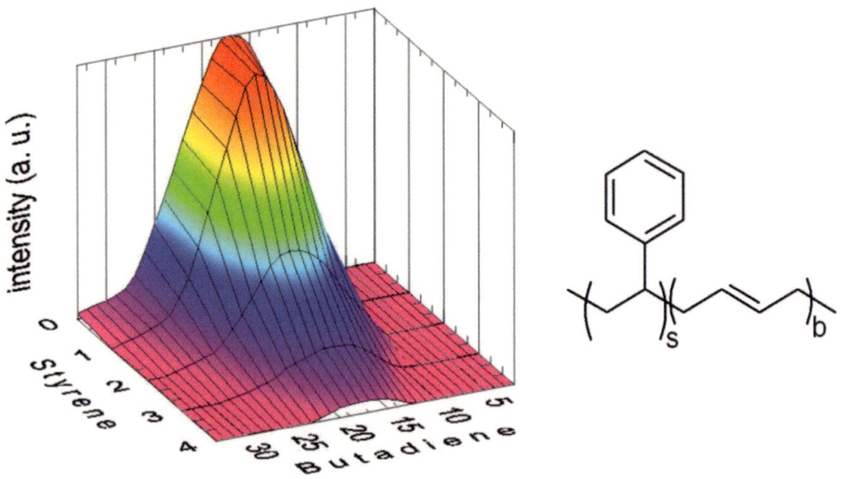

Figure 6. MALDI-TOF cumulative diagram of a poly(butadiene-*co*-styrene) oligomer initiated with butyl groups and functionalized with benzhydrylidene anthrone.

around 50–60%. With all the catalysts, the polymerization is significantly slowed down with BEM excesses *vs* 1 equiv. magnesium dialkyl, as observed for butadiene [62]. The excess of BEM has another consequence to the polymerization process: the transmetalation is accompanied by a modification of the selectivity of the reaction, from 98.5% 1,4-*trans* with 1 BEM to up to 46% 3,4-polyisoprene using 10 equiv. chain transfer agent. This simply appears as a way to tune the microstructure of the polyisoprene just by adjusting the quantity of chain transfer agent [63]. Actually, a gradual decrease of the 1,4-*trans* stereoselectivity of the reaction, leading to a great variety of poly(1,4-*trans*-isoprene) based materials, is observed at the benefit of 3,4-selectivity with increasing quantities of magnesium dialkyl. By combining dialkylmagnesium and trialkylaluminum, we disclosed that the *trans*-selectivity can be preserved: a 1,4-*trans* stereoselective oligomerization of isoprene with a CCTP character leading to the growth of several poly(1,4-*trans*-isoprene) chain per catalyst metal is reached using the half-lanthanocene $Cp^*La(BH_4)_2(THF)_2$ in combination with BEM and Al^iBu_3 in 1/1/9, 1/1/19, or 1/1/39 quantities, respectively [64].

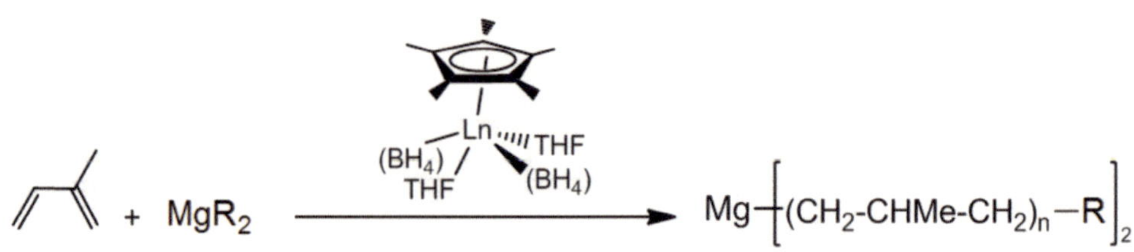

Scheme 15. Rare earth catalyzed polyisoprene chain growth on magnesium. Ln=Nd, La

http://dx.doi.org/10.5772/58217

An example of functionalization has been assessed on magnesium end-capped oligomers of isoprene, by oxidation with oxygen and hydrolysis to primary alcohols. Analysis by MALDI-TOF mass spectrometry (see figure 7) was in accordance with the expected structure for half of the observed peaks. The other half witnessed the presence of unfunctionalized species which may arise from difficulties to avoid hydrolysis or coupling reactions, before or during the oxidation.

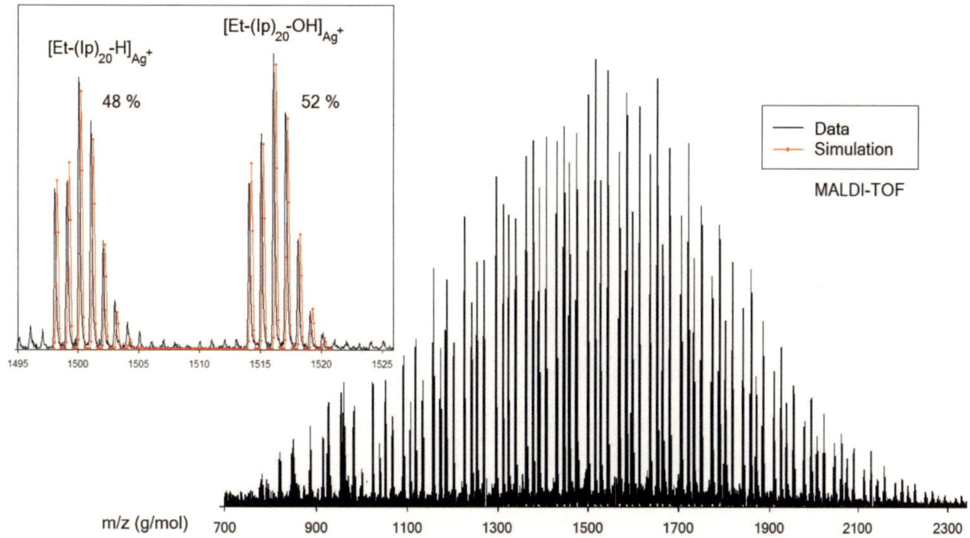

Figure 7. MALDI-TOF spectrum of isoprene oligomers functionalized in primary alcohols.

3.5. Myrcene

As expected from our previous results with isoprene and butadiene, polymerization of β-myrcene with neodymium borohydride-based coordination catalysts ($Cp^*Nd(BH_4)_2(THF)_2$ and $Nd(BH_4)_3(THF)_3$) in the presence of gradual excesses of BEM (1 to 20 equiv) shows high level of transfer reactions efficiency between neodymium and magnesium [65]. For 1-5 Mg co-catalyst equiv, the measured molecular weights (by SEC and end-group integration NMR) match quite well with calculated values for the growth of two chains per magnesium atom. As the BEM quantity increase, dispersities become more narrow, which accounts for rapid and reversible polymer chain transfer. In addition, the selectivity turns from > 90 %-1,4-*trans*[1] (1 BEM) to 3,4-rich (64%, 20 BEM), thus illustrating the "tuning ability" of the BEM concentration in such processes, as already observed with isoprene and butadiene (see scheme 16).

1 The selectivity was initially claimed as cis but recent additional experiments led us to reconsider it as trans. This will be published soon.

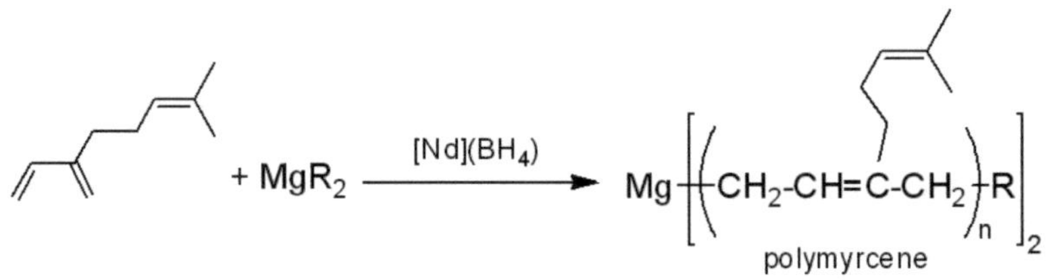

Scheme 16. Oligomerization of myrcene by CCTP with rare earth / magnesium system. $[Nd](BH_4)$=$Nd(BH_4)_3(THF)_3$, $Cp^*Nd(BH_4)_2(THF)_2$, MgR_2=BEM.

4. Conclusion

Readily applicable syntheses of magnesium end-capped oligomers of olefinic unsaturated monomers leading to tailor-made macromolecular objects were proposed. The strategy implemented to achieve this goal consisted in Coordinative Chain Transfer Polymerization involving rare earths precatalysts combined to a dialkylmagnesium reagent. After work-up, organo-functionalized oligomers bearing an hydroxyl, anthrone, rhodamine group, etc, or a second polymer sequence could be prepared.

The rare earth precatalyst was unprecedentedly and advantageously prepared via an *in situ* methodology, which allowed to start with basic compounds as dialkylmagnesium, commercial rare earth salts, and a cyclopentadiene (typically the permethylated one C_5Me_5H), to afford a highly efficient metallocene-based catalytic combination.

A complete panel of analyses was carried out on the oligomer samples, thus allowing a well-defined knowledge of the mechanism of reaction and of the active species involved, along with the ideal operative conditions in order to accurately monitor the process, in particular in the case of ethylene monomer. Basically, the growing length of these oligomers could be tuned all along the reaction just by checking the ethylene intake. After consumption of the monomer, the reaction was stopped by addition of a functionalization reagent, affording a rich variety of end-capped oligomers of ethylene. Similar processes were applied to others monomers including olefins, styrene, conjugated dienes, and including also copolymers of these monomers.

Applications of these compounds include copolymers and formulation additives for fine chemistry, plasturgy and electronics, pharmaceutics and cosmetics, dyes and adhesives etc.

Acknowledgements

Pr. André Mortreux, as former leader of the polymerization group in Lille, is greatly acknowledged for its essential contribution to initiate the basics of ethylene CCTP and related applications

http://dx.doi.org/10.5772/58217

Author details

Thomas Chenal* and Marc Visseaux

University of Science and Technology of Lille, France

National Center for Scientific Research, Unit of Catalysis and Solid State Chemistry, Villeneuve d'Ascq, France

National School of Chemistry of Lille (ENSCL), France

References

[1] Gladyz J. A. Frontiers in Metal-Catalyzed Polymerization: Designer Metallocenes, Designs on New Monomers, Demystifying MAO, Metathesis Déshabillé. Chem. Rev., 2000, 100(4), 1167-1168, and all articles in this special issue.

[2] Coates, G. W.; Hustad, P. D.; Reinartz, S. Catalysts for the Living Insertion Polymerization of Alkenes: Access to New Polyolefin Architectures Using Ziegler–Natta Chemistry. Angew. Chem., Int. Ed., 2002, 41, 2236-2257.

[3] Guan, Z; Cotts, P. M.; McCord, E. F.; McLain, S. J. Chain Walking: A New Strategy to Control Polymer Topology. Science, 1999, 283, 2059-2062.

[4] Arriola, D. J.; Carnahan, E. M.; Hustad, P. D.; Kuhlman, R. L.; Wenzel, T. T. Catalytic Production of Olefin Block Copolymers via Chain Shuttling Polymerization. Science, 2006, 312, 714-719.

[5] Britovsek, G. J. P.; Cohen S. A.; Gibson V. C.; van Meurs M. Iron Catalyzed Polyethylene Chain Growth on Zinc: A Study of the Factors Delineating Chain Transfer versus Catalyzed Chain Growth in Zinc and Related Metal Alkyl Systems. J. Am. Chem. Soc., 2004, 126, 10701-10712.

[6] Chenal, T.; Mortreux, A.; Visseaux, M. Method for Preparing Dialkyl Magnesium Compounds by Ethylene Polymerization and Uses Thereof. Patent WO2013/014383A1.

[7] Pelletier, J. F.; Bujadoux, K.; Olonde, X.; Adisson, E.; Mortreux, A.; Chenal, T. Long-Chain Dialkylmagnesium, its Preparation Process and Applications. Patent US005779942A, 1996.

[8] Pelletier, J. F.; Mortreux, A.; Olonde, X.; Bujadoux, K. Synthesis of New Dialkylmagnesium Compounds by Living Transfer Ethylene Oligo-and Polymerization with Lanthanocene Catalysts. Angew. Chem. Int. Ed. Engl., 1996, 35(16), 1854-1856.

[9] Chenal, T.; Olonde, X.; Pelletier, J. F.; Bujadoux, K.; Mortreux, A. Controlled polyethylene chain growth on magnesium catalyzed by lanthanidocene: A living transfer

polymerization for the synthesis of higher dialkyl-magnesium. Polymer, 2007, 48, 1844-1856.

[10] Carpentier, J.-F.; Sarazin, Y. Alkaline-earth metal complexes in homogeneous polymerization catalysis. Topics in Organometallic Chemistry, 2013, 45, 141-189.

[11] Mazzolini, J.; Espinosa, E.; D'Agosto, F.; Boisson, C. Catalyzed Chain Growth (CCG) on a Main Group Metal: an Efficient Tool to Functionalize Polyethylene. Polym. Chem., 2010, 1, 793-800.

[12] German, I.; Kelhifi, W.; Norsic, S.; Boisson, C.; D'Agosto F. Telechelic Polyethylene from Catalyzed Chain-Growth Polymerization. Angew. Chem. Int. Ed., 2013, 52(12), 3438-3441.

[13] Espinosa, E.; Charleux, B.; D'Agosto, F.; Boisson, C.; Tripathy, R.; Faust, R.; Soulié-Ziakovic C. Di-and Triblock Copolymers Based on Polyethylene and Polyisobutene Blocks. Toward New Thermoplastic Elastomers. Macromolecules, 2013, 46, 3417-3424.

[14] Bieligmeyer, M.; Mehdizadeh Taheri, S.; German, I.; Boisson, C.; Probst, C.; Milius,W.; Altstadt, V.; Breu,J.; Schmidt, H.-W.; D'Agosto, F.; Forster, S. Completely Miscible Polyethylene Nanocomposites. J. Am. Chem. Soc., 2012, 134, 18157-18160.

[15] Mazzolini, J.; Boyron, O.; Monteil, V.; D'Agosto, F.; Boisson, C.; Sanders, G. C.; Heuts, J. P. A.; Duchateau, R.; Gigmes, D.; Bertin D. Polyethylene End Functionalization using Thia-Michael Addition Chemistry. Polym. Chem., 2012, 3, 2383-2392.

[16] Lefay, C.; Glé D.; Rollet, M.; Mazzolini, J.; Bertin, D.; Viel, S.; Schmid, C.; Boisson, C.; D'Agosto, F.; Gigmes, D.; Barner-Kowollik, C. Block Copolymers via Macromercaptan Initiated Ring Opening Polymerization. J. Polym. Sci. Part A: Polym. Chem., 2011, 49(3), 803-813.

[17] Akbar, S.; Beyou, E.; Chaumont, P.; Mazzolini, J.; Espinosa, E.; D'Agosto F.; Boisson C. Synthesis of Polyethylene-Grafted Multiwalled Carbon Nanotubes via a Peroxide-Initiating Radical Coupling Reaction and by Using Well-Defined TEMPO and Thiol End-Functionalized Polyethylenes. J. Polym. Sci. Part A: Polym. Chem., 2011, 49(4), 957-965.

[18] Espinosa, E.; Glassner, M.; Boisson, C.; Barner-Kowollik, C.; D'Agosto, F. Synthesis of Cyclopentadienyl Capped Polyethylene and Subsequent Block Copolymer Formation Via Hetero Diels-Alder (HDA) Chemistry. Macromol. Rapid Commun., 2011, 32, 1447-1453.

[19] Mazzolini, J.; Mokthari, I.; Briquel, R.; Boyron, O.; Delolme, F.; Monteil, V.; Bertin,D.; Gigmes, D.; D'Agosto, F.; Boisson, C. Thiol-End-Functionalized Polyethylenes. Macromolecules, 2010, 43, 7495-7503.

[20] Cortial, G.; Le Floch, P.; Nief, F.; Thuilliez, J. Novel Organometallic Compounds Containing a Metal Belonging to the Second Column of the Periodic Table, and Method for Preparing Same. Patent WO2010/139449A1.

[21] Jeske, G.; Lauke, H.; Mauermann, H.; Swepston, P. N.; Schumann, H.; Marks T. J. Highly Reactive Organolanthanides. Systematic Routes to and Olefin Chemistry of Early and Late Bis(pentamethylcyclopentadienyl) 4f Hydrocarbyl and Hydride Complexes. J. Am. Chem. Soc., 1985, 107, 8091-8103.

[22] Watson, P.L.; Herskovitz, T. Homogeneous Lanthanide Complexes as Polymerization and Oligomerization Catalysts: Mechanistic Studies. ACS Symposium Series., 1983, 459-479

[23] Cendrowski-Guillaume, S.M.; Le Gland, G.; Nierlich, M.; Ephritikhine, M. Lanthanide Borohydrides as Precursors to Organometallic Compounds. Mono(cyclooctatetraenyl) Neodymium Complexes. Organometallics, 2000, 19(26), 5654-5660.

[24] Ballard, D.G.H.; Courtis, A.; Holton, J.; McMeeking, J.; Pearce, R. Alkyl bridged complexes of the group 3A and lanthanoid metals as homogeneous ethylene polymerisation catalysts. Journal of the Chemical Society, Chemical Communications., 1978, 22, 994-995

[25] Olonde, X.; Bujadoux, K.; Mortreux, A.; Petit, F. Catalysts and Process for the Preparation of Same for Use in the Polymerization of Ethylene. Patent WO 93/07180.

[26] Pettijohn, T. M. Olefin Polymerization Process and Polymer Produced. Patent US005350816A, 1994.

[27] Zinck, P.; Valente, A.; Terrier, M.; Mortreux, A.; Visseaux, M. Half-Lanthanidocenes Catalysts via the "Borohydride/Alkyl" Route: A simple Approach of Ligand Screening for the Controlled Polymerization of Styrene. C. R. Chimie, 2008, 11, 595-602.

[28] Visseaux, M.; Terrier, M.; Mortreux, A.; Roussel, P. Facile Synthesis of Lanthanidocenes by the "Borohydride/Alkyl Route" and Their Application in Isoprene Polymerization. Eur. J. Inorg. Chem., 2010, 2867-2876.

[29] Ziegler, K.; Gellert, H. G.; Külhorn, H.; Martin, H.; Meyer, K.; Nagel, K.; Sauer, H.; Zosel, K. Aluminium-organische Synthese im Bereich olefinischer Kohlenwasserstoffe. Angewandte Chemie, 1952, 64(12), 323-329.

[30] Bergbreiter, D. E.; Blanton J. R.; Chandran R.; Hein, M. D.; Huang, K.-J.; Treadwell, D. R.; Walker, S. A. Anionic Syntheses of Terminally Functionalized Ethylene Oligomers. J. Polym. Sci. Part A: Polym. Chem., 1989, 27(12), 4205-4226.

[31] Valente, A.; Mortreux, A.; Visseaux, M.; Zinck, P. Coordinative Chain Transfer Polymerization. Chem. Rev., 2013, 113, 3836-3857.

[32] Sita, L. R. Ex Uno Plures ("Out of One, Many"): New Paradigms for Expanding the Range of Polyolefins through Reversible Group Transfers. Angew. Chem. Int. Ed., 2009, 48, 2464-2472.

[33] Guan, Z.; Cotts, P. M.; McCord, E. F.; McLain, S. J. Chain Walking: A New Strategy to Control Polymer Topology Science, 1999, 283, 2059-2062.

[34] D'Agosto, F.; Boisson, C. A RAFT Analogue Olefin Polymerization Technique using Coordination Chemistry. Australian Journal of Chemistry, 2010, 63(8), 1155-1158.

[35] Makio, H.; Ochiai, T.; Mohri, J.-I.; Takeda, K.; Shimazaki, T.; Usui, Y.; Matsuura,S.; Fujita, T. Synthesis of Telechelic Olefin Polymers via Catalyzed Chain Growth on Multinuclear Alkylene Zinc Compounds. J. Am. Chem. Soc., 2013, 135, 8177-8180.

[36] Kretschmer, W. P.; Meetsma, A.; Hessen, B.; Schmalz, T.; Qayyum, S.; Kempe, R. Reversible Chain Transfer between Organoyttrium Cations and Aluminum: Synthesis of Aluminum-Terminated Polyethylene with Extremely Narrow Molecular-Weight Distribution. Chem. Eur. J., 2006, 12, 8969-8978.

[37] Ganesan, M.; Gabbaï, F. P. [Cp*Cr(C_6F_5)(Me)(Py)] as a Living Chromium(III) Catalyst for the "Aufbaureaktion". Organometallics, 2004, 23 (20), 4608-4613.

[38] Bazan, G. C.; Rogers, J. S.; Fang C. C. Catalytic Insertion of Ethylene into Al–C Bonds with Pentamethylcyclopentadienyl–Chromium(III) Complexes. Organometallics, 2001, 20 (10), 2059-2064.

[39] Chenal, T.; Visseaux, M. Combining Polyethylene CCG and Stereoregular Isoprene Polymerization: First Synthesis of Poly(ethylene)-b-(trans-isoprene) by Neodymium Catalyzed Sequenced Copolymerization. Macromolecules, 2012, 45, 5718-5727.

[40] Lin-Gibson, S.; Brunner, L.; Vanderhart, D. L.; Bauer, B. J.; Fanconi, B. M.; Guttman, C. M.; Wallace W. E. Optimizing the Covalent Cationization Method for the Mass Spectrometry of Polyolefins. Macromolecules, 2002, 35(18), 7149-7156.

[41] Lennon, P. J.; Mack D. P.; Thompson Q. E. Nucleophilic Catalysis of Organosilicon Substitution Reactions. Organometallics, 1989, 8, 1121-1122

[42] Amin, S. B.; Marks T. J. Versatile Pathways for In Situ Polyolefin Functionalization with Heteroatoms: Catalytic Chain Transfer. Angew. Chem. Int. Ed., 2008, 47, 2006-2025.

[43] Bogaert, S.; Chenal, T.; Mortreux, A.; Nowogrocki, G.; Lehmann, C. W.; Carpentier, J. –F. ansa-Bis(cyclopentadienyl) Ligands: Synthesis and Use in Olefin Oligomerization. Organometallics, 2001, 20, 199-205.

[44] Bogaert, S.; Chenal, T.; Mortreux, A.; Carpentier, J.-F. Unusual product distribution in ethylene oligomerization promoted by in situ ansa-chloroneodymocene-dialkylmagnesium systems. Journal of Molecular Catalysis A: Chemical, 2002, 190(1-2), 207-214.

[45] Bogaert, S.; Carpentier, J. –F. Chenal, T.; Mortreux, A.; Ricart, G. Chlorolanthanocene-dialkylmagnesium systems for styrene bulk polymerization and styrene-ethylene block copolymerization. Macromolecular Chemistry and Physics, 2000, 201(14), 1813-1822.

[46] Sarazin, Y.; Chenal, T.; Mortreux, A.; Vezin, H.; Carpentier, J.-F. Binary cerium(IV) tert-butoxides-dialkylmagnesium systems: Radical versus coordinative polymerization of styrene. Journal of Molecular Catalysis A: Chemical, 2005, 238, 207-214.

[47] Menoret, S.; Carlotti, S.; Fontanille, M.; Deffieux, A.; Desbois, P.; Schade, C.; Schrepp, W.; Warzelhan, V. Retarded anionic polymerization, 5 influence of the structure of dialkylmagnesium additives on the reactivity of polystyryllithium species. Macromolecular Chemistry and Physics, 2001, 202(16), 3219-3227.

[48] Carlotti, S.; Desbois, P.; Warzelhan, V.; Deffieux, A. Retarded anionic polymerization (RAP) of styrene and dienes. Polymer, 2007, 48(15), 4322-4327.

[49] Sarazin, Y.; de Fremont, P.; Annunziata, L.; Duc, M.; Carpentier, J.-F. Syndio-and Isoselective Coordinative Chain Transfer Polymerization of Styrene Promoted by ansa-Lanthanidocene/Dialkylmagnesium Systems. Advanced Synthesis & Catalysis, 2011, 353(8), 1367-1374.

[50] Annunziata, L.; Duc, M.; Carpentier, J.-F. Chain Growth Polymerization of Isoprene and Stereoselective Isoprene-Styrene Copolymerization Promoted by an ansa-Bis(indenyl) allyl-Yttrium Complex. Macromolecules, 2011, 44(18), 7158-7166.

[51] Rodrigues, A.-S.; Kirillov, E.; Vuillemin, B.; Razavi, A.; Carpentier, J.-F. Binary ansa-lanthanidocenes/dialkylmagnesium systems versus single-component catalyst: Controlled synthesis of end-capped syndiotactic oligostyrenes. Journal of Molecular Catalysis A: Chemical, 2007, 273(1-2), 87-91.

[52] Gromada, J.; le Pichon, L.; Mortreux, A.; Leising, F.; Carpentier, J.-F. Neodymium alk(aryl)oxides-dialkylmagnesium systems for butadiene polymerization and copolymerization with styrene and glycidyl methacrylate. Journal of Organometallic Chemistry, 2003, 683(1), 44-55.

[53] Zinck, P.; Valente, A.; Mortreux, A.; Visseaux, M. In situ generated half-lanthanidocene based catalysts for the controlled oligomerisation of styrene: Selectivity, block copolymerization and chain transfer. Polymer 2007, 48, 4609-4614.

[54] Zinck, P.; Valente, A.; Bonnet, F.; Violante, A.; Mortreux, A.; Visseaux, M.; Ilinca, S.; Duchateau, R.; Roussel, P. Reversible coordinative chain transfer polymerization of styrene by rare earth borohydrides, chlorides/dialkylmagnesium systems. J. Polym. Sci. Part A: Polym. Chem., 2010, 48, 802-814.

[55] Valente, A.; Zinck, P.; Mortreux, A.; Bria, M.; Visseaux, M. Half-lanthanocene/dialkylmagnesium-mediated coordinative chain transfer copolymerization of styrene and hexane. J. Polym. Sci. Part A: Polym. Chem., 2011, 49, 3778-3782.

[56] Jenkins, D. K. Polymerization of Conjugated Dienes. Patent US004619982, 1986.

[57] Jenkins, D. K. Butadiene polymerization with a rare earth compound using a magnesium alkyl cocatalyst: 1. Polymer, 1985 26(1), 147-151.

[58] Jenkins, D. K. Butadiene polymerization with a rare earth compound using a magnesium alkyl cocatalyst: 2. Polymer, 1985 26(1), 152-158.

[59] Monakov, Y..B.; Duvakina, N.V.; Ionova, I.A. Polymerization of Butadiene with a Halogen-containing Trans-regulating Neodymium-magnesium Catalytic System in the Presence of Carbon Tetrachloride. Polymer Science-Series B, 2008, 50 (5-6), 134-138.

[60] Jenkins, D. K. Sequential Formation of Trans and Cis Butadiene Homopolymers with One Catalyst System. Polymer, 1994, 35(13), 2897-2898.

[61] Ventura, A.; Chenal, T.; Bria, M.; Bonnet, F.; Zinck, P.; Ngono-Ravache, Y.; Balanzat, E.; Visseaux, M. Trans-stereospecific polymerization of butadiene and random copolymerization with styrene using borohydrido rare earths / magnesium dialkyl catalysts. European Polymer Journal, 2013, 49, 4130-4140.

[62] Terrier, M.; Visseaux, M.; Chenal, T.; Mortreux, A. Controlled Trans-stereospecific Polymerization of Isoprene with Lanthanide(III) Borohydride/Dialkylmagnesium Systems: the Improvement of the Activity and Selectivity, Kinetic Studies, and Mechanistic Aspects. Journal of Polymer Science, Part A: Polymer Chemistry, 2007, 45 (12), 2400-2409.

[63] Valente, A.; Zinck, P.; Mortreux, A.; Visseaux, M. Catalytic Chain Transfer (co-)Polymerization: Unprecedented Polyisoprene CCG and a New Concept to Tune the Composition of a Statistical Copolymer Macromol. Rapid Commun., 2009, 30, 528-531.

[64] Valente, A.; Zinck, P.; Vitorino, M.J.; Mortreux, A.; Visseaux, M. Rare earths/main group metal alkyls catalytic systems for the 1,4-trans stereoselective coordinative chain transfer polymerization of isoprene. Journal of Polymer Science, Part A: Polymer Chemistry., 2010, 48 (21), 4640-4647.

[65] Loughmari, S.; Hafid, A.; Bouazza, A.; El Bouadili, A.; Zinck, P.; Visseaux, M. Highly Stereoselective Coordination Polymerization of β-Myrcene from a Lanthanide-based Catalyst: Access to Bio-sourced Elastomers. Journal of Polymer Science, Part A: Polymer Chemistry, 2012, 50 (14), 2898-2905.

Chapter 2

The Use of Ionic Liquids in the Oligomerization of Alkenes

Csaba Fehér, Eszter Kriván, Zoltán Eller,
Jenő Hancsók and Rita Skoda-Földes

Additional information is available at the end of the chapter

http://dx.doi.org/10.5772/57478

1. Introduction

The more and more stringent quality standards of motor fuels, targeting the reduction of harmful material emission of the vehicles, require the development of new technologies for fuel production or improvement of the existing ones. The need for environmentally friendly, relatively clean-burning and practically heteroatom-free, high *n*-and *i*-paraffin containing blending fractions has been increasing.

From the options to convert light hydrocarbons (C_3-C_6 paraffins and olefins) of lower value to high quality blending components, oligomerization is one of the most promising methodologies. It provides extra flexibility to respond to changes in market demands with regards to the required gasoline:diesel ratio. C_3–C_5 olefins obtained by fluid catalytic cracking can be oligomerized to produce branched products which have higher octane numbers. The need for an increased overall diesel fuel yield can be addressed by C_3–C_5 olefin oligomerization operated in the trimer or tetramer mode, followed by hydrogenation of the products.

Beside the application as blending components in diesel fuels, triisobutenes are considered to be highly useful for the synthesis of specialty chemicals including dodecylbenzene, base oils and solidifying agents for epoxy resins. Oligomerization, and further hydrogenation of other olefins, such as 1-octene and 1-decene results in the formation of high viscosity index (good lubricating properties) synthetic base oils with low pour point and good oxidative stability.

The main challenges in the design of catalysts for oligomerization are focused to reach high conversion and high selectivity. Consequently, several catalysts have been developed for the oligomerization of lower alkenes. [1] Both Brønsted and Lewis acids have been used in either homogeneous or heterogeneous phase. Transition-metal complexes, such as Ni-based homogeneous catalysts containing phosphine ligands have also been developed. Ionic liquids were introduced first, as highly polar media, to ensure catalyst recovery due to biphasic conditions.

In recent years the use of room-temperature molten salts, or ionic liquids, has received increasing attention. The ionic liquids are good solvents for inorganic complexes (catalysts) while being immiscible with most hydrocarbons. As such, they provide a nonaqueous alternative for two-phase catalysis. Ionic liquid solvents eliminate the need to use volatile organic solvents. Furthermore, the nearly infinite combinations of suitable cations and anions lead to the possibility of tailoring their properties.

However, ionic liquids may serve not only as solvents, but also as catalysts during the oligomerization reaction. Both Lewis and Brønsted acidic ionic liquids have been used efficiently. Another possibility is the development of supported ionic liquid phases to combine the advantages of the ionic liquid with heterogeneous catalysis.

In the present chapter, after a short review of the importance and mechanism of light olefin oligomerization, the use of ionic liquids as solvents and catalysts in oligomerization reactions is presented. The effect of the composition of ionic liquids on the catalytic activity and selectivity is discussed in detail.

2. Industrial relevance of oligomerization of lower alkenes

Oligomerization of light olefins is an important alternative for the production of higher molecular weight hydrocarbon mixtures useful as fuels (e.g. gasoline or diesel). [2] It makes possible the upgrading of low value components of plant process streams, *e.g.* from fluid catalytic cracking, from cracking of polymers or wastes or from Fischer-Tropsch C_3-C_6 olefins.

During oligomerization of light olefins, different boiling point range isoolefins can be produced. Gasoline, jet fuel, diesel gas oil and base oil boiling point range components can be formed depending on the level of oligomerization. They can be transformed to isoparaffins by hydrogenation. Because of the constant growth of the energy demand of transportation, there is an increasing need for the production of isoparaffins.

Inland, aerial and waterway mobility is a keystone of sustainable development. To operate the equipment of mobility, internal combustion engines will be used (dominant share ≥85%) based on forecasts until 2030-2035. [3]

Intermittent-duty engines are used in inland and waterway transport, while aircrafts are operated by continuous jet engines. From the former, the Otto-and Diesel-engines are the most prevalent. The operational materials of these engines are the motor fuels and, in a wider sense, lubricants and other materials (*e.g.* oxygen source, refrigerants), too. [4]

http://dx.doi.org/10.5772/57478

The quantity of the fuel used is in accordance with the type of the drive. [5,6]

Based on the fuel demand (Figure 1), liquid hydrocarbons (engine gasoline, diesel gas oil) are used in the widest range in inland transport. They must satisfy the most important general and special (see also Table 1 and Table 2) requirements, such as:

- long-time availability of quality and quantity all over the world
- good combustion properties
  - high octane number (gasoline)
  - high cetane number (Diesel fuels)
  - relatively low combustion temperature (lower amount of NO_x formation; lower load of construction materials)
  - lower total harmful material emission, *etc.*
- very low sulphur content (≤ 10 mg/kg)
- reduced aromatic content
  - gasoline: benzene content ≤ 1.0 v/v%; total aromatic content ≤ 35 v/v%
  - Diesel fuels: polyaromatic content ≤ 8 %
- reduced olefin content
- gasoline ≤ 18 v/v%
- good blending ability with alternative components (not only with hydrocarbons but also with e.g. oxygen containing compounds)
- good additive sensibility and compatibility with additives (e.g. nozzle, valve and combustion chamber clearness)
- compatibility with motor oils
- fewer harmful matters in the exhaust gas
- user-friendly
- secure utilisation
- easy biodegradation
- environmentally friendly (non toxic)
- low book-cost, etc.

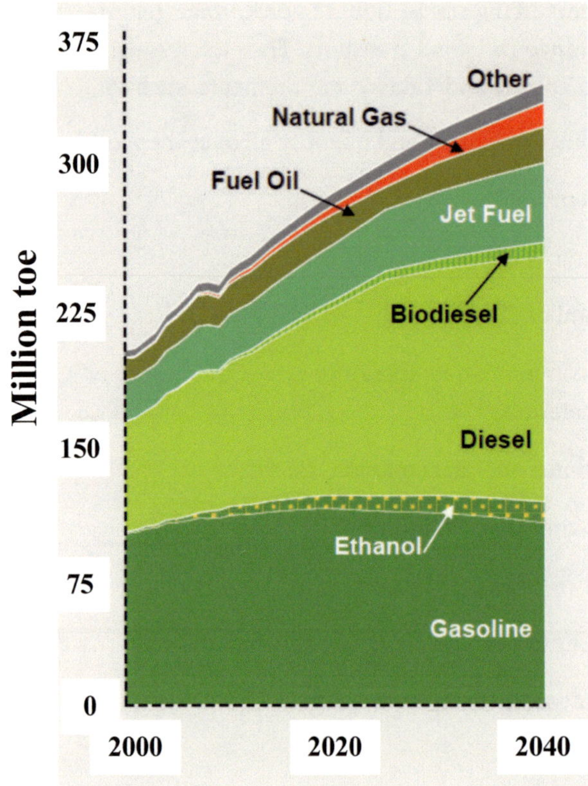

Figure 1. Changing demand of fuel by type (toe: tonne of oil equivalent)

Based on these requirements, it can be unequivocally stated that it is only the quantity of paraffins (mainly iso-and cycloparaffins and, with less importance, *n*-paraffins) that is not limited directly or indirectly in fuels. The concentration of cycloparaffins with suitable boiling points is low in crude oil and their synthesis is still expensive. [4] As a consequence, the most important components of gasoline and diesel gas oils are mixtures of different carbon number (different boiling point) isoparaffins or high isoparaffin content fractions, from the aspect of performance properties, environment protection and health. For example, gasoline boiling point range isoparaffins have high octane number and energy content (Figure 2), [4,7] their sensibility is low, they are practically sulphur and aromatic free, less toxic and because of their 'cleaner ignition', lead to the formation of less harmful material. Isoparaffins in diesel gas oil boiling point range have high energy content and cetane number (Figure 3) and they have good flow properties even at low temperature (low freezing point) (Figure 4). [8,9] Moreover, they are the most suitable hydrocarbons from the aspect of environmental protection. These properties are due to their high hydrogen content (C_nH_{2n+2}) and consequently low carbon content, their relatively easier biodegradation and low toxicity.

property	European Union				United States			WWFC
	EN 228 (1993)	EN 228 (2000)	EN 228 (2005)	EN 228 (2013)	RFG Phase II. 2012	CARB Phase 3 (2004/ 2006)	CARB Phase 3 (2009)	5. category (2012)
maximum sulphur content, mg/kg	500	150	50/10	10	80/30	60/30	30/20/15	10
maximum aromatic content, v/v%	-	42	35	35	-	35	35	35
maximum olefincontent, v/v%	-	18	18	18	-	10	10	10
maximum benzene content, v/v%	5.0	1.0	1.0	1.0	0.62/1.3[a]	1.1	0.7	1.0
maximum oxygen content, %	-	2.7	2.7	2.7/3.7	2.7/3.5	1.8-3.5[b]	1.8-3.5[b]	2.7
maximum ethanol content v/v%	-	5.0	5.0	5.0/10.0	10.0	-	-	5.0/10.0
maximum Reidvapour pressure, kPa	35-100	60/70	60/70	45-105	44-69	44-50	41-50	45-105

CARB: California Air Resources Board, WWFC: World Wide Fuel a: Commitment of crude oil refinery, b: Prohibited since 01. 01. 2004.

Table 1. Quality requirement changes of gasoline

Property	European Union				United States		WWFC
	EN 590 (1999)	EN 590 (2000)	EN 590 (2005)	EN 590 (2009)	Federal ASTM D 975-13 No. 2-D; S-15	CARB (2008)	5. category (2012)
minimum cetane number	48	51	51	51	40	53	55
maximum density at 15°C, kg/m3	820-860	820–845	820–845	820–845		-	820-840
maximum sulphur content, mg/kg	500	350	50.0/10.0	10.0	15	15	10.0
maximum total aromatic content, %	-	-	-	-	35.0 (m/m%)	10	15
maximum polyaromatic content, %	-	11	11	8	-	3.5	2.0
maximum distillation recovery (95%) temperature, °C	370	360	360	360	338 (90 v/v%)	-	350
maximum biodiesel (FAME-) content, v/v%,			5.0	7.0	5.0		not allowed

CARB: California Air Resources Board, WWFC: World Wide Fuel Charter

Table 2. Quality requirement changes of diesel gas oils

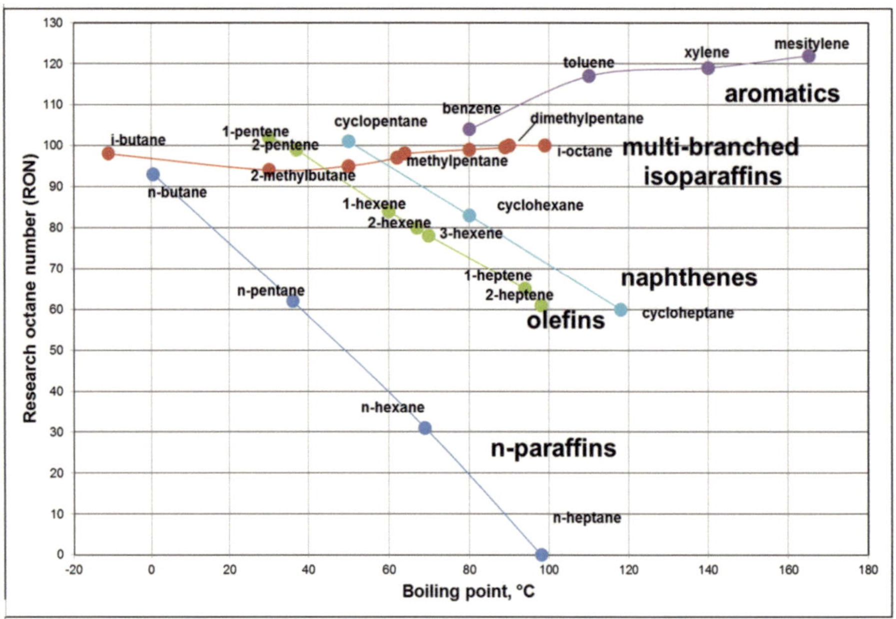

Figure 2. Research octane numbers of different hydrocarbons

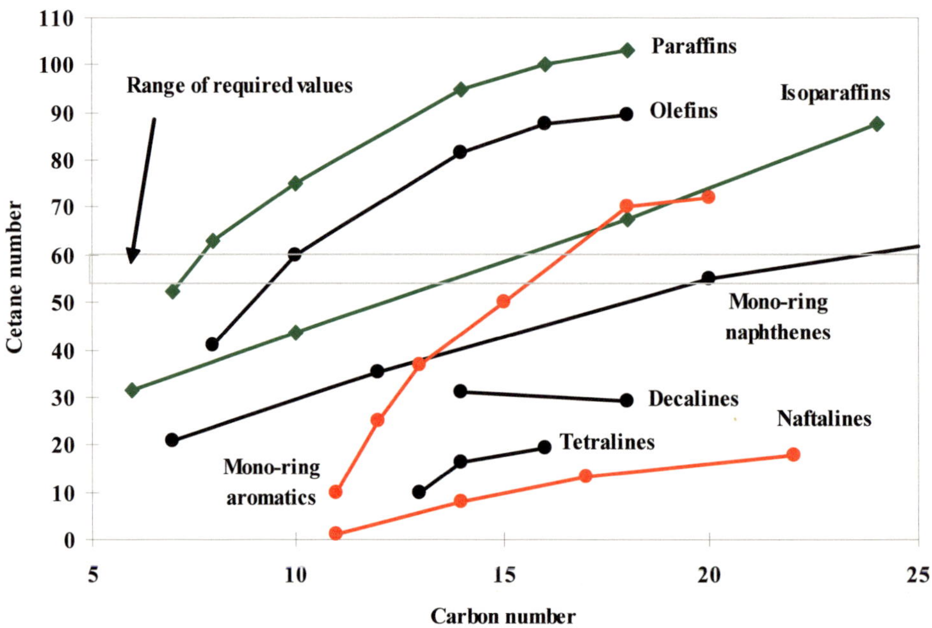

Figure 3. Cetane numbers of different hydrocarbons

http://dx.doi.org/10.5772/57478

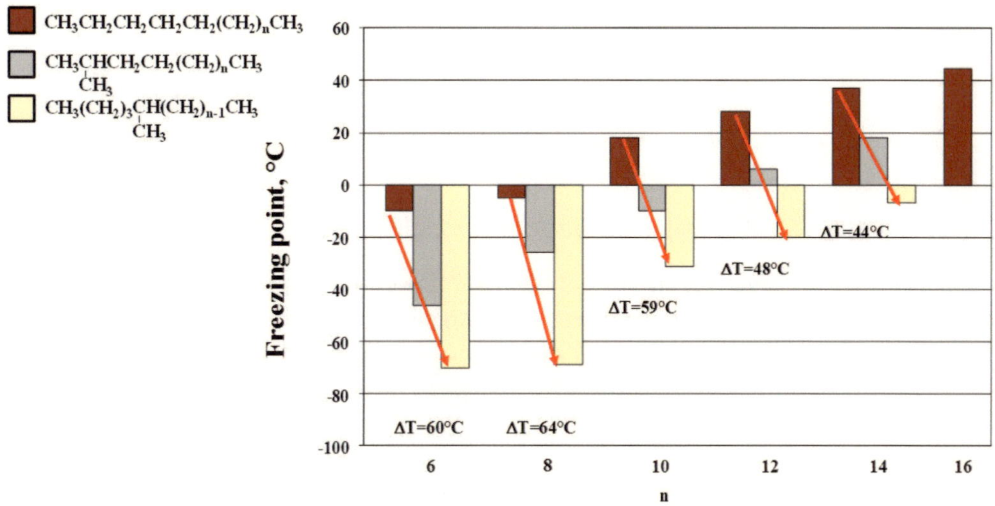

Figure 4. Freezing points of different hydrocarbons

Isoparaffin hydrocarbons are of key importance not only as energy carriers but also as suitable lubricants in case of engine oils. From the main components of engine oils (base oil and additives), hydrocarbon based lubricating oils have 65-80% share. From the aspect of performance properties, paraffin hydrocarbons and especially isoparaffins with a proper boiling point range are the most suitable constituents (Table 3.). [10, 11] These hydrocarbons have excellent lubricant properties (high viscosity index), good or suitable flow properties (-65°C — -20°C), low evaporization loss, moreover, they are aromatic and sulphur free [12].

property	base oil					
	conventional	HC-1	HC-2	HC-3	HC-4	PAO
kinematic viscosity, (100°C) mm^2/s	4	4	4	4	4	4
viscosity index	100	105	125	130	140	125
volatility (NOACK), %	23	18	14	13	11	12
flowpoint, °C	-15	-15	-18	-20	-30	-65
n- and i-paraffins, %	25	30	55	75	100	96
aromatics, %	24	05	0.3	0.1	0	-
sulphur content, %	<0.3	<0.1	<0.1	<100	<10	<1

HC-1:hydrocracking base oil; HC-2: rigorous hydrocracking base oil; HC-3: hydroisomerized paraffinic base oil, HC-4: Fischer-Tropsch paraffinic base oil, PAO: poly(alpha-olefins)

Table 3. Properties of different base oils

There are a lot of possibilities to produce isoparaffin rich fractions with different boiling point range:

- alkylation of isobutene with olefins (naphtha blending components)
- isomerisation of *n*-paraffins with suitable carbon number (C_5-C_7 naphtha fractions, gas oil fraction)
- hydrocracking of higher molecular weight hydrocarbon mixtures
- suitable level oligomerization of light olefins followed by hydrogenation to isoparaffins (indirect alkylation in case of naphthas, gas oil targeted oligomerization and hydrogenation; base oil targeted oligomerization and hydrogenation) (Scheme 1, Scheme 2).

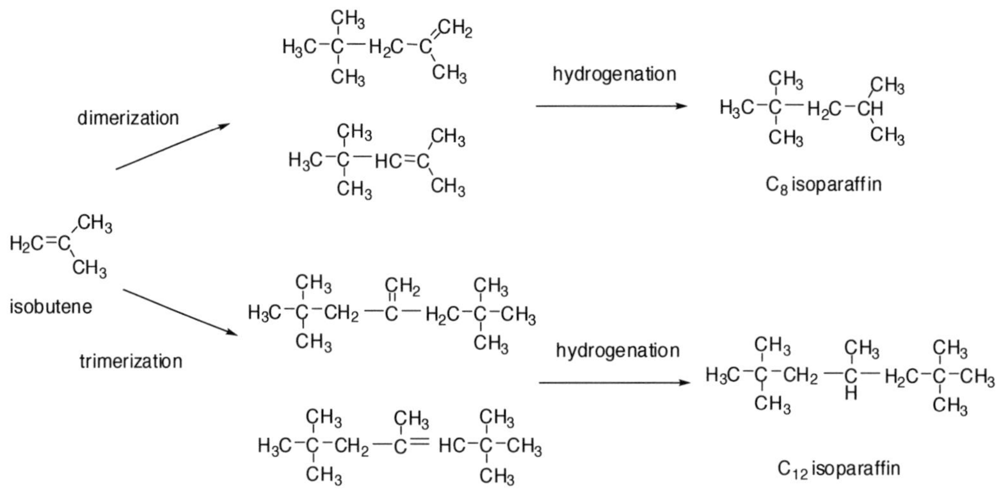

Scheme 1. Synthesis of naphtha boiling point range (C_8) and jet fuel/gas oil boiling point range (C_{12}) isoparaffins by an oligomerization — hydrogenation reaction sequence, starting from isobutene

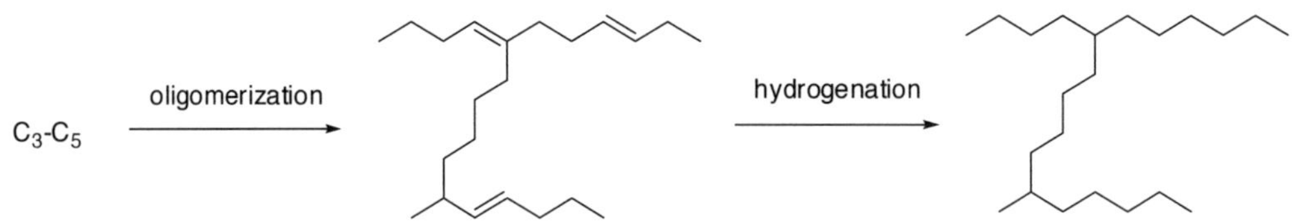

Scheme 2. Base oil targeted oligomerization and hydrogenation of light olefins

Poly-α-olefins (PAOs), used as lubricant base oils, are synthesized by a two-step reaction sequence from linear α-olefins derived from ethene (Scheme 3.). PAOs have good flow properties at low temperatures, relatively high thermal and oxidative stability, low evaporation losses at high temperatures, higher viscosity index, good friction behavior, good hydrolytic stability and good erosion resistance.

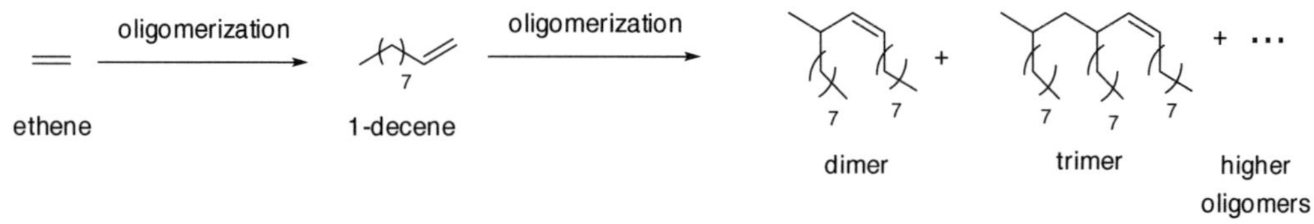

Scheme 3. Synthesis of poly-α-olefins (PAOs)

In summary, oligomerization of olefin hydrocarbons has an important role in the production of operational materials of internal combustion engines, and its importance is growing further. A great advantage of this method is that the quantity of each product can be controlled with the level of oligomerization. It makes the flexible adaptation to the market demands possible.

Beside the production of valuable blending components for fuels in petroleum refineries, oligomerization can be used for the large scale synthesis of fine chemicals and intermediates in the petrochemical industry. Linear C_8-olefin dimers are highly-desirable intermediates for the production of C_9-plasticizers, exhibiting better thermal properties than branched isomers. Oligomerization of ethene producing higher α-olefins also represents an important industrial process, as the products, depending on the chain length, can be used as intermediates for plastics, plasticizers, surfactants.

During the quest for sustainable technologies for the production of oligomers, research concerning the use of ionic liquids as solvents and/or catalysts came into focus.

3. Main features of ionic liquids used in oligomerization reactions

Ionic liquids (ILs) are salts consisting of bulky organic cations and inorganic or organic anions. (Figure 5 shows the general formulas of the most commonly used ILs, as well as the structures of the compounds mentioned in this chapter.) They melt at relatively low temperature, usually below 100 °C. They are good solvents for polar organic molecules and inorganic salts but they do not dissolve apolar compounds. These properties, together with their ability to stabilize transition metal complexes in low oxidation states, make them ideal solvents for transition metal catalyzed reactions. [13] When the polarity of the products are sufficiently low, biphasic reactions take place. After the completion of the reaction, the products can be separated by simple decantation and the metal catalyst remains in the IL phase that can be reused. ILs have negligible vapor pressure and they are not flammable which makes them very easy and safe to handle. Mainly because of their low volatility, they are considered to be 'green solvents'. However, their toxicity, investigated more thoroughly only recently, [14] should also be taken into account. Because of the great variety of anion – cation pairs, and the diversity in the side chains of the cations, an almost infinite IL combinations can be produced. Task-specific ILs are developed by the fine-tune of their physical and chemical properties through a careful choice of the structure of the cation-anion pair.

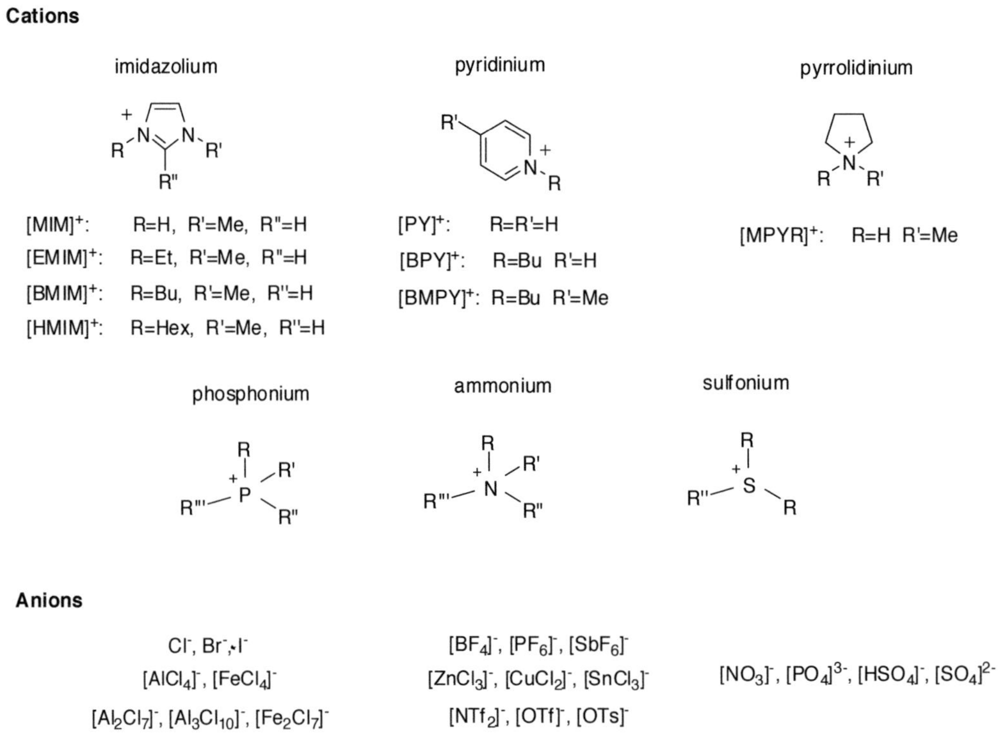

Figure 5. Common cations and anions of ILs

From the point of view of oligomerization of alkenes, mainly acidic ILs (Figure 6) are of interest. [15] Acidity may be due either to the anion or to the cation of the IL. [16]

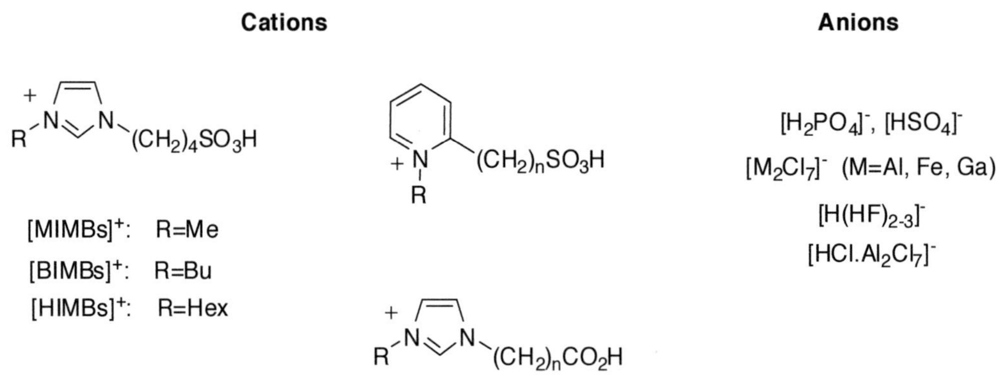

Figure 6. Cations and anions of acidic ILs

ILs with polynuclear metallic anions, such as chloroaluminate, chloroferrate or chlorozincate ions, show Lewis acidity [16] and, in the presence of protons, superacidity. [17] Brønsted acidity can be achieved by the addition of Brønsted acids such as HF or HCl into halide based ILs. The advantage of these systems lies in the fact that by supporting the acid in the ILs, its volatility can be reduced.

Acidity of the ILs depends greatly on the structure of the cations, too. [MIM][BF_4] shows higher acidity compared to dialkylimidazolium ILs. [18] At the same time, Brønsted acidic ILs are usually prepared by the introduction of alkane sulfonic acid or carboxylate acid groups as side chains of the cations (Figure 6).

4. Oligomerization mechanisms relevant in IL solvents

In oligomerization of alkenes, the most widely used methods involve cationic oligomerization in the presence of acids and transition metal catalyzed oligomerization. [19] In the first case, a carbocation intermediate is formed by the transfer of a proton from the acid catalyst to the alkene (for oligomerization of isobutene, see Scheme 4). Then the carbocation acts as an electrophile and reacts with the alkene to form another carbocation produced from two monomers. Termination happens when the dimeric carbocation loses a proton. When the carbocation is reasonably stable, this termination reaction is slower than chain elongation and the reaction leads to a trimer then a tetramer and finally a polymer.

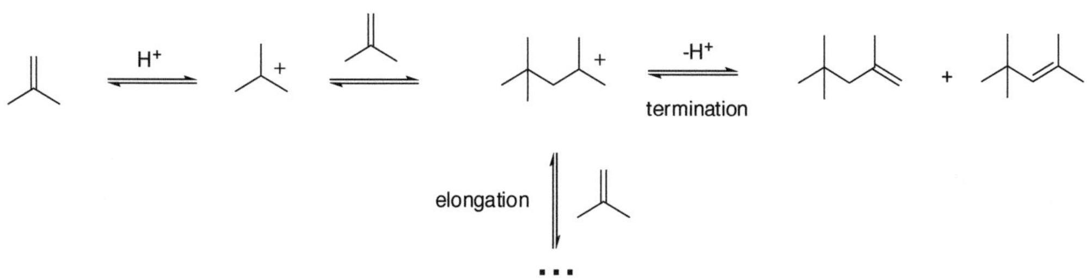

Scheme 4. Mechanism of cationic oligomerization of isobutene

In the transition metal catalyzed reaction the catalytically active species is regenerated after coordination of the olefin, chain growth and termination by β-elimination, producing the primary oligomerization product. Dimerization of alkenes can be described as two successive insertions of the olefin in the catalytic species. [20]

In the industrial Dimersol^{TM} process, the active catalyst is formed in situ by the reaction of a nickel (II) salt with an ethylaluminumhalide derivative. In the first step, insertion of either the C1 or C2 carbon of the alkene can occur, leading to various linear or branched dimers (for the mechanism of propene dimerization, see Scheme 5). At the same time, the nickel-complexes are active catalysts in isomerization, leading to further isomeric products. [19] The regioselectivity of dimerization can be directed by the addition of appropriate ligands. For example, in nickel-catalyzed oligomerizations sterically demanding phosphines favor the formation of highly branched oligomers.

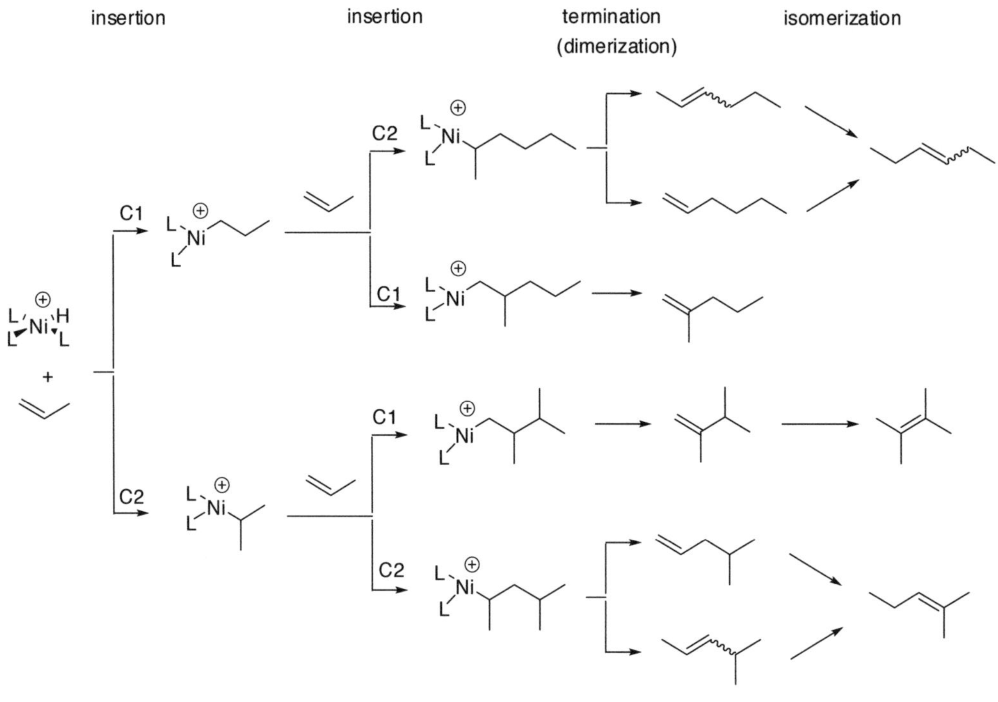

Scheme 5. Mechanism of nickel-catalyzed dimerization of propene

The main challenges in the design of catalysts for oligomerization are focused to reach high conversion and high selectivity. Although transition metal catalysts can efficiently be fine-tuned to achieve these goals, a great drawback of the homogeneous systems lies in the problem of catalyst recovery and recycle. The use of a two-phase solvent mixture is an attractive alternative, as it enables catalyst separation and reuse. ILs were introduced first in oligomerization reactions as co-catalysts, as well as polar solvents ensuring biphasic nickel-catalyzed reactions. [21]

5. The use of ILs as solvents in oligomerization catalyzed by organometallic complexes

5.1. Oligomerization with Ni catalysts

Chloroaluminate ILs, composed of imidazolium [22-34] or pyridinium [33, 35] halides, $AlCl_3$ and, in most cases, an alkylaluminum compound ($AlEtCl_2$ or $AlEt_2Cl$), act both as a medium for catalyst immobilization and nickel activator (Table 4). The cationic nickel active species is immobilized in the ionic phase without the need of a special ligand. Because of the high solubility of the Ni-complex but poor solubility of the products in the IL, complete separation of the catalyst can be achieved by simple decantation. In principle, the recovered catalyst/IL phase can continuously be reused maintaining its catalytic activity and selectivity.

http://dx.doi.org/10.5772/57478

Activities of catalysts in organic solvents were found to be considerably lower than those obtained from the same nickel precursors immobilized in an IL. [22]. This can be explained by an increase in the electrophilic nature of the nickel metal center by the weak coordinating ability of chloroaluminate anions.

The efficiency of the IL/catalyst system depends noticeably on the composition of the chloroaluminate IL. In ILs with an aluminum molar fraction lower than 0.50 ('basic ionic liquids'), the presence of an excess of coordinating chloride anions inhibits catalytic activity due to the formation of stable anionic species such as $NiCl_4^{2-}$ and $NiCl_3L^-$.[23, 24]

$$NiCl_2(PR_3)_2 + 2Cl^- \rightarrow NiCl_4^{2-} + 2PR_3 \tag{1}$$

Acidic chloroaluminates, with an aluminum molar fraction higher than 0.50, show enhanced activity. At the same time, in acidic ILs composed of [BMIM]Cl / $AlCl_3$, the formation of multinuclear species, such as $[Al_2Cl_7]^-$ or $[Al_3C_{10}]^-$ were detected. In the presence of $[Al_2Cl_7]^-$, an abstraction of the phosphine ligand, ensuring high branching in the product mixture, from the coordination sphere of the metal occurs [24] that alters the selectivity of the reaction.

$$[HNiPR_3]^+ + [Al_2Cl_7]^- \rightarrow [HNi]^+ + AlCl_3.PR_3 + [AlCl_4]^- \tag{2}$$

To avoid this, a weak competitive base, an aromatic hydrocarbon [24, 25] that did not interfere with the cationic nickel active species, was added to the reaction mixture. Thus, the acidity and the distribution of aluminum anionic species could be controlled due to the coordination of $AlCl_3$ to the aromatic ring.

$$[Al_2Cl_7]^- + ArH \rightarrow AlCl_3.ArH + [AlCl_4]^- \tag{3}$$

As another disadvantage, the presence of $[Al_2Cl_7]^-$ leads to the formation of yellow highly viscous heavy oligomers, characteristic of a cationic oligomerization, even in the absence of a Ni-precursor, [23, 26] arising from the superacidity of proton contamination of the ionic liquid.

$$HCl + [Al_2Cl_7]^- \rightarrow [H]^+_{nonsolvated} + 2[AlCl_4]^- \tag{4}$$

First, these problems were eliminated by the use of $AlEtCl_2$ instead of $AlCl_3$. In these ILs, both the formation of higher oligomers, and the loss of phosphine ligands could be avoided. The latter statement was based on the fact, that selectivity of the reaction was shifted towards 2,3-dimethylbutenes during dimerization of propene, characteristic in the presence of catalysts with bulky ligands. [23]

In [BMIM]Cl / $AlEtCl_2$ / $AlCl_3$ systems, ethylaluminum species such as $[AlEtCl_3]^-$, $[Al_2Et_2Cl_5]^-$, $[Al_2EtCl_6]^-$ and $[Al_3Et_3Cl_7]^-$ [36] are formed, depending on the amount of the alkylaluminum compound. The composition of the mixture greatly affects the acidity of the final IL, so an adjustment of the $AlEtCl_2$ to $AlCl_3$ ratio is necessary in order to optimize the efficiency of the catalyst/ionic liquid system.[24, 26]

Cationic side reactions in [BMIM]Cl / $AlCl_3$ ILs could totally be suppressed by the addition of $AlEtCl_2$, even in a molar fraction as low as 0.05. [26] This behavior was explained by a hydride abstraction on the olefin by $[Al_2Cl_7]^-$, generating an allylic cation (Scheme 6), with a subsequent alkylation consuming the co-catalyst.

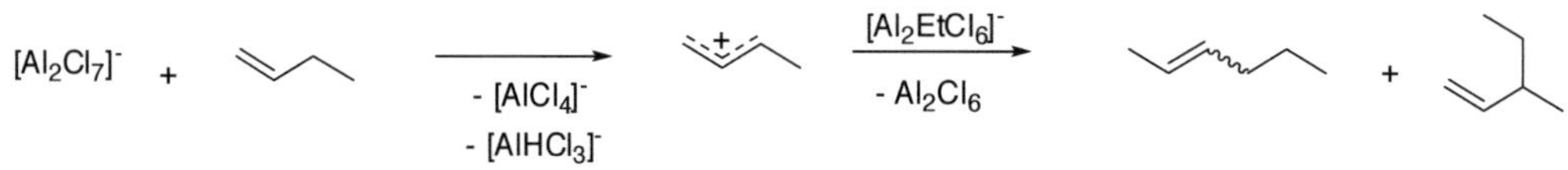

Scheme 6. The role of alkylaluminum derivatives in the suppression of cationic oligomerization

However, the loss of $AlEtCl_2$ may decrease the possibility of precatalyst activation in nickel-containing systems and consequently decreases their activity. This explanation is in agreement with the observation that increasing the $AlCl_3$ content of the ILs decreases the activity, due to the lower availability of the co-catalyst, necessary for the alkylation of the nickel precursor.

Generally, the use of slightly acidic [BMIM]Cl / $AlEtCl_2$ / $AlCl_3$ mixtures is the most favorable. Catalytic activity of nickel precursors was found to be higher due to the more acidic nature of these ionic liquids compared to [BMIM]Cl / $AlEtCl_2$. [24]

When an IL containing $AlEtCl_2$ is contacted with a hydrocarbon layer, the dissociation equilibrium of polynuclear chloroethylaluminum anions is shifted to the formation of lower nuclear chloroaluminum anions and neutral chloroethylaluminum compounds (Eqs 5, 6). Because of the total miscibility of the latter aluminum derivatives and hydrocarbons, ethylaluminium species are leached into the organic phase and this leads to a change in the composition of the ionic liquid phase. [24] The extraction process, as well as the nature of the alkylaluminum species formed in various chloroaluminate ILs was investigated in detail by Gilbert *et al.* by Raman spectroscopy [37] in order to optimize acidity and catalytic activity of the system as well as to minimize the loss of $AlEtCl_2$.

$$2\left[Al_2Et_2Cl_5\right]^- \rightarrow 2\left[AlCl_4\right]^- + Al_2Et_4Cl_2 \qquad (5)$$

$$2\left[Al_2EtCl_6\right]^- \rightarrow 2\left[AlCl_4\right]^- + Al_2Et_2Cl_4 \qquad (6)$$

Ni salts (such as NiF_2 [25]), neutral (**1** (Figure 7) [23, 34], **2** [35], **3** [33, 34], **4** [34]) and cationic Ni-complexes ($[Ni(MeCN)_6][BF_4]_2$ [25-27, 32], ($[Ni(MeCN)_6][AlCl_4]_2$, ($[Ni(MeCN)_6][ZnCl_4]$ [27]) as well as preformed (*e. g.* $NiCl_2(P(i\text{-}Pr)_3)_2$ (**5**) [23, 24], $NiBr_2(PPh_3)_2$ (**6**) [22], $NiCl_2(PCy_3)_2$ (**7**) [25]) or *in situ* produced Ni-phosphine complexes (*e. g.* $NiCl_2$+PBu_3, $NiCl_2$+PPh_3 $NiCl_2$+PCy_3 [28])

and Ni-carbene complexes (**8**) [30] were found to be suitable catalyst precursors in chloroaluminate ILs (Table 4).

Various ligands/additives were found to affect greatly the activity and/or selectivity of the catalytic systems.

A decisive effect of the counter-anion of cationic Ni-complexes on the catalytic activity was observed. [27] Precursors with $[ZnCl_4]^{2-}$ and $[AlCl_4]^-$ anions led to enhanced catalytic activity (TOF: 6480 h^{-1}) compared to $[Ni(MeCN)_6][BF_4]_2$ (TOF: 2412 h^{-1}). This phenomenon was attributed to the highly reactive behavior of the $[BF_4]^-$ ligand towards alkyl aluminum compounds. It can be assumed that a single active nickel hydride was formed in the ionic liquid, regardless of the precursor used, but in varying amounts depending on the counter ion for the different salts.

In case of $[Ni(MeCN)_6][BF_4]_2$, the use of $PCy_3.CS_2$ as an additive led to considerably higher reaction rate (TOF: 6840 h^{-1}) in oligomerization of 2-butene carried out in acidic ILs, than the addition of the phosphine PCy_3 itself (TOF: 3960 h^{-1}). [28] A systematic study of the effect of $PCy_3.CS_2$ was carried out by de Souza. [29] According to the NMR investigations of different IL— $PCy_3.CS_2$ systems, the improvement of catalytic activity is a consequence of a new anionic IL species, formed by the reaction of $[Al_2EtCl_6]^-$ with the zwitterionic $PCy_3.CS_2$, that is able to coordinate to the nickel center and has a strong electron-withdrawing effect on the metal.

Various Ni-carbene complexes (**8**, Figure 7) were found to be more active catalysts of 1-butene oligomerization (with TOFs 3820 h^{-1} – 7020 h^{-1}) in chloroaluminate ILs than $NiCl_2(PCy_3)_2$ (**7**, TOF=2950 h^{-1}) under the same conditions. [30]

At the same time, because of a similar selectivity obtained in the presence of $[Ni(MeCN)_6][BF_4]_2$, a Ni-carbene complex and a nickel complex in an imidazolium IL with an imidazolium ion blocked with a methyl group in position 2, the active species are thought to be similar. As the formation of a Ni-carbene complex is not possible in the last case, it was concluded that the active catalyst was formed by the coordination of aluminum species to the nickel center in all cases. [26]

In the [BMIM]Cl / $AlEtCl_2$ / $AlCl_3$ systems usually selective dimerization of ethene [25, 31, 38], propene [23, 24, 32] and butenes [26-29] could be observed due to the inhibition of cationic oligomerization/polymerization by the $AlEtCl_2$ co-catalyst. [26]

It should be mentioned however, that by a proper change in the reaction temperature and alkylaluminum co-catalyst/Ni ratio, oligomerization of ethene could be shifted towards the formation of trimers. Both TOF values and C_6 selectivity showed a curve with a maximum value with an increase in the $AlEt_2Cl$/Ni ratio. [22]

Oligomerization in acidic ILs was reported to lead to highly branched products. Besides, isomerization of both products and starting material was observed in most cases. For example, oligomerization of 1-butene and 2-butene led to the same product distribution, showing that isomerization was not a limiting step. [28]

Isomerization of the oligomerization products could be minimized by the proper choice of temperature and pressure during dimerization of propene: good selectivity for 1-hexene, up to 63%, could be achieved. [32]

As linear C_8-alkenes are highly desirable precursors of plasticizers, great efforts were made to develop a catalyst system producing linear oligomers. Complexes **2** and **3** (Figure 7), showing high selectivity in linear oligomerization in organic solvents, did not give satisfactory results in [BMPY]Cl / $AlEt_2Cl$ / $AlCl_3$ ILs. To avoid cationic side reactions, the use of buffered IL systems, consisting of [BMPY]Cl / $AlCl_3$ and an organic base, was investigated. [33, 35] The function of the base is to trap any free acidic species in the IL which may initiate cationic side reactions. Besides suitable basicity, the base should be non-coordinating with respect to the catalytically active nickel center. Depending on the nature of the base, 32-72% linear selectivity could be achieved in 1-butene oligomerization. Catalytic activity and selectivity was found to be optimal in the presence of *N*-methylpyrrole (TOF: 2100 h^{-1}, linear selectivity: 51%) or chinoline (TOF: 1240 h^{-1}, linear selectivity: 64%).

In oligomerization of propene, the same catalyst showed higher activity and dimer selectivity but lower linearity of the products in the buffered IL [EMIM]Cl /$AlCl_3$ / N-methylpyrrole than in toluene. [34] Selective dimerization was due to the 22 times lower solubility of the product hexenes in the IL than the starting material. The low linearity was attributed to ligand degradation in the presence of $[Al_2Cl_7]^-$ anions present in the ionic liquid.

Highly selective dimerization of ethene, propene, 1-butene and 1-hexene could be carried out in the presence of **10** (Figure 7) in triphenylbismuth buffered chloroluminate ILs, such as [MPYR]$[Al_2Cl_7]$/ $BiPh_3$. [39]

As chloroaluminate ILs promote isomerization of the double bond giving internal olefins as the main products, the possibility of the use of neutral ILs was also investigated. Hexafluorophosphate ILs with imidazolium cations were found to be suitable solvents for ethylene oligomerization with a cationic nickel-phosphine complex leading to 91-95% linear hexene selectivity, together with a 89-94% 1-hexene selectivity in the C_6 fraction. [40] The experiments showed enhanced activity in ILs (TOFs: 2058-12712 h^{-1}) compared to the reaction in CH_2Cl_2 (1852 h^{-1}). Decreasing activity was observed with increasing alkyl chain length of the imidazolium cation. This phenomenon was explained by inhibition of the cationic nickel catalyst by the oligomers formed: oligomerization activity was reduced by both the monophasic reaction in CH_2Cl_2 and increasing solubility of the products in the ILs with increasing alkyl chain length. Excellent 1-hexene selectivity is due to the low solubility of the oligomerization products in the IL. Since the primarily formed 1-olefins are quickly extracted into the organic layer, consecutive isomerization of these products at the Ni-center is suppressed. [41] A decisive negative effect of solvent impurities, such as chloride and water, on the outcome of the reaction was also revealed.

Active catalysts were generated from $Ni(COD)_2$ and the Brønsted acid, $H(Et_2O)_2B[3,5\text{-}(CF_3)_2C_6H_3]_4$ in ILs such as [BMIM]$[NTf_2]$ or [BMIM]$[SbF_6]$. [38] Contrary to organic solvents, the IL was able to stabilize and immobilize the active Ni species even in the absence of a coordinative ligand. In ethene oligomerization, high selectivity towards C_4-C_8 olefins (above

http://dx.doi.org/10.5772/57478

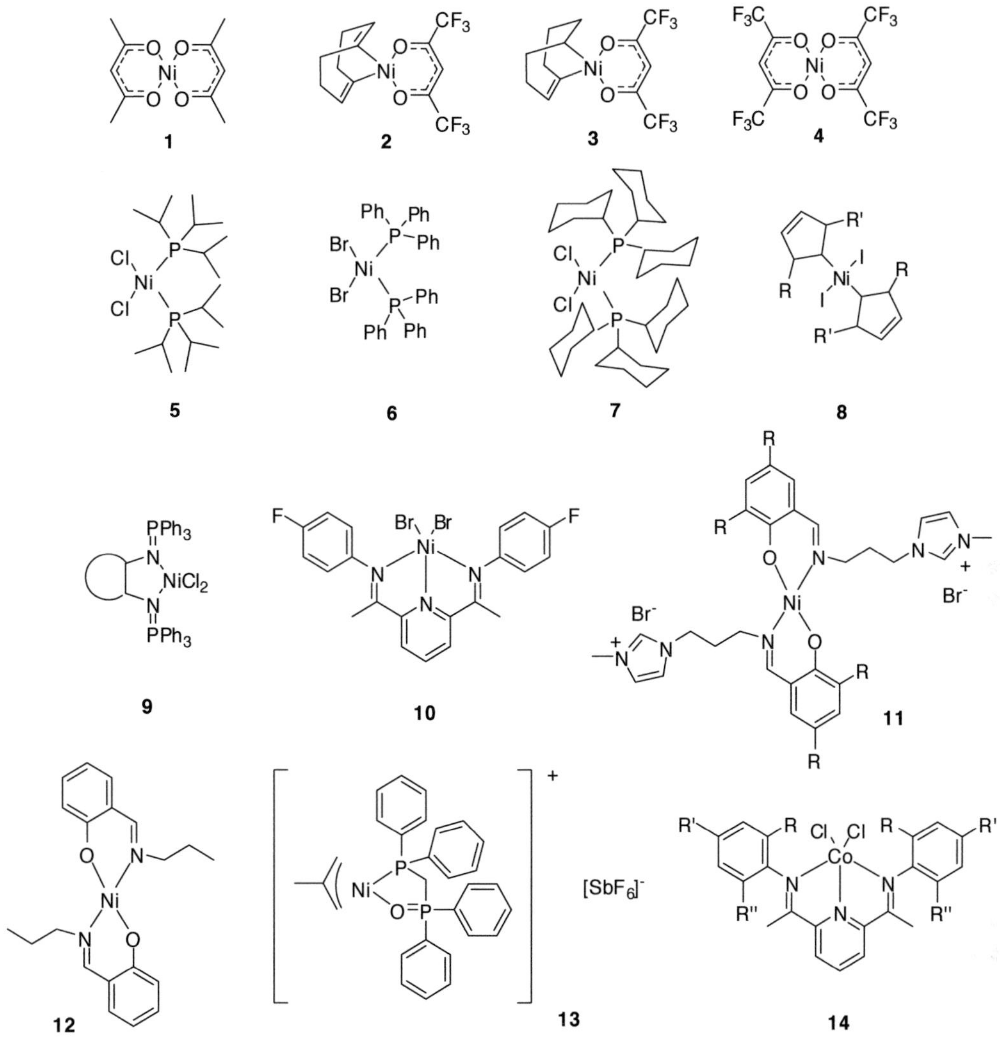

Figure 7. Some transition metal complexes used in oligomerization reactions

90%), together with good activity, with TOFs between 1400-12000 h^{-1}, were achieved in most cases. Selectivity for 1-butene was around 35% in the presence of IL/catalyst systems leading to C_4 products in good yields.

The main advantage of the use of the ionic liquid solvents in oligomerization is the possibility of catalyst reuse.

In butene dimerization the catalytic system, a [BMIM]Cl / $AlCl_3$ / $AlEtCl_2$ ionic liquid with an aluminum molar fraction of 0.57 and a cationic nickel complex, was reused for six times without any significant changes in the catalytic activity or selectivity. [22] The dimer distribution was almost constant throughout the catalytic runs. These indicate that the composition of the catalytic species in the ionic liquid phase is not affected by the removal of the organic phase.

On the contrary, a drop of activity with constant selectivity was observed in the second run in chloroaluminate ILs composed of [BMIM]Cl / $AlCl_3$ / $AlEtCl_2$=0.48/0.5/0.02. [26] At the same time, an increase in the TOFs from 5700 h^{-1} to 6500 h^{-1} and from 6100 h^{-1} to 7600 h^{-1} took place when the $AlEtCl_2$ content of the IL was raised to 5% and 10%, respectively. This shows that a minimal amount of co-catalyst is needed, probably to maintain a significant amount of active species through alkylation of the nickel species after the recycle.

Almost constant activity and selectivity could be ensured by the addition of a fresh supply of $AlEt_2Cl$ together with the *n*-heptane solvent after each run to the IL phase consisting of [BMIM]Cl / $AlCl_3$ / $AlEt_2Cl$ in ethene oligomerization. After three cycles, 98% of the original amount of nickel remained in the IL phase.[22]

The stability of the catalyst could be enhanced by the use of ligands with imidazolium tags. [42] Recycling experiments showed that bis-(salicylaldimine) Ni(II) complexes bearing imidazolium moieties in the side chains (**11**, Figure 7) could be recycled more efficiently, retaining their catalytic activity (with TOF values of 49 h^{-1} — 41 h^{-1}) at least in three cycles of ethene oligomerization. In contrast, activity of complex **12** dropped considerably upon reuse with TOFs of 48 h^{-1} and 30 h^{-1} in the first and second batch, respectively.

The catalytic system, the cationic Ni-complex (**13**, Figure 7) in [BMIM][PF_6], was used successfully for biphasic oligomerization of ethene to higher α-olefins. It was proved to be recyclable with little change in selectivity, although with somewhat lower activity. [40] The drop of TOF from 12712 h^{-1} to 7952 h^{-1} in the third run was attributed to the practical problem of quantitative transfer of the IL — catalyst mixture back into the autoclave under completely inert conditions.

Good results of catalyst reuse were obtained in ethene oligomerization in the presence of a catalyst generated from a Ni(0) complex and a Brønsted acid in [BMIM][SbF_6]. [38]

The use of the IL solvent makes it possible to carry out the oligomerization reaction in a continuous mode. A loop reactor was designed where the biphasic reaction mixture was circulated by a pump with high flow rates. [33] Two static mixers in the reactor loop provided an efficient dispersion of the ionic catalyst solution in the organic phase. The product was separated from the IL/catalyst mixture with a gravity separator integrated in the reactor loop. After 3 h reaction time the [BMIM]Cl / $AlCl_3$ / N-methylpyrrole / **3** catalytic system showed good activity (TOF=2700 h^{-1}) and selectivity (selectivity to C_8-product >98%, selectivity to linear C_8-product=52%).

5.2. Oligomerization with other transition metal catalysts

To avoid high carbon–carbon double bond isomerization rate, observed for nickel-catalyzed reactions, oligomerization catalysts based on other transition metals have also been tested (Table 4).

http://dx.doi.org/10.5772/57478

Cationic complexes of iron and cobalt, $[Fe(MeCN)_6][BF_4]_2$ and $[Co(MeCN)_6][BF_4]_2$, showed catalytic activities *ca.* one order of magnitude lower in chloroaluminate ILs than the analogous nickel catalyst in ethene oligomerization. [43] At the same time, they exerted higher dimer selectivity (in the range of 79-100% for iron and 66-99% for cobalt) than the nickel derivative (43-68%) and excellent selectivity for 1-butene (up to 100% and 99% within the dimer fraction for the iron and cobalt complex, respectively.) $FeCl_2$ and $CoCl_2$ turned out to be more active but the absence of ligands resulted in considerably lower selectivity to α-olefins.

Similarly to the nickel-catalyzed reactions, the addition of the ionic liquid greatly increased catalytic activity of the cationic complexes. Also, systems with more acidic co-catalysts, such as $AlEtCl_2$ and $AlEt_2Cl$, showed higher activity than $AlEt_3$ or methylaluminoxane.

Cyclodimerization of 1,3-butadiene, producing 4-vinyl-1-cylohexene, was carried out in $[BMIM][BF_4]$ in the presence of the catalyst precursor $[Fe(NO)_2Cl]_2$ and a reducing agent, such as Zn or $AlEtCl_2$. [44] In the latter case the conversion and turnover frequencies were similar to those obtained using Zn as reducing agent, but formation of some linear oligomers was observed. The use of the IL solvent greatly enhanced catalytic activity: the TOF achieved in the two-phase system (1404 h^{-1}) was higher than that under homogeneous conditions (253 h^{-1}). The catalytic system could be recovered and used in three successive experiments without any loss in the catalytic activity or selectivity.

Bis(imino)pyridine cobalt (II) catalysts (**14**, Figure 7) activated by methylaluminoxane were also found to be active in ethene oligomerization in chloroaluminate ILs. [45] They showed activities for ethene conversion in the range of 4000–15300 h^{-1}. The presence of electron withdrawing groups in the ligands, such as CF_3 and F, Cl or Br, enhanced catalytic activity. The catalysts exhibited high selectivity (above 90%) for the dimerization of ethene. The major product was 1-butene for all the catalysts.

Tungsten complexes, generally applied to induce metathesis reactions, could also be used for the dimerization of ethene in chloroaluminate ILs. [46] The tungsten-imido complex $Cl_2W{=}NPh(PMe_3)_3$ was found to be an active catalyst in slightly acidic [BMIM]Cl / $AlCl_3$ even without the addition of $AlEtCl_2$. At the same time, the presence of the alkylaluminum co-catalyst was necessary in propene dimerization with the $Cl_4W{=}NPh$ catalyst. In the latter case, contrary to the experiment starting from the phosphine complex, some leaching of the tungsten species to the organic phase was observed.

Palladium — Lewis acid systems, such as $Pd(OAc)_2/Cu(OTf)_2$ and $Pd(OAc)_2/In(OTf)_3$ were found to be very efficient catalysts for the selective dimerization of styrene to 1,3-diphenyl-1-butene in $[BMIM][PF_6]$. [47] Linear dimerization of 1,3-butadiene was achieved with $PdCl_2$ in $[BMIM][BF_4]$ and $[BMIM][PF_6]$. [48] Butadiene conversion was significantly higher in ILs than under homogeneous conditions. According to atomic absorption analysis, the palladium complex was almost completely retained in the ionic phase.

alkene	catalyst[a]	ionic liquid	reference
oligomerization with Ni catalysts			
ethene	NiF_2, $NiCl_2(PCy_3)_2$ (**7**), $[Ni(MeCN)_6][BF_4]_2$	[BMIM]Cl / $AlEtCl_2$ / $AlCl_3$	25
ethene	$NiBr_2(PPh_3)_2$ (**6**)	[BMIM]Cl / $AlEt_2Cl$ / $AlCl_3$	22
ethene	**9**	[BMIM]Cl / $AlEt_2Cl$ / $AlCl_3$	31
ethene	**11, 12**	[BMIM]Cl / $AlEt_2Cl$ / $AlCl_3$	42
ethene	$Ni(COD)_2$ / $H(Et_2O)_2B[3,5-(CF_3)_2C_6H_3]_4$	[BMIM]$[NTf_2]$, [BMIM]$[SbF_6]$	38
ethene	**13**	[BMIM]$[PF_6]$	40, 41
propene	$[Ni(MeCN)_6][BF_4]_2$	[BMIM]Cl / $AlEtCl_2$ / $AlCl_3$	32
propene	$NiCl_2(P(i\text{-}Pr)_3)_2$ (**5**) $NiCl_2(PBu_3)_2$ $NiCl_2(PPh_3)_2$ $NiCl_2(PBz_3)_2$ $NiCl_2(PCy_3)_2$ (**7**)	[BMIM]Cl / $AlEtCl_2$ / $AlCl_3$	23, 24
propene	$Ni(acac)_2$ (**1**)	[BMIM]Cl / $AlEtCl_2$ / $AlCl_3$	23
propene	**8**	[BMIM]Cl / *N*-methylpyrrole / $AlCl_3$	30
propene	$Ni(acac)_2$ (**1**), **3, 4**	[EMIM]Cl / *N*-methylpyrrole / $AlCl_3$	34
propene	**10**	[MPYR]$[Al_2Cl_7]$/ $BiPh_3$	39
1-butene	$[Ni(MeCN)_6][BF_4]_2$ $Ni(MeCN)_6][AlCl_4]_2$, $[Ni(MeCN)_6][ZnCl_4$	[BMIM]Cl / $AlEtCl_2$ / $AlCl_3$	26, 27
1-butene	$[Ni(MeCN)_6][BF_4]_2 + PCy_3CS_2$	[BMIM]Cl / $AlEtCl_2$ / $AlCl_3$	29
1-butene, 2-butene	$NiCl_2$+ PBu_3, $NiCl_2$+ PPh_3 $NiCl_2$+ PCy_3 $NiCl_2$+ PCy_3CS_2	[BMIM]Cl / $AlEtCl_2$ / $AlCl_3$	28
1-butene	**8**	[BMIM]Cl / *N*-methylpyrrole / $AlCl_3$	30
1-butene	**3**	[BMPY]Cl / LiCl / $AlCl_3$ [BMPY]Cl / $AlEtCl_2$ / $AlCl_3$ [BMPY]Cl / *N*-methylpyrrole / $AlCl_3$ [BMIM]Cl / *N*-methylpyrrole / $AlCl_3$	33
1-butene	**2**	[BMPY]Cl / $AlEtCl_2$ / $AlCl_3$ [BMPY]Cl / *N*-methylpyrrol / $AlCl_3$	35
oligomerization with other transition metal catalysts			
ethene	$[Fe(MeCN)_6][BF_4]_2$	[BMIM]Cl / $AlEtCl_2$ / $AlCl_3$	43

http://dx.doi.org/10.5772/57478

alkene	catalyst [a]	ionic liquid	reference
		$[BMIM]Cl / AlEt_2Cl / AlCl_3$ $[BMIM]Cl / AlEt_3 / AlCl_3$ $[BMIM]Cl / MAO / AlCl_3$	
1,3-butadiene	$[Fe(NO)_2Cl]_2$ / Zn or $AlEtCl_2$	$[BMIM][BF_4]$	44
ethene	$[Co(MeCN)_6][BF_4]_2$	$[[BMIM]Cl / AlEtCl_2 / AlCl_3$ $[BMIM]Cl / AlEt_2Cl / AlCl_3$ $[BMIM]Cl / AlEt_3 / AlCl_3$ $[BMIM]Cl / MAO / AlCl_3$	43
ethene	**14**	$[BMIM]Cl / MAO / AlCl_3$	45
ethene	$Cl_2W{=}NPh(PMe_3)_3$	$[BMIM]Cl / AlCl_3$	46
propene	$Cl_4W{=}NPh$	$[BMIM]Cl / AlEtCl_2 / AlCl_3$	46
styrene	$Pd(OAc)_2/Cu(OTf)_2$ $Pd(OAc)_2/In(OTf)_3$	$[BMIM][PF_6]$	47
1,3-butadiene	$PdCl_2$	$[BMIM][BF_4]$ $[BMIM][PF_6]$	48

[a]: for structures of catalysts **1**-**14** see Figure 7.

Table 4. Oligomerization of alkenes in ILs with transition metal catalysts

5.3. The use of ILs as solvents in acid-catalyzed oligomerizations

Air-stable nonchloroaluminate ILs can be used as solvents for catalytic amounts of protic acids (Table 5). In this case the IL is used to immobilize the acid catalyst and to ensure a biphasic reaction. The acidity of the system can be tuned by the nature of the acid/IL composition. The acidity level of these systems can be evaluated by the determination of the Hammett acidity functions using UV-Visible spectroscopy. [49, 50] (For a review, see [51].)These results were in good agreement with the catalytic data obtained for oligomerization of isobutene in most ILs except for $[BMIM][NTf_2]$. [52] The choice of the anion of the IL and the nature and concentration of the Brønsted acid were the main factors influencing catalytic performance. Although excellent conversions were obtained with $HNTf_2$ or HOTf, dimer selectivity remained moderate and trimers were formed in a large amount. Under the same acid concentrations, no reaction was observed either in organic solvents, such as heptane, or in water. Due to the solvating effect of water with respect to the proton, even addition of water to an IL/acid system decreased considerably the acidity level of the proton.

The catalytic system composed of $[HMIM][BF_4]$ and HBF_4 showed high catalytic activity in dimerization of α-methylstyrene (Scheme 7) to produce 4-methyl-2,4-diphenyl-1-pentene (**17**) at 60 °C. [53] The catalyst could be reused in four cycles without a loss of activity or selectivity. HBF_4 alone was found to be a considerably less active catalyst leading to a 6% conversion of α-methylstyrene under identical conditions. The IL $[HMIM][BF_4]$ exerted no catalytic activity without the addition of an acid. The use of ILs with other anions or the addition of other acids than HBF_4 led to lower activity and/or lower selectivity. At the same time, the cation of the IL had no significant influence, $[BMIM][BF_4]$–HBF_4, $[MIM][BF_4]$–HBF_4 and $[EMIM][BF_4]$–HBF_4 were similarly suitable compositions. The reaction temperature had a great effect on the selectivity of the reaction, at 120 °C selective formation of **16** was observed.

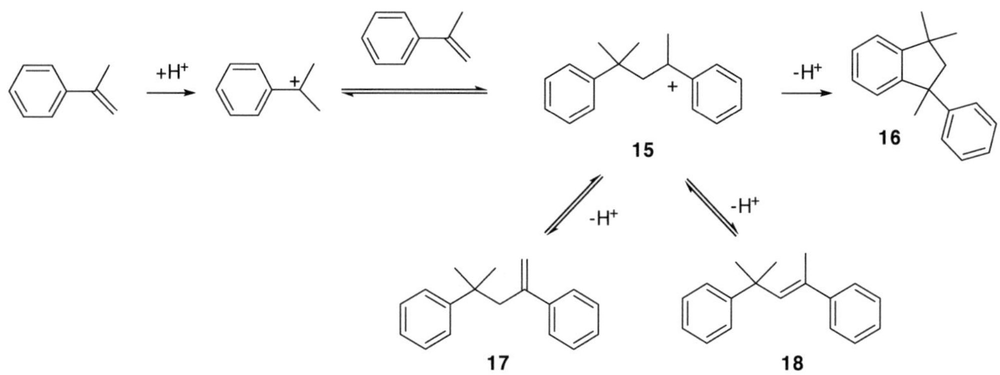

Scheme 7. Oligomerization of α-methylstyrene

alkene	acidic catalyst	ionic liquid	reference
isobutene	$HNTf_2$	$[BMIM][NTf_2]$ [BMIM][OTf] $[BMIM][PF_6]$ $[BMIM][BF_4]$ $[BMIM][SbF_6]$	52
isobutene	HOTf	$[BMIM][NTf_2]$	52
isobutene	CH_3SO_3H	$[BMIM][NTf_2]$	52
isobutene	CF_3CO_2H	$[BMIM][NTf_2]$	52
α-methylstyrene	HBF_4	$[HMIM][BF_4]$ $[BMIM][BF_4]$ $[EMIM][BF_4]$ $[MIM][BF_4]$	53

Table 5. Acid catalyzed oligomerization in ILs

6. Ionic liquids as catalysts

The first example for an IL catalyzed oligomerization of low molecular weight olefins was described in 1993 (Table 6). [54] Ethene or propene were converted to a mixture of oligomeric saturated and unsaturated hydrocarbons. The [BMIM]Cl / $AlCl_3$ system was shown to give unsaturated C_4-C_6 hydrocarbons with better selectivity than the more acidic [BPY]Cl / $AlCl_3$.

As it was described in the previous section, the formation of highly viscous oligomeric products were found to be formed in acidic chloroaluminate ILs even in the absence of a nickel catalyst [23, 26] but according to the reports, this reaction could be suppressed in the presence of the alkylaluminum co-catalyst. At the same time, Stenzel et al. reported on the oligomerization of 1-alkenes in the IL with a composition of [BMIM]Cl / $AlCl_3$ / $AlEtCl_2$=1/1.1/0.1. [55] The reaction was much slower than the transition metal catalyzed oligomerization: 67% ethene conversion was achieved in 16h at 60 °C. The total yield of oligomers decreased with increasing chain length of the monomer (to 7% in the oligomerization of 1-hexene).

Oligomerization of 1-hexene, [56, 57] 1-octene [58] and 1-decene [58] led to oligoalkylnaphtenic oils in chloroaluminate ionic liquids, such as [Et_3NH]Cl / $AlCl_3$, [PY]Cl / $AlCl_3$ and 2,6-bis(morpholinylethyl)-4-methylphenol/$AlCl_3$. The products were formed via an oligomerization – cyclization reaction sequence. The introduction of a titanium containing modifier into the chloroaluminate IL leads to the formation of new catalytic centers that mediate the oligomerization process toward the formation of oligomers with a higher molecular mass. [56]

Chloroaluminate ionic liquids are extremely moisture-sensitive and hydrolyze to release hydrogen chloride in contact with traces of water. Besides, they promote not only cationic oligomerization but also isomerization. As a consequence, the catalytic activity of chloroferrate(III)-[59] and chlorogallate(III) ILs [60], less sensitive to hydrolysis, were also tested.

Acidic compositions of [Et_3NH]Cl / $FeCl_3$ and [$C_{13}H_{22}$N]Cl / $FeCl_3$ showed high activity in isobutene oligomerization leading to a mixture of diisobutene and triisobutene with high selectivity. [59] The conversion of isobutene increased with increasing reaction temperature, but at high temperature cracking reactions occurred. The addition of CuCl to iron(III) chloride ionic liquids increased catalytic activity and the selectivity for diisobutene plus triisobutene up to 90 % using a composition of CuCl / [Et_3NH]Cl / $FeCl_3$=0.25/1/1.5.

[EMIM][Ga_2Cl_7] was successfully used in the oligomerization of 1-pentene to produce a C_{20}–C_{50} fraction, that can be used as a base for synthetic automotive lubricants, with high selectivity. [60]

A considerably different activity and selectivity of chloroaluminate and chloroferrate ILs were observed in oligomerization of α-methylstyrene (Scheme 7). [61] With [Et_3NH]Cl / 2$AlCl_3$ 100% conversion was achieved in 5 minutes. The selectivity for 1,1,3-trimethyl-3-phenylindan (**16**) was as high as about 97% in the absence of organic solvents. [Et_3NH]Cl / 2$FeCl_3$, [BMIM]Br / 2$AlCl_3$ [BMIM]Br / 2$FeCl_3$ turned out to be similarly active but less selective catalysts producing 22-32% trimers as side products. In [BMIM]Cl / 2$AlCl_3$ complete selectivity towards **16** was observed. [53] The high activity of the chloroaluminate ionic liquid was attributed to the strong

alkene	ionic liquid	reference
ethene	[BMIM]Cl / $AlCl_3$ [BPY]Cl / $AlCl_3$	54
ethene	[BMIM]Cl / $AlEtCl_2$ / $AlCl_3$	55
propene	[BMIM]Cl / $AlCl_3$ [BPY]Cl / $AlCl_3$	54
propene	[BMIM]Cl / $AlEtCl_2$ / $AlCl_3$	55
1-butene	[BMIM]Cl / $AlEtCl_2$ / $AlCl_3$	55
isobutene	[Et_3NH]Cl / $FeCl_3$ [$C_{13}H_{22}N$]Cl / $FeCl_3$ [Et_3NH]Cl / $FeCl_3$ / CuCl	59
isobutene	[MIMBs][OTf] [HIMBs][OTf]	64
isobutene	[MIMBs][OTf] / [BMIM][OTf]	52
1-pentene	[BMIM]Cl / $AlEtCl_2$ / $AlCl_3$	55
1-pentene	[EMIM][Ga_2Cl_7]	60
1-hexene	[BMIM]Cl / $AlEtCl_2$ / $AlCl_3$	55
1-hexene	[Et_3NH]Cl / $AlCl_3$ [PY]Cl / $AlCl_3$ 2,6-bis(morpholinylethyl)-4-methylphenol/$AlCl_3$ 2,6-bis(morpholinylethyl)-4-methylphenol/$TiCl_4$	56, 57
1-octene	[Et_3NH]Cl / $AlCl_3$ [PY]Cl / $AlCl_3$	58
1-decene	[Et_3NH]Cl / $AlCl_3$ [PY]Cl / $AlCl_3$	58
α-methylstyrene	[BMIM]Cl / $AlCl_3$	53
α-methylstyrene	[Et_3NH]Cl / $AlCl_3$ [Et_3NH]Cl / $FeCl_3$ [BMIM]Br / $AlCl_3$ [BMIM]Br / $FeCl_3$	61
α-methylstyrene	[MIM][BF_4]	62
α-methylstyrene	[MIMBs][OTf] [HIMBs][OTf]	64

Table 6. Oligomerization of alkenes with IL catalysts

Lewis acid, as well as superacidic protons, existing in the system due to the release of hydrochloric acid in the presence of trace amounts of water (see Eq. 4).

http://dx.doi.org/10.5772/57478

Stabilization of the intermediate carbenium ion (**15**, Scheme 7) by the IL with great polarity may make it possible for the positively charged carbon to attack the aromatic ring leading to the cyclodimer **16**. At the same time, in tertiary-amylalcohol as solvent, $[Et_3NH]Cl$ / $2FeCl_3$ oligomerizes α-methylstyrene to unsaturated dimer **17** with high selectivity. Interestingly, activity of $[Et_3NH]Cl$ / $2AlCl_3$ was much lower under these conditions, probably due to the better solvation of Al^{3+} than that of Fe^{3+} in the organic solvent.

The Brønsted acidic IL, $[MIM][BF_4]$ was shown to catalyze selective dimerization of α-methylstyrene, without the addition of an acidic co-catalyst. [62] This can be explained by the higher acidity of $[MIM][BF_4]$ compared to dialkylimidazolium ILs. [63] In the dimerization reaction, a great temperature dependence on the selectivity was observed: at 60 °C 4-methyl-2,4-diphenyl-1-pentene (**17**) was formed with 93% selectivity at 92% conversion of the substrate, while indan **16** could be obtained with 100% selectivity when the reaction temperature was increased to 170 °C. The ionic liquid was successfully recycled six times.

Other Brønsted acidic ILs, consisting of imidazolium cations with alkane sulfonic acid side chains, were found to be equivally active but less selective catalysts in the dimerization of α-methylstyrene. [64] At the same time, in oligomerization of isobutene, 68% and 94 % conversion and 99% selectivity to $C_{8+}C_{12}$ products, starting material for the production of high-octane gasoline blending components, could be achieved with [MIMBs][OTf] and [HIMBs][OTf] (Figure 6), respectively. The selectivity of the reaction was greatly dependent on the length of the side chain of the imidazolium cation. The use of an ionic liquid with smaller side chain led to higher selectivity for dimeric products. The increase in the catalytic activity of [HIMBs][OTf] was attributed to the higher solubility of isobutene in this IL because of the greater lipophilicity of the imidazolium cation with the longer alkyl side chain. It should be mentioned that using 1-alkenes as substrates, isomerization, instead of oligomerization, was found to be the main reaction.

A Brønsted acidic IL could also be used in catalytic amounts in a neutral IL as solvent. [MIMBs] [OTf] (Figure 6) was proved to be a more efficient acidic catalyst for isobutene dimerization in [BMIM][OTf] solvent than $HNTf_2$ or HOTf regarding dimer selectivity (up to 88 % at 70% isobutene conversion). [52] The [MIMBs][OTf] / [BMIM][OTf] mixture retained its activity and selectivity in ten subsequent cycles.

7. Supported catalysts based on ILs

Despite the several advantages of ILs discussed in the previous sections, they also have some drawbacks including the difficulties in handling because of the high viscosity of some ILs and the problems for application in fixed bed reactors. Also, biphasic IL—organic systems require large amounts of the expensive ILs, which hinders industrial applications. These difficulties can be overcome by the use of supported ionic liquid phases (SILPs) prepared by the immobilization of ILs on solid supports. [65]

Supported versions of the transition metal catalyzed processes, easy to use in a continuous mode reactor, were described. A buffered IL immobilized on a support material together with an organometallic complex of the type **19** (Figure 8) was used in selective dimerization reactions. The OH groups of the support were modified with an aluminum halide or alkylaluminum halide (Figure 8). The immobilized buffered catalyst was formed by mixing the organometallic catalyst, the IL composition of [BMIM]Cl/$AlCl_3$ and the coated support. The methodology was successfully used for the dimerization of propene in a fix bed reactor. [66]

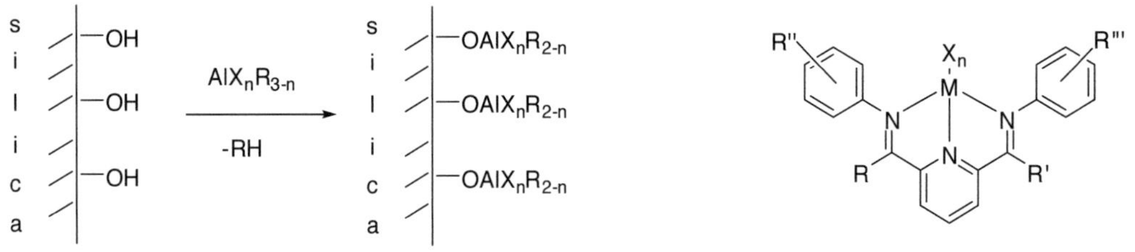

19

Figure 8. Modification of silica with an aluminumhalide/alkylaluminumhalide, and complex **19** (M=Ni, Fe, Co, Ti, V) used as the catalyst precursor

Immobilization of metallocene catalysts for the same reaction was achieved by similar methodology. [67]

Silica-supported SO_3H-functionalized IL catalysts were used in the oligomerization of isobutene. [68] The catalysts were obtained by impregnation of silica supports with Brønsted acidic ILs with $[MIMBs]^+$or $[BIMBs]^+$cations resulting in the formation of a solid material. It was shown that various factors, such as the properties and pretreatment of the solid support, the choice of cation and anion, as well as the reaction time and temperature affected the outcome of the reaction considerably. Oligomerization of isobutene, carried out at 60 °C, led to the formation of C_8 products with very good selectivity in each case. An increase in the temperature and/or reaction time led to an increase in the ratios of higher oligomers with selectivities for C_{12}+C_{16} products up to 85%. The catalysts comprising a trifluoromethanesulfonate anion could be reused several times without loss of activity. Total catalyst leaching of eight successive runs was 2.0% of the original load. At the same time, a quick inactivation of the catalysts obtained from ILs with hydrogensulfate anions was observed.

The TON and TOF values of the SILP catalysts were found to be ten times higher than those of the same ILs. When using the SILP catalysts, a smaller amount of the relatively expensive ionic liquid was sufficient for the reaction. The mass transport into the ionic liquid phase can be rate limiting due to the high viscosity of the ILs. This drawback can be circumvented by dispersing the ILs on support materials. Furthermore, the solid catalysts are easier to handle than the ionic liquids themselves, so separation and recovery of even small amounts of catalysts are simple.

Both mesoporous and microporous silica materials were used as solid supports. [69] A close relationship between catalytic activity and catalyst morphology was observed. Silica with mesoporous structure was able to adsorb a higher amount of the IL and produced SILP materials with higher catalytic activity at identical catalyst loadings. The formation of a stable film was a prerequisite for an unvarying selectivity of the catalyst, so impregnation at a high temperature was necessary to obtain a suitable composition, especially in case of microporous support materials.

As it could be expected, immobilization of the IL led to a loss of the BET surface of the supports. From the experimental data of the nitrogen adsorption/desorption isotherms of the supports and the solid catalysts, it could be concluded that the pores of mesoporous supports retained their shape during the catalyst preparation process, although the total pore volume values decreased because of the active IL film on the wall of the pores. On the other hand, the microporous supports contained narrow shaped pores that became almost totally filled with IL film during the catalyst preparation. The diffusion of isobutene to the micropores was blocked by the IL that filled the pore, so the contact surface between the IL phase and the organic phase was lower than that of SILP catalysts with mesopores. Also, acid capacity of microporous material was found to be lower, in accordance with the different amounts of adsorbed ionic liquid. Lower contact surface and lower acidity resulted in lower catalytic activity. At the same time, using a proper pretreatment, a stable catalyst with excellent C_{12} selectivity, exceeding even that of the mesoporous material, could be obtained with ILs supported on microporous silica.

Supported chloroaluminate ILs were used for the trimerization of isobutene in a feed contaning C_4 mixture. [70] Both the support and immobilization methodology were found to play a crucial role in the reaction. Catalysts obtained by mixing the IL with the silica support showed excellent oligomerization activity. At the same time, ILs immobilized on glass or molecular sieves induced only isobutane/butene alkylation. Exclusive alkylation took place also in the presence of catalysts prepared by grafting the IL on the silica. According to ^{29}Si CP/MAS NMR measurements, in the latter case a covalent bond was formed between the cation and the support. On the contrary, the spectrum of the catalyst with oligomerization activity indicated the formation of a covalent bond between the anion of the Lewis acid and the silanol group on the surface of the support (Figure 9).

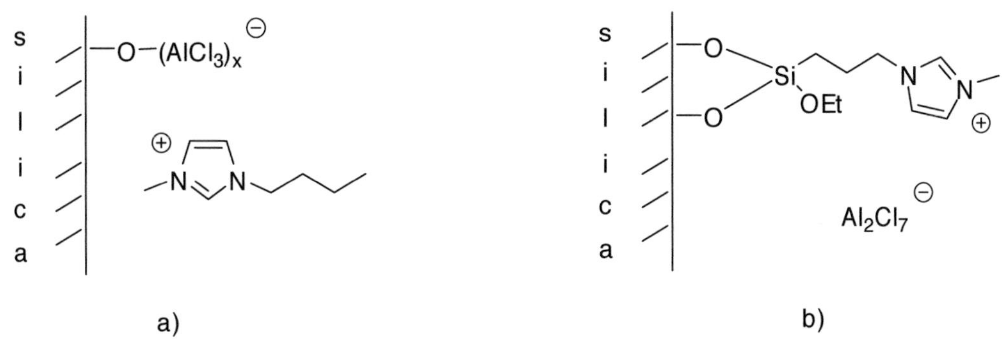

Figure 9. Structures of immobilized IL on silica support prepared by different methods (a) by mixing the IL and the silica support, b) by grafting)

[BMIM]Cl was used as a template during the preparation of a β-zeolite, a support for nickel-β-diimine catalysts that were highly active in oligomerization of ethene. [71] The high regularity of the microspherical agglomerates, obtained during the preparation of the support, was attributed to the formation of micellar aggregates due to the IL. The nickel complex incorporated into the β-zeolite framework showed higher activity, higher C_4 (up to 93.8%) and 1-butene selectivities (85.7% of the C4 fraction) than the homogeneous catalyst. The β-zeolite structure was found to work as a shape-selective support that inhibited re-coordination of 1-butene, thereby preventing isomerization and growth of the oligomer chain.

8. Industrial processes and patents based on the use of ionic liquids

Some of the catalyst compositions, described in the previous sections, were also patented.

Researchers of the Institue Français du Petrole (IFP) described the use of chloroaluminate ILs, composed of imidazolium or pyridinium halides and $AlCl_3$ as solvents for nickel catalyzed oligomerizations as early as 1987. [21] An improvement of the system by the addition of an alkylaluminum chloride co-catalyst, [72] and the use of an aromatic hydrocarbon to control the acidity of the system were patented soon afterwards. [73]

In 1998, an industrial process based on chloroaluminate ILs and nickel catalysts, known as the Difasol™ technology, was commercialized. It can be considered as a biphasic variant of the original homogeneous Dimersol X™ process that converts butenes to dimers. [19] As conversion is dependent on the initial concentration of butenes, the use of the latter technology is limited to C_4 feed with a minimum of 60% butene content. The liquid-liquid biphasic process can convert dilute feeds and can produce dimers with high selectivity that does not depend on conversion due to the low solubility of the product dimers in the ionic phase. The Difasol system is ideally suited for use as the finishing reaction section for a conventional Dimersol unit and importantly, it can be fitted into existing Dimersol plants to give improved yields and lower catalyst consumption. The sequence increases the relative octene gain by 22-41% over the traditional single process. [74] By the use of an appropriate ligand with a coordinating nitrogen, the catalyst system containing a zerovalent nickel complex, an acid and the IL can be tuned to obtain either dimers, oligomers or polymers with good selectivity. [75]

BP Chemicals also patented catalytic systems containing nickel-complexes in buffered ILs with a composition of R_nMCl_{3-n} or $R_mM_2Cl_{6-m}$ (M: Al, Ga, B or Fe(III)), an organic halide and a weak base. [76]

Catalysts composed of a Lewis acid and an IL [77] as well as the catalytic activity of acidic chloroaluminate(III) and alkylchloroaluminate(III) ILs [78] in oligomerization of light olefins were patented in 1995. Using the latter system, the olefinic feedstock could simply be bubbled through the ionic liquid catalyst, or alternatively, in a batch process, the IL was injected into a charged autoclave. With the use of imidazolium ions with alkyl groups longer than C_5, the reaction could be pushed towards polymerization.

Good trimer selectivity was obtained in oligomerization of isobutene in the presence of chloroindate ILs. Catalysts with ILs supported on silica were also used efficiently. [79]

Synthesis of polyalphaolefins, starting from 1-decene or 1-dodecene, was described in imidazolium, pyridinium, phosphonium, ammonium or sulfonium halides combined with an aluminum-or gallium halide or-alkylhalide. [80] A continuous mode of operation, using the same composition, was also patented by Chevron. [81] Not only the proper choice of the catalytic system but also the use of an appropriate operational mode is important to achieve optimal results. It was shown that the activity of the IL catalyst could be increased by emulsifying the IL with one or more liquid components. [82]

Efficient processes can be developed by the combination of oligomerization in ILs with other methodologies. An acidic IL catalyst consisting of an organic salt and a Lewis acid was used for the oligomerization of a light olefinic by-product fraction from a metallocene-catalyzed polyalphaolefin oligomerization process. [83]

Oligomerization of olefins present in the condensate recovered from the Fischer-Tropsch reactor can lead to high quality lube base oils. [84] However, the presence of oxygenates was found to interfere with the oligomerization of olefins catalyzed by Lewis acidic ILs. The amount of oxygenates could be reduced by contacting the feed with a hydrotreating catalyst prior to oligomerization. [85] The olefin stream formed by the Fisher-Tropsch synthesis and containing 1-alkenes with 5-18 carbon atoms could be converted into lubricating oils having a viscosity index of at least 120 and a pour point of -45 °C or less. [86]

Oligomerization of olefins together with an alkylation in the presence of isoparaffins could be carried out using the same IL catalysts. This may provide an efficient way to reduce the concentration of double bonds and at the same time to enhance the quality of the product. [87]

A Lewis acidic IL was supported on a porous solid and served as an adsorbent and activator for a Brønsted acid catalyst. The catalytic system was used in a fixed bed reactor, constructed for oligomerization or combined oligomerization /alkylation reactions. [88] In order to be able to respond to changing market demands, a process adjusting operational mode between two levels, a low level that favors the synthesis of C_{5+} products boiling at 137.8 °C or below or another C_{5+} product boiling above this temperature, was developed combining alkylation/ oligomerization in ILs. [89]

9. Conclusions

The use of ILs in catalytic reactions is a quickly expanding area of chemical research that has a great potential for industrial applications. Oligomerization of alkenes with ILs is one of the most promising methodologies to be implemented as industrial processes, although until now only one example, the Difasol™ technology, exists. [19]

The results presented in the previous sections show that ILs can really be used efficiently either as solvents or catalysts in oligomerization reactions. Catalytic activity of transition metal complexes was shown to be enhanced in ILs compared to organic solvents and ILs were proved to ensure easy product separation and efficient catalyst recycling. Technologies using ILs can convert not only pure alkenes but also process streams containing olefins to oligomers and can also be combined with other hydrocarbon conversion processes.

The effects of chloroaluminate ILs on nickel-catalyzed reactions are relatively well explored and from the investigations it can be concluded that a careful fine-tune of IL composition is necessary to obtain satisfactory results. However, in most cases little is known about the active transition metal species and the choice of the cations restricted mainly to imidazolium or pyridinium ions. There are still plenty of possibilities for the design of new IL compositions with better performance. This is even more true for ILs with oligomerization activity. To obtain an IL composition with optimal acidity, in order to achieve high conversion and at the same time to avoid polymerization or formation of higher oligomers, is a key issue in future research.

However, one of the main drawbacks that hinder industrial applications of ILs is the still high cost of these materials. In this respect, the use of supported catalysts opens new possibilities. Besides reducing the quantity of ILs necessary for the catalytic reactions, immobilization on solid supports makes it possible to use the catalysts in fixed bed reactors in continuous mode. At the same time, recent observations clearly show that compared to the non-immobilized IL catalysts, there are numerous other factors here that influence catalytic performance. Not only the IL composition but also the choice of the support material, porosity of the support and immobilization methodology affects catalytic activity and selectivity.

Abbreviations

BIMBs: 1-(4-sulfobutyl)-3-butylimidazolium

BMIM: 1-butyl-3-methylimidazolium

BMPY: 1-butyl-4-methylpyridinium

BPY: 1-butyl-pyridinium

Bz: benzyl

Cy: cyclohexyl

EMIM: 1-ethyl-3-methylimidazolium

HIMBs: 1-(4-sulfobutyl)-3-hexylimidazolium

HMIM: 1-hexyl-3-methylimidazolium

IL: ionic liquid

MAO: methylaluminoxane

MIM: 1-methyl-3H-imidazolium

MIMBs: 1-(4-sulfobutyl)-3-methylimidazolium

MPYR:1-methylpyrrolidinium

NTf_2: bistrifluoromethylsulfonimide

OTf: trifluoromethanesulfonate

PAO: poly-α-olefin

PY: pyridinium

RON: research octane number

toe: tonne of oil equivalent

Acknowledgements

The present work was funded by the project TÁMOP-4.2.2.A-11/1/KONV-2012-0071, realized with the support of the Hungarian Government and the European Union, with the co-funding of the European Social Fund.

Author details

Csaba Fehér[1], Eszter Kriván[2], Zoltán Eller[2], Jenő Hancsók[2] and Rita Skoda-Földes[1]

1 University of Pannonia, Institute of Chemistry, Department of Organic Chemistry, Hungary

2 University of Pannonia, Department of MOL Hydrocarbon and Coal Processing, Hungary

References

[1] Olah GA, Molnár A, Oligomerization and Polymerization. In: Hydrocarbon Chemistry (2nd edition). Hoboken, New Jersey: Wiley; 2003. p723-806.

[2] Olivier-Bourbigou H, Forestiere A, Saussine L, Magna L, Favre F, Hugues F. Olefin oligomerization for the production of fuels and petrochemicals. OIL GAS European Magazine 2010;2:97-102.

[3] Megatrends matter, Infineum Insight, 2012;(56):5

[4] Hancsók J, Kasza T. The importance of isoparaffins at the modern engine fuel production. In: W. Bartz (ed) Proceedings of 8th International Colloquium Fuels 2011, 8th International Colloquium Fuels, 19-20 January 2011, Stuttgart/Ostfildern, Germany: Esslingen: TAE, 2011. p.361-73.

[5] Bridsal C. The Outlook for Energy: A View to 204. Mineral Oil Congress, Stuttgart, Germany April 9-10. 2013.

[6] BP Statistical Review, June 2013.

[7] Hancsók J, Magyar S, Holló A. Importance of isoparaffins in the crude oil refining industry. Chemical Engineering Transactions 2007;11:41-7.

[8] Hancsók J, Krár M, Magyar Sz, Boda L, Holló A, Kalló D. Investigation of the production of high cetane number biogasoil from pre-hydrogenated vegetable oils over Pt/HZSM-22/Al2O3. Microporous and Mesoporous Materials 2007;101(1-2):148-52.

[9] Krár M, Thernesz A, Tóth Cs, Kasza T, Hancsók J. Investigation of catalytic conversion of vegetable oil/gas oil mixtures. In: Halász I. (ed) Silica and Silicates in Modern Catalysis, Kerala, India:Transworld Research Network; 2010. p 435-55.

[10] Pölczmann G, Hancsók J. Production of high stability base oil from Fischer-Tropsch wax, 18th International Colloquium Tribology, January 10-12.2012, Book of Synopsis. Stuttgart/Ostfildern, Germany, 2012. p71.

[11] Pölczmann G, Valyon J, Szegedi Á, Mihályi RM, Hancsók J.Hydroisomerization of Fischer–Tropsch wax on Pt/AlSBA-15 and Pt/SAPO-11 catalysts. Topics in Catalysis 2011;54:1079–83.

[12] Auer J, Borsi Z, Hancsók J, Lakics L, Lenti M, Nemesnyik Á, Valasek I. Tribology 2. Lubricants and the investigations. Budapest:Tribotechnik Ltd.; 2003.

[13] Olivier-Bourbigou H, Magna L, Morvan D. Ionic liquids and catalysis: Recent progress from knowledge to applications. Applied Catalysis A 2010;373:1–56.

[14] Pham TPT, Cho CW, Yun YS. Environmental fate and toxicity of ionic liquids: A review. Water Research 2010;44:352–72.

[15] Nasirov FA, Novruzova FM, Aslanbeili AM, Azizov AG. Ionic liquids in catalytic processes of transformation of olefins and dienes (Review). Petroleum Chemistry 2007;47(5):309–17.

[16] Chiappe C, Rajamani S. Structural effects on the physico-chemical and catalytic properties of acidic ionic liquids: An overview. European Journal of Organic Chemistry 2011;(28):5517-39.

[17] Johnson KE, Pagni RM, Bartmess J. Brønsted Acids in Ionic liquids: fundamentals, organic reactions, and comparisons. Monatshefte für Chemie 207;138:1077–101.

[18] Cui X, Zhang S, Shi F, Ma X, Lu L, Deng Y. The influence of the acidity of ionic liquids on catalysis. ChemSusChem 2010;3:1043-7.

[19] Forestière A, Olivier-Bourbigou H, Saussine L. Oligomerization of monoolefins by homogeneous catalysts. Oil & Gas Science and Technology – Rev. IFP 2009;64(6): 649-67.

[20] Nikiforidis I, Görling A, Hieringer W. On the regioselectivity of the insertion step in nickel complex catalyzed dimerization of butene: A density-functional study. Journal of Molecular Catalysis A 2011;341:63–7.

[21] Chauvin Y, Commereuc D, Hirschauer A, Hugues F, Saussine, L. Procédé de dimérisation ou de codimérisation d'oléfines. FP 2611700, 1987.

[22] Pei L, Liu X, Gao H, Wu Q. Biphasic oligomerization of ethylene with nickel complexes immobilized in organochloroaluminate ionic liquids. Applied Organometallic Chemistry 2009,23:455–59.

[23] Chauvin Y, Gilbert B, Guibard I. Catalytic dimerization of alkenes by nickel complexes in organochloroaluminate molten salts. Journal of the Chemical Society, Chemical Communications 1990;1715-6.

[24] Chauvin Y, Einloft S, Olivier H. Catalytic dimerization of propene by nickel-phosphine complexes in 1-butyl-3-methylimidazolium chloride/$AlEt_xCl_{3-x}$, (x=0, 1) ionic liquids. Industrial & Engineering Chemistry Research 1995;34:1149-55.

[25] Einloft S, Dietrich FK, de Souza RF, Dupont J. Selective two-phase catalytic ethylene dimerization by NiII complexes/AlEtCl2 dissolved in organoaluminate ionic liquids. Polyhedron 1996;15(19):3257-59.

[26] Thiele D, de Souza RF. The role of aluminum species in biphasic butene dimerization catalyzed by nickel complexes. Journal of Molecular Catalysis A 2007;264:293–8.

[27] Simon LC, Dupont J, de Souza RF. Two-phase n-butenes dimerization by nickel complexes in molten salt media. Applied Catalysis A 1998;175:215-20.

[28] Chauvin Y, Olivier H, Wyrvalski CN, Simon LC, de Souza RF. Oligomerization of n-butenes catalyzed by nickel complexes dissolved in organochloroaluminate ionic liquids. Journal of Catalysis 1997;165:275-8.

[29] de Souza RF, Thiele D, Monteiro AL. Effect of phosphine–CS2 adducts on the nickel-catalyzed butenes oligomerization in organochloroaluminate imidazolium ionic liquids. Journal of Catalysis 2006;241:232–4.

[30] McGuinness DS, Mueller W, Wasserscheid P, Cavell KJ, Skelton BW, White AH, Englert U. Nickel(II) heterocyclic carbene complexes as catalysts for olefin dimerization in an imidazolium chloroaluminate ionic liquid. Organometallics 2002;21:175-81.

[31] Bernardo-Gusmão K, Trevisan Queiroz LF, de Souza RF, Leca F, Loup C, Réau R. Biphasic oligomerization of ethylene with nickel–1,2-diiminophosphorane complexes immobilized in 1-n-butyl-3-methylimidazolium organochloroaluminate. Journal of Catalysis 2003;219:59–62.

[32] de Souza RF, Leal BC, de Souza MO, Thiele D. Nickel-catalyzed propylene dimerization in organochloroaluminate ionic liquids: Control of the isomerization reaction. Journal of Molecular Catalysis A 2007;272:6–10.

[33] Wasserscheid P, Eichmann M. Selective dimerisation of 1-butene in biphasic mode using buffered chloroaluminate ionic liquid solvents —design and application of a continuous loop reactor. Catalysis Today 2001;66:309–16.

[34] Eichmann M, Keim W, Haumann M, Melcher BU, Wasserscheid P. Nickel catalyzed dimerization of propene in chloroaluminate ionic liquids: Detailed kinetic studies in a batch reactor. Journal of Molecular Catalysis A 2009,314: 42–8.

[35] Ellis B, Keim W, Wasserscheid P. Linear dimerisation of but-1-ene in biphasic mode using buffered chloroaluminate ionic liquid solvents. Chemical Communications 1999;337–8.

[36] Chauvin Y, Di Marco-Van Tiggelen F, Olivier H. Determination of aluminium electronegativity in new ambient temperature acidic molten salts based on 3-butyl-1-methylimidazolium chloride and AlEt3-xClx (x=0-3) by 1H nuclear magnetic resonance spectroscopy. Journal of the Chemical Society, Dalton Transactions 1993;1009–11.

[37] Gilbert B, Olivier-Bourbigou H, Favre F. Chloroaluminate ionic liquids: from their structural properties to their applications in process intensification. Oil & Gas Science and Technology – Rev. IFP 2007; 62(6):745-59.

[38] Lecocq V, Olivier-Bourbigou H. Biphasic Ni-catalyzed ethylene oligomerization in ionic liquids. Oil & Gas Science and Technology – Rev. IFP 2007; 62(6):761-73.

[39] Dötterl M, Alt HG. Heavy metal with a heavy impact: olefin dimerization reactions in triphenylbismuth buffered chloroaluminate ionic liquids. ChemCatChem 2011;3(11):1799-804.

[40] Wasserscheid P, Gordon CM, Hilgers C, Muldoon MJ, Dunkin IR. Ionic liquids: polar, but weakly coordinating solvents for the first biphasic oligomerisation of ethene to higher α-olefins with cationic Ni complexes. Chemical Communications 2001;1186–7.

[41] Wasserscheid P, Hilgers C, Keim W. Ionic liquids—weakly-coordinating solvents for the biphasic ethylene oligomerization to α-olefins using cationic Ni-complexes. Journal of Molecular Catalysis A 2004;214:83–90.

[42] Song KM, Gao HY, Liu FS, Pan J, Guo LH, Zai SB, Wu Q. Ionic liquid-supported bis-(salicylaldimine) nickel complexes: robust and recyclable catalysts for ethylene oligomerization in biphasic solvent system. Catalysis Letters 2009;131:566–73.

[43] Thiele D, de Souza RF. Oligomerization of ethylene catalyzed by iron and cobalt in organoaluminate dialkylimidazolium ionic liquids. Catalysis Letters 2010;138:50–5.

[44] Ligabue RA, Dupont J, de Souza RF. Liquid–liquid two-phase cyclodimerization of 1,3-dienes by iron-nitrosyl dissolved in ionic liquids. Journal of Molecular Catalysis A 2001;169:11–7.

[45] Thiele D, de Souza RF. Biphasic ethylene oligomerization using bis(imino)pyridine cobalt complexes in methyl-butylimidazolium organochloroaluminate ionic liquids. Journal of Molecular Catalysis A 2011;340:83–8.

[46] Olivier H, Laurent-Gérot P. Homogeneous and two-phase dimerization of olefins catalyzed by tungsten complexes. The role of imido ligands and Lewis acids. Journal of Molecular Catalysis A 1999;148:43–8.

[47] Peng J, Li J, Qiu H, Jiang J, Jiang K, Mao J, Lai G. Dimerization of styrene to 1,3-diphenyl-1-butene catalyzed by palladium–Lewis acid in ionic liquid. Journal of Molecular Catalysis A 2006;255:16–8.

[48] Silva SM, Suarez PAZ, de Souza RF, Dupont J. Selective linear dimerization of 1,3-butadiene by palladium compounds immobilized into 1-n-butyl-3-methyl imidazolium ionic liquids. Polymer Bulletin 1998;40:401–5.

[49] Crowhurst L, Welton T. Brønsted acidity in ionic liquids. EUCHEM Conference on Molten Salts and Ionic Liquids: 16-22 September 2006, Hammamet, Tunisia Abstract No. 1980.

[50] Robert T, Magna L, Olivier-Bourbigou H, Gilbert B. A comparison of the acidity levels in room-temperature ionic liquids. Journal of the Electrochemical Society 2009;156: F115-F121.

[51] Mihichuk LM, Driver GW, Johnson KE. Brønsted acidity and the medium: fundamentals with a focus on ionic liquids. ChemPhysChem 2011;12:1622–32.

[52] Magna L, Bildé J, Olivier-Bourbigou H, Robert T, Gilbert B. About the acidity-catalytic activity relationship in ionic liquids: Application to the selective isobutene dimerization. Oil & Gas Science and Technology – Rev. IFP 2009;64(6):669-679.

[53] Shen ZL, He XJ, Hu BX, Sun N, Mo WM. Study on selective dimerization of α-methylstyrenes promoted by ionic liquids. Catalysis Letters 2008;122:359–65.

[54] Goledzinowski M, Birss VI. Oligomerization of low molecular weight olefins in ambient temperature molten salts. Industrial & Engineering Chemistry Research 1993;32:1795-7.

[55] Stenzel O, Brüll R, Wahner UM, Sanderson RD, Raubenheimer HG. Oligomerization of olefins in a chloroaluminate ionic liquid. Journal of Molecular Catalysis A 2003;192:217–22.

[56] Azizov AG, Kalbalieva ES, Alieva RV, Bektashi NR, Aliev BM. Oligomerization of hexene-1 in the presence of catalytic systems on the basis of ionic liquids. Petroleum Chemistry 2010;50(1):56–64.

[57] Azizov AH, Aliyeva RV, Kalbaliyeva ES, Ibrahimova MJ. Selective synthesis and the mechanism of formation of the oligoalkylnaphthenic oils by oligocyclization of 1-hexene in the presence of ionic-liquid catalysts. Applied Catalysis A 2010;375:70–7.

[58] Ibragimova MD, Samedova FI, Gasanova RZ, Azmamedov NG, Eivazov EZ. Synthesis of oligooctene and oligodecene oils in the presence of chloroaluminate ionic liquids. Petroleum Chemistry 2007;47(1):61–6.

[59] Yang S, Liu Z, Meng X, Xu C. Oligomerization of isobutene catalyzed by iron(iii) chloride ionic liquids. Energy & Fuels 2009;23:70–3.

[60] Atkins MP, Seddon KR, Swadźba-Kwaśny M. Oligomerisation of linear 1-olefins using a chlorogallate(III) ionic liquid. Pure and Applied Chemistry 2011;83(7):1391–406.

[61] Cai Q, Li J, Bao F, Shan Y. Tunable dimerization of a-methylstyrene catalyzed by acidic ionic liquids. Applied Catalysis A 2005;279:139–43.

[62] Wang H, Cui P, Zou G, Yang F, Tang J. Temperature-controlled highly selective dimerization of α-methylstyrene catalyzed by Bro°nsted acidic ionic liquid under solvent-free conditions. Tetrahedron 2006;62:3985–8.

[63] Cui X, Zhang S, Shi F, Ma X, Lu L, Deng Y. The influence of the acidity of ionic liquids on catalysis. ChemSusChem 2010;3:1043-1047.

[64] Gu Y, Shi F, Deng Y. SO3H-functionalized ionic liquid as efficient, green and reusable acidic catalyst system for oligomerization of olefins. Catalysis Communications 2003;4:597–601.

[65] Riisager A, Fehrmann R, Haumann M, Wasserscheid P. Supported ionic liquids: versatile reaction and separation media. Topics in Catalysis 2006;40: 91-102.

[66] Dötterl M, Schmidt R, Engelmann T, Denner C, Alt HG. Selective olefin dimerizaton with supported metal complexes activated by alkylaluminum compounds or ionic liquids. US 2011/0004039, 2011.

[67] Engelmann T, Denner C, Alt HG, Schmidt R. Heterogeneous dimerization of alpha-olefins with activated metallocene complexes. US 2011/0004036, 2011.

[68] Fehér C, Kriván E, Hancsók J, Skoda-Földes R. Oligomerisation of isobutene with silica supported ionic liquid catalysts. Green Chemistry 2012;14:403-409.

[69] Fehér C, Kriván E, Kovács J, Hancsók J, Skoda-Földes R. Support effect on the catalytic activity and selectivity of SILP catalysts in isobutene trimerization. Journal of Molecular Catalysis A2013;372:51–57.

[70] Liu S, Shang J, Zhang S, Yang B, Deng Y. Highly efficient trimerization of isobutene over silica supported chloroaluminate ionic liquid using C4 feed. Catalysis Today 2013,200:41– 48.

[71] Mignoni ML, de Souza MO, Pergher SBC, de Souza RF, Bernardo-Gusmão K. Nickel oligomerization catalysts heterogenized on zeolites obtained using ionic liquids as templates. Applied Catalysis A 2010;374:26–30.

[72] Chauvin Y, Commereuc D, Guibard I, Hirschauer A, Olivier H, Saussin L. Non-aqueous liquid composition with an ionic character and its use as a solvent. US 5,104,840, 1992.

[73] Chauvin Y, Einloft S, Olivier H. Catalytic process for the dimerization of olefins. US 5,550,306, 1996.

[74] Commereuc D, Forestiere A, Hughes F, Olivier-Bourbigou H. Sequence of processes for olefin oligomerization. US 6,444,866, 2002.

[75] Lecocq V, Olivier Bourbigou H. Catalyts composition for dimerizing, co-dimerizing, oligomerizing and polymerizing olefins. US 6,951,831, 2005.

[76] Wasserscheid P, Keim W. Catalyst comprising a buffered IL and hydrocarbon conversion process, e.g. oligomerisation. WO9847616, 1998.

[77] Goledzinowski M, Birss VI. Oligomerization of low molecular weight olefins in ambient temperature melts. US 5,463,158, 1995.

[78] Abdul-Sada AK, Ambler PW, Hodgson PKG, Seddon KR, Stewart NJ, Ionic liquids. WO9521871, 1995.

[79] Earle MJ, Karkkainen J, Plechkova NV, Tomaszowska A, Seddon KR. Oligomerisation. US 2008/0306319, 2008.

[80] KD Hope, MS Driver, TV Harris, High viscosity polyalphaolefins prepared with ionic liquid catalyst. US 6,395,948, 2002.

[81] Hope KD, Stern DA, Twomey DW, Collins JB. Method for manufacturing high viscosity polyalphaolefins using ionic liquid catalysts. US 7,351, 780, 2008.

[82] Bergmann LH, Hope KD, Benham EA, Stern DA. Method and system to add high shear to improve an ionic liquid catalyzed chemical reaction. US 8,163,856, 2012.

[83] Patil AO, BodigeS Process for synthetic lubricant production. US 2009/0156874, 2009.

[84] Harris TV, Cheng, Lei GD. Process for the oligomerization of olefins in Fischer-Tropsch derived feeds. WO 2005/005352, 2005.

[85] Johnson DR, HarrisTV, DriverMS Hydrotreating of Fischer-tropsch derived feeds prior to oligomerization using an ionic liquid catalyst. WO 2005/005353, 2005.

[86] Atkins MP, Smith MR, Ellis B. Lubricating oils. EP 0 791 643, 1997.

[87] Elomari S, Krug R. Process for the formation of a superior lubricant or fuel blendstock by ionic liquid oligomerization of olefins in the presence of isoparaffins. WO 2007/078595, 2007.

[88] Hommeltoft SI, Zhou Z. Supported ionic liquid reactor. US 2011/0318233, 2011.

[89] Timken K, Winter S, Lacheen HS, Hommeltoft SI. Market driven alkylation or oligomerization process. US 2011/0226669, 2011.

Chapter 3

Silk Fiber — Molecular Formation Mechanism, Structure-Property Relationship and Advanced Applications

Xinfang Liu and Ke-Qin Zhang

Additional information is available at the end of the chapter

http://dx.doi.org/10.5772/57611

1. Introduction

Silk fibers spun by several species of arthropods have existed naturally for hundreds of millions of years. The ecological functions of the silk fibers are closely related to their properties. For example, orb-weaving spiders produce a variety of different silks with diverse properties, each tailored to achieve a certain task (Figure 1) [1]. Most arthropod species produce silks used for building structures to capture prey and protect their offspring against environmental hazards [2]. The most investigated categories that have piqued the greatest amount of interest are spider silk and dragline silk in particular, produced by major ampullate glands and the cocoon silk of *Bombyx mori* (*B. mori*). The ongoing evolutionary optimization of silks from silkworms and spiders exhibit outstanding mechanical properties, such as strength and extensibility, as well as toughness, which outperform most other natural and man-made silk fibers (Table 1) [3, 4]. Due to its smooth texture, luster and strength, silks from natural silkworms have been extensively used in apparel and fashion applications for thousands of years [5]. Silks from spiders have also been utilized throughout history, such as sutures and fishing equipment in ancient Greece and Australasia.

In contrast petrochemical-based synthetic polymers commonly used today, such as polyethylene, which is formed by polymerization of ethylene at high temperature and pressure, or under the presence of some metal-based catalysis, *B. mori* and spider spin fibers from a highly concentrated, water-based protein solution under mild conditions [6, 7]. Due to current trends in exploration of natural biological materials and the demand for environmentally friendly (green) materials, investigation of the applications of silk fibers has steadily gained prominence. Silk fibers are emerging as candidates for applications in even non-apparel areas due in part to recognition for their extraordinary mechanical properties, as well as their biocompatibility and biodegradability. Currently, the promotion of silkworm as bio-factory to produce silk fibers fitting for innovative and advanced functional biological applications is a big trend. Compared to silkworm silks, the potential commercial applications for many spider silks are still extremely finite due to reasons such as difficulty of high-density spider farming, which is limited by the cannibalistic nature of most spiders. Additionally, only ~12 m of silk can be obtained from a complete spider web, this is extremely small in comparison to the 600 to 900 m of silk that is yielded by one silkworm cocoon [8].

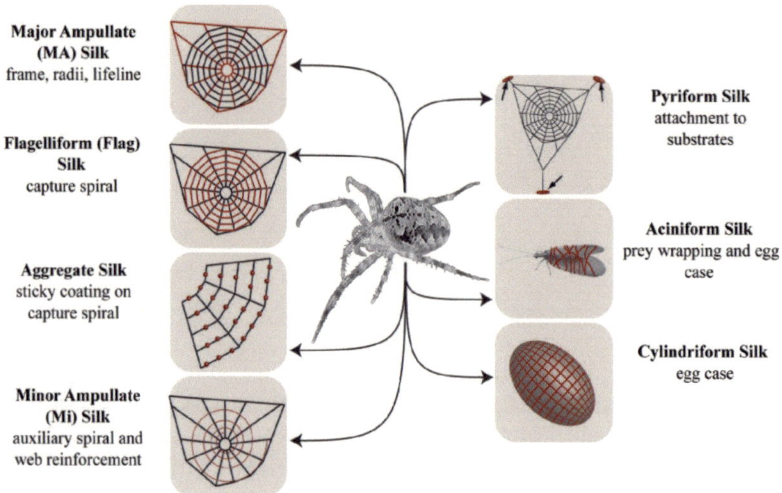

Figure 1. Schematic overview of different silk types produced by female orb-weaving spiders (*Araneae*). Each silk type (highlighted in red) is tailored for a specific purpose. (Reprinted from Ref. [1]. Copyright 2011, with permission from Elsevier.)

Fibers	Stiffness (GPa)	Strength (GPa)	Extensibility (%)	Toughness (MJ•m^{-3})
B. mori cocoon silk	7	0.6	18	70
B. mori reeled silk	15	0.7	28	150
A. Diadematus silk (dragline)	10	1.1	27	180
A. Diadematus silk (flagelliform)	0.003	0.5	270	150
Wool (at 100% RH[b])	0.5	0.2	5	60
Elastin	0.001	0.002	15	2
Nylon fiber	5	0.95	18	80
Kevlar 49 fiber	130	306	2.7	50
Carbon fiber	300	4	1.3	25
High-tensile steel	200	1.5	0.8	6

Table 1. Comparison of mechanical properties of natural silks and other synthetic fibers[a]. ([a] Data taken from refs. [3, 4]. [b] RH, relative humidity.)

Artificial spinning is the most promising method of promoting the application of silk fibers, as it can output sufficient man-made fibers cost-effectively and with specific tailored properties. Remarkable efforts for silk fiber reproduction via reconstituted/recombinant silk fibroin are currently underway [9, 10]. Reconstituted silk protein is derived from *B. mori* cocoons and the degummed *B. mori* silk fibers, they are soluble in concentrated LiBr aqueous solutions, yielding reconstituted silk protein solutions after dialysis [9]. Recent advances in transgenic technology enable the high level expression of recombinant proteins. Spider silk proteins have been produced by other organisms so that recombinant spider silk protein might be suitable for creation of artificial silk threads or other applications. Host organisms include bacteria, yeasts, animal cells and plants [3]. In order to biomimic native silks, the reconstituted or recombinant proteins used to spin artificial silks should possess amino acid similar to the ones found in native sites for sequence and composition. However, the properties of synthetic silk fibers currently do not meet the standards of native silks yet, as the composition, hierarchical structure and production conditions of natural silks all reportedly affect their mechanical properties [11]. Consequently, a profound knowledge of the natural formation process, chemical composition, relationships of structure and properties of silk fibers seems therefor imperative.

Thanks to recent developments in modern analytical techniques, significant progress has been made with respect to the structural characterization of silk. These techniques can provide molecular information about silk, including microscopic methods (atomic force microscopy (AFM), scanning and transmission electron microscopy (SEM and TEM), and scanning transmission x-ray microscopy (STXM)) and synchrotron x-ray diffraction (wide-angle x-ray diffraction (WAXD) and small-angle x-ray scattering (SAXS) combined with synchrotron radiation). Solid-state nuclear magnetic resonance (SS-NMR) is a powerful technique because it allows for the study of molecular structure and dynamics of semi-crystalline and amorphous materials. Raman and FTIR spectroscopy can provide the dominant conformational contents of a fiber. Raman microspectroscopy can be used to determine quantitative parameters characterizing the molecular structure (orientation and conformation, amino acid composition) of micrometer-sized biological samples. In this chapter, we will provide an overview of the current understanding of the silk fibers' structure taken advantage of these analytic methods, then describe in detail the structure-property relationships and the formation processes of silk fiber. Additionally, we will explore material morphologies and applications of these silk fibers.

2. The structure-property relationship of silk fibers

The structure-property relationship is one of the most intriguing 'mysteries' of silk fibers. Various studies have suggested that there is a strong connection between the structures of silk fibers and their physical (e. g., mechanical) properties. An understanding of the structure-property relationship requires background knowledge of local structure, including the component and composition of silk fiber, the conformation and orientation of constitutive units with respect to the fiber, and so on.

2.1. The structure and composition of *B. mori* and spider dragline silk fibers

In principle, the full range of properties of silk fibers can be calculated from their structural morphology and chemical composition. On the macroscopic level, the morphological structure of *B. mori* silk and spider dragline silk are very similar, as both possess a core-shell structure (Figure 2) [12]. The silk thread diameter varies across types and species. For example, coating the two core brins of *B. mori* silk fiber with sericin yields fibers about 20 μm width. Spider dragline silks have a diameter of 3-5 μm and to date, have been described to contain only one protein monofilament.

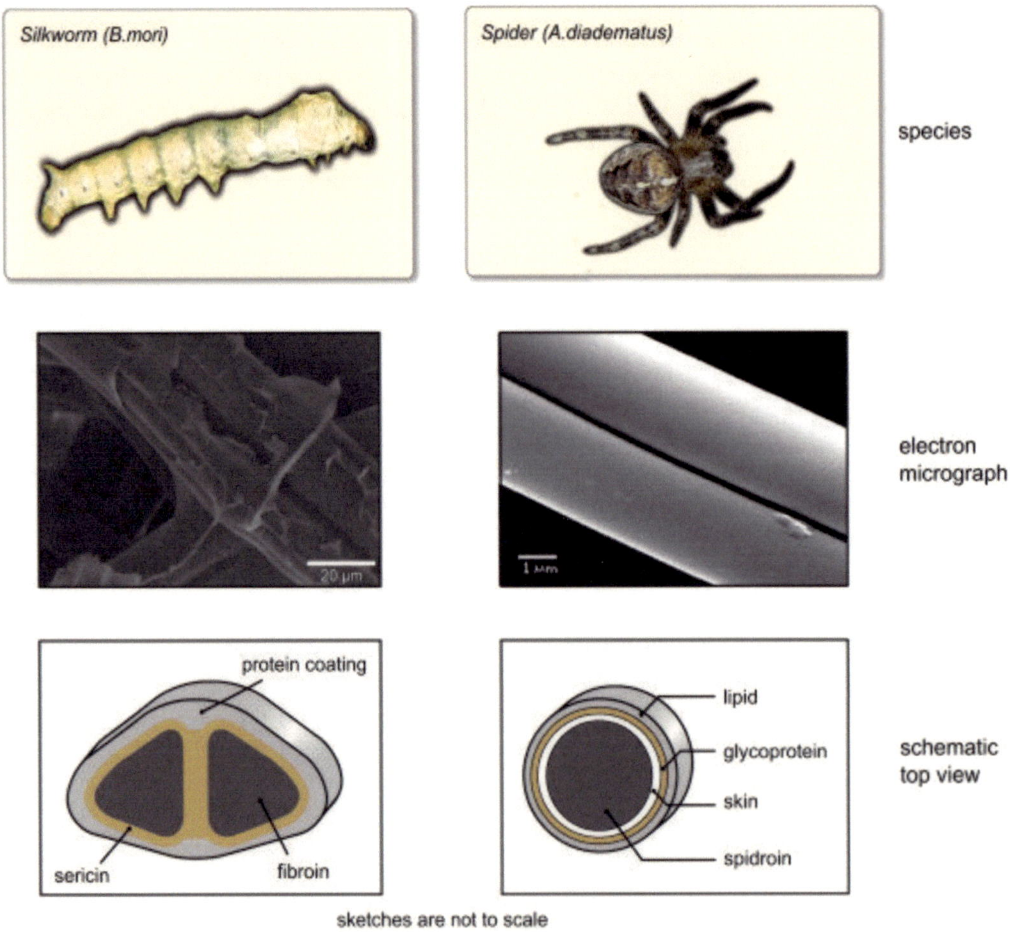

Figure 2. Examples of silk fibers produced by silkworms and spiders and a schematic illustration. (Reprinted from Ref. [12]. Copyright 2008, with permission from Elsevier.)

Silk fibers are normally polyamino acid-based fibrous proteins. In contrast, the synthetic polymers, which are usually homopolymers or copolymers consisting of one or several simpler monomer, the biopolymers – silk fibers, the primary sequence and linkage between the

http://dx.doi.org/10.5772/57611

monomers are arranged in a strictly controlled manner and are responsible for the formation of well-defined structure [13]. A range of microscopy methods, including SEM, TEM, and AFM, have been used to investigate the microstructure of silk fiber [14-19]. The results confirmed that silk fibers are composed of well-oriented bundles of nanofibrils. Generally, the coatings of silk fibers function as glue. The sericin coating, which occupies 25-30% of the weight of *B. mori* silk fiber, glues the two core brins together. However, recent studies have provided evidences that the coating may act as a fungicidal or bactericidal agent [20]. It may also have a role in the complex spinning process. Studies have demonstrated the presence of microvoids for both silk fibers [21, 22]. Microvoids are thought to develop during the final stages of the spinning process, in which viscous protein aqueous is stretched or loaded into the fibers.

As two major families of silk proteins, fibroin is the chief component of silkworm silk fiber, while spidroin (also named spider fibroin) is the analogue in spider silk fiber. The *B. mori* silk fibroin is composed of two protein chains, heavy-chain (H-fibroin) with the molecular weight of approximately 350 kDa and light chain (L-fibroin, M_w ~ 26 kDa) covalently linked by a disulfide bond at the carboxy-terminus of the two subunits [23-25] (Figure 3a). The main proteinaceous constituents of spider dragline silk are typically two major ampullate spidroins, MaSp1 and MaSp2, which are estimated to range from 250-350 kDa or larger [26-29]. A common feature of fibroins is the high content of alanine and glycine residues.

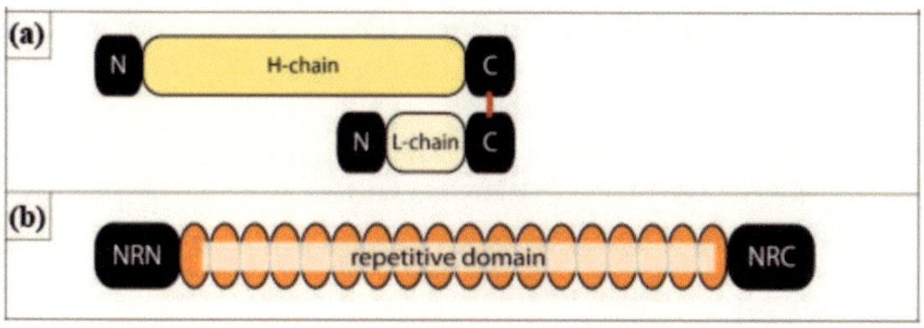

Figure 3. (a) Silkworm fibroin consisting of a covalently linked highly repetitive heavy and non-repetitive light chain. (b) Spider silk spidroins consist of a large repetitive core domain flanked by non-repetitive amino-(NRN) and carboxy-terminal (NRC) domains. (Figure slightly modified with permission from Ref. [30]. Copyright 2011 Wiley Periodicals, Inc.)

2.2. Hierarchical structure of fibroin in *B. mori* and spider silk fibers

The primary sequence plays an important role in defining basic materials. Despite being quite different in their primary structure, *B. mori* fibroin heavy chain and spider spidroins share fundamental similarities. Both have large central core of repeated modular units (Figure 4), flanked by nonrepetitive amino- (NRN) [31, 32] and carboxy- (NRC) [29] terminal domains (Figure 3). The light chain of *B. mori* fibroin, has a standard amino acid composition and a nonrepeating sequence. It plays only a marginal role in the fiber [33]. The organization of the repeating modular units can differ significantly, as seen in the sequences of different protein types. As the major component of *B. mori* fibroin, the complete amino acid sequence of the *B. mori* fibroin heavy chain is composed of a highly repetitive

(Gly-Ala)$_n$ sequence motif and tyrosine-rich domains [34]. In MaSp1, the modular units mainly consist of a subset of the sequence motifs (Ala)$_n$ followed by several GGX motifs, with X representing a variable amino acid. In MaSp2, the GGX motif is replaced by the GPGXX motif, which contains more proline residues [26, 27]. The modular units are repeated up to several hundred times in the central core of *B. mori* fibroin heavy chain and spider spidroins such that they largely determine the macroscopic properties of the fibers. The highly conserved sequence of nonrepetitive amino- and carboxy- terminal domains are essential for fiber formation and expected to be of functional relevance [35-39]. Moreover, the analysis of the hydropathicity of these fibroins reveals a pair of hydrophobic and hydrophilic counterparts. The central region of the protein is mostly hydrophobic, while the nonrepetitive amino- and carboxy- terminal domains are more hydrophilic [40].

***Bombyx mori* fibroin:**

GAGAGSGAGAGSGAGAGSGAGAGSGAGAGSGAGAGYGAGVGVGYGAGYGAGAG
AGYGAGAGSGAASGAGAGSGAGAGSGAGAGSGAGAGSGAGAGSGAGAGSGAGA
GSGAGAGSGAGAGSGAGAGSGAGVGSGAGAGSGAGAGVGYGAGAGVGYGAGAG
SGAASGAGAGSGAGAGSGAGAGSGAGAGSGAGAGSGAGAGSGAGAGSGAGAGSG

***Nephila clavipes* MaSp1:**
AAAAAAGGAGQGGYGGLGSQGAGRGGLGGQGAGAAAAAAGGAGQGGYGGLGG
QGAGQGGYGGLGSQGAGRGGLGGQGAGAAAAAAA

***Nephila clavipes* MaSp2:**
AAAAAAAASGPGQQGPGGYGPGQQGPGGYGPGQQGPSGPGSAAAAAAAASGPGQ
QGPGGYGPGQQGPGGYGPGQQGLSGPGSAAAAAAA

***Araneus diadematus* MaSp1:**
AAAAAAAAVGAGGGGQGGLGSGGAGQGYGAGLGGQGGASAAAAAAGGQGGQG
GQGGYGGLGSQGAGGAGQLGYGAGQESAAAAAAAA

***Araneus diadematus* MaSp2:**
AAAAAAGGYGPGSGQQGPSQQGPGQQGPGGQGPYGPGASAAAAAAGGYGPGSGQ
QGPGGQGPYGPGSSAAAAAA

Figure 4. Typical amino acid sequences of repetitive core of *B. mori* fibroin heavy chain, minor ampullate spidroins and major ampullate spidroins. The highly repetitive (Gly-Ala)$_n$ and (Ala)$_n$ sequence motifs are highlighted in red. The accession numbers for the sequences are P05790, P19837, P46804, AAC47009 and AAC47010 respectively.

The primary structural motifs have a preferred secondary structure and give rise to structures higher up the hierarchy. NMR, circular dichroism (CD), IR and Raman spectroscopy were usually used to examine the chemical, conformational, and orientational information of secondary structures for silk proteins [41-51]. There are three major conformations of silk proteins: the random coil, the α-helix and the β-sheet [52-54]. Using the approach of Porter, and reducing the complex secondary structure of silk proteins into fractions of ordered and

http://dx.doi.org/10.5772/57611

disordered material, they are roughly equivalent to crystalline and non-crystalline phase of silk proteins, respectively [55, 56].

The solid threads are characteristic of well-oriented β-sheet, the dominant secondary structure in silk fibers [57-59]. The first Raman spectrum of *B. mori* silk fiber has clearly shown the predominance of β-sheet, matching the results previously obtained from other techniques [60]. The total amount of β-sheet is around 50% for *B. mori* silk, which matched the proportion of the $(Gly\text{-}Ala)_n$ motif [42, 43, 46]. Therefore, it is widely accepted that the *B. mori* fibroin is composed of a highly repetitive $(Gly\text{-}Ala)_n$ sequence motif adopting antiparallel β-sheet conformation, namely silk II of the crystalline form. The β-sheet crystallite is the molecular network constructed by crosslinking β-sheet conformation of the molecular structures within several neighboring silk protein molecules [61]. It can be indexed as a monoclinic space group with a rectangular unit cell parameter of a=0.938 nm, b=0.949 nm and c=0.698 nm for *B. mori* silk [62]. Drummy *et al.* investigated *B. mori* silk fiber bundles using wide angle X-ray scattering (WAXS) [63]. The amorphous halo was also investigated from the WAXS pattern. The results concluded that silk fiber is made up of crystalline regions and connected by regions of amorphous or non-crystalline regions, each comprising of approximately 50% of the total structure. These features are in agreement with structural model proposed before [64].

It is quite firmly believed that the $(Ala)_n$ domains in spider dragline silk fibers adopt a β-sheet conformation, which are predominantly found to be antiparallel folding and organized into crystallite [41, 61, 65]. Based on the x-ray diffraction pattern, the β-sheet crystallites can be indexed as an orthogonal unit with cell parameters of a=1.03 nm, b=0.944 nm and c=0.695 nm for *Nephila* spider dragline silk [66]. From the data presented, the content of β-sheet conformation in *Nephila* spider dragline silks ranging from 30% to 40% [46, 48]. This value is much higher than the 20% average for degree of crystallinity (amount of β-sheet crystallites), as reported by XRD [67-69]. The lower crystallinity of spider dragline seems to closely correlate with less highly ordered β-sheet content for diffraction [61].

It is expected that the mechanical properties of silk fibers will critically depend on the characters of β-sheet crystallites, significant properties include crystallinity, size (aspect ratio, distribution) and dispersion of β-sheet crystallites, the intercrystallite distance, and the degree of orientation in the silk fiber. The size of β-sheet crystallite in *B. mori* silk fiber, determined via quantitative examination of the dark field TEM images, were revealed to be 20 to 170 nm in the axial direction and 1 to 20 nm in the lateral direction. And all the crystallites were uniformly distributed in the whole fiber matrix [14]. The smaller crystallite sizes, as measured from LVTEM and WAXS images, are a reasonable match to those calculated from Scherrer analysis of x-ray fiber pattern [63]. It has a large distribution of sizes range from only a few nanometers to tens of nanometers in length. The average crystallite size of *Nephila* spider dragline silks is calculated to be approximately $2 \times 5 \times 7$ nm based on X-ray diffraction patterns of single and bundles of fiber [70]. The correlation lengths related to intercrystallite distance along the fiber axis is ~ 13-18 nm, measured from SAXS for *Nephila* spider dragline silk. This value is agreeable with the distance between the β-sheets from the MaSp1 sequence [19]. The quantitative determination of orientation of the secondary structure for silk protein molecules shows that the β-sheets are aligned parallel to the fiber axis [57] and the β-sheets crystallites representing the highly ordered fraction are well-oriented along the silk fiber [70-72]. Additionally, the β-sheets of *B. mori* silks are slightly better oriented than those of dragline silks, corresponding to the fact that they are more crystalline than spider dragline silk [46].

The non-crystalline regions are often described as amorphous, poorly orientated, or randomly coiled sections of the peptide. The structural organization in the amorphous phase is not well understood yet. The existence of β-turn or β-spiral and helical conformations has been suggested for amorphous domains [42, 46, 65, 73, 74]. Tyrosine residue, on average, may form distorted β-turns and distorted β-sheets, which is characterized by ^{13}C solid-state NMR in the amorphous matrix of *B. mori* silk [42, 75]. The Gly-rich regions in spider dragline silk have been described as the amorphous rubber based on X-ray diffraction studies [76]. The precise structure of the GlyGlyX motif in MaSp1 has been somewhat controversial. Recent NMR studies provided evidence of the presence of less ordered helical type structure or distorted β-sheets adopted by the GlyGlyX motif [48, 73]. However, for MaSp2, ADF 3 and 4 (the fibroin of major ampullate dragline silk for spider *Araneus diadematus*), the structure of the GlyProGlyXX repeat has been proposed to be a β-turns or spiral structure. The stability of these structures is given by the interchain hydrogen bonding [77]. The molecular chains in the amorphous phase are often considered to be randomly oriented. Studies from Raman [46] and SS-NMR [73] reported that the protein backbones in the amorphous regions of silk fibers are not randomly oriented but exhibit certain degree of orientation along the fiber axis, albeit much less oriented than β-sheet crystallites. Meanwhile, the higher level of orientation of the amorphous phase for the spider silks than that for *B. mori* silk.

Recent computational approaches have been useful in modeling nanostructure of silk. Molecular modeling integrated the information known about the structures, and has been used to characterize the nanostructure of the silk. Based on a bottom-up molecular computational approach using replica exchange molecular dynamic, Keten *et al.* reported atomic-level structures of MaSp1 and MaSp2 proteins from the *Nephila Clavipes* spider dragline silk sequence. It showed that poly-alanine segments in silk have an extremely high propensity for forming distinct and orderly β-sheet crystallites. Previous molecular dynamic simulations on poly-alanine aggregation also suggested that anti-parallel orientations in the hydrogen bonding direction and parallel stacking in the side – chain direction leads to stable β-sheets [78]. Glycine-rich regions are less orderly, predominantly forming helical type structures and β-turns in amorphous domains. The density of hydrogen bonds in amorphous regions is lower than in β-sheet crystallites [79, 80]. All of the results are excellently consistent with available experiment evidence and may contribute towards an improved understanding of the source of silk's strength and toughness.

According to the prevalent characterizations mentioned above, silk fiber is considered a semicrystalline polymer with a hierarchical structure in which highly oriented β-sheets crystallites connecting with an amorphous matrix are organized in nanofibrils or fibrillar entities [81] (Figure 5). However, it has been proposed that there exists a third phase, or interphase consisting of weakly oriented β-sheets regions [68, 82-84] or oriented amorphous domains [85, 86] in silk. Recent NMR and IR studies performed on silk used hydrogen-deuterium (H-D) exchange to differentiate among three structures. The data revealed that the

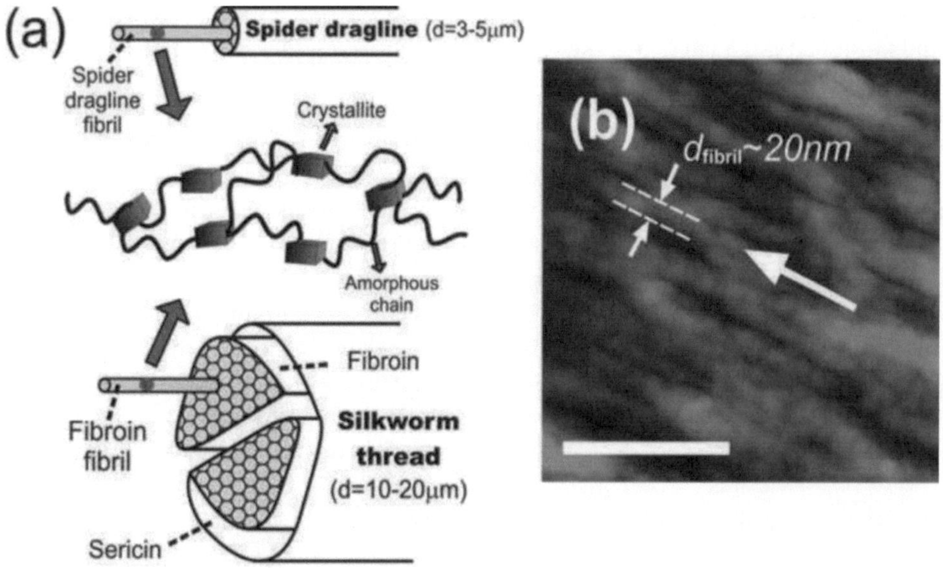

Figure 5. (a) The hierarchical structure of spider dragline and silkworm silk fiber. Both spider dragline and fibroin are composed of numerous minute fibrils, which are separated into crystalline and amorphous segments. (b) The minute fibrils in silkworm *B. mori* silk as revealed in an AFM image (scale bar: 150 nm). The silk fiber direction is indicated by the arrow. (Adapted with permission from Ref. [81]. Copyright 2011, WILEY-VCH.)

D_2O-inaccessible β-sheets are associated with crystallites, while D_2O-accessible ones are composed of amorphous domains and interphase β-sheets [82-84]. In the case of spider dragline silk, crystalline component larger in size and poorer in orientation are detectable beside the ~ 2 nm sized β-sheets crystallite that are commonly observed [69, 87, 88]. These observations of larger ordered regions have been explained as *'Non Periodic Lattice (NPL)'* crystals which form as a result of statistical matches between compatible sequences on adjacent molecular chains. It was revealed that the border where β-sheet crystallite regions and amorphous domains do not have any discrete phase boundaries. The presence of the interphase has also been deduced from STXM studies on *Nephila* dragline silk. It indicates that highly oriented and unoriented domains are surrounded by a moderately oriented matrix [85]. In summary, there is ongoing debate on the molecular structure of silk at the nanoscopic level.

2.3. The physical (mechanical) properties of silk fibers

Spider silk and *B. mori* silk feature unique physical properties – such as superior mechanical properties in terms of toughness (the amount of energy absorbed before breakage) (Table 1). So far, the maximum strength of spider dragline silk (dragline of *Caerostris darwini*) up to 1.7 GPa, which exceeds that of steel (1.5 GPa), is in the range of high-tech materials [1]. Due to its great extensibility, spider dragline silks have three times of toughness of man-made synthetic fibers like Kevlar 49 [3, 89]. Typical *B. mori* silk is presumed to be weaker and less extensible than spider dragline silk. However, when forcibly silking from immobilized silkworms artificially at certain spinning speed, the mechanical properties of the specific *B. mori* silk have greatly improved to a level that is comparable the toughest spider silk [4].

The mechanical properties of silk fibers can be described by stress-strain curve profiles, which are generated by stretching the fibers at a specific strain rate. The stress is expressed as force per cross-sectional area and the strain is defined as a normalized extensibility. Typical stress-strain curves for *B. mori* silkworm silk and spider dragline silk show both elastic behavior followed by plastic deformation [90]. The linear portion of the curve, up to the yield point, is the elastic region. The slope is defined as Young's modulus [91], a measure of the stiffness of the fiber. After the yield point, the fiber buffers the plastic deformation and the stress-strain profiles are subjected to sudden slope changes. This behavior indicates that major structural transition from rubberlike to glassy state occurs in the fiber [92-94]. These characteristics have driven scientists to explore the structural origin of the high-performance silk fiber, with the goal of obtaining templates for designing novel materials with comparable properties.

2.4. The structure-property relationship of silk fibers

Evidently, the attractive macroscopic mechanical properties of silk fiber can be ascribed to the structural effects. Most of the attention has focused on the nanometer scale: predominantly, primary and secondary structure, as well as organization and arrangement of protein molecules. In terms of primary structure of silk proteins, amino acid composition, sequential order and the number of the motifs in each module are important for the mechanical properties of the final fibers. For example, the primary structure of *Antheraea pernyi* (*A. pernyi*) silk fibroin produced by the 'wild' silkworm, especially the motif, is more like that of major ampullate spidroins than that of *B. mori* fibroin [34, 95]. It has been found that such 'wild' silkworm silk displays similar mechanical properties as spider dragline silk [96]. In addition, six novel silk proteins from *Mygalomorphae* (terantulas) do not possess high tensile strength and elasticity, due to the absence of the four motifs found in major ampullate spidroins [97]. Several studies have tried to establish correlations between specific peptide segments and the mechanical functions of the silk fibers [81, 89, 96, 98-100]. For example, glycine and proline play important specific roles in silk, as they modulate the backbone hydration and conformational order of peptides to govern the behavior of the fibers [101]. The proline-containing motif, GPGXX, was hypothesized to account for the elasticity of silk [102]. However, the primary structure of silk proteins alone does not explain the properties of silk fibers. With merely the protein of the right primary structure, the artificial spun silk fiber is far inferior to the native one [103].

The mechanical properties of silk fibers, also depend crucially on spinning conductions, such as humidity, temperature, and reeling speed, and so on [19, 104]. Variations in crystallinity and alignment can be found within the silk fiber due to variations in reeling speed of the collected sample. These variations have been mapped to mechanical properties by affecting the formation of the β-sheet crystals. As reeling speed is increased, the content of β-sheet structures rise in silk, with increasing orientation of both crystalline and amorphous fractions [105, 106]. Additionally, the tensile properties (the breaking stress and modulus) of silk fibers increase while breaking strain decreases [4, 19, 107]. It has also shown that reducing the crystal size by increasing the reeling speed has a significant influence on the toughness and ultimate

http://dx.doi.org/10.5772/57611

strength of the fiber [19]. Recently, Buehler *et al.* investigated β-sheet nanocrystals using the sequence from *B. mori* silk as a model system. They examined the key mechanical parameters of the silk β-sheet nanocrystals as a function of size. It concluded that small nanocrystals are predominantly loaded in uniform shear so that the hydrogen bonds in β-sheet strands break by means of stick-slip motion with enhanced energy dissipation and leading to greater stiffness and fracture resistance of silk [108]. Molecular models of silk protein β-sheet crystals with variation in their β-strand length were mechanically tested in molecular dynamics simulations. It was found that β-strands of around eight residues in length were optimal [109]. In summary, primary structure and spinning conditions both contribute to the observed structures higher up the hierarchy associated with the mechanical properties of the silk fiber.

Modern analytical technologies and tools have steadily contributed to the progress in experimental studies of the structure of silk fibers, as described above. However, it is still no consensus on the hierarchical structure of silk at the nanometer scale. Some models have been proposed to interpret the structure-property relationship of silk fibers. The first such model was Termonia's early model [64]. The model hypothesized that silk is a hydrogen-bonded amorphous phase with embedded stiff crystal domains acting as multifunctional cross-links and creating a thin layer of high modulus in the amorphous regions. The stiff hydrogen bonds are first broken to give the fiber its high initial modulus. Meanwhile, it allows the dynamic rubber phase to redistribute the deformation field for prediction of the nonlinear large strain deformation. The simulated properties based on the theoretical model properly reproduce the combination of high initial modulus, strength and toughness of dragline silk fiber. However, in this model, a theoretical modulus of 160 GPa for rigid β-sheets crystals, assuming fully extended crystals, is much higher than the moduli for β-sheets crystals obtained from experiments and molecular dynamic simulation [110-112]. Porter and Vollrath *et al.* includes morphological parameters by simplifying complex structural arrangements of silk fiber into ordered and disordered fractions which are best quantified by the number of amide-amide hydrogen bonds between adjacent chains. These fractions can each impart individual attributes to the property profile (such as stiffness and energy dissipation). This model predicts the range of silk tensile properties in good agreement with the experimental observation [56, 113]. Krasnov *et al.* established a viscoelatic model for *B. mori* silk in the form of standard three-parameter Maxwell model, where the elastic modulus is split into amorphous and crystalline elastances. Subsequently, the elastic modulus is parallelly connected to the elements standing for the relaxation processes of the amorphous regions which are observed through cyclic tensile stretching measurements on a single silkworm silk fiber. They separated the mechanical properties of the crystalline and amorphous phases, as well as the interplay between mechanical properties and morphology, of silk. The model fits well with the reports of testing experiment [111]. Buehler *et al.* presented a simple coarse-grained model in which a combination of β-sheet nanocrystal and semi-amorphous region is modeled by beads connected via multilinear springs in a serial arrangement, representing the fundamental unit building block of the silk fiber. The mechanical behavior of these domains was simulated based on this model and the resulting stress-strain curve displays the characteristic shape observed in silk. It develops a fundamental understanding of silk's mechanics. In general, amorphous regions contribute to the elasticity of the material. The amorphous regions unravel first when silk is being stretched, leading to its large extensibility. Conversely, highly ordered, crystalline regions play a major role in determining the strength and stiffness of silks [114].

Multi-scale experimental and simulation analyses are the key to improve our systematic understanding of how structure and properties are linked. The mechanical mechanism at the macroscopic scale, namely, the fibril, including morphology and its consequence for mechanical behavior and the mechanistic interplay with nanostructure of silk, has also been elucidated [115-117]. At the same time, many experiments have been employed to assess the effect of structural changes on the mechanical deformation of silk [118-121]. When mechanical load are applied to the fibers, conformation, reorientation, crystallite size, and some other structural characters are monitored to explain the structure-property relationships.

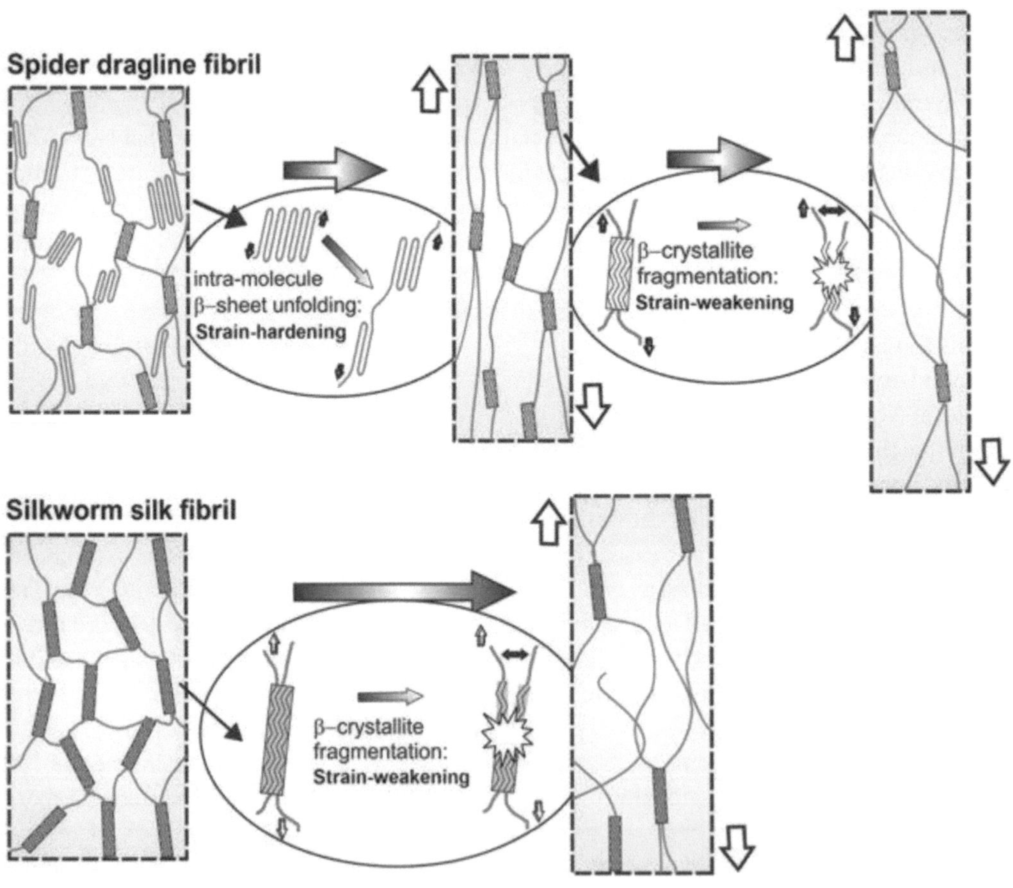

Figure 6. A schematic model demonstrating how the silkworm and spider dragline fibers respond when they are subjected to stretching. There are two components in the alanine-rich regions of spider dragline silk: β-crystallites and intramolecular β-sheets. (Adapted with permission from Ref. [81]. Copyright 2011, WILEY-VCH.)

The experimental and computational investigations shown above have explored mechanical properties of *B. mori* silkworm silks and spider draglines at different structural hierarchies

http://dx.doi.org/10.5772/57611

from sequence to crystallites to fibrils to fibers, as well as the effect of structural changes on the overall mechanical behavior of silk fibers. In particular, a closer analysis of the mechanical response, spider draglines behave typically the strain-hardening process in the post-yield region [81]. Based on the 'β-sheet splitting' mechanism [122], the occurrence of strain-hardening in spider dragline as response to the structural factors has been clearly addressed. Spider dragline silk can acquire extra toughness as a strain-hardening material by breaking intramolecular β-sheets. On the other hand, *B. mori* silkworm silk has far fewer intramolecular β-sheets in the amorphous region, therefore it is less extensible and only exhibits strain-weakening after yield point (Figure 6).

Furthermore, unlike *B. mori* silk and the other types of spider silks, the mechanical properties of spider dragline silks are greatly influenced by water. When an unconstrained dragline silk fiber is immersed in water or comes in contact with a relative humidity greater than 60%, the thread starts to swell radially, doubling in diameter and a shrinking to half of its original length [123]. This process is known as supercontraction, which is another interesting property of spider silk. A number of research groups have used different experimental techniques to understand supercontraction and the underlying mechanisms [124-126]. It is assumed that supercontraction is a result of reorientation of hydrogen bonds within the chains of protein molecules and is accompanied with the release of the prestress [127-130]. Some researchers attributed supercontraction mainly on the proline content of MaSp2 protein. Notably, the content of proline does not correlate with the mechanical performance of spider dragline silk fibers from different species, but influences the mechanical properties of wetted dragline silks [99, 126, 131, 132]. Recently, Guan *et al.* discussed the role of the two MaSp1 and MaSp2 proteins in supercontraction and quantified a contraction of about 13% maximum, linked to the disordered component of MaSp1 protein. Thus the remaining supercontraction to a total of about 30% is linked to the intrinsically disordered proline-containing fraction of MaSp2 protein [133]. After supercontraction, the silk is called supercontracted fibers, and are usually to be employed to study the structure-property relationship of silk fiber [123, 134].

3. Silk protein assembly and silk fiber formation mechanism

The remarkable mechanical properties of silk fibers have spawned great interests in determination of their origin. Systematic studies of the natural spinning process of silk fibers have shown a highly sophisticated hierarchical process, allowing for the transformation of soluble silk protein into solid fibers with specific mechanical and functional properties. Although much is already known about the characteristics of the silk proteins and silk fibers themselves, the process for silk assembly and spinning into fibers is yet to be resolved. A detailed knowledge of silk fiber formation is critical for the biomimetic production of tough silk-like fibers.

3.1. Natural spinning process for *B. mori* silk and spider draglines

In nature, silk proteins are secreted and stored in the glands until they are processed into fibers. Morphological and histological studies demonstrate that the silk glands of *B. mori* silkworm are a pair of tubes and the two tubal glands are connecting before the spinneret. The gland of *B. mori* silkworm can be divided into three parts: posterior, middle and anterior [135]. The fibrion protein is synthesized and present in a weak gel in the posterior division. The secreted proteins are transported to the middle division, where the sericin is synthesized, accumulating as a shell around the fibroin. Due to the water going out through the cell wall of the gland in the middle division, highly concentrated gel-like fibroin begins to undergo a gel-sol transition and serves as a concentrated protein solution of 30 wt% [136]. Notably, the highly concentrated liquid protein, often referred to as the spinning dope, displays nematic liquid crystal properties [137-139]. The spinning dope is exposed to the elongational flow and moves forward in the anterior division. The shear force increases along the anterior division and the spinneret, leading to the orientation of the liquid crystallinity. The crystalline spinning dope is converted into a fiber containing water-insoluble silk II. This process is accompanied by extrusion through the spinneret into air, evaporating the residual water. Another remarkable feature of the fiber spinning is the stretching force, which is brought about by the repeated drawing back of the silkworm's head, causing the orientation of protein molecules along with the silk fiber.

The major gland responsible for the dragline silk of *Nephila clavipes* spider contains the following components: a long tail, a wider sac, named ampulla, and spinning duct approaching the spinneret [140]. Each division of the gland possesses a unique function in fiber formation. For instance, a highly viscous silk protein solution of ~ 50% (w/v) is secreted from the A-zone of the gland, which is comprised of the tail and two thirds part of the sac. Further compounds forming the shell of the fiber may arise in the B-zone of the gland, which occupies the rest part of the sac. Like in the *B. mori* silkworm, the viscosity of the spider's liquid crystalline protein becomes lower and the spinning dope moves forward in the spinning duct where the orientation of liquid crystalline protein into a fiber begins. Due to its tapering, the shear force is increasing along the spinning duct and the stress forces generated in the drawdown process bring the protein molecules into alignment. Hence the protein molecules join together with hydrogen bonds to give the final fiber with anti-parallel β-sheet structure. As the silk protein molecules aggregate and crystallize, they become more hydrophobic, inducing the loss of water from the surface of the silk fiber [141-143].

3.2. Silk protein assembly and silk fiber formation mechanism on a structural view

The formation of a solid fiber from soluble silk proteins is a remarkable process owing to complex biochemical and physical changes. For silk spinning, several assembly models, such as liquid spinning theory [136] and micelle theory [144] have been proposed for the fiber formation, whereas the details remain to be elucidated. In order to understand the mechanisms of silk proteins assembly and fiber formation, the structure of proteins stored in *B. mori* silkworm gland and the major ampullate gland of spider should be clarified. In *vivo*, freshly secreted fibroin first adopts silk I (the crystalline form of *B. mori* silk fibrion found before the spinning process) and random-coil conformation [145]. Silk I is less stable as shown by attempts to study the secondary structure of silk I form using x-ray diffraction, electron diffraction or SS-NMR have caused the silk I to convert to silk II easily. Silk I remains poorly understood. Most investigations on the structure of the silk I form have been based on model building of peptides such as (Gly-Ala)$_n$ [146, 147]. The comparison of these models with limited experimental data, resulting in a number of conflicting models describing the structure of silk I. Recently, the structure of silk I has been proposed as a repeated β-turn type II-like structure [148, 149]. The secreted dragline proteins are mainly natively unfolded within the gland and consist of random-coil and polyproline-II with helix-like structure [150-152]. There is evidence indicates that the polyalanine motifs form polyproline-II with a helix-like structure. Particularly, the polyproline-II conformation may be important for maintaining the highly concentrated spinning dope, since the extended polyproline-II structure could prevent the formation of intramolecular hydrogen bonds. Additionally, the polyproline-II helix in spider fibroin favors transforming into a β-sheet structure due to their similarity of dihedral angles.

It has reported that *B. mori* fibroins and spider spidroins usually show micellar-like structure [16, 144] with an amphiphilic sequence, implying short alternating hydrophilic and hydrophobic amino acid stretches flanked by larger hydrophilic terminal regions [153, 154]. The intervening hydrophilic blocks located among the hydrophobic blocks in the protein prevent premature β-sheet formation, thus maintaining the solubility of the solution. Hence, the silk fiber formation involves shear force inducing the conversion of silk protein with specific structural conformations into β-sheet structure. This conversion occurs in the spinning ducts followed by drawing down into fibrillar structure. It has shown upon passage through the gland and spinning duct, the proteins encounter remarkable changes in their solvent environment, such as extensional flow, protein concentration, pH and metal ion concentrations, which are thought to be contributing factors in silk processing and affecting structural conformations [3, 142, 155-157]. The changes include removal of some water, slight acidification. In addition, the concentration of calcium ions (Ca^{2+}) are increased as the silk protein flowed through the gland in *B. mori*. Unlike in the silkworm *B. mori*, the elemental composition of the silk dope in the spider suggested that Ca^{2+} ions concentration stayed constant while potassium ions (K^+) concentration increased. Meanwhile, sodium ions (Na^+), chlorine ions (Cl^-) are removed from the major ampullate gland.

Experiments made *in vitro* can provide several relevant insights into the process of silk protein assembly and the formation mechanism of the silk fiber. To unravel the assembly mechanism, the reconstituted/recombinant proteins were applied for fiber assembly under specific condition [158-163]. Jin *et al.* characterized the change of the supramolecular structure of silk fibroin with a reduction of the pH, which demonstrates the self-assembly of silk fibroin as a function of pH. When the pH is reduced from 6.8 to 4.8, a morphological transition of silk fibroin from spherical micelles to nanofibrils and the conformational transition of silk fibroin from random coil to β-sheet were observed [158]. It may be driven by the stretching entropy effect related to the hydrophobic block in the protein. Shniepp *et al.* [160] deposited the silk protein solution on mica substrates without and with shearing by spin-coating. Only when shear force was applied during deposition, fibrillar structures were obtained. A microfluidic device was employed in which the ion concentrations and pH value could be controlled, and simultaneously, physical stress could be applied by channel design [163]. Silk fibers formed after addition of phosphate, application of an elongational flow, and a pH change from 8 to pH 6.

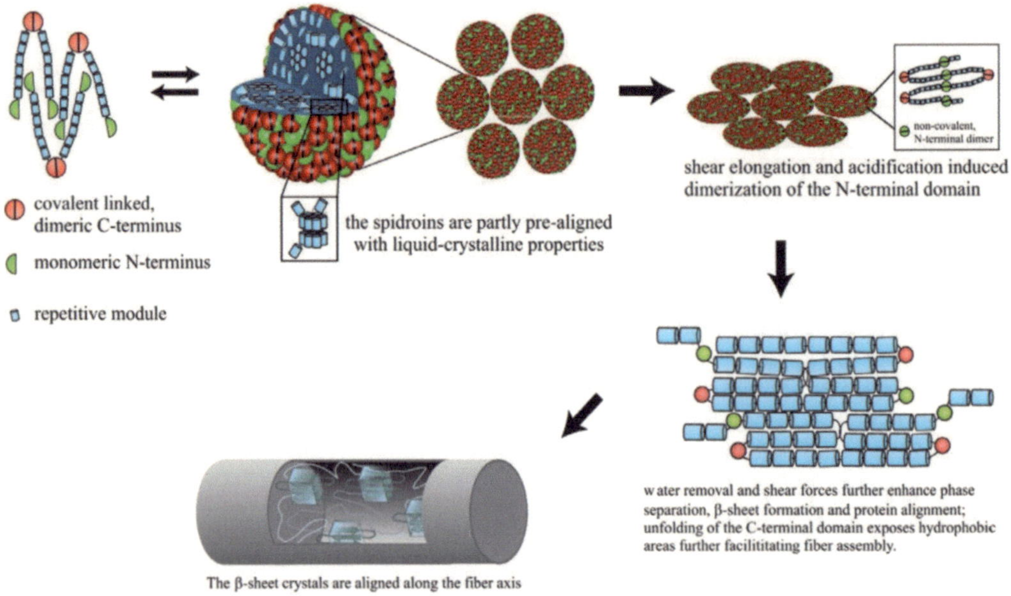

Figure 7. Schematic formation mechanism of the hierarchical assembly from molecular silk fibroin to microfibers.(Reprinted from Ref. [1]. Copyright 2011, with permission from Elsevier.)

Actually, the chemical and mechanical stimuli together are likely to influence the fold of nonrepetitive amino-terminal and carboxy-terminal and the hydrophilic spacers within the hydrophobic core domain [37, 135, 158, 164-166]. Due to the larger hydrophilic blocks at the chain ends of the protein molecules having charged groups, it is possible that they might play an important role in the molecular assembly and conformational transition at a specific pH through decreased electrostatic repulsion. A significant step towards understanding the effect of the terminal domains in assembly was the determination of atomistic structures of the nonrepetitive terminal regions of MaSp proteins. Kessler and Scheibel's group reported the structure of carboxy-terminal domain of *Araneus diadematus* ADF 3 by NMR spectroscopy [36]. And Johannson, Knight reported the structures of amino-terminal domain of *Euprosthenops australis* MaSp1 by X-ray scattering [35]. Interestingly, both terminal domains are mainly composed of α-helical barrels but with different folds. The carboxy-terminal domain mediates homodimerization via a disulfide bond [36] and forms a clamp-like structural arrangement. It seems to be implicated in a number of different functions, including control of solubility and fiber formation. The carboxy-terminal domains are able to form supramolecular assemblies resembling micellar-like structure, which is stabilized by the chaotropic ions (Figure 7). The amino-terminal domains are monomeric at pH 6.8 and above [35]. And recent NMR and light scattering studies on the amino-terminal nonrepetitive domain of *Latrodectus hesperus* confirmed that a combination of pH and salt concentration controlled the dimerization. The monomer was clearly stabilized at neutral pH in the presence of salt. While the lower pH and/ or the reduced salt concentration causes the amino-terminal nonrepetitive domain to dimerize in an antiparallel fashion to create head-to-tail dimmers to dimmers [165]. The pH dependence of silk fiber formation showed that an oligomerization is greatly increased with a drop in pH at about 6.0, which is triggered by the amino-terminal domain [35, 165]. The structural changes of the amino-terminal nonrepetitive domains rearrange the position of the core regions within the micellar-like structure, together with mechanical stimuli supports the β-sheet formation. With the exchange of chaotropic ions for the relatively kosmotropic ions, exposing hydrophobic patches can enhance the assembly process. The amino-terminal and carboxy-terminal domains sense changes in salt, pH, and shear force. And the fine-tuned interplay between these parameters enables the silkworm and spider to efficiently produce a stable very tough fiber under mild conditions

4. Advanced applications of silk fibers

Traditionally, silk has been utilized in the construction of textiles. Current research in silk fibers involves their innovative trends and advanced applications. Basically, the rich proportion of essential amino acids in silk fibers indicates high nutritive value, meaning that silk fibroin can be used as a dietary additive [167-169]. Furthermore, the amino acids, glycine, alanine, serine and tyrosine are of vital for nourishing the skin. The crystalline structure of silk protein reflects UV radiation, acting as protective buffer between the skin and environment. The extracts of silk protein are used in soap making, personal care and cosmetic products. The silk protein is also applied to enhance glossy, brightness, and softness of products. In addition, the production of advanced man made super-fibers such as Kevlar involves petrochemical processing, which contributes to pollution. Interest in silk fibers is mainly due to the combination of the mechanical properties and eco friendly way in which they are made. Spider silk fibers have been envisioned to be applied in a variety of technical textiles, including parachute cords, protective clothing and composite materials in aircrafts, which demand high toughness in combination with sleaziness.

Silks are biocompatible, biodegradable and have implant ability, as well as morphologic flexibility. Silk fiber has been used as extremely thin suture for eye or nerve surgery for long history [170]. Nowadays, one attractive application of silk fibers is act as a source of novel biomaterials. Recent progress with processing of silk fibers into various material forms, usually via the formation of the fibroin/spidroin solution, including thread, hydrogels, tubes, sponges, microspheres, particles and films [9, 171], promotes the field of applications for silk fibers in general (Figure 8) [172]. Silk protein can be modified by chemical treatment or used in combination with other materials and the silk-based biomaterials have been transformed for high-technology uses, with promising futures in the fields of biomedicine and material engineering. Numerous studies have demonstrated that fibroin supports cell attachment and proliferation for a variety of cell types [173-178]. Studies have established a potential for silk-based biomaterials use as tissue engineering scaffolds, such as skeletal tissue like bone [179], ligaments [180], and cartilage [181, 182], as well as skin [183], blood vessels [184] and nerve [185]. Silks can be designed and offer another biomedical applications, such as delivery of small molecule drugs, proteins and genes [186]. Silk fibroin possesses remarkable optical properties, such as near-perfect transparency in a visible range. It has been identified as a suitable material

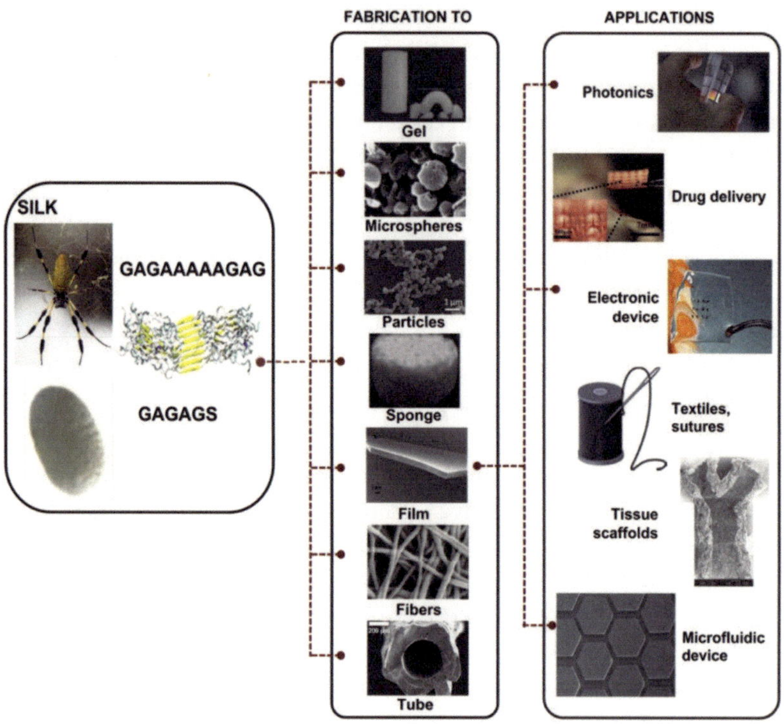

Figure 8. Possible structure and technical applications of the silk fibers. The dotted line shows an example of the versatility of silk and the multiple possible applications. (Figure slightly modified with permission from Ref. [172]. Copyright 2012, Elsevier.)

for the development of biophotonic components [187-189] in biomedical device performing electronics or sensors [190-198]. Surely, these impressive biopolymers are extremely promising for their potential applications in material science and engineering.

5. Conclusions

Our review in current chapter concentrated on *B. mori* silk and spider dragline. It links the physical and mechanical properties of native silk to the molecular make up, assembly and formation process in *B. mori* silkworm and spider. Over the last decade, there has been considerable progress in understanding the molecular structure of silk, which has inspired us a range of research utilizing the repeating modules of silk in combination with other chemical motifs to develop novel materials. The physical properties of silks highlight the potential for threads to act as high performance fibers. The advances in genetic engineering and gene sequencing enable the production of recombinant proteins in large amount and the exploration of various applications foreseeable in industry. The key point for this area is the development of suitable spinning technologies to reproducibly form threads with properties similar to that of the natural silk. The molecular assembly process of silk has provided us concepts to copy and mimic the silkworm's or spider's ways of making and processing silks with tunable properties. Such knowledge is beneficial for further improvement of synthetic polymer-based materials. Combined with the discovery of new bio-inspired materials, the future application space seems more and more broad. As a whole, further development of related yield is underway.

Acknowledgements

The authors thank the financial support from the National Science Foundation of China (NSFC) under Grant 51073113, 91027039 and the Natural Science Foundation of the Jiangsu Higher Education Institutions of China under Grant 10KJA540046. This work was also supported by the Priority Academic Program Development of Jiangsu Higher Education Institutions (PAPD). We also acknowledge support from the Priority Academic Program Development of Jiangsu Higher Education Institutions (PAPD), Qing Lan Project for Excellent Scientific and Technological Innovation Team of Jiangsu Province (2012) and Project for Jiangsu Scientific and Technological Innovation Team (2013). The author, Xinfang Liu, especially thank the support of the Postdoctoral Science Foundation of Jiangsu province (No. 1201030B).

Author details

Xinfang Liu and Ke-Qin Zhang*

*Address all correspondence to: kqzhang@suda.edu.cn

National Engineering Laboratory for Modern Silk, College of Textile and Clothing Engineering, Soochow University, Suzhou, China

References

[1] Eisoldt L., Smith A., Scheibel T. Decoding the secrets of spider silk. Materials Today 2011; 14(3): 80–86.

[2] Craig C L. Evolution of arthropod silks. Annual Review of Entomology 1997; 42: 231–267.

[3] Heim M., Keerl D., Scheibel T. Spider silk: From soluble protein to extraordinary fiber. Angewandte Chemie, International Edition 2009; 48(20): 3584–3596.

[4] Shao Z., Vollrath F. Surprising strength of silkworm silk. Nature 2002; 418(15): 741–741.

[5] Iizuka E. Silk thread: Mechanism of spinning and its mechanical properties. Journal of Applied Polymer Science: Applied Polymer Symposium 1985; 41: 173–185.

[6] Heslot H. Artifical fibrous proteins: A review. Biochimie 1998; 80(1): 19–31.

[7] Asakura T., Kaplan D L. Silk production and processing. Encyclopedia of Agricultural Science 1994; 4: 1–11.

[8] Lewis R. Unraveling the weave of spider silk. Bioscience 1996; 46(9): 636–638.

[9] Rockwood D N., Preda R C., Yücel T., Wang X Q., Lovett M L., Kaplan D L. Materials fabrication from *Bomyx mori* silk fibroin. Nature Protocols 2011; 6(10): 1612–1631.

[10] Humenik M., Smith A M., Scheibel T. Recombinant spider silks – Biopolymers with potential for future applications. Polymer 2011; 3(1): 640–661.

[11] Fu C., Shao Z., Fritz V. Animal silks: their structures, properties and artificial production. Chemical Communication 2009; 43: 6515–6529.

[12] Hardy J G., Römer L M., Scheibel T R. Polymeric materials based on silk proteins. Polymer 2008; 49(30): 4309–4327.

[13] Gührs K H., Weisshart K., Grosse F. Lessons from nature – protein fibers. Reviews in Molecular Biotechnology 2000; 74(2): 121–134.

[14] Shen Y., Johnson M A., Martin D C. Microstructural characterization of *Bombyx mori* silk fibers. Macromolecules 1998; 31(25): 8857–8864.

[15] Li S F., McGhie A J., Tang S L. New internal structure of spider dragline silk revealed by atomic force microscopy. Biophysical Journal 1994; 66(4): 1209–1212.

[16] Oroudjev E., Soares J., Arcdiacono S., Thompson J B., Fossey S A., Hansma H G. Segmented nanofibers of spider dragline silk: Atom force microscopy and single-molecule force spectroscopy. Proceedings of the National Academy of Science of the United States of America 2002; 99(9, suppl.2): 6460–6465.

[17] Augsten K., Muehlig P., Hermann C. Glycoproteins and skin-core structure in *Nephila clavipes* spider silk observed by light and electron microscopy. Scanning 2000; 22(1): 12–15.

[18] Putthanarat S., Stribeck N., Fossey S A., Eby R K., Adams W W. Investigation of the nanofibrils of silk fibers. Polymer 2000; 41(21): 7735–7747.

[19] Du N., Liu X Y., Narayanan J., Li L., Lim M L M., Li D. Design of superior spider silk: From nanostructure to mechanical properties. Biophysical Journal 2006; 91(12): 4528–4535.

[20] Padamwar M N., Pawar A P. Silk sericin and its applications: A review. Journal of Scientific & Industrial Research 2004; 63(4): 323–329.

[21] Robson R M. Microvoids in *Bombyx mori* silk. An electron microscope study. International Journal of Biological Macromolecules 1999; 24(2-3): 145–150.

[22] Frische S., Maunsbach A B., Vollrath F. Elongate cavities and skin-core structure in *Nephila* spider silk observed by electron microscopy. Journal of Microscopy 1998; 189(1): 64–70.

[23] Takei F., Kikuchi Y., Kikuchi A., Mizuno S., Shimura K. Further evidence for importance of the subunit combination of silk fibroin in its efficient secretion from the posterior silk gland cells. The Journal of Cell Biology 1987; 105(1): 175–180.

[24] Tanaka K., Mori K., Mizuno S. Immunological identification of the major disulfide-linked light component of silk fibroin. Journal of Biochemistry (Tokyo) 1993; 114(1): 1–4.

[25] Tanaka K., Kajiyama N., Ishikura K., Waga S., Kikuchi A., Ohtomo K., Takagi T., Mizuno S. Determination of the site of disulfide linkage between heavy and light chains of silk fibroin produced by *Bombyx mori*. Biochimica et Biophysica Act (BBA) – Protein Structure and Molecular Enzymology 1999; 1432(1): 92–103.

[26] Xu M., Lewis R V. Structure of a protein superfiber: spider dragline silk. Proceedings of the National Academy of Science of the United States of America 1990; 87(18): 7120–7124.

[27] Hinman M B., Lewis R V. Isolation of a clone encoding a second dragline silk fibroin. *Nephila clavipes* dragline silk is a two-protein fiber. The Journal of Biological Chemistry 1992; 267(27): 19320–19324.

[28] Beckwitt R., Arcidiacono S. Sequence conservation in the C-terminal region of spider silk proteins (spidroin) from *Nephila clavipes* (Tetragnathidae) and *Araneus bicentenarius* (Araneidae). The Journal of Biological Chemistry 1994; 269(9): 6661–6663.

[29] Sponner A., Vater W., Rommerskirch W., Vollrath F., Unger E., Grosse F., Weisshart K. The conserved C-termini contribute to the properties of spider silk fibroins. Biochemical and Biophysical Research Communications 2005; 338(2): 897–902.

[30] Eisoldt L., Thamm C., Scheibel T. The role of terminal domains during storage and assembly of spider silk proteins. Biopolymers 2012; 97(6): 355–361.

[31] Rising A., Hjälm G., Engström W., Johansson J. N-terminal nonrepetitive domain common to dragline, flagelliform, and cylindriform spider silk proteins. Biomacromolecules 2006; 7(11): 3120–3124.

[32] Motriuk-Smith D., Smith A., Hayashi C Y., Lewis R V. Analysis of the conserved N-terminal domains in major ampullate spider silk proteins. Biomacromolecules 2005; 6(6): 3152–3159.

[33] Zhou C Z., Confalonieri F., Jacquet M., Perasso R., Li Z G., Janin J. Silk fibroin: structural implications of a remarkable amino acid sequence. Proteins: Structure, Function, and Genetics 2001; 44(2): 119–122.

[34] Zhou C Z., Confalonieri F., Medina N., Zivanovic Y., Esnault C., Yang T., Jacquet M., Janin J., Duguet M., Perasso R., Li Z G. Fine organization of *Bombyx mori* fibroin heavy chain gene. Nucleic Acid Research 2000; 28(12): 2413–2419.

[35] Askarieh G., Hedhammar M., Nordling K., Saenz A., Casals C., Rising A., Johansson J., Knight S D. Self-assembly of spider silk proteins is controlled by a pH-sensitive relay. Nature 2010; 465(13): 236–238.

[36] Hagn F., Eisoldt L., Hardy J G., Vendrely C., Coles M., Scheibel T., Kessler, H. A conserved spider silk domain acts as a molecular switch that controls fibre assembly. Nature 2010; 465(13): 239–242.

[37] He Y X., Zhang N N., Li W F., Jia N., Chen B Y., Zhou K., Zhang J., Chen Y., Zhou C Z. N-terminal domain of *Bombyx mori* fibroin mediates the assembly of silk in response to pH decrease. Journal of Molecular Biology 2012; 418(3-4): 197–207.

[38] Hagn F. A structural view on spider silk proteins and their role in fiber assembly. Journal of Peptide Science 2012; 18(6): 357–365.

[39] Ittah S., Cohen S., Garty S., Cohn D., Gat U. An essential role for the C-terminal domain of a dragline spider silk protein in directing fiber formation. Biomacromolecules 2006; 7(6): 1790–1795.

[40] Bini E., Knight D P., Kaplan D L. Mapping domain structures in silks from insects and spiders related to protein assembly. Journal of Molecular Biology 2004; 335(1-2): 27–40.

[41] Simmons A., Ray E., Jelinski L. W. Solid-state ^{13}C NMR of *Nephila clavipes* dragline silk establishes structure and identity of crystalline regions. Macromolecules 1994; 27(18): 5235–5237.

[42] Asakura T., Yao J. ^{13}C CP/MAS NMR study on structural heterogeneity in *Bombyx mori* silk fiber and their generation by stretching. Protein Science 2002; 11(11): 2706–2713.

[43] Boulet-Audet M., Lefèvre T., Buffeteau T., Pézolet M. Attenuated total reflection infrared spectroscopy: An efficient technique to quantitatively determine the orientation and conformation of proteins in single silk fibers. Applied Spectroscopy 2008; 62(9): 956–962.

[44] Holland G P., Creager M S., Jenkins J E., Lewis R V., Yarger J L. Determining secondary structure in spider dragline silk by carbon-carbon correlation solid-state NMR spectroscopy. Journal of the American Chemistry Society 2008; 130(30): 9871–9877.

[45] Colomban P., Dinh H M., Riand J., Prinsloo L C., Mauchamp B. Nanomechanics of single silkworm and spider fibres: a Raman and micro-mechanical *in situ* study of the

conformation change with stress. Journal of Raman Spectroscopy 2008; 39(12): 1749–1764.

[46] Lefèvre T., Rousseau M E., Pézolet M. Protein secondary structure and orientation in silk as revealed by Raman spectromicroscopy. Biophysical Journal 2007; 92(8): 2885–2895.

[47] Holland G P., Jenkins J E., Creager M S., Lewis R V., Yarger J L. Quantifying the fraction of glycine and alanine in β-sheet and helical conformations in spider dragline silk using solid-state NMR. Chemical Communication 2008; 43: 5568–5570.

[48] Jenkins J E., Creager M S., Lewis R V., Holland G P., Yarger J L. Quantitative correlation between the protein primary sequences and secondary structures in spider dragline silks. Biomacromolecules 2010; 11(1): 192–200.

[49] Ling S., Qi Z., Knight D P., Shao Z., Chen X. Synchrotron FTIR microspectroscopy of single natural silk fibers. Biomacromolecules 2011; 12(9): 3344–3349.

[50] Dicko C., Knight D., Kenney J M., Vollrath F. Structural conformation of spidroin in solution: A synchrotron radiation circular dichroism study. Biomacromolecules 2004; 5(3): 758–767.

[51] Lefèvre T., Paquet-Mercier F., Rioux-Dubé J F., Pézolet M. Structure of silk by Raman spectromicroscopy: From the spinning glands to the fibers. Biopolymers 2012; 97(6): 322–336.

[52] Kaplan D.; Adams W W., Farmer B., Viney C., editors. Silk polymers: materials science and biotechnology. American Chemical Society Symposium Series 1994.

[53] Marsh R E., Corey R B., Pauling L. An investigation of the structure of silk fibroin. Biochimica et Biophysica Acta 1955; 16: 1–34.

[54] Warwicker J O. Comparative studies of fibroins: II. The crystal structures of various fibroins. Journal of Molecular Biology 1960; 2(6): 350–362.

[55] Dekker M. In: Porter D. (ed.) Group interaction modeling of polymer properties. New York. 1995. P499.

[56] Porter D., Vollrath F., Shao Z. Predicting the mechanical properties of spider silk as a model nanostructured polymer. The European Physical Journal E 2005; 16(2): 199–206.

[57] Rousseau M E., Lefèvre T., Beaulieu L., Asakura L., Pézolet M. Study of protein conformation and orientation in silkworm and spider silk fibers using Raman microspectroscopy. Biomacromolecules 2004; 5(6): 2247–2257.

[58] Krimm S., Bandekar J. Vibrational Spectroscopy and conformation of peptides, polypeptides, and proteins. Advances in Protein Chemistry 1986; 38: 181–364.

[59] Dong J., Wan Z., Popov M., Carey P R., Weiss M. A. Insulin assembly damps conformational fluctuations: Raman analysis of amide I line-widths in native states and fibrils. Journal of Molecular Biology 2003; 330(2): 431–442.

[60] Zheng S., Li G., Yao W., Yu T. Raman spectroscopic investigation of the denaturation process of silk fibroin. Applied Spectroscopy 1989; 43(7): 1269–1272.

[61] Simmons A H., Michal C A., Jelinski L W. Molecular orientation and two-component nature of the crystalline fraction of spider dragline silk. Science 1996; 271(5245): 84–87.

[62] Takahashi Y. Crystal structure of silk of *Bombyx mori*. In: Kaplan D.; Adams W W., Farmer B., Viney C. (eds.) Silk polymers: materials science and biotechnology. American Chemical Society Symposium Series 1994. 544: P168–175.

[63] Drummy L F., Farmer B L., Naik R R. Correlation of the β-sheet crystal size in silk fibers with the protein amino acid sequence. Soft Matter 2007; 3(7): 877–882.

[64] Termonia Y. Molecular modeling of spider silk elasticity. Macromolecules 1994; 27(25): 7378–7381.

[65] Kümmerlen J., van Beek J D., Vollrath F., Meier B H. Local structure in spider dragline silk investigated by two-dimensional spin-diffusion nuclear magnetic resonance. Macromolecules 1996; 29(8): 2920–2928.

[66] Becker M A., Tuross N. Initial degradation changes found in *Bombyx mori* silk fibroin. In: Kaplan D.; Adams W W., Farmer B., Viney C. (eds.) Silk polymers: materials science and biotechnology. American Chemical Society Symposium Series 1994. 544: P252–269.

[67] Yang Z., Grubb D T., Jelinski L W. Small-angle X-ray scattering of spider dragline silk. Macromolecules 1997; 30(26): 8254–8261.

[68] Plaza G R., Pérez-Rigueiro J., Riekel C., Perea G B., Agulló-Rueda F., Burghammer M., Guinea G V., Elices M. Relationship between microstructure and mechanical properties in spider silk fibers: identification of two regimes in the microstructural changes. Soft Matter 2012; 8(22): 6015–6020.

[69] Sampath S., Isdebski T., Jenkins J E., Ayon J V., Henning R W., Orgel J P R O., Antipoa O., Yarger J L. X-ray diffraction study of nanocrystalline and amorphous structure within major and minor ampullate dragline spider silks. Soft Matter 2012; 8(25): 6713–6722.

[70] Grubb D T., Jelinski L W. Fiber morphology of spider silk: The effects of tensile deformation. Macromolecules 1997; 30(10): 2860–2867.

[71] Grubb D T., Ji G. Molecular chain orientation in supercontracted and re-extended spider silk. International Journal of Biological Macromolecules 1999; 24(2-3): 203–210.

[72] Cruz D H., Rousseau M E., West M M., Pézolet M., Hitchcock A P. Quantitative mapping of the orientation of fibroin β-sheets in *B. mori* cocoon fibers by scanning transmission X-ray microscopy. Biomacromolecules 2006; 7(3): 836–843.

[73] van Beek J D., Hess S., Vollrath F., Meier B H. The molecular structure of spider dragline silk: Folding and orientation of the protein backbone. Proceedings of the National Academy of Science of the United States of America 2002; 99(16): 10266–10271.

[74] Marcotte I., van Beek J D., Meier B H. Molecular disorder and structure of spider dragline silk investigated by two-dimensional solid-state NMR spectroscopy. Macromolecules 2007; 40(6): 1995–2001.

[75] Asakura T., Yao J., Yamane T., Umemura K., Ulrich A S. Heterogeneous structure of silk fibers from *Bombyx mori* resolved by ^{13}C solid-state NMR spectroscopy. Journal of the American Chemical Society 2002; 124(30): 8794–8795.

[76] Gosline J M., Denny M W., DeMont M E. Spider silk as rubber. Nature 1984; 309(5968): 551–552.

[77] Gatesy J., Hayashi C., Motriuk D., Woods J., Lewis R. Extreme diversity, conservation, and convergence of spider silk fibroin sequence. Science 2001; 291(5513): 2603–2605.

[78] Ma B Y., Nussinov R. Molecular dynamics simulations of alanine rich β-sheet oligomers: Insight into amyloid formation. Protein Science 2002; 11(10): 2335–2350.

[79] Keten S., Buehler M J. Atomistic model of the spider silk nanostructure. Applied Physics Letters 2010; 96(15): 153701–153703.

[80] Keten S., Buehler M J. Nanostructure and molecular mechanics of spider dragline silk protein assemblies. Journal of the Royal Society Interface 2010; 7(53): 1709–1721.

[81] Du N., Yang Z., Liu X Y., Li Y., Xu H Y. Structural origin of the strain-hardening of spider silk. Advanced Functional Materials 2011; 21(4): 772–778.

[82] Li X., Eles P T., Michal C A. Water permeability of spider dragline silk. Biomacromolecules 2009; 10(5): 1270–1275.

[83] Ene R., Papadopoulos P., Kremer F. Partial deuteration probing structural changes in supercontracted spider silk. Polymer 2010; 51(21): 4784–4789.

[84] Paquet-Mercier F., Lefèvre T., Auger M., Pézolet M. Evidence by infrared spectroscopy of the presence of two type of β-sheets in major ampullate spider silk and silkworm silk. Soft Matter 2013; 9(1): 208–215.

[85] Rousseau M E., Cruz D H., West M M., Hitchcock A P., Pézolet M. Nephila clavipes spider dragline silk microstructure studies by scanning transmission X-ray microscopy. Journal of the American Chemical Society 2007; 129(13): 3897–3905.

[86] Fossey S A., Tripathy S. Atomistic modeling of interphases in spider silk fibers. International Journal of Biological Macromolecules 1999; 24(2-3): 119–125.

[87] Thiel B L., Guess K B., Viney C. Non-periodic lattice crystals in the hierarchical microstructure of spider (major ampullate) silk. Biopolymers 1997; 41(7): 703–719.

[88] Trancik J E., Czernuszka J T., Bell F I., Viney C. Nanostructural features of a spider dragline silk as revealed by electron and X-ray diffraction studies. Polymer 2006; 47(15): 5633–5642.

[89] Swanson B O., Blackledge T A., Beltran J., Hayashi C Y. Variation in the material properties of spider dragline silk across species. Applied Physics A: Materials Science & Processing 2006; 82(2): 213–218.

[90] Sirichaisit J., Brookes V L., Young R J., Vollrath F. Analysis of structure/property relationships in silkworm (*Bombyx mori*) and spider dragline (*Nephila edulis*) silks using Raman spectroscopy. Biomacromolecules 2003; 4(2): 387–394.

[91] Denny M. The physical properties of spider's silk and their role in the design of orb-webs. The Journal of Experimental Biology 1976; 65(2): 483–506.

[92] Gosline J M., Guerette P A., Ortlepp C S., Savage K N. The mechanical design of spider silks: From fibroin sequence to mechanical function. The Journal of Experimental Biology 1999; 202(23): 3295–3303.

[93] Hu X., Vasanthavada K., Kohler K., McNary S., Moore A M F., Vierra C A. Molecular mechanisms of spider silk. Cellular and Molecular Life Sciences 2006; 63(17): 1986–1999.

[94] Ko F K., Jovicic J. Modeling of mechanical properties and structural design of spider web. Biomacromolecules 2004; 5(3): 780–785.

[95] Sezutsu H., Yukuhiro K. Dynamic rearrangement within the *Antheraea pernyi* silk fibroin gene is associated with four types of repetitive units. Journal of Molecular Evolution 2000; 51(4): 329–338.

[96] Fu C., Porter D., Chen X., Vollrath F., Shao Z. Understanding the mechanical properties of *Antheraea pernyi* silk – From primary structure to condensed structure of the protein. Advanced Functional Materials 2011; 21(4): 729–737.

[97] Garb J E., Dimauro T., Lewis R V., Hayashi C Y. Expansion and intragenic homogenization of spider silk since the triassic: Evidence from mygalomorphae (tarantulas and their kin) spidroins. Molecular Biology and Evolution 2007; 24(11): 2454–2464.

[98] Brooks A E., Steinkraus H B., Nelson S R., Lewis R V. An investigation of the divergence of major ampullate silk fibers from *Nephila clavipes* and *Argiope aurantia*. Biomacromolecules 2005; 6(6): 3095–3099.

[99] Liu Y., Sponner A., Porter D., Vollrath F. Proline and processing of spider silks. Biomacromolecules 2008; 9(1): 116–121.

[100] Brooks A E., Stricker S M., Joshi S B., Kamerzell T J., Middaugh C R., Lewis R V. Properties of synthetic spider silk fibers based on *Argiope aurantia* MaSp2. Biomacromolecules 2008; 9(6): 1506–1510.

[101] Rauscher S., Baud S., Miao M., Keeley F., Pomes R. Proline and glycine control protein self-organization into elastomeric or amyloid fibrils. Structure 2006; 14(11): 1667–1676.

[102] Hayashi C Y., Shipley N H., Lewis R V. Hypotheses that correlate the sequence, structure, and mechanical properties of spider silk proteins. International Journal of Biological Macromolecules 1999; 24(2-3): 271–275.

[103] Shao Z., Vollrath F., Yang Y., Thøgersen H C. Structure and behavior of regenerated spider silk. Macromolecules 2003; 36(4): 1157–1161.

[104] Lee S M., Pippel E., Göesele U., Dresbach C., Qin Y., Chandran C V., Bräeuniger T., Hause G., Knez M. Greatly increased toughness of infiltrated spider silk. Science 2009; 324(5926): 488–492.

[105] Riekel C., Müller M. In situ X-ray diffraction during forced silking of spider silk. Macromolecules 1999; 32(13): 4464–4466.

[106] Riekel C., Vollrath F. Spider silk fibre extrusion: combined wide-and small-angle X-ray microdiffraction experiments. International Journal of Biological Macromolecules. 2001; 29(3): 203–210.

[107] Khan M R., Morikawa H., Gotoh Y., Miura M., Ming Z., Sato Y., Iwasa M. Structural characteristics and properties of *Bombyx mori* silk fiber obtained by different artificial forcibly silking speeds. Internation Journal of Biological Macromolecules 2008; 42(3): 264–270.

[108] Keten S., Xu Z., Ihle B., Buehler M J. Nanoconfinement controls stiffness, strength and mechanical toughness of β-sheet crystals in silk. Nature Materials 2010; 9(4): 359–367.

[109] Xiao S., Stacklies W., Debes C., Gräter F. Force distribution determines optimal length of β-sheet crystals for mechanical robustness. Soft Matter 2011; 7(4): 1308–1311.

[110] Sinsawat A., Putthanarat S., Magoshi Y., Pachter R., Eby R K. X-ray diffraction and computational studies of the modulus of silk (*Bombyx mori*). Polymer 2002; 43(4): 1323–1330.

[111] Krasnov I., Diddens I., Hauptmann N., Helms G., Ogurreck M., Seydel T., Funari S S., Muller M. Mechanical properties of silk: Interplay of deformation on macroscopic and molecular length scales. Physical Review Letters 2008; 100(4): 048104/1–048104/4.

[112] Sinsawat A., Putthanarat S., Magoshi Y., Pachter R., Eby R K. The crystal modulus of silk (*Bombyx mori*). Polymer 2003; 44(3): 909–910.

[113] Vollrath F., Porter D. Spider silk as a model biomaterial. Applied Physics A: Materials Science & Processing 2006; 82(2): 205–212.

[114] Nova A., Keten S., Pugno N M., Redaelli A., Buehler M J. Molecular and nanostructural mechanisms of deformation, strength and toughness of spider silk fibrils. Nano Letters 2010; 10(7): 2626–2634.

[115] Papadopoulos P., Sölter J., Kremer F. Hierarchies in the structural organization of spider silk – a quantitative model. Colloid and Polymer Science 2009; 287(2): 231–236.

[116] Giesa T., Arslan M., Pugno N M., Buehler M J. Nanoconfinement of spider silk fibrils begets superior strength, extensibility, and toughness. Nano Letters 2011; 11(11): 5038–5046.

[117] Brown C P., Harnegea C., Gill H S., Price A J., Traversa E., Licoccia S., Rosei F. Rough fibrils provide a toughening mechanism in biological fibers. ACS Nano 2012; 6(3): 1961–1969.

[118] Lefèvre T., Paquet-Mercier F., Lesage S., Rousseau M E., Bédard S., Pézolet M. Study by Raman spectromicroscopy of the effect of tensile deformation on the molecular structure of *Bombyx mori* silk. Vibrational Spectroscopy 2009; 51(1): 136–141.

[119] Eles P T., Michal C A. A DECODER NMR study of backbone orientation in *Nephila clavipes* dragline silk under varying strain and draw rate. Biomacromolecules 2004; 5(3): 661–665.

[120] Seydel T., Kölln K., Krasnov I., Diddens I., Hauptmann N., Helms G., Ogurreck M., Kang S G., Koza M M., Müller M. Silkworm silk under tensile strain investigated by synchrotron X-ray diffraction and neutron spectroscopy. Macromolecules 2007, 40(4): 1035–1042.

[121] Brookes V L., Young R J., Vollrath F J. Deformation micromechanics of spider silk. Journal of Materials Science 2008; 43(10): 3728–3732.

[122] Wu X., Liu X Y., Du N., Xu G., Li B. Unraveled mechanism in silk engineering: Fast reeling induced silk toughening. Applied Physics Letters 2009; 95: 093703/1–093703/4.

[123] Jelinski L W., Blye A., Liivak O., Michal C., LaVerde G., Seidel A., Shah N., Yang Z. Orientation, structure, wet-spinning, and molecular basis for supercontraction of spider dragline silk. International Journal of Biological Macromolecules 1999; 24(2-3): 197–201.

[124] Holland G P., Jenkins J E., Creager M S., Lewis R V., Yarger J L. Solid-state NMR investigation of major and minor ampullate spider silk in the native and hydrated states. Biomacromolecules 2008; 9(2): 651–657.

[125] Shao Z., Vollrath F., Sirichaisit J., Young R J. Analysis of spider silk in native and supercontracted states using Raman spectroscopy. Polymer 1999; 40(10): 2493–2500.

[126] Savage K N., Gosline J M. The effect of proline on the network structure of major ampullate silks as inferred from their mechanical and optical properties. The Journal of Experimental Biology 2008; 211(12): 1937–1947.

[127] Liu Y., Shao Z., Vollrath F. Relationships between supercontraction and mechanical properties of spider silk. Nature Materials 2005; 4(12): 901–905.

[128] Bell F I., McEwen I J., Viney C. Supercontraction stress in wet spider dragline. Nature 2002; 416(6876): 37–37.

[129] Papadopoulos P., Ene R., Weidner I., Kremer F. Similarities in the structural organization of major and minor ampullate spider silk. Macromolecular Rapid Communications 2009; 30(9-10): 851–857.

[130] Blackledge T A., Boutry C., Wong S C., Baji A., Dhinojwala A., Sahni V., Agnarsson, I. How super is supercontraction? Persistent versus cyclic responses to humidity in spider dragline silk. The Journal of Experimental Biology 2009; 212(13): 1981–1988.

[131] Liu Y., Shao Z., Vollrath F. Elasticity of spider silks. Biomacromolecules 2008; 9(7): 1782–1786.

[132] Boutry C., Blackledge T A. Evolution of supercontraction in spider silk: structure – function relationship from tarantulas to orb-weavers. The Journal of Experimental Biology 2010, 213(20): 3505–3514.

[133] Guan J., Vollrath F., Porter D. Two mechanisms for supercontraction in *Nephila* spider dragline silk. Biomacromolecules 2011; 12(11): 4030–4035.

[134] Glišović A., Vehoff T., Davies R J., Salditt T. Strain dependent structural changes of spider dragline silk. Macromolecules 2008; 41(2): 390–398.

[135] Wong Po Foo C., Bini E., Hensman J., Knight D P., Lewis R V., Kaplan D L. Role of pH and charge on silk protein assembly in insects and spiders. Applied Physics A: Materials Science and Processing 2006; 82(2): 223–233.

[136] Vollrath F., Knight D. P. Liquid crystalline spinning of spider silk. Nature 2001; 410(6828): 541–548.

[137] Knight D P., Vollrath F. Liquid crystals and flow elongation in a spider's silk production line. Proceedings of the Royal Society B: Biological Science 1999; 266(1418): 519–523.

[138] Kerkam K., Viney C., Kaplan D L., Lombardi S. Liquid crystallinity of natural silk secretions. Nature 1991; 349(6310): 596–598.

[139] Willcox P J., Gido S P. Muller W. Kaplan D L. Evidence of a cholesteric liquid crystalline phase in natural silk spinning processes. Macromolecules 1996; 29(15): 5106–5110.

[140] Lefevre T, Boudreault S, Cloutier C, Pezolet M. Conformational and orientational transformation of silk proteins in the major ampullate gland of *Nephila clavipes* spiders. Biomacromolecules 2008; 9(9): 2399–2407.

[141] Vollrath F., Knight D P. Structure and function of the silk production pathway in the spider *Nephila edulis*. International Journal of Biological Macromolecules 1999; 24(2-3): 243–249.

[142] Knight D P., Vollrath F. Changes in element composition along the spinning duct in a *Nephila* spider. Naturwissenschaften 2001; 88(4): 179–182.

[143] Terry A E., Knight D P., Porter D., Vollrath F. pH induced changes in the rheology of silk fibroin solution from the middle division of *Bombyx mori* silkworm. Biomacromolecules 2004; 5(3): 768–772.

[144] Jin H J., Kaplan D L. Mechanism of silk processing in insects and spiders. Nature 2003; 424(28): 1057–1061.

[145] Monti P., Taddei P., Freddi G., Asakura T., Tsukada M. Raman spectroscopic characterization of *Bombyx mori* silk fibroin: Raman spectrum of silk I. Journal of Raman Spectroscopy 2001; 32(2): 103–107.

[146] Asakura T., Yamane T., Nakazawa Y., Kameda T., Ando K. Structure of *Bombyx mori* silk fibroin before spinning in solid state studied with wide angle X-ray scattering and ^{13}C cross-polarization/magic angle spinning NMR. Biopolymers 2001; 58(5): 521–525.

[147] Taddei P., Asakura T., Yao J., Monti P. Raman study of poly(alanine-glycine)-based peptides containing tyrosine, valine, and serine as model for the semicrystalline domains of *Bombyx mori* silk fibroin. Biopolymers 2004; 75(4): 314–324.

[148] Asakura T., Ashida, J., Yamane T., kameda T., Nakazawa, Y., Ohgo K., Komatsu K. A repeated β-turn structure in poly(Ala-Gly) as a model for silk I of *Bombyx mori* silk fibroin studied with two-dimensional spin-diffusion NMR under off magic angle spinning and rotational echo double resonance. Journal of Molecular Biology 2001; 306(2): 291–305.

[149] Asakura T., Ohgo K., Komatsu K., Kanenari M., Okuyama K. Refinement of repeated β-turn structure for silk I conformation of *Bombyx mori* silk fibroin using ^{13}C solid-state NMR and X-ray methods. Macromolecules 2005; 38(17): 7397–7403.

[150] Hirijida D H., Do K G., Wong S., Zax D., Jelinski L W. ^{13}C NMR of *Nephila clavipes* major ampullate silk gland. Biophysical Journal 1996; 71(6): 3442–3447.

[151] Hronska M., van Beek J D., Willimason P T F., Vollrath F., Meier B H. NMR characterization of native liquid spider dragline silk from *Nephila edulis*. Biomacromolecules 2004; 5(3): 834–839.

[152] Lefèvre T., Leclerc J., Rioux-Dubé J F., Buffeteau T., Paquin M C., Rousseau M E., Cloutier I., Auger M., Gagné S M., Boudreault S., Cloutier C., Pézolet M. Conforma-

tion of spider silk proteins in situ in the intact major ampullate gland and in solution. Biomacromolecules 2007; 8(8): 2342–2344.

[153] Roemer L., Scheibel T. In: Scheibel T. (ed.) Fibrous proteins. Austin: Landes Bioscience; 2008. P137–151.

[154] Exler J H., Hummerich D., Scheibel T. The amphiphilic properties of spider silks are important for spinning. Angewandte Chemie International Edition 2007; 46(19): 3559–3562.

[155] Magoshi J., Magoshi Y., Becker M A., Nakamura S. Biospinning (silk fiber formation, multiple spinning mechanisms). In: Salamone J C. (ed.) Polymeric Materials Encyclopedia. CRC Press: New York. 1996. vol. 1: P667–679.

[156] Vollrath F., Knight D P., Hu X W. Silk production in a spider involves acid bath treatment. Proceedings of the Royal Society B: Biological Science 1998; 265(1398): 817–820.

[157] Hardy J G., Scheibel T R. Silk-inspired polymers and proteins. Biochemical Society Transactions 2009; 37(4): 677–681.

[158] Chen P., Kim H S., Park C Y., Kim H S., Chin I J., Jin H J. pH-triggered transition of silk fibroin from spherical micelles to nanofibrils in water. Macromolecular Research 2008; 16(6): 539–543.

[159] Dicko C., Kenney J M., Knight D., Vollrath F. Transition to a β-sheet-rich structure in spidroin in vitro: The effects of pH and cations. Biochemistry 2004; 43(44): 14080–14087.

[160] Greving I., Cai M., Vollrath F., Schniepp H C. Shear-induced self-assembly of native silk proteins into fibrils studied by atomic force microscopy. Biomacromolecules 2012; 13(3): 676–682.

[161] Xie F., Zhang H., Shao H., Hu X. Effect of shearing on formation of silk fibers from regenerated *Bombyx mori* silk fibroin aqueous solution. International Journal of Biological Macromolecules 2006; 38(3-5): 284–288.

[162] Kinahan M E., Filippidi E., Köster S., Hu X., Evans H M., Pfohl T., Kaplan D L., Wong J. Tunable silk: Using microfluidics to fabricate silk fibers with controllable properties. Biomacromolecules 2011; 12(5): 1504–1511.

[163] Rammensee S., Slotta U., Scheibel T., Bausch A R. Assembly mechanism of recombinant spider silk proteins. Proceedings of the National Academy of Science of the United States of America 2008; 105(18): 6590–6595.

[164] Gronau G., Qin Z., Buehler M J. Effect of sodium chloride on the structure and stability of spider silk's N-terminal protein domain. Biomaterial Science 2013; 1(3): 276–284.

[165] Hagn F., Thamm C., Scheibel T., Kessler H. pH-Dependent dimerization and salt-dependent stabilization of the N-terminal domain of spider dragline silk – implications for fiber formation. Angewandte Chemie International Edition 2011; 50(1): 310–313.

[166] Landreh M., Askarieh G., Nordling K., Hedhammar M., Rising A., Casals C., Astorga-Wells J., Alvelius G., Knight S D., Johansson J., Jornvall H., Bergman T. A pH-dependent dimer lock in spider silk protein. Journal of Molecular Biology 2010; 404(2): 328–336.

[167] Dandin S B., Nirmal Kumar S. Bio-medical uses of silk and its derivative. Indian Silk 2006; 45(9):5–8.

[168] Manohar Reddy R. Value addition span of silkworm cocoon – Time for utility optimization. International Journal of Industrial Entomology 2008; 17(1): 109–113.

[169] Michael B. Food management. Proquest Agriculture Journals 2004; 60: 39.

[170] Altman G H., Diaz F., Jakuba C., Calabro T., Horan R L., Chen J., Lu H., Richmond J., Kaplan D L. Silk-based biomaterials. Biomaterials 2003; 24(3): 401–416.

[171] Vepari C., Kaplan D L. Silk as a biomaterial. Progress in Polymer Science 2007; 32(8-9): 991–1007.

[172] Gronau G., Krishnaji S T., Kinahan M E., Giesa T., Wong J Y., Kaplan D L., Buehler M J. A review of combined experimental and computational procedures for assessing biopolymer structure-process-property relationships. Biomaterials 2012; 33(33): 8240–8255.

[173] Meinel L., Hofmann S., Karageorgiou V., Kirker-Head C., McCool J., Gronowicz G., Zichner L., Langer R., Vunjak-Novakovic G., Kaplan D L. The inflammatory responses to silk films in vitro and in vivo. Biomaterials 2005; 26(2): 147–155.

[174] Gupta M K., Khokhar S K., Phillips D M., Sowards L A., Drummy L F., Kadakia M P., Naik R R. Patterned silk films cast from ionic liquid solubilized fibroin as scaffolds for cell growth. Langmuir 2007; 23(3): 1315–1319.

[175] Servoli E., Maniglio D., Motta A., Predazzer R., Migliaresi C. Surface properties of silk fibroin films and their interaction with fibroblasts. Macromolecular Bioscience 2005; 5(12): 1175–1183.

[176] Min B M., Jeong L., Lee K Y., Park W H. Regenerated silk fibroin nanofibers: water vapor-induced structural changes and their effects on the behavior of normal human cells. Macromolecular Bioscience 2006; 6(4): 285–292.

[177] Unger R E., Wolf M., Peters K., Motta A., Migliaresi C., James Kirkpatrick C. Growth of human cells on a non-woven silk fibroin net: a potential for use in tissue engineering. Biomaterials 2004; 25(6): 1069–1075.

[178] Jin H J., Chen J., Karageorgiou V., Altman G H., Kaplan D L. Human bone marrow stromal cell responses on electrospun silk fibroin mats. Biomaterials 2004; 25(6): 1039–1047.

[179] Wharram S E., Zhang X., Kaplan D L., McCarthy S P. Electrospun silk material systems for wound healing. Macromolecular Bioscience 2010; 10(3): 246–257.

[180] Fan H., Liu H., Wong E J W., Toh S L., Goh J C H. In vivo study of anterior cruciate ligament regeneration using mesenchymal stem cells and silk scaffold. Biomaterials 2008; 29(23): 3324–3337.

[181] Mandal B B., Priya A S., Kundu S C. Novel silk sericin/gelation 3-D scaffolds and 2-D films: Fabrication and characterization for potential tissue engineering applications. Acta Biomaterialia 2009; 5(8): 3007–3020.

[182] Kasoju N., Bhonde R R., Bora U. Preparation and characterization of *Antheraea assama* silk fibroin based novel non-woven scaffolds for tissue engineering applications. Journal of Tissue Engineering and Regenerative Medicine 2009; 3(7): 539–552.

[183] Sugihara A., Sugiura K., Morita H., Ninagawa T., Tubouchi K., Tobe R., Izumiya M., Horio T., Abraham N G., Ikehara S. Promotive effects of a silk film on epidermal recovery from full-thickness skin wounds. Proceedings of the Society for Experimental Biology and Medicine 2000; 225(1): 58–64.

[184] Unger R E., Sartoris A., Peters K., Motta A., Migliaresi C., Kunkel M., Bulnheim U., Rychly J., Kirkpatrick C J. Tissue-like self-assembly in cocultures of endothelial cells and osteoblasts and the formation of microcapillary-like structures on three-dimensional porous biomaterials. Biomaterials 2007; 28(27): 3965–3976.

[185] Yang Y., Chen X., Ding F., Zhang P., Liu J., Gu X. Biocompatibility evaluation of silk fibroin with peripheral nerve tissues and cells in vitro. Biomaterials 2007; 28(9): 1643–1652.

[186] Numata K., Kaplan D L., Silk-based delivery systems of bioactive molecules. Advanced Drug Delivery Reviews 2010; 62(15): 1497–1508.

[187] Lawrence B D., Cronin-Golomb M., Georgakoudi I., Kaplan D L., Omenetto F G. Bioactive silk protein biomaterial systems for optical devices. Biomacromolecules 2008; 9(4): 1214–1220.

[188] Perry H., Gopinath A., Kaplan D L., Negro L D., Omenetto F G. Nano- and micropatterning of optically transparent, mechanically robust, biocompatible silk fibroin films. Advanced Materials 2008; 20(16): 3070–3072.

[189] Omenetto F G., Kaplan D L. A new route for silk. Nature Photonics 2008; 2(11): 641–643.

[190] Amsden J J., Perry H., Boriskina S V., Gopinath A., Kaplan D L., Negro L D., Omenetto F G. Spectral analysis of induced color change on periodically nanopatterned silk films. Optics Express 2009; 17(23): 21271–21279.

[191] Kim D H., Kim Y S., Amsden J., Panilaitis B., Kaplan D L., Omenetto F G., Zakin M R., Rogers J A. Silicon electronics on silk as a path to bioresorbable, implantable devices. Applied Physics Letters 2009; 95(13): 133701/1–133701/3.

[192] Kim D H., Viventi J., Amsden J. J., Xiao J., Vigeland L., Kim Y S., Blanco J A., Panilaitis B., Frechette E. S., Contreras D., Kaplan D L., Omenetto F G., Huang Y., Hwang K C., Zakin M R., Litt B., Rogers J A. Dissolvable films of silk fibroin for ultrathin conformal bio-integrated electronics. Nature Materials 2010; 9(6): 511–517.

[193] Parker S T., Domachuk P., Amsden J., Bressner J., Lewis J A., Kaplan D L., Omenetto F G. Biocompatible silk printed optical waveguides. Advanced Materials 2009; 21(13): 2411–2415.

[194] Kharlampieva E., Kozlovskaya V., Wallet B., Shevchenko V V., Naik R R., Vaia R., Kaplan D L., Tsukruk V V. Co-cross-linking silk matrices with silica nanostructures for robust ultrathin nanocomposities. ACS Nano 2010; 4(12): 7053–7063.

[195] Steven E., Park J G., Paravastu A., Lopes E B., Brooks J S., Englander O., Siegrist T., Kaner P., Alamo R G. Physical characterization of functionalized spider silk: electronic and sensing properties. Science and Technology of Advanced Materials 2011; 12(5): 0055002/1–005002/13.

[196] Huby N., Vié V., Renault A., Beaufils S., Lefèvre T., Paquet-Mercier F., Pézolet M., Bêche B. Native spider silk as a biological optical fiber. Applied Physics Letters 2013; 102(12): 123702/1–123702/3.

[197] Kim S., Mitropoulos A N., Spitzberg J D., Kaplan D L., Omenetto F G. Silk protein based hybrid photonic-plasmatic crystal. Optics Express 2013; 21(7): 8897–8903.

[198] Huang X., Liu G., Wang X. New secrets of spider silk: Exceptionally high thermal conductivity and its abnormal change under stretching. Advanced Materials 2012; 24(11): 1482–1486.

Chapter 4

Ethylene Oxide Homogeneous Heterobifunctional Acyclic Oligomers

Calin Jianu

Additional information is available at the end of the chapter

http://dx.doi.org/10.5772/57610

1. Introduction

It is difficult, if not hazardous and partisan, to rank the top advances of organic chemistry in the last approx. 20 years for their role in the development of this science and adjacent, complementary others. It can, however, be said with certainty that the field of homogeneous (HOHAO) and heterogeneous heterobifunctional acyclic oligomers (HEHAO) [1] of ethylene oxide, through their diversity of macromolecular architectures and effects, turns out to be one of the most significant achievements. Their importance resides mainly in the resolve by affordable means of some fundamental problems of organic synthesis by the transfer to the same phase of reaction partners with different polarities (organic substrate and water-soluble, usually inorganic, reactant).

Pioneering attempts to structure homogeneous PEO chains with n=3-20 are recorded between the fourth and sixth decades of the XXth century and found in technical bulletins issued by large corporations (Hülles, Henkel, Union Carbide, Shell Oil etc.). For reasons of intellectual property protection (in the absence of patents) the technological information recorded within these publications are summary in terms of processing parameters, conversions, or secondary products.

Mulley, B.A. [2] has the merit of the first to de-centralize and systematize the efforts to structure the first proper homogeneous PEO chains in the true meaning of "homogeneous heterobifunctional acyclic oligomers of ethylene oxide" (HOHAO). The compounds reported by him are really just "homogeneous monofunctional acyclic oligomers of ethylene oxide".

The material presented below introduces the reader to the field of homogeneous heterobifunctional acyclic oligomers of ethylene oxide (HOHAO).

HOHAO, "tailor-made macromolecules" ("designer macromolecules") with the general structure (Figure 1), fall into the category of "niche" unitary organic compounds (derivatives of polyethylene glycols PEG_n).

$$R_1\text{-O-(-}CH_2CH_2O\text{-)}_n\text{-}R_2$$

n=homogeneous oligomerization degree (strictly monitored value); R_1, R_2=aliphatic, aromatic or mixed derivatization terminals

Figure 1. General structure of homogeneous heterobifunctional acyclic oligomers of ethylene oxide

Why have unitary polyoxyethylene chains PEO imposed themselves? On the one hand to eliminate the cumulative manifestation of the colloidal physico-chemical behavior of the group of chain homologues in the polydisperse heterogeneous structures (technical products) obtained by anionic polymerization; on the other hand to definitely delineate the colloidal physico-chemical competences of each homologue.

2. Ethylene oxide (EO) — Structure, properties, consequences

The origin of the steadily increasing interest that has fascinated for over a century the scientific effort of many researchers, laboratories and concerns for polyethylene glycols (PEG_n) and polyoxyethylene chains (PEO), is largely due to the specific structure and properties of ethylene oxide [3,4]. A slightly colored gas at 25°C, with a sweetish odor and taste characteristic of ethers particularly at concentrations above 500 ppm in air, it is readily soluble in water, ethanol and other organic solvents.

It is relatively thermally stable. In the absence of catalysts up to 300°C it does not dissociate, but above 570°C the major exothermic decomposition is recorded.

Union Carbide at the beginning of the 20th century inaugurated the first production plant of EO by the air oxidation of ethylene in the presence of catalytic metallic silver. Later Shell Oil Co. replaced air with high-purity oxygen and processed EO at 200-300°C and 1-3 MPa, respectively, with an oxidation yield between 63-75% and 75-82%.

The reactivity of the three-atom (two carbon, one oxygen) ether heterocycle (oxirane) (Figure 2a), also founded on the "ring tension theory", favors the nucleophilic attack of organic compounds with hydroxyl, thiol, primary and/or secondary amine, etc., function with breaking of the C-O bond of the oxirane ring.

Its typical reactions are with nucleophiles, which proceed via the SN2 mechanism, both in acidic (weak nucleophiles: water, alcohols) and alkaline media (strong nucleophiles: OH^-, RO^-, NH_3, RNH_2, RR'NH, etc.). The general reaction scheme is presented in Figure 2b.

Reactions of ethylene oxide with fatty alcohols proceed in the presence of sodium metal, sodium hydroxide or boron trifluoride and are used for the synthesis of surfactants.

http://dx.doi.org/10.5772/57610

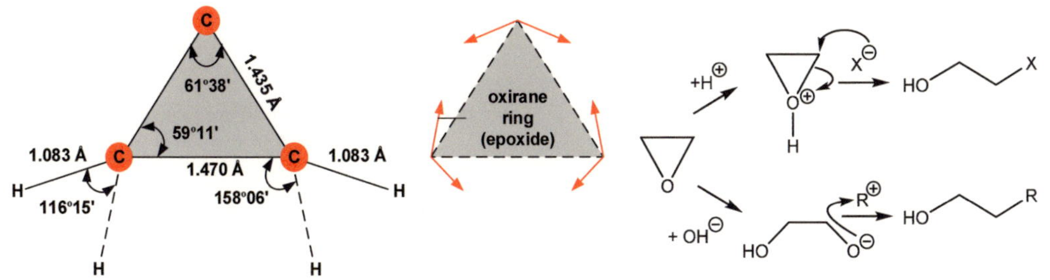

Figure 2. The detailed structural geometry of ethylene oxide (EO) [3,4]

The reaction is carried out in a special insulated reactor under inert atmosphere (nitrogen) to prevent the possible explosion of ethylene oxide. Finally, the reaction mixture is neutralized, degassed and purified.

A narrow-range ethoxylated alcohol also called "peaked ethoxylated" alcohol has a distribution curve that is narrower than the equivalent standard alcohol ethoxylate and a considerably lower content of unreacted alcohol. This gives the nonionic surfactant focused properties, a very low odor, even if based on a short-chain alcohol and avoids the formulation problems often associated with standard alcohol ethoxylates [3].

The lower the degree of ethoxylation, the higher the amount of free alcohol [3].

Polyethylene glycol is produced by the interaction of ethylene oxide with water, ethylene glycol, or ethylene glycol oligomers [5]. The reaction is catalyzed by acid or base catalysts.

The size distribution can be characterized statistically by its weight average molecular weight (M_w) and its number average molecular weight (M_n), the ratio of which is called the polydispersity index (M_w/M_n). M_w and M_n can be measured by mass spectrometry.

The different oligomeric and/or polymeric structures of ethylene oxide depend on the polymerization initiators. Ethylene glycol and its oligomers are preferable as starting materials ("lead compounds") instead of water, because they allow the creation of polymers with low polydispersity ("narrow molecular weight distribution") (NMWD).

A number of recent reviews [5,6] have also covered PEG_n chemistry and its applications in biotechnology and medicine, supported catalysis, aqueous two-phase systems in bioconversion, and solvent and phase transfer catalyst (PTC) in organic synthesis.

PEGylation [7] represents the covalent coupling of a PEG_n to a macromolecule (*e.g.*, lipids, therapeutic proteins, etc.). The effect of PEGylation is prolongation of the biological effect (produces a larger molecule with a prolonged half-life). PEGylation is similar to the structuring processes of heterogeneous heterobifunctional acyclic oligomers of ethylene oxide (HEHAO) (Figure 3).

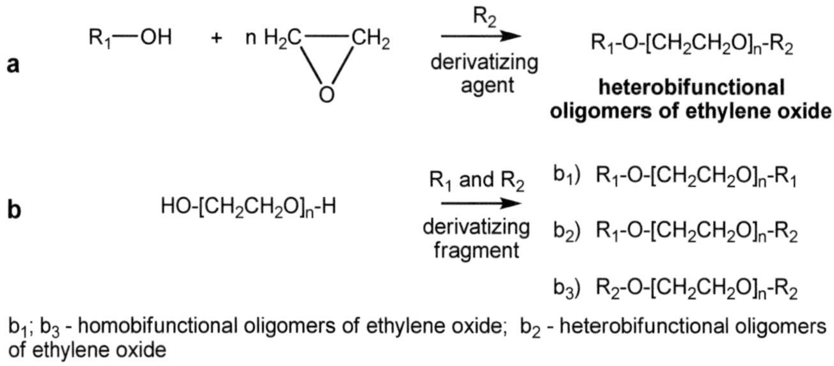

Figure 3. Synthetic methods for the preparation of heterogeneous heterobifunctional acyclic oligomers of ethylene oxide (HEHAO) a) direct preparation; b) end group derivatization of PEG_n diols

With the decision "no observed adverse effect level" (NOAEL) higher homologues PEG_n-1500 for doses of 600 mg/kg have also been advised.

PEG_n, PEO or POE refer to an oligomer or polymer of ethylene oxide. The three names are chemically synonymous.

Five concepts frequently accessed in the specialized literature on polyethylene glycols as such and/or derivatized have also cumulatively fueled the steadily increasing interest in homogeneous heterobifunctional acyclic oligomers of ethylene oxide (HOHAO):

- the particular physical properties of aqueous solutions of PEG_n;
- the unique solvent properties and the coordination competence of cations present in solution;
- the solvent competences of liquid low-molecular-weight PEG_n in chemical reactions;
- the employment of PEG_n as such and/or derivatized as alternative PTC (phase-transfer catalysis);
- the acceptance of aqueous biphasic reactive extraction (ABRE) as a present phenomenon in the development of alternative processes for wood pulping and green catalytic oxidation systems.

In addition to their own phase-transfer activity, PEG_n have also been employed as polymer support for other phase-transfer catalysts (PTCs). PEG_n have been modified with some typical PTCs such as crown ethers, ammonium salts, cryptands, and polypodants to enhance the phase-transfer in two-phase reactions.

After 1960 the coordination competences of alkali cations in the matrix of polyoxyethylene chains (PEO), as such and/or derivatized were also explicitly recognized. The striking analogy with crown polyethers could not fail to impose questions and provide answers pertinent to their capacity, stability and coordination geometry. Certainties that appeared after the series

http://dx.doi.org/10.5772/57610

of scientific speculations and subsequently of experimental evidence, preceded by several decades the conformational and geometrical interpretation of polyoxyethylene chains.

The infrared (IR) absorption characteristics of PEO-MY complexes differ from the individual species, but are similar to those seen in MY-cyclic polyethers systems reported earlier, thus suggesting that the coordination effect may be due to ion-dipole-type interactions in these systems.

This hypothesis becomes credible if the helical conformation (helix) of oligomeric PEO chains with minimum 7 units EO/coil and the electron-donor character (Lewis base) of the oxygen in the PEO chain are accepted.

3. Anionic polymerization of ethylene oxide

Heterogeneous polyoxyethylene chains obtained by the anionic polymerization of ethylene oxide ("anionic ring-opening polymerization") with a polydispersity degree $M_w/M_n < 1.1$ are hydrophilic, flexible (specific spatial conformation), biocompatible "bridges".

Generically, the term "oligomer", *oligos* being Greek for "a few", while *mer* with the meaning of primary structural unit (ethylene oxide) which is repeating, can be defined as a molecular assembly composed of a small number of monomeric units covalently grafted (ethylene oxide for HOHAO). Since ethylene oxide (EO) is accessed exclusively, the macromolecular structure formed is a homo-oligomer (homomer).

Oligomerization in the casuistry analyzed is a process of monitored attachment in a macromolecular architecture of (n) primary structural units of ethylene oxide. The demarcation of n between oligomerization and polymerization is undecided in the specialized literature. For HOHAO the range 3-20 is accepted compared to 10-100 in general. In the case of heterogeneous polyoxyethylene chains (polydisperse polymerization degree $n_{average}$) it is natural to assume the existence of homologous (mixtures of oligomers and/or polymers) oligomeric (polymeric) series with the same strictly defined structure, with molecular weights different from homogeneous oligomeric chains (definite oligomerization degree n).

For heterogeneous polyoxyethylene (PEO) chains the statistical distribution quantitatively expressed through the equations: Natta, Weilbull/Nycander/Gold, Natta/Mantica, Poisson, etc., is accepted [8]

4. Homogeneous polyoxyethylene chains — Preparation, structure, competences

Although the structuring of homogeneous PEO chains represented a scientific challenge for more than six decades, today we still can not say that there is a single, rapid procedure for their synthesis due to preparative difficulties, complex process flow diagrams for purification and characterization, consequently due to the high cost of processing in industrial quantities.

During our research the results obtained contributed decisively to the confirmation of the direct participation of PEO chains in nucleophilic addition reactions (cyanoethylation, amidoethylation) (Figure 10) of polyethoxylated higher alcohols purified of free higher alcohols, polyethylene glycols (PEG_n) and water, when the processing yields under similar conditions increase proportionally with the oligomerization degree, (n), of the PEO chain [9,10].

After almost a century of investigations, similar to other classes of macromolecular compounds for polyethylene glycols (PEG_n) and their derivatives (glyme, oligoglymes, PEGylated compounds), respectively, it can also be reasonedly claimed that in addition to a primary structure there are a secondary (conformational) and a tertiary structure (micellar macromolecular architectures).

The main qualities of these tridimensional macromolecular architectures with consequences in the study of HOHAO are dimensional flexibility, transfer mobility, the existence of "meander", "zig-zag", and "helix" conformations of variable geometry, free coaxial C-C/C-O rotation, and the absence of "ring tensions" specific to rigid structures (crown polyethers).

With few exceptions, the preparation of homogeneous PEO chains as such and derivatized is reported with yields ranging between 60-80% for relatively small oligomerization degrees (n=3-6) [2]. The laborious, difficult to accomplish purification, associated with the presence of "neighboring effects" ("sympathy effects") between two or more hydrophilic (polyoxyethylene) chain homologues with close physico-chemical constants, limited the extension of synthetic efforts [10].

The main colloidal characteristics of HEHAO and HOHAO, respectively, depend on their structure and heterogeneous or homogeneous composition. As mixtures with wide distribution of hydrophobic R_1,R_2 and hydrophilic PEO chain homologues, respectively, HEHAO manifest cumulatively through the individual colloidal behavior of each homologue present in the mixture, but also through mutual interdependences. That is why the experimental values of the main basic colloidal characteristics evaluated in the research carried out [9,10] were preliminarily only indicative, even though they were the result of the mathematical processing of a considerable number of measurements.

The actual distribution of the PEO chain homologues and hence of the oligomerization degrees ($n_{average}$) changes in the series $n_{average}$=3-18 from advanced symmetry for $n_{average} \leq 8$ to pronounced asymmetry for $n_{average}$=9-18.

The steadily increasing interest in the definite explanation of colloidal properties of heterogeneous (polydisperse) polyoxyethylene chains is evident in the literature after the sixth decade of the 20th century and by the research related to the obtaining, purification and characterization of homogeneous polyoxyethylene chains. Two types of PEO conformations are postulated during this period ("zig-zag" and "meander") (Figure 4).

Today one can draw a unitary conclusion that would eliminate earlier partial speculative assumptions. Major contributions in this area were due to the accessing of modern instrumental methods: "short distance diffraction techniques"; "wide-angle-diffraction techniques" and "low-angle-diffraction techniques".

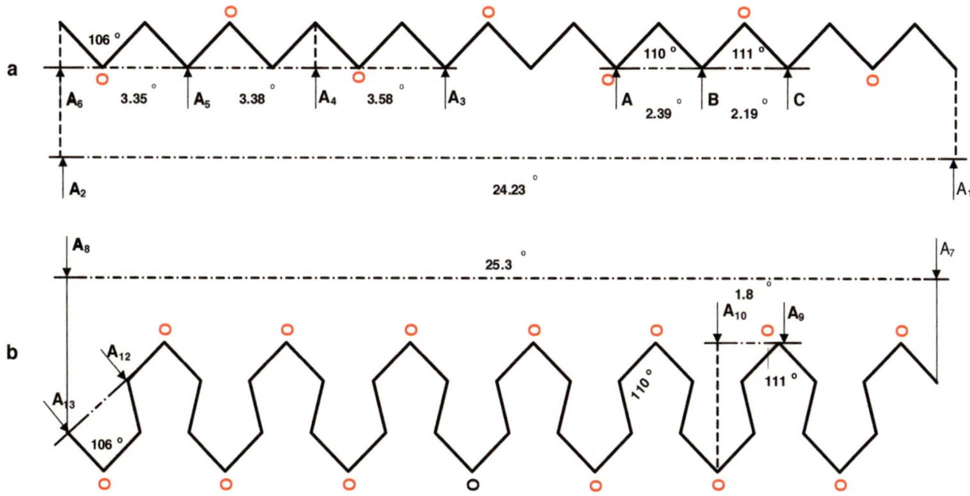

Figure 4. The "zig-zag" (a) and meander (b) conformations of the polyoxyethylene chain [11-15]

These constructive details about the PEO chain ("strain-free polyoxyethylene chain") and the manner of "packing" in the "macromolecular lattice" are reasonedly argued with the specifications:

- the "monoclinic unit cell" appears at four chain meanders;
- the meander has in its structure nine oxyethylene structural units ($-CH_2CH_2-O-$), *i.e.*, 4 x 9=36 total oxyethylene units in a "monoclinic unit cell";
- the "repeat period" is identical (19.5 Å);
- each oxyethylene unit is "twisted" to the neighboring structural unit, such that the main PEO chain returns to the original position at every tenth "lead/turn" (19.5 Å);
- in the "meander" conformation one oxyethylene structural unit has a length of 1.9 Å and a diameter of 4 Å compared to the "zig-zag" conformation where the same geometrical parameters are 3.5Å and 2.5 Å, respectively.

Today it is accepted that the "zig-zag" conformational form is specific to PEO with low oligomerization degrees, and the "meander" conformation to PEO chains with high oligomerization degrees.

Be mentioned that differences between some experimental results are due to the study of polyoxyethylene chains: in solution; in the solid state, in the range of average molecular weights between 2,400 and 100,000 (n=55-2300), and in the solid state, derivatized.

Experimental evidence favorable to the concepts expressed, associated with continuously developing advances in instrumental investigation, have decisively stimulated the theoretical and practical interest for the synthesis of HOHAO in general, and for mono-and diderivatized homogeneous and heterogeneous polyoxyethylene chains in particular.

The rediscovery of crown polyethers and their role as phase-transfer catalysts, and the pronounced mutual structural similarity constituted an additional major impetus for the theoretical and practical conformational study of acyclic PEO chains, suggested changes in terminology and their recognition as biomacromolecules with major physiological role.

The homologous series of dimethylated polyethylene glycols [$CH_3(OCH_2CH_2)_nOCH_3$; ($n \geq 3$)] also suggested the term *glyme* (oligoglyme) (*met*hylated *gly*cols). Although not yet widely accepted, there is a tendency to generalize this concept also to derivatized acyclic polyether (polyoxyethylene) chains.

From a simple working hypothesis (speculative nature) able to explain phenomena or processes, subsequent experimental studies based on X-ray investigations, electronic microscopy and diffraction confirmed their ability of intra-and interchain contraction, dependent on the structure and medium and the formation of "cavities" ("cage" of variable geometry) at oligomerization degrees (n, $n_{average}$) below and above nine ethylene oxide (EO) units, and "sandwich" below three structural units (EO), respectively [16].

The synergistic cumulation in a unitary structure of the conformational and colloidal qualities of homogeneous PEO chains with the possibility of controlled modification of the HLB (hydrophilic/hydrophobic balance) was and still is of wide theoretical and practical interest.

Technologies based on grafted PEG_n conjugates launched products of major biological importance, for research and diagnosis (PEG_n-modified proteins and liposomes, food, medical and analytical matrices), based on the accumulated knowledge on PEG_n and the prospects of PEG_n as biomaterial.

These considerations suggested the idea of structuring the HOHAO of the PEG_n-L(R';2R)(2R';R) structured-lipids-type (Figure 5).

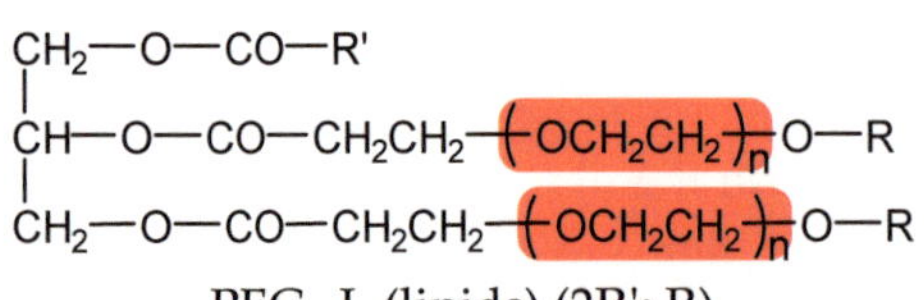

PEGn-L (lipids) (2R'; R)

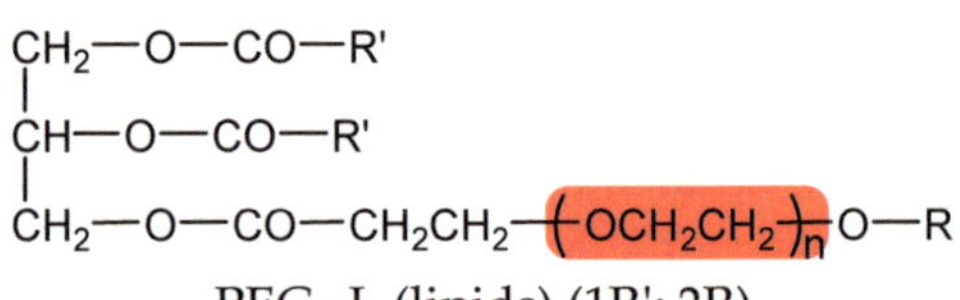

PEGn-L (lipids) (1R'; 2R)

Figure 5. The structure of homogeneous heterobifunctional acyclic oligomers of ethylene oxide of the tailor-made-lipids type [10]

In Figures 6-9 are presented the main process flow charts which formed the basis of the synthesis of homogeneous polyoxyethylene chains in HOHAO. It is noted that in the stages with strongly ionic character were accessed PTC with homogeneous PEO chains, able to activate nucleophilic agents by coordinating the alkaline cation [9,10].

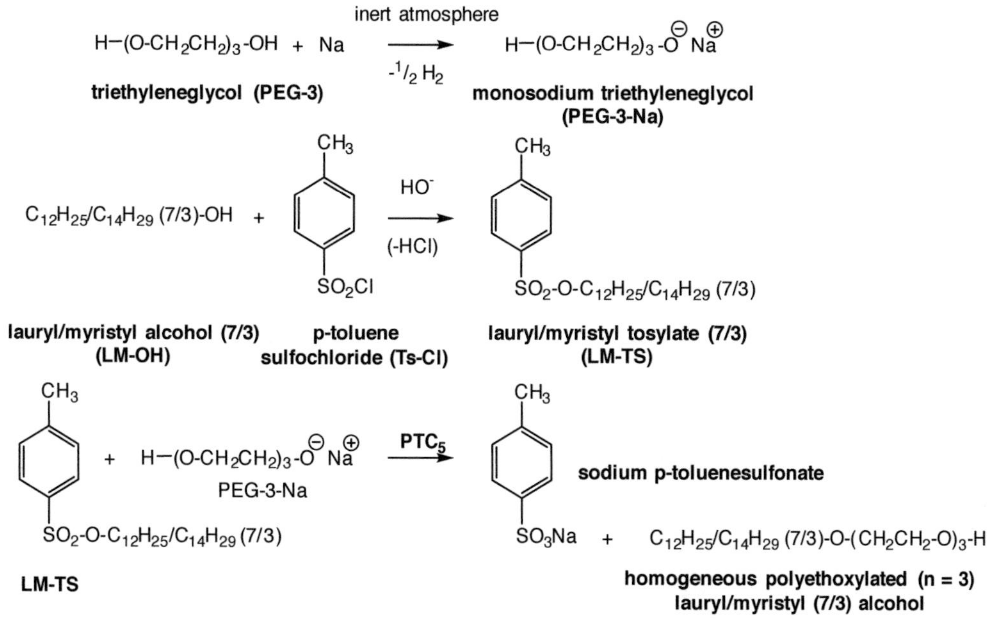

Figure 6. Reaction scheme for the production of homogeneous polyethoxylated (n=3) lauryl/myristyl (7/3) alcohol [LM(EO)$_3$H]

Homogeneous heterobifunctional acyclic oligomers of ethylene oxide (HOHAO) fall into the category of "niche" unitary organic compounds (derivatives of homogeneous polyethylene glycols PEG$_n$), whose synthesis, purification, chemical and physico-chemical characterization detaches from the classic heterodisperse character specific to the oligomerization and polymerization products of ethylene oxide (EO), through the unitary (homogeneous) structure of the polyoxyethylene chain (PEO) "constructed" through controlled successive covalent grafting of lower oxyethylene units, diethyleneglycol (DEO) or triethyleneglycol (TEO) (adapted Williamson synthesis), followed by the derivatization of the two terminals ("end groups") with various fragments R$_1$, R$_2$ (Table 1) [9,10].

Aromatic sulphochlorides (tosyl chloride) react with higher alcohols (C$_8$-C$_{18}$), diols (PEG$_n$), and alkyl (C$_8$-C$_{12}$) phenols (NF), respectively, in the presence of a base, pyridine, forming sulphonate esters, effective alkylating agents (Figures 6, 7, 9). Higher alcohols (C$_8$-C$_{18}$), diols (DEO, TEO, PEG$_n$), and alkyl (C$_8$-C$_{12}$) phenols, respectively, react with thionyl chloride ($SOCl_2$) in the presence of pyridine as organic base, generating the corresponding chloro derivatives (PEG$_2$-2Cl; PEG$_3$-2Cl; PEG$_6$-2Cl) (Figures 8, 9). In order to direct (control) the course of the reaction in the direction of mono-or dichlorination, one hydroxyl terminal can be protected by acetylation (Ac) with acetic anhydride (Ac_2O) with the formation of PEG$_n$-Ac; PEG$_n$-2Ac; PEG$_n$-Ac,Cl [9,10].

Higher alcohols (C$_8$-C$_{18}$), alkyl (C$_8$-C$_{12}$) phenols and diols (DEO, TEO, PEG$_n$) also form under controlled inert atmosphere, without oxygen, CO_2 or traces of water (Figures 6, 9), alcoholates or phenolates (PEG$_2$-Na; PEG$_3$-Na; PEG$_6$-Na; PEG$_9$-Na; PEG$_{12}$-Na; PEG$_{18}$-Na), who subsequently participate in a directed manner (Figures 7-9) in the nucleophilic substitution of chloride in the mono-and/or dichloro derivatives (PEG_2-2Cl; PEG_3-2Cl; PEG_6-2Cl) for the elongation of the homogeneous PEO chain [9,10].

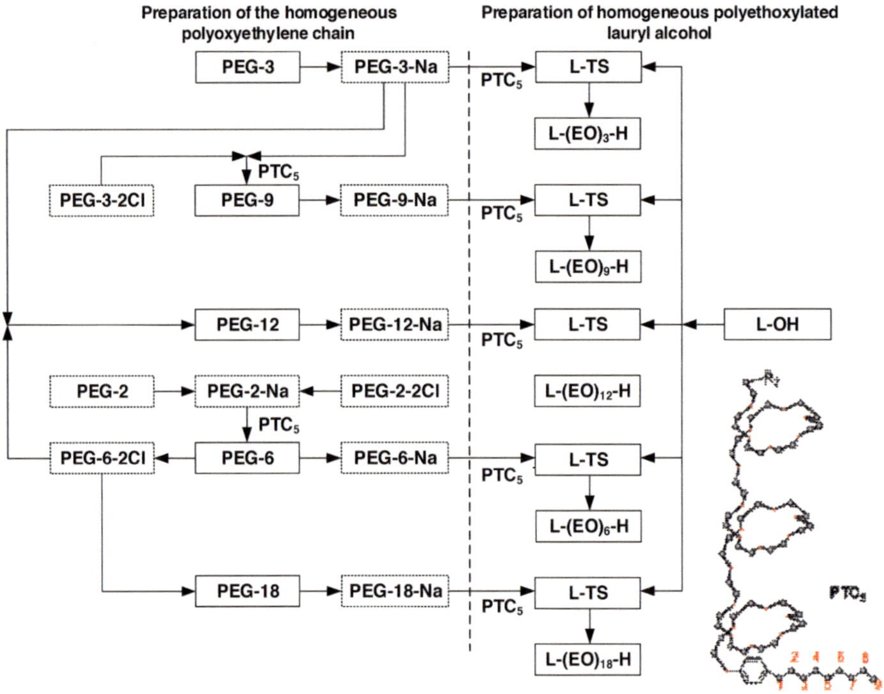

Figure 7. The preparation scheme of homogeneous polyethoxylated lauryl alcohol n=3, 6,9,12,18 under phase-transfer catalysis conditions (PTC_5 – homogeneous β-nonylphenoxy polyethyleneoxy (n=24) methyl ether)

The process with a pronounced polar (ionic) character is additionally favored by the presence of micellar phase-transfer catalysts (PTC_{1-5}) (Figures 6-9, 13), which "sequester" the alkaline cation (sodium) in the "cavity" with flexible geometry of the homogeneous PEO chain's helix for n≥8-9 (PTC_1, PTC_4, PTC_5) or in the interchain space ("sandwich" type) for n≤3-4 (PTC_2, PTC_3).

5. Homogeneous heterobifunctional acyclic oligomers of ethylene oxide

a. Synthesis of homogeneous heterobifunctional acyclic oligomers of ethylene oxide from the category of surface-active compounds

An overview of the colloidal characteristics of two major categories of surface-active structures (ionic and nonionic) suggested the idea of creating a new class of hybrid surface-active compounds, of the nonionic-ionic type, with synergistically cumulated colloidal effects. After confirming the real possibilities of approaching the study it was decided upon the structuring of HOHAO proper (Table 1). The common element of all the molecular architectures obtained was the directed synthesis of the homogenous polyoxyethylene chains (n) (Figure 7) (n=3,6,9,12,18), following the adapted Williamson variant [9,10]. Subsequently there was the monitored grafting (PEGylation) with fragments R_1 (Figures 7, 8), and the derivatization of the second hydroxyl terminal with fragments R_2 accessing a scheme of adapted classical reactions (Figures 10, 11). Literature reports similar structures of the nonionic-ionic type with heterogeneous polyoxyethylene chains ($n_{average}$=3) [9,10]. The study extends the range of HOHAO with surface-active competences using higher homogenous polyoxyethylene chains (n=3-18).

To facilitate the presentation of synthesized HOHAO (Figure 10) their chemical names have been encoded. The main organic functions were symbolized by the initials of the chemical names (*e.g.*:-propionitrile, PN;-primary ethylamine, EP; etc.), the homologues of the base hydrocarbon chain with the initials of the trivial names (*e.g.*: lauryl/myristyl, LM; cetyl/stearyl, CS) followed in parentheses by the ratio (7/3) signifying their mutual relative distribution. The hydrocarbon chains attached later by synthesis were symbolized by the number of carbon atoms contained (*e.g.*: in the cationic structure – EC-1.1.16., we find two methyl groups symbolized as (1.1.), and a hexadecyl chain indicated by the number (16.), respectively), and within the class in the natural order. For example: *N,N*-dimethyl-*N*-dodecyl (lauryl)-*N*-β-lauryl/myristyl (7/13) polyethyleneoxy (n=9) ethylammonium chloride has the symbol LM-$(EO)_9$-EC-1.1.12.

Similarly was done with the structures of structured lipids HOHAO (Figure 11) assimilated with PEG_n-L (lipids) conjugates.

Under equimolar AN/LM-OH conditions, increasing the temperature in the range 25-35°C determines the increase of the cyanoethylation yields, then between 40-60°C the yields drop. With excess acrylic monomer the evolution of yields follows the same trend. In these conditions between 25-40°C the amount of acrylic oligomers formed is below 1%, independently of the excess of monomer introduced, while between 40-60°C it increases dramatically. For cetyl/stearyl alcohol, under equimolar conditions or excess of monomer AN/CS-OH, increasing the temperature between 45-55°C (below this range the yields are low) favors the nucleophilic addition yields, the more so as the excess of acrylic monomer is higher. In the range 55-70°C the yields drop under the same conditions, and the content of acrylic oligomers is higher than in the case of lauryl/myristyl alcohol. At the cyanoethylation of homogeneous polyethoxylated (n=3) lauryl/myristyl alcohol under equimolar conditions or excess of acryl monomer, the addition yields increase between 25-35°C, then between 45-60°C decrease appreciably. The amount of acrylic oligomers formed follows roughly the same evolution as in the case of lauryl/myristyl alcohol over the entire temperature range. For homogeneous polyethoxylated (n=3) cetyl/stearyl alcohol under equimolar conditions or excess of monomer, the addition yields increase between 30-40°C then drop between 45-65°C. The content of acrylic oligomers increases in proportion to the temperature and excess of monomer. Compared with cetyl/stearyl alcohol it is noted that the cyanoethylation yields are higher at the same value of temperature even below 45°C. These suggest that the cyanoethylation reaction is reversible in character (Figure 12), the polyoxyethylene chain favors the addition, while increasing the hydrocarbon one reduces the yields.

No.	Structure R_1-$(PEO)_n$-R_2			Name	Competences	References
	R_1	n	R_2			
	HO	3,6,9, 12,18	OH	homogeneous polyethylene glycols	controlled PEGylation structures	9,10
	Alk-O- Alk-Aryl- O-	3,6,9, 12,18	OH	monoderivatized homogeneous polyethylene glycols	homogeneous polyethoxylated (C_1-C_{20}) alcohols	9,10
	Alk-O- Alk-Aryl- O-	3,6,9, 12,18	$-O-CH_2CH_2CN$	homogeneous β-alkyl (C_8-C_{18})/alkyl (C_8-C_{12}) aryl polyethyleneoxy propionitriles	PEGylation intermediates	9,10,17,28
	Alk-O- Alk-Aryl- O-	3,6,9, 12,18	$-O-CH_2CH_2CONH_2$	homogeneous β-alkyl (C_8-C_{18})/alkyl (C_8-C_{12}) aryl polyethyleneoxy propionamides	PEGylation intermediates	9,10,29
	Alk-O- Alk-Aryl- O-	3,6,9, 12,18	$-O-CH_2(CH_2)_mNH_2$ m=1,2	homogeneous β-alkyl (C_8-C_{18})/alkyl (C_8-C_{12}) aryl polyethyleneoxy ethyl/propyl amines	PEGylation intermediates	9,10,18
	Alk-O- Alk-Aryl- O-	3,6,9, 12,18	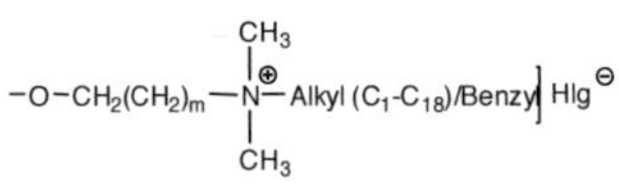 (m = 1,2)	homogeneous *N,N*-dimethyl-*N*-alkyl (C_8-C_{18})/benzyl-*N*-β-alkyl (C_1-C_{18})/ polyethyleneoxy (n = 3-18) ethyl/propyl ammonium	flotation agents/phase-transfer catalysts	9,24
	Alk-O- Alk-Aryl- O-	3,6,9, 12,18	$-O-CH_2(CH_2)_mNH_3{}^+]X^-$ (m = 1,2)	salified homogeneous β/γ-alkyl (C_8-C_{18})/alkyl (C_8-C_{12}) aryl polyethyleneoxy ethyl/ propyl amines	micellar flotation agents	9,18,20
	Alk-O- Alk-Aryl- O-	3,6,9, 12,18	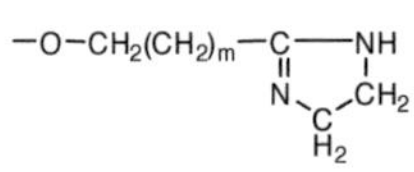 (m = 1,2)	homogeneous β-alkyl (C_8-C_{18})/alkyl (C_8-C_{12}) aryl polyethyleneoxy 2-imidazolines	structuring intermediates of amphoteric surface-active compounds	22,23,26
	Alk-O- Alk-Aryl- O-	3,6,9, 12,18	the structure is presented in detail on figure 5	glycerol esters containing carboxy ethyl/propyl polyethyleneoxy alkyl ether groups	tailor-made lipids, designer lipids, fat mimetics	10,21,35

Table 1. Homogeneous heterobifunctional acyclic oligomers R_1-$(PEO)_n$-R_2 of ethylene oxide (selective exemplification)

http://dx.doi.org/10.5772/57610

These conclusions suggest a greater reactivity of the lauryl chain compared to the myristyl one with the increase of the homogeneous polyether chain.

In the cyanoethylation process of higher alcohols, the reaction time favors the formation of β-alkyl-oxy-propionitriles up to 180 minutes, and the oligomerization of the acrylic monomer throughout the process. After this period the cyanoethylation yields decrease, further confirming the reversible character of the nucleophilic addition under prolonged contact between reactants (Figure 12).

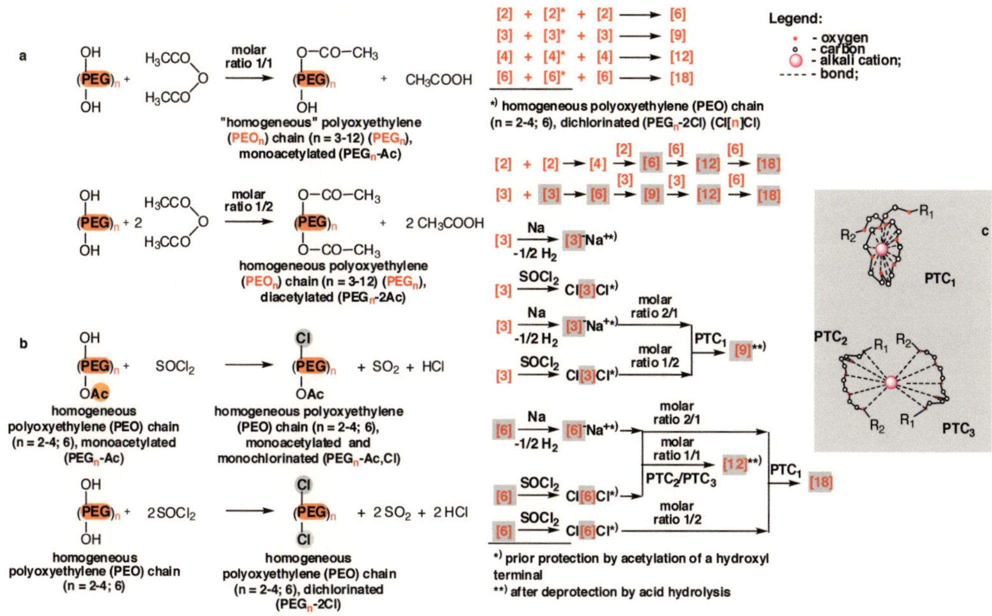

Figure 8. Operations flow chart of the monitored structuring of homogeneous polyoxyethylene (PEO) chains (n=6-18) as such accessing structural units of diethyleneglycol PEG_2 (DEO) (2), triethyleneglycol PEG_3 (TEO) (3) and/or homogeneous polyethylene glycols PEG_n (n). a) mono-and diacetylation, respectively, of the homogeneous chain (PEO) (n=3) (protection); b) mono-and dichlorination, respectively, of the PEO chains protected by acetylation; c) schemes of (directed) structuring of homogeneous polyoxyethylene (PEO) chains (n=6,9,12,18) by phase-transfer catalysis (PTC_1, PTC_2, PTC_3); PTC_1 – homogeneous dimethylglyme; PTC_2 – homogeneous β-alkyl (L/M) polyethyleneoxy (n=4) methyl ether; PTC_3 – dicyanoethyl triethyleneglycol.

Similar trends are observed in the cyanoethylation of nonylphenol and homogeneous polyethoxylated nonylphenols, respectively, for the entire series of homogeneous and/or heterogeneous PEO chain homologues.

In the series of homogeneous polyethoxylated lauryl/myristyl alcohols (n=3-18) the maximum value of the cyanoethylation yield is obtained at lower processing periods, which may suggest the favorable intervention of the polyoxyethylene chain in the cyanoethylation process (Figure 13).

On the overall process, increasing the reaction time up to approx. 180 minutes favors all reaction processes, including the formation of acrylic oligomers.

Increasing the amount of catalyst above the optimum value (4-5 × 10^{-3} mol/L) increases the alkalinity of the medium and the oligomerization reactions of the acrylic monomer. In the concentration range 15-50 × 10^{-3} mol/L, the content of homogeneous β-lauryl/myristyl (7/3) polyethyleneoxy (n=3) propionitrile decreases, in parallel with the sharp rise of the acrylic oligomers content. In the series of polyoxyethylene chain homologues, the maximum cyanoethylation yield is achieved at higher values of the catalyst concentration. At the same catalyst concentration, increasing the polyether chain determines a significant increase of the nucleophilic addition yields, but also a reduction in the amount of acrylic oligomers formed, probably due to the solvation of the acrylic monomer in the polyether chain [9,10].

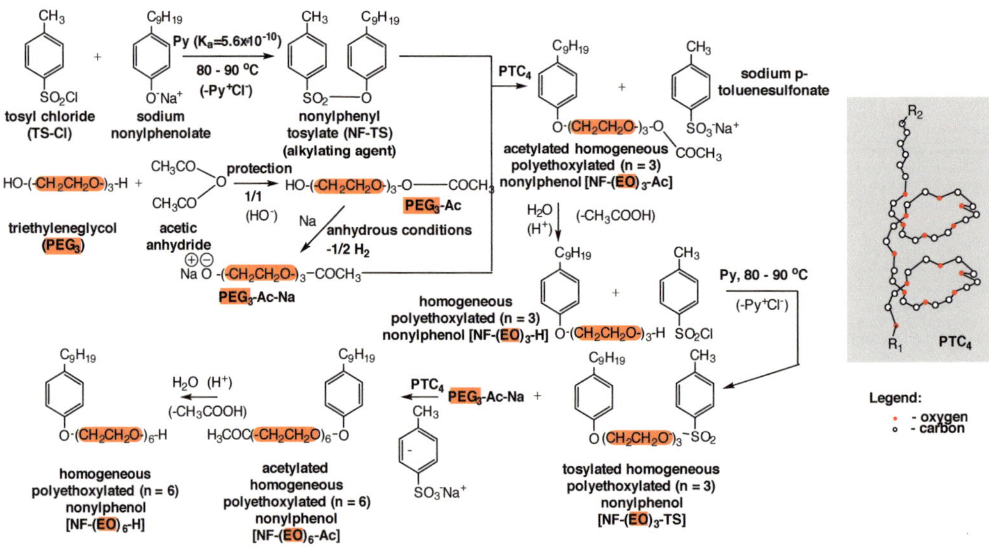

Figure 9. Flow chart of the processing and purification of homogeneous polyoxyethylene (PEO) chains (n=3, 6) monoderivatized with nonylphenol (NF) by condensation of tosylated nonylphenol (NF-TS) with monosodium protected (acetylated) diethyleneglycol under phase-transfer catalysis conditions; PTC_4 – β-alkyl (L/M) polyethyleneoxy (n=16) ethylamine

The formation of acrylic oligomers was avoided by the introduction of ferrous cations, in the present case of anhydrous ferrous sulphate ($FeSO_4$), as polymerization inhibitors of the acrylic monomer. For 1%, the cyanoethylation yield of homogeneous lauryl/myristyl (7/3) alcohol increases by more than 10%, without the formation of acrylic oligomers. Similar results are obtained in the homologous series of homogeneous (C_{12}-C_{18}) polyethoxylated (n=3-18) alcohols, and also of nonylphenols as such and polyethoxylated nonylphenols, respectively, for the entire series of (PEO) chain homologues [9,10]. Increasing the length of the hydrocarbon chain, for the same size of the homogeneous polyoxyethylene chain, reduces the cyanoethylation yields through unfavorable steric effects.

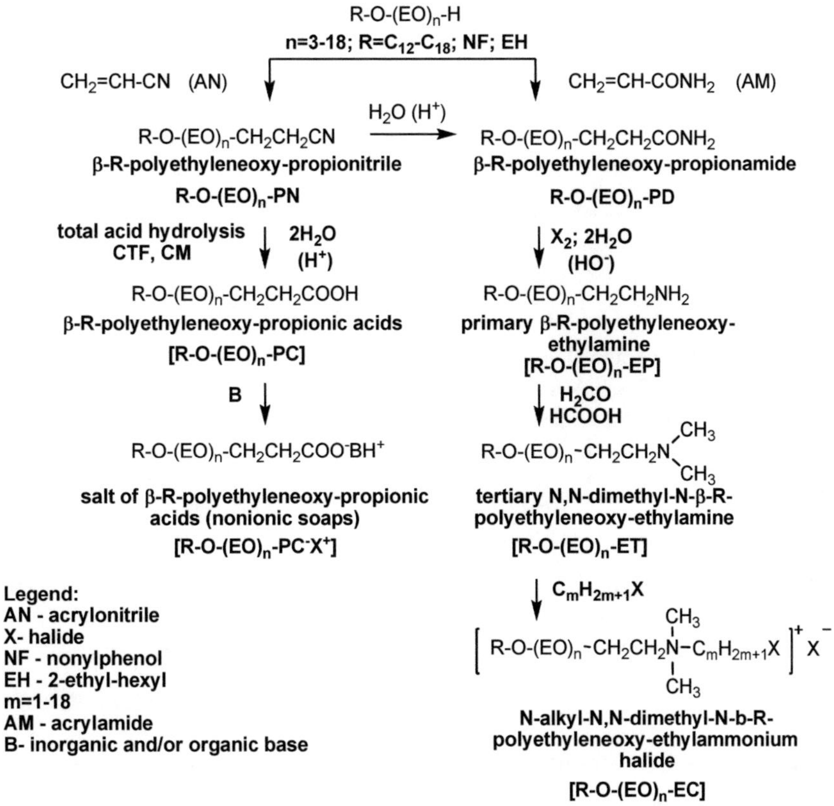

Figure 10. Reaction scheme of the processing of surface-active structures HOHAO [9,19]

The polyoxyethylene chain with its specific conformation interferes in the cyanoethylation process in non-polar reaction media (toluene, etc.) by activating the nucleophile, so that the cyanoethylation yields for higher polyethoxylated alcohols with the same hydrocarbon chain, but with a variable (n=3-18) polyoxyethylene chain increase.

The hydrocarbon chains in higher alcohols or in general β-R-oxy-propionitriles present in the process, through their length or tridimensional arrangement, respectively, generate "steric hindrance" phenomena (Figure 14), which reduce the overall rate the more so as the length is greater. The polyoxyethylene chains through their conformation favor the formation of non-solvated nucleophiles, which accelerates the cyanoethylation. For this reason cyanoethylation is also favored by the presence of phase-transfer catalysts which activate the nucleophile in the "reverse" micelle [9,25] medium system. Unprotected glymes, polyethylene glycols (n=3-30), in comparison with protected ones: dicyanoethylated homogeneous polyethylene glycols (n=3,6,9,12,18), homogeneous β-lauryl-polyethyleneoxy (n=3,6,9,12,18) propionitriles, together with the reaction rate increase the monomer consumption. The determination of the partial reaction order with respect to glymes allowed the indirect estimation of the size of the elementary coordination "cell" of the alkaline cation at the value of 8-9, and the thermody-

(a)

(b)

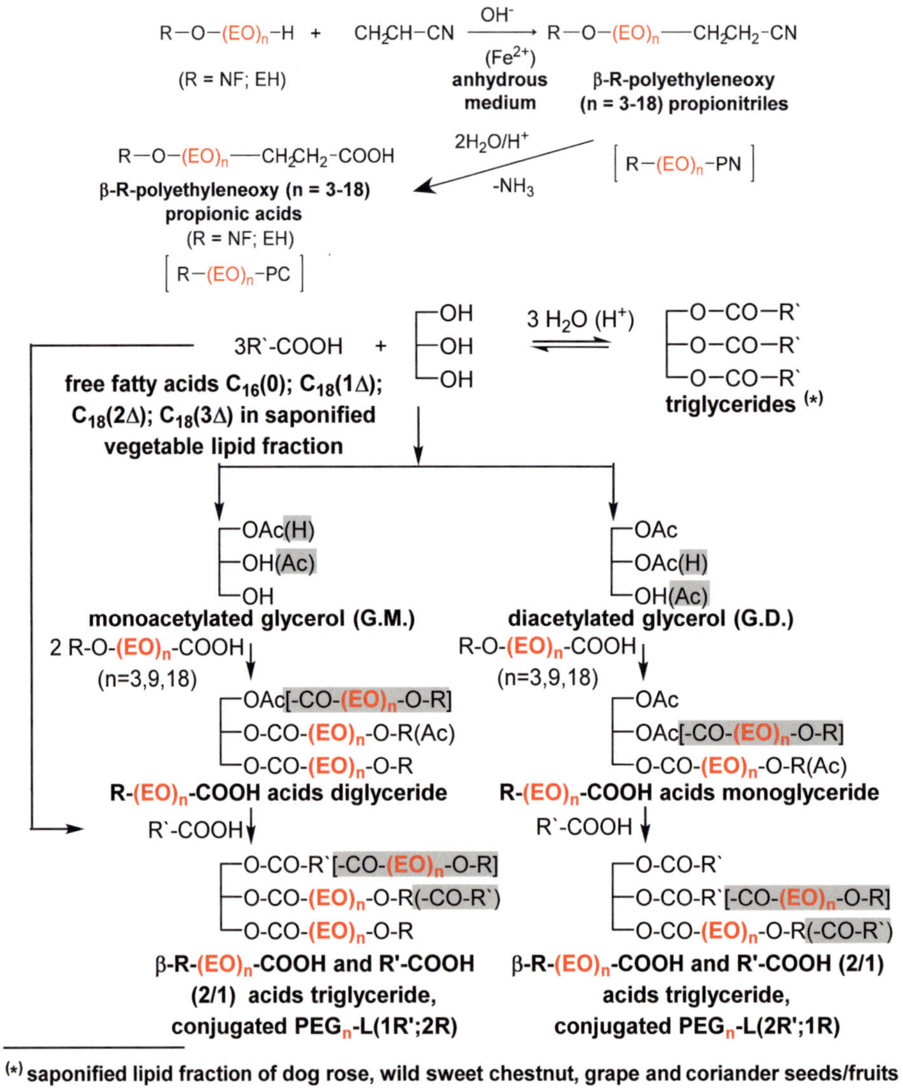

(a) controlled modification of homogeneous polyoxyethylene (PEO) chains (n=3, 6, 9, 12, 18), monoderivatized R(NF;EH) by cyanoethylation and total acid hydrolysis [$R(EO)_nPC$];
(b) actual processing of conjugates PEG_n-L (2R';R) (R';2R)

Figure 11. Overall processing scheme of conjugates PEG_n-L(2R';R) (R';2R) [10]

namic expression of the rate constant did the same for the calculation of the "reverse" micelle-processing medium phase-transfer free energy [9].

A reversible reaction, cyanoethylation is influenced by temperature, time, excess reagent (monomer) and addition products. Secondary products existent in the unpurified technical raw materials (higher alcohols, polyethylene glycols, traces of water) and oligomers of AN affect the yields of nucleophilic addition in the synthesis of HEHAO through the consumption

http://dx.doi.org/10.5772/57610

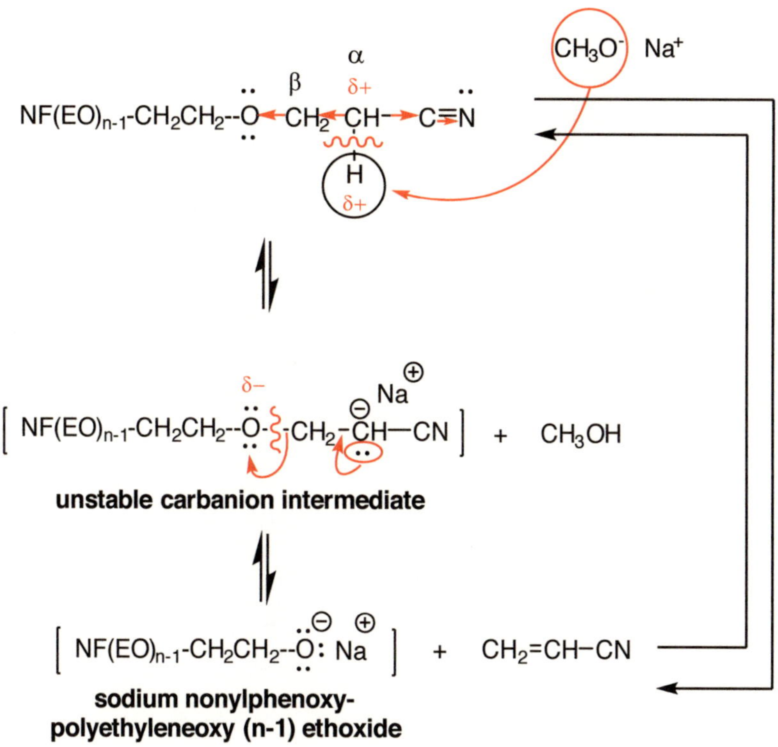

Figure 12. Mechanisms of the cleavage of the "ether bridge" formed in β-nonylphenoxypolyethyleneoxy (n=3-18) propionitriles under prolonged contact with the basic catalyst [CH_3O-Na^+]

of monomer and further purification complications. The presence of oligomerization inhibitors favors the yields.

The partial hydrolysis of homogeneous β-R-polyethyleneoxy-propionitriles (HOHAO) is a heterogeneous process due to the limited solubility in water of homogeneous polyethyleneoxy (n=0-18) nitriles and their corresponding amides, respectively. The low reaction temperature along with the reduced solubility and the waxy, consistent appearance of homogeneous β-R-polyethyleneoxy (n=0-18) propionamides, constitute serious impediments in obtaining high hydrolysis yields. The use of large amounts of water or high reaction temperatures increase the total hydrolysis yields with the formation of the corresponding homogeneous β-R-polyethyleneoxy (n=0-18) propionic acids. In the research carried out it was proceeded to the partial hydrolysis of nitriles with 90% concentrated sulphuric acid in the temperature range 0-15°C.

Homogeneous β-R-polyethyleneoxy-propionic acids were also obtained through the acid-catalyzed exhaustive hydrolysis of homogeneous β-R-polyethyleneoxy-propionitriles. Depending on the processing conditions of the reaction products two classes of HOHAO are obtained: the free acids or their salts (soaps), R-$(EO)_n$-PC, with confirmed surface-active properties [19,22,23]. The evolution of the total hydrolysis yield is determined by the hetero-

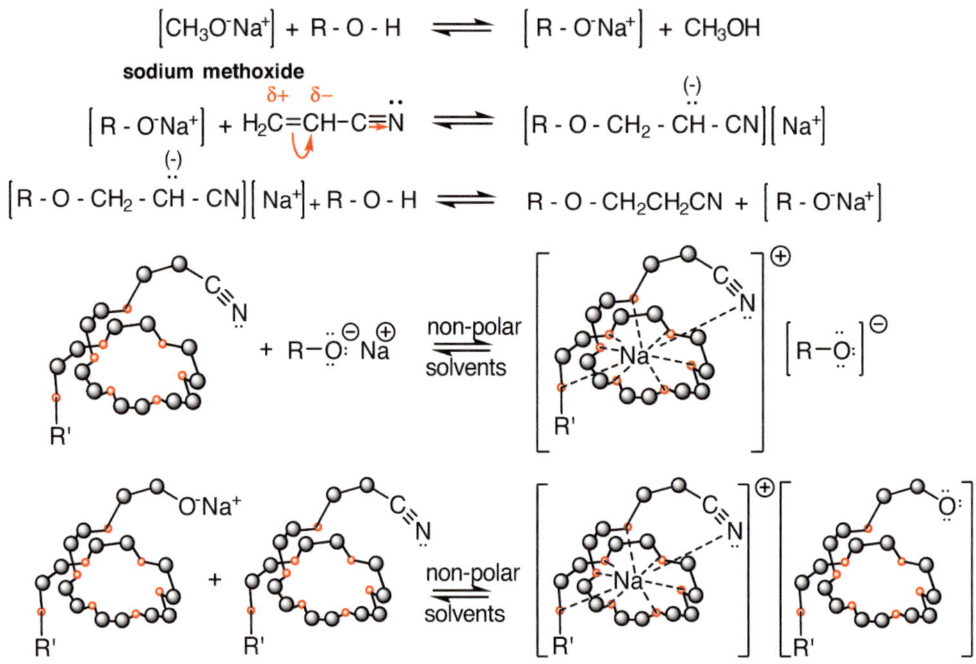

$R = C_9H_{19} - C_6H_5$ -; or R';
$R' = C_9H_{19} - C_6H_5 - O - (CH_2CH_2 - O)_{n-1} - CH_2CH_2$ -;

Figure 13. Scheme of the principle of coordination in non-polar solvents of alkali cations in the cavity of the polyoxyethylene chain with 8-9 oxygen atoms of β-R-polyethyleneoxy-propionitriles

geneity of the reaction medium (biphasic system); phenomena of micellar catalysis in emulsion or of phase-transfer play a major role in the development of the processes [22].

The nature of the acid catalyst (HY) Y=Cl^-; $HOSO_3^-$; $CH_3C_6H_4SO_3^-$, paratoluenesulphonic acid (TS); $C_{12}H_{25}C_6H_4SO_3^-$, dodecylbenzenesulphonic acid (DBSH), influences favorably the hydrolysis yields through the acid's strength. While in the presence of concentrated sulphuric acid at low temperatures are obtained predominantly homogeneous β-substituted propionamides, in the presence of hydrochloric acid above 80°C the corresponding propionic acids are formed. Under paratoluenesulphonic and/or dodecylbenzenesulphonic acid catalysis the yields in homogeneous β-R-polyethyleneoxy-propionamides increase at low temperatures for low molar ratios HY/R-$(EO)_n$PN, also on the account of the homogenizing effect of these structures with surface-active properties [22].

Increasing the temperature and the molar ratio HY/propionitrile favors the total hydrolysis yields. The amount of acid selectively influences the hydrolysis of nitriles. Thus, excess hydrochloric acid favors the formation of homogeneous β-substituted propionic acids, alongside their corresponding esters with higher alcohols either present as impurities or originating in the cleavage in acid medium of "ether bridges" formed in the cyanoethylation or amidoethylation process or existing initially in the homogeneous polyoxyethylene chain,

http://dx.doi.org/10.5772/57610

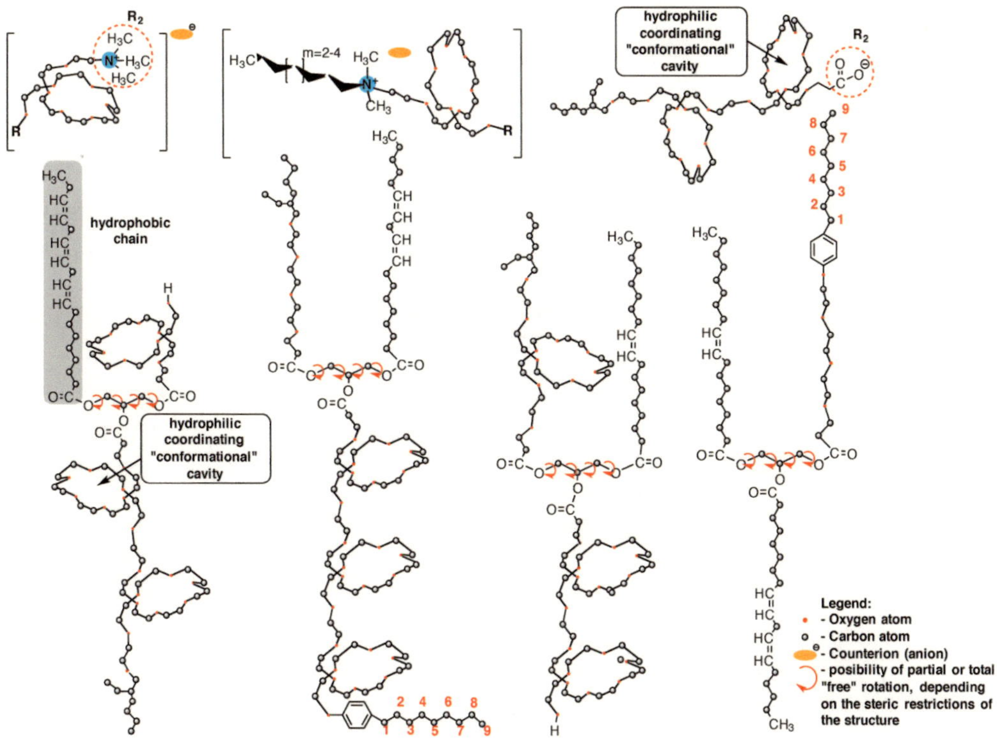

Figure 14. Three-dimensional conformation of HOHAO with structures generating difficulties of "packing" at hydrophilic and/or hydrophobic interfaces

while excess *n*-dodecylbenzenesulphonic acid favors preferentially the formation of propionamides and less of propionic acids, without the formation of the corresponding esters [28,29].

In all cases, above 80°C the content of homogeneous β-R-polyethyleneoxy (n=0-18) propionitriles decreases markedly regardless of the acid catalyst used and the presence of propionate esters is observed. Above 110°C in the presence of hydrochloric acid increasing amounts of higher alcohol are observed due to the cleavage of "ether bridges". Increasing the reaction time favors the total acid hydrolysis. At high temperatures (over 80°C) in the presence of hydrochloric acid, after approx. 60 minutes the β-substituted propionitrile disappears from the reaction mixture, and after 90 minutes also does the propionamide formed intermediately. In the presence of TS, at the same temperature and amount of water, propionitriles and propionamides can be found in traces even after 180 minutes and total conversions are generally not reached.

Hydrolysis of homogeneous β-R-polyethyleneoxy (n=3-9) propionitriles in the presence of free homogeneous polyethoxylated (n=3-9) higher alcohols (C_{12}-C_{18}) ensures the obtention of high yields of hydrolysis. In parallel the content of propionic ester is increasing, the more so as the length of the homogeneous polyoxyethylene chain in the homogeneous polyethoxylated (n=3-9) higher alcohol introduced is smaller.

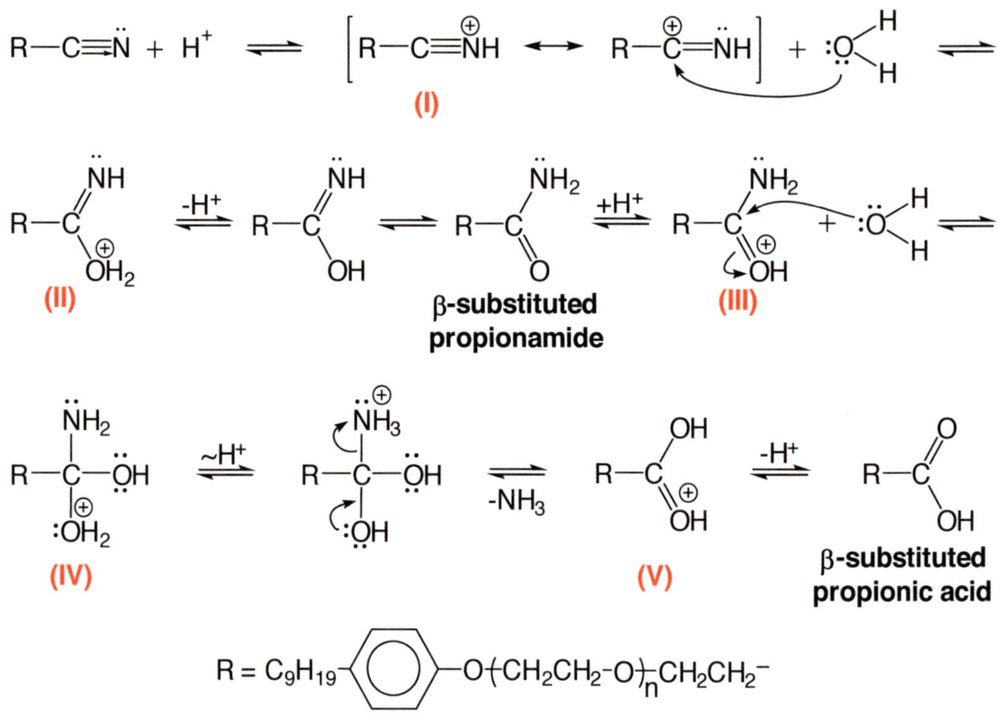

Figure 15. Probable mechanism for the exhaustive acid hydrolysis of β-nonylphenoxy polyethyleneoxy (n=3-18) propionitriles (β-substituted propionamide; β-substituted propionic acid) [22,23]

The favorable effect of homogeneous polyethoxylated (n=3-9) higher alcohols (C_{12}-C_{18}) with increasing polyoxyethylene chain on the process of total acid hydrolysis of propionitriles implies the existence of micellar catalysis (n≥9) or in emulsion phenomena (for n=3-9) [9].

Under the same conditions, increasing the homogeneous polyoxyethylene chain's length in the nitrile subjected to hydrolysis favors the yields, due to the stabilization effect on the hydrolysis intermediates. Increasing the hydrocarbon chain influences negatively, through its length, the total acid hydrolysis yield, probably due to steric reasons similar to cyanoethylation.

Because the hydrolysis of β-substituted propionitriles is a heterogeneous process which takes place both at the separation interface of the two phases (water/organic) and inside the two phases due to the mutual solubility of the two reagents, the use of the most diverse structures of the aforementioned phase-transfer catalysts in the class of the cationic HOHAO synthesized [*N,N,N*-trimethyl-*N*-β-lauryl/myristyl (7/3) oxy-ethylammonium chloride], LM-O-EC-1.1.1, determines the increase of total hydrolysis yields, without the cleavage of "ether bridges" even under mild reaction conditions [28,29].

b. Synthesis of homogeneous heterobifunctional acyclic oligomers of ethylene oxide in the category of customized (structured) lipids.

http://dx.doi.org/10.5772/57610

In the series of efforts to diversify the macromolecular architectures HOHAO structured lipid similar to conjugates PEG_nL (Figure 5) were obtained with composition, physico-chemical and functional characteristics specific to the physiological benefits of cell membranes.

An adapted classical reaction scheme was accessed following these steps (Figure 14):

- synthesis, purification and characterization of the homologous series of homogeneous β-alkyl (EH) alkyl aryl (NF) polyethyleneoxy (n=0-18) propionic acids, HOHAO presented above $R(EO)_nPC$ (Figure 10);
- solid/liquid extraction in petroleum ether (b.p.=30-60°C) of glycerides in the divided seed material of coriander (*Coriandrum sativum*, R'_{co}), grapes (*Vitis vinifera*, R'_s), dog rose (*Rosa canina*, R'_m), and nuts (fruits) of wild sweet chestnut (*Castanea vesca*, R'_{ca}), respectively.

In all the variants studied unsaturation in variable proportions is confirmed, by higher acids: oleic (C_{18};1Δ); linoleic (C_{18};2Δ) and linolenic (C_{18};3Δ).

The content of saturated higher acids, approx. 10.1% C_{16} in grapes (s), 10.9% C_{16} in wild sweet chestnut (ca), 3.18% C_{16} in dog rose hips (m) and 2.7% C_{16} in coriander (co), does not change the "fingerprint" of the vegetable lipid fractions rich in ω_3 acids of major interest in the composition of functional lipids reported in the last decades.

It is noted that in the saponified lipid fraction from grapes and dog rose hips predominate the acids C_{18};2Δ with a share of 57% and 83%, respectively. In continuation of the preparation scheme of PEG_n-L follow:

- separation of unsaponifiables, followed by the exhaustive acid hydrolysis (HCl) of free higher acids in R'_{co}; R'_m; R'_{ca}; R'_s, their purification and gas-chromatographic characterization;
- directed esterification of glycerol mono-and/or diprotected with $R(EO)_nPC$ or free higher acids R'_{co}; R'_m; R'_{ca}; R'_s, respectively.

c. Fundamental colloidal competences of surface-active HOHAO

Two basic surface-active properties were evaluated suggesting potential directions for the exploitation of synthesized HOHAO depending on the structural elements of the respective homologous series (hydrophilic-hydrophobic index, HLB, surface tension and critical micelle concentration, respectively) [33]. The determinations allowed the formulation of structure-surface activity correlations [30].

Surface tension as a form of free energy (expressed in N/m or dyne/cm) independent of the shape of the interface separating two phases in a system, is a function of temperature, time, and the structural characteristics of the HOHAO considered. Because the interface equilibrium is established within a short period, the existence of static and dynamic surface tension is accepted [34]. The latter manifests itself in aqueous floats of HOHAO and subsequently acquires great practical significance in the process of their actual use. Two aspects can be distinguished:

- the capacity of HOHAO to reduce the surface tension, expressed as the concentration required to achieve a certain effect of reducing the surface tension;

- the effectiveness of HOHAO, expressed by the minimum value which is capable of reducing the surface tension.

It will be proceeded to selectively comment some of the colloidal characteristics of the class. Following the evolution of the surface tension in the homologous series of cationic HOHAO synthesized, the following can be stated:

- for the same n and R_2, R_2=-$N(CH_3)_2$ C_mH_{2m+1}, m=1-18, the capacity to reduce the surface tension increases in the series LM<CS, and the effectiveness of reduction decreases in the order LM>CS, as a result of the increase of the length of the hydrocarbon chain R_1;

- for the same n and R_1, the capacity to reduce the surface tension decreases in the series of chain homologues R_2, and the effectiveness of reduction increases with the R_2 chain's length due to the movement towards the center of the hydrophilic ionic polar group (Figure 14) ;

- for the same R_1 and R_2, the effectiveness of reducing the surface tension increases in the order $(EO)_0 < (EO)_3 < (EO)_6 < (EO)_9 ... < (EO)_{20}$, while the capacity to reduce the surface tension decreases in the same order: $(EO)_0 > (EO)_6 ... > (EO)_{20}$, due to the increased hydrophilic (polar) character of the cationic structure and its solubility.

The most pronounced capacity to reduce the surface tension in the homologous series of *N,N*-dimethyl-*N*-alkyl (C_1-C_4)-*N*-β-cetyl/stearyl (7/3) oxy-ethylammonium chlorides and N,N-dimethyl-*N*-alkyl (C_1-C_4)-N-β-cetyl/stearyl (7/3) polyethyleneoxy (n=3-6) ethylammonium chlorides, in comparison with *N,N*-dimethyl-*N*-alkyl (C_8-C_{16})-N-β-lauryl/myristyl (7/3) oxy-ethylammonium chlorides and *N,N*-dimethyl-*N*-alkyl (C_1-C_4)-N-β-lauryl/myristyl (7/3) polyethyleneoxy (n=3-18) ethylammonium chlorides can be explained by the difference in the intensity of repulsion between the ionic and nonionic polar hydrophilic groups identically oriented at the separation interface [9]. In the first case the role of entropic factor lowers significantly on account of the increase of the micellization free energy, while in the second case it increases due to the reduction of the micellization free energy at the separation interface with the consequence of accumulation of increased numbers of surface-active cationic molecules (micelles). Another explanation can also be found in the relationship between the capacity to reduce the surface tension and the concentration of the surfactant at the liquid-air separation interface, the latter being decisively influenced by the free energy of the diffusion process from the aqueous float to the interface and the free energy of formation of cationic HOHAO micelles. The decrease of the micelle formation entropy corresponds to an advanced ordering in the micelle, that accumulates around it a greater number of water molecules as the polarity of the ionic hydrophilic group is higher.

The movement of the ionic polar group or the nonionic one towards the center of the cationic structure is followed by an advanced reduction of the entropic effect in solution (the effective length decreases along with the degree of "packing" at the interface), therefore of the capacity to reduce the surface tension [9].

In the homologous series of *N,N*-dimethyl-*N*-alkyl (C_1-C_{16})-N-β-alkyl (C_{12}-C_{18}) polyethyleneoxy (n=0-20) ethylammonium chlorides, the value of the critical micelle concentration (CMC) for the same environmental conditions is influenced by:

- the increase of the hydrocarbon chain's R_1 length (R_2 and n identical), which lowers the CMC value by reducing the solubility;
- the increase of the hydrocarbon chain's R_1 length (R_2 and n identical) lowers the CMC value due to the reduction of the degree of "packing" at the interface;
- the increase of the hydrocarbon chain's length (R_1 and R_2 identical) lowers the CMC value through the high hydrophilicity of the HOHAO structures.

d. Fundamental colloidal competences and coordination (sequestration) competences of homogeneous heterobifunctional acyclic oligomers of ethylene oxide from the category of customized (structured) lipids

The numerous hopes of obtainment and technological implementation of the conjugates PEG_n-L(2R';R)(R';2R) due primarily to their diversified structure, were based on their potential colloidal qualities: colloidal (micellar) solubility, interface phenomena, interfacial surface tension (σ), correlated with the critical micelle concentrations (CMC).

In the structure of customized lipids the independent variable, the homogeneous oligomerization degree (n), can provide in a homologous series the controlled modification of the hydrophilic/hydrophobic balance (HLB index), and implicitly the range of later practical applications.

From the evaluation of HLB values the following can be stated:

- increasing the homogeneous oligomerization degree (n) for the same hydrocarbon chain R and/or R', either for the PEG_n-L (R';2R) or PEG_n-L (2R';R) conjugates, respectively, shows a more prominent increase of the HLB index for the first case regardless of the nature of R and/or R';
- increasing the share of hydrocarbon chain R' against R also determines the reduction of the HLB index value;
- modifying the chain R(NF;EH) for the same structure of conjugate PEG_n-L and the same homogeneous oligomerization degree (n) causes variations in the HLB index due to the structural differences between the two chains;
- the modification of the hydrocarbon chain R' (R'_{ca}; R'_s; R'_m; R'_{co}) for the same homologous series of conjugate PEG_n-L and homogeneous oligomerization degree (n) does not attract significant variations of the HLB index either for the same reasons;
- we find a good agreement between the HLB values determined [10] and calculated [30], which justifies the premises and operating colloidal strategies;
- in the series of the 48 conjugated PEG_n-L studied the overall range of variation of the HLB index is between 2.5-14, which allowed their indicative grouping in structures: HLB=1-4: lipophilic (insoluble in water) non-dispersible; HLB=4-6: partially dispersible (partially soluble in water); HLB=6-8: unstable microemulsions (after vigorous mechanical stirring); HLB=8-10: stable microemulsions (after gentle mechanical stirring); HLB=10-13: translucent (opalescent) to clear in the upper area of the domain; HLB>13: soluble, transparent).

Surface tension (σ) is a function of temperature, duration and structural characteristics also in the series of conjugates PEG_n-L (2R';1R) (R';2R).

The evaluation of this colloidal characteristic in close connection with the critical micelle concentration (CMC) sought to assess the ability of these structures to reduce the surface tension overall, and also the effectiveness of providing a minimum surface tension for a given concentration.

From the comparative interpretation of the experimental data the following correlations can be formulated:

- for the same homogeneous oligomerization degree (n) and the same hydrocarbon chain (R') the capacity to reduce the surface tension (σ) increases in the series R(EH)→R(NF), while the effectiveness of reduction decreases in the order R(NF)→R(EH), probably due to differences in chain length C_8 (EH) and C_{15} (NF), respectively;
- for the same homogeneous oligomerization degree (n) and the same hydrocarbon chain (R)) the capacity to reduce the surface tension (σ) decreases significantly in the hydrocarbon chain series R' ($R'_{co} \geq R'_m > R'_{ca} > R'_s$), while the effectiveness of reduction of (CMC) increases in the same order, probably due to the movement of the hydrophilic polar groups towards the center of the structure of conjugates PEG_n-L (Figure 14);
- for the same hydrocarbon chain R and R' the capacity to reduce the surface tension (σ) decreases with increasing the homogeneous oligomerization degree (n), while the effectiveness of reduction of (CMC) increases in the same homologous series of polyoxyethylene (PEO) chain (n=3-18), probably due to the intensification of the hydrophilic character of conjugates PEG_n-L, and of the solubility in the polar medium (water).

From the comparative interpretation of the experimental data on the coordination characteristics the following can be stated [10]:

- the sequestration (coordination) competences of the conjugates PEG_n-L(2R';1R) (R';2R) depend on their concentration (below and/or above CMC) in the processing environment;
- the coordination competences in the homologous series of derivatized homogeneous polyoxyethylene (PEO) chains (n=3,9,18) studied can be correlated with the values of the main colloidal parameters: HLB index, cumulative homogeneous oligomerization degree (n_c), surface tension (σ);
- the coordination competences depend on the ionic radius (Å) in the series of metal cations studied: Mg^{2+}(0.65), Na^+(0.95), Ca^{2+}(0.99), K^+(1.33) [10], below and above the CMC values of the respective conjugates PEG_n-L(2R';1R) (R';2R); the premises formulated in the specialized literature on the similarity of the geometric coordinates of the conformational "host site" (diameter, radius, area etc.) of the homogeneous polyoxyethylene (PEO) chains (n≥9), with the geometric coordinates of the "guest" (ionic radius, diameter of metallic cations) are confirmed;
- the coordination competences follows the same trend in the case of coordination of transitional metal cations Ni^{2+}(0.69), Co^{2+}(0.72), Mn^{2+}(0.80);

- the coordination competences can also be extrapolated through the structural parameter ionic radius (Å) to a coordination number (N.C.). The correlation relative to the sodium cation (centered cubic lattice with N.C.=6) took into account that in the case of transitional metal cations the evaluated interatomic distances are smaller due to the polarizability of the anions under the influence of proper metal cations, and that the atomic volume of alkali and alkaline-earth cations is larger than that of transitional cations;

- the preliminary study of the coordination competences for certain transitional metal cations (known promoters of lipid autooxidation) has technological importance in avoiding and/or eliminating autooxidation processes due to the high degree of unsaturation of the hydrocarbon chains (R'_{ca}; R'_{s}; R'_{m}; R'_{co}) in conjugates PEG_n-L(2R';1R) (R';2R);

- the phase-transfer competences assessed comparatively through the partition coefficient values K_D (1,2-dichloroethane) and/or K'_D (isooctane) suggest that the hydrophilicity of conjugates PEG_n-L(2R';1R) (R';2R) favors the interphase distribution of the "host-guest" (PEO chain-metal cation) systems in industrial processing;

- because there are no significant differences between the K_D and K'_D values, the evaluated experimental data confirm the reality of the phase-transfer processes studied;

- the phase-transfer competences for the same cumulative homogeneous oligomerization degree (n_C) are lower for the conjugates PEG_n-L(2R';1R) (R';2R) with hydrocarbon chains R(EH) compared with R(NF) due to differences in hydrophobicity [EH(C_8) < NF(C_{15})] and [2EH($2C_8$) < 2NF($2C_{15}$)], respectively.

The comparative evaluation of the colloidal experimental data of conjugates PEG_n-L (2R';1R) (R';2R) themselves revealed that not all the structures obtained offer directly potential colloidal and coordination competences due to the different degree of dispersibility in aqueous media (HLB=1-8) (increased lipophilicity).

Their exclusion from further tests does not represent the acceptable technical solution, which suggested two future work strategies:

- the study of cumulated colloidal competences in mixed systems PEG_n-L (2R';1R) (R';2R) (HLB=1-8) with homogeneous polyoxyethylene chains (n=9,18), monoderivatized R(EH;NF);

- the study of cumulated colloidal competences in mixed systems PEG_n-L (2R';1R) (R';2R) (HLB=1-8) / PEG_n-L (2R';1R) (R';2R) (HLB=9-13) (HLB>13).

In both variants it was counted on the cumulative properties recognized in the specialized literature, similar for mixtures with different proportions of colloidal (surface-active) compounds, but also of the structural units composing them. Thus, in order to shift the cumulative hydrophilic/hydrophobic balance (HLB_c) towards increasing hydrophilic character, it was originally resorted to homogeneous polyoxyethylene (PEO) chains (n=3-18) monoderivatized R(EH;NF) with high HLB values (HLB_{EH9}=15.35; HLB_{NF9}=12.86; HLB_{EH18}=17.37; HLB_{NF18}=14.24). In this respect we can also admit the existence of a cumulated homogeneous oligomerization degree (n_c), exemplified randomly by PEG_3-L (R_s;2NF) (n_c=6) or PEG_{18}-L (R_s;2NF) (n_c=36).

Three categories of conjugates PEG_n-L (2R';1R) (R';2R) have been selected in the intervals HLB=1-4, HLB=3-6, HLB=6-8, respectively, each with eight representatives to which were controlledly added 1%, 10% and 20%, respectively, homogeneous polyoxyethylene (PEO) chain (n=3-18) monoderivatized R(EH;NF), corresponding to the hydrocarbon chain R in the conjugates PEG_n-L (2R';1R) (R';2R) (HLB=1-8) evaluated. The HLB_c values (Pearson rule) experimentally verified by sampling led to results falling below the error limit of the confidence interval (± 1%).

The typical colloidal characteristics were correlated and interpreted: surface tension (σ), critical micelle concentration (CMC), hydrophilic-hydrophobic balance (HLB), with the structural parameters for the 18 conjugates PEG_n-L(2R';1R) (R';2R) nominated and subsequently associated in binary and ternary systems, overall finding that:

- increasing the homogeneous oligomerization degree (n) and the unsaturation in the chains R' (R'_s; R'_{ca}; R'_m; R'_{co}) 1Δ→2Δ→3Δ favors the hydrophilicity of conjugates PEG_n-L(2R';1R) (R'; 2R);

- between the chains R(EH) and R(NF) differences in the colloidal behavior exist and manifest themselves sensibly in favor of increasing the hydrophobic character of conjugates [R(EH)(C_8) < R(NF)(C_{15})], due to the movement of the homogeneous polyoxyethylene (PEO) chains (n=3-18) towards the center of the structure of conjugates [R(NF)(C_{15}) < R(EH)(C_8)];

- the exclusive modification of the ratio R'/R affected the interface and transfer colloidal characteristics (R' > R decrease R' < R increase), probably due to "steric restrictions" adversely affecting "the degree of packing" of conjugates PEG_n-L(2R';R) (R';2R) at interfaces;

- a high degree of unsaturation affects the colloidal characteristics, the more so as the share of higher unsaturated acids 2Δ and 3Δ is higher, due to their possibilities of spatial arrangement as geometric isomers, but also to the reduction of the capacity of free rotation (C-C) in the structure of conjugates PEG_n-L(2R';R) (R';2R);

- the colloidal phenomena in the structural category of systems PEG_n-L(2R';R) (R';2R) studied differ for concentration values below CMC [conformational competences are present exclusively due to the homogeneous polyoxyethylene (PEO) chains (n=3,9,18)] and values above CMC, both the conformational competences and the ones of micellar solubilization of homogeneous polyoxyethylene (PEO) chains (n=3,9,18) occur cumulatively and simultaneously;

- the gradual presence of homogeneous polyoxyethylene (PEO) chains (n=9, 18) monoderivatized R(EH;NF) along with conjugates PEG_n-L(2R';R) (R';2R) with HLB < 6 in binary and/or ternary associated systems favors the structuring of systems PEG_n-L with HLB' n_c ≥ 8, which widens the range of structural variants capable of interface and transfer phenomena, beneficial to the technological practice.

6. Conclusions and perspectives

Overall it can be stated that the obtainment and characterization for the first time of homogeneous heterobifunctional oligomers of ethylene oxide (HOHAO) as "niche" structures represents a "challenge" with real future perspectives.

The potential applications envisaged are based primarily on their structure but also their varied composition which allow the expression in perspective of colloidal phenomena [wetting, foaming/defoaming with the three components (strength, density, stability) [31,32], softening agent, micellar solubilization, controlled emulsification, adsorption at the interface in normal micelles (aqueous medium) and reverse micelles (non-aqueous media), cleaning in the most various hypostases (including resistance to hard water), chemical interface processes (chemisorption), coordination/sequestration and phase transfer, etc.]. The diversity of lipid conjugates and their competences recommends them as specialized structural units of the cell membrane walls.

Preliminary tests carried out and ongoing support these assertions.

Acknowledgements

My entire gratitude to my mentor, Prof. Eng. Ionel Jianu, Ph.D., the founder of agrifood school of Timisoara, in whose team I started and trained as a researcher and discovered the hitherto unsuspected potential of homogeneous and heterogeneous polyoxyethylene chains.

Author details

Calin Jianu*

Address all correspondence to: calin.jianu@gmail.com

Department of Food Science, Faculty of Food Processing Technology, Banat's University of Agricultural Sciences and Veterinary Medicine, Timișoara, Romania

References

[1] Thompson MS, Vadala TP, Vadala ML, Lin Y, Riffle JS. Synthesis and applications of heterobifunctional poly(ethylene oxide) oligomers. Polymer 2008; 49(2) 345-373.

[2] Mulley BA. Synthesis of homogeneous nonionic surfactants. In: Schick MJ. (ed.) Nonionic Surfactants (Chapter 13). New York: Marcel Dekker; 1967. p421-439.

[3] Edwards CL. Polyoxyethylene alcohols. In: van Os NM. (ed.) Nonionic Surfactants: Organic Chemistry; Surfactant Science Series. Volume 72. New York: Marcel Dekker; 1998. p87-121.

[4] Edwards CL. Chemistry and Handling of Ethylene Oxide. In: van Os NM. (ed.) Nonionic Surfactants: Organic Chemistry Science Series (Chapter 1). New York: Marcel Dekker; 1998. p2-4.

[5] Harris J. Poly(ethylene) glycol Chemistry: Biotechnical and Biomedical Applications. New York: Plenum Press; 1992.

[6] Chen J, Spear SK, Huddleston JG, Rogers RD. Polyethylene glycol and solutions of polyethylene glycol as green reaction media. Green Chemistry 2005; 7(2) 64-82.

[7] Abuchowski A, McCoy JR, Palczuk NC, van Es T, Davis FF. Effect of covalent attachment of polyethylene glycol on immunogenicity and circulating live of bovine liver catalase. Journal of Biological Chemistry 1977; 252(11) 3582-3586.

[8] Edwards CL. Polyoxyethylene Chain Lenght Distribution. In: van Os NM. (ed.) Nonionic Surfactants: Organic Chemistry Science Series (Chapter 1). New York: Marcel Dekker; 1998. p4-34.

[9] Jianu I. Aminoethers surfaceactive agents. Cationic surfaceactive agents. PhD thesis. Polytehnic University; 1984.

[10] Jianu C. Research concerning the potential of some metal ion complexing additives to improve food value in horticultural raw matter. PhD thesis. Banat's University of Agricultural Sciences and Veterinary Medicine; 2006.

[11] Yokogama Y, Hirajima R, Morigaki K, Yamaguchi Y, Ueda K. Alkali-cation affinities of polyoxyethylene dodecylethers and helical conformation of their cationized molecules studied by electrospray mass spectrometry. Journal of the American Society for Mass Spectrometry 2007; 18(11) 1914-1920.

[12] Jackson AT, Scrivens JH, Williams JP, Baker ES, Gidden J, Bowers MT. Microstructural and Conformational studies of polyether copolymers. International Journal of Mass Spectrometry 2004; 238(3) 287-297.

[13] Gidden J, Wyttenbach T, Jackson AT, Scrivens JH, Bowers MT. Gas-Phase Conformations of Synthetic Poymers: Poly(Ethylene Glycol), Poly(Propylene Glycol) and Poly (Tetramethylene Glycol). Journal of the American Chemical Society 2000; 122(19) 4692-4699.

[14] Bailey FE, Koleske JV. Configuration and Hydrodynamic Proprieties of Polyoxyethylene Chain in Solution. In: Schick MJ. (ed.) Nonionic Surfactants: Physical Chemistry (Chapter 16). New York: Marcel Dekker; 1998. p927-971.

[15] Gejji SP, Tegenfeldt J, Lindgren J. Conformational analysis of poly (ethylene oxide) oligomers: diglyme. Chemical Physics Letters 1994; 226(3-4) 427-432.

[16] Rösch M. The configuration of the polyethyleneoxide chain of nonionic surfactants (part 1). Tenside Detergents 1971; 8 302-313. (in german)

[17] Jianu, I., Process for the cyanoethylation of higher alcohols C12-C18, RO Patent 63637** (A2), International Classification: C07C41/00; C07C43/00; C07C41/00; C07C43/00; (IPC1-7): C07C41/00; C07C43/00.

[18] Drugarin C. and Jianu I., Process for the preparation of new amphoteric surface-active agents, RO Patent 78634 (A2), International Classification: (IPC1-7): C07C87/30.

[19] Jianu, I., Process for obtaining polyethoxylated soaps, RO Patent 85437 (A2), International Classification: C11D1/02; C11D1/02; (IPC1-7): C11D1/02.

[20] Drugarin C. and Jianu I., Process for the preparation of surface-active agents, RO Patent 77367 (A2), International Classification: C07D233/08; C07D233/00; (IPC1-7): C07D233/08.

[21] Jianu C, Jianu I. Colloidal competences of new tailor-made lipids. Journal of Food Agriculture and Environment 2010; 8(3-4) 148-155.

[22] Jianu C. Synthesis of nonionic-anionic colloidal systems based on alkaline and ammonium beta-nonylphenol polyethyleneoxy (n=3-20) propionates/dodecylbenzene sulfonates with prospects for food hygiene. Chemistry Central Journal 2012; 6:95, doi: 10.1186/1752-153X-6-95.

[23] Jianu C. Colloidal competences of some food cleaning agents based on alkaline and ammonium nonionic soaps. Journal of Food Agriculture and Environment 2012; 10(1) 10-15.

[24] Drugarin C. and Jianu, I., Process for the preparation of new cationic agents, RO Patent 82439* (A3), International Classification: (IPC1-7): C07C87/30.

[25] Jianu I. Beitrage zum kinetischen und thermodynamischen studium der cyanoethylierungs reaktionen höherer alkohole (C10-C18). In: Proceeding World Surfactants Congress. Surfactants in our Word Today and Tomorrow, vol. II 1984; p188-196. (in German).

[26] Drugarin C. and Jianu I., Imidazoline derivatives and process for the preparation thereof, RO Patent 82335** (A2), C.A. 102, 26882p, 1985, International Classification: C07D233/20; C07D233/00; (IPC1-7): C07D233/20.

[27] Weinheimer RM, Varineau PT. Polyoxyethylene alkylphenols. In: van Os NM. (ed.) Nonionic Surfactants: Organic Chemistry. New York: Marcel Dekker; 1998.

[28] Drugarin C, Jianu I. The hydrolysis of β-[(p)-Alkylphenyloxy polyethyleneoxy]-propionitrils under phase transfer catalysis-Part I: The synthesis of β-[(p)-Alkylphenyloxy polyethyleneoxy]-propionitrils. Tenside Detergents 1983; 20 128-129.

[29] Drugarin C, Jianu I. The hydrolysis of β-[(p)-Alkylphenyloxy polyethyleneoxy]-propionitrils under phase transfer catalysis-Part II: The synthesis of β-[(p)-Alkylphenyloxy polyethyleneoxy]-propionic acids. Tenside Detergents 1983; 20 130-131.

[30] Marszall L. HLB of nonionic surfactants: PIT and EIP methods. In: Schick MJ. (ed.) Nonionic Surfactants, Physical Chemistry; Surfactant Science Series. Volume 23. New York: Marcel Dekker; 1998. p493-547.

[31] Drugarin C, Jianu I. Correlation between molecular structure and surface activity, Part I: The foaming power of the class of β-Alkyloxyethyl and β-Alkylpoly(ethylenoxy)-ethyl-alkyl-dimethylammonium halides. Tenside Detergents 1983; 20 76-77.

[32] Drugarin C, Jianu I. Correlation between molecular structure and surface activity, Part II: Stability and density of the foam of β-Alkyloxyethyl and β-Alkylpoly(ethylenoxy)-ethyl-alkyl-dimethylammonium halides. Tenside Detergents 1983; 20 78-79.

[33] SR ISO/TR 896:1995 Surface Active Agents-Scientific Clasification identical with the International ISO/TR 893:1977.

[34] SR EN 14370:2005 Surface active agents. Determination of surface tension. Ring variante (identical with the European Standard EN 14370:2004).

[35] Drugarin C. and Jianu, I., Process for obtaining synthetic triglycerides, RO Patent 82041 (A2), C.A. 103, 55720k, 1985. International Classification: C07C67/22; C07C69/003; C07C67/00; C07C69/00; (IPC1-7): C07C67/22; C07C69/003.

Chapter 5

Higher Oligomeric Surfactants — From Fundamentals to Applications

D. Jurašin and M. Dutour Sikirić

Additional information is available at the end of the chapter

http://dx.doi.org/10.5772/57655

1. Introduction

Surfactants (shorten for surface-active agents) are organic compounds containing in one molecule both lyophobic (hydrophobic) and lyophilic (hydrophilic) parts (Fig. 1.). Due to such amphiphilic structure surfactants exhibit specific properties in solutions, as well as in solid state. When present at low concentrations in solutions they adsorb at all available interfaces (liquid/gas, liquid /liquid, liquid /solid,) and as a consequence dramatically change their free energy. At higher concentrations, above so called critical micellization concentration (cmc), when all the interfaces are occupied, surfactants self-assemble in the bulk in various aggregates: micelles, vesicles and liquid crystals. The type of formed supramolecular structure depends upon the structure and concentration of the surfactant, presence of the electrolyte, temperature, etc. [1, 2].

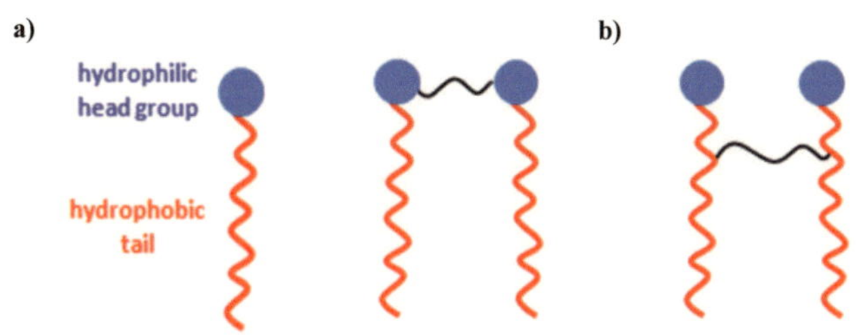

Figure 1. Schematic representation of a) monomeric and b) dimeric surfactant molecule.

Surfactants versatile phase behavior and ability to form different structures, with sizes from nano to micro-scale, is a reason why they are widely used in various industrial processes, ranging from classical (paints, cosmetics, pharmaceuticals, foods) to modern technologies (synthesis of advanced materials, environmental protection). Moreover, surfactants have important roles in living organisms. Examples are pulmonary surfactants, proteins, biological membranes which can be considered to be self-assembled bilayers of surface active compounds (phospholipids), etc. [1, 2].

In constant search for more efficient and environmentally friendly surfactants, both academic and industrial interest has been focused on design and preparation of novel, complex, surfactant structures, with improved properties in comparison with conventional surfactants, i.e. those containing only one hydrophilic and one hydrophobic group.

Novel surfactants which have attracted considerable interest in last two decades are oligomeric surfactants. These compounds are made up of two (dimeric surfactants) or more (higher oligomeric surfactants) amphiphilic moieties covalently linked at the level of the head groups or very close to them by a spacer group [3]. This means that, in theory, it is possible to synthesize oligomeric surfactants using two or more molecules of identical and/or different conventional surfactant and connecting them with a spacer group varying in chemical nature, length, hydrophobicity and rigidity. The number of possible architectures, and thus properties, is vast [3].

First to report about dimeric or gemini surfactant (Fig. 1. b) in scientific literature were Bunton and collaborators in 1971 [4] They have synthetized bisquaternary ammonium surfactants and studied rate of nucleofilic substitution in their micellar solutions. This work was followed by Devinsky and collaborators in 1985, who synthesized the great variety of bisquaternary ammonium surfactants and investigated their surface activity and micellization [5]. In 1990 Okahara and collaborators synthesized first anionic gemini surfactants, with two sulphate groups and two alkyl chains [6]. In 1990s work by Zana's, and latter Esumi's group, on bisquaternary ammonium surfactants in which they have shown that these surfactants posses unique properties and various self-assembly behaviors compared to the corresponding monomeric surfactants, motivated the investigation of different dimeric surfactants [7-17].

Constantly growing interest for the investigation and synthesis of novel gemini surfactants is a consequence of their superior properties in comparison to the conventional ones [3]:

- their cmcs are one or two order of magnitude lower than for the corresponding monomeric surfactants.
- they are more efficient in lowering surface tension.
- their aqueous solution can have a very high viscosity or even show viscoelastic properties at relatively low surfactant concentrations, whereas the solutions of corresponding monomers remain low viscous as water,
- also, they have better: solubilizing, wetting and foaming properties.

In addition, the Krafft temperatures of dimeric surfactants with hydrophilic spacers are generally very low giving to these surfactants the capacity to be used in cold water [3]. Some cationic dimeric surfactants even have interesting biological activity [5, 18].

http://dx.doi.org/10.5772/57655

The fact that the dimeric surfactants posses properties superior to the corresponding monomers, was motivation to extend the concept of gemini surfactants to higher oligomers (degree of oligomerization ≥ 3), expecting that their properties would be even better. This expectations were further supported by theoretical considerations predicting that the critical micellization concentration decreases continuously with increasing degree of oligomerization, while preferentially small spherical micelles form at low concentration and wormlike or threadlike micelles at high concentration [19,20]. In addition, since higher oligomeric surfactants represent transitional structures between conventional and polymeric surfactants, investigation of their properties can give insight in the behavior of the latter.

First synthesized and investigated higher oligomeric surfactants were trimeric [9, 11, 16, 17] and tetrameric [13] linear quaternary ammonium bromides with alkyl spacers. Up to now different cationic, anionic and nonionic oligomeric surfactants, with degree of oligomerization up to 7 [21] and linear, ring-typed or star-like topology [22] of the molecule have been investigated.

However, the efforts in investigation of higher oligomeric surfactants are hindered by the more complex synthesis and purification. Recently progress in this area has been reported. White and Warr have synthesized oligomeric alkylpyridinium surfactants by a simple elimination–addition reaction [23]. Also Feng and collaborators have reported a new method of synthesis for oligomeric surfactants by atom transfer radical polymerization (ATRP) [24, 25].

In this chapter the influence of the oligomerization degree, length of hydrophobic chains, nature of the spacer and topology of the molecule on the properties of higher oligomeric surfactants[1] in the solution and at the interfaces will be discussed. Comparison will be made with behavior of corresponding monomeric and dimeric surfactants. Possible applications of higher oligomeric surfactants will also be discussed.

2. Clasification of the oligomeric surfactants

Oligomeric surfactants, like conventional ones, are most commonly classified depending on the type of the hydrophylic headgroup surfactants in four major groups:

- anionic – hydrophilic group is negatively charged,
- cationic – hydrophilic group is positively charged,
- nonionic – hydrophilic group bears no charge, but solubility in water is a consequence of its high polarity,
- zwitterionic – molecule contain both positively and negatively charged hydrophylic groups.

So far cationic, anionic and nonionic higher oligomeric surfactants have been synthesized, but no synthesis of zwitterionic has been reported.

1 The term oligomeric surfactants as used in this chapter does not include surfactants molecules in whose structure only part of the moiety, i.e. head group or tail, is repeated. Example of such amphiphilic molecules are nonionic oligomeric surfactants like Brij or polyetheramine surfactants.

Since oligomeric surfactants are made up of two or more amphiphilic moieties covalently linked at the level of head groups or very close to them (Fig. 1. b) by a hydrophylic or hydrophobic, flexible or rigid spacer group, they can also be classified based on:

- number of amphiphilic moieties present in the molecule – dimeric (gemini), trimeric, tetrameric, etc.
- molecular structure (features of their spacer groups)-linear, ring-type, and star-shaped.

The structures of typical examples of the higher oligomeric surfacatants, reviewed in this chapter, are given in Fig. 2-4. They include:

- **cationic** – different derivates of quaternary ammonium salts were prepared, like:
  - *quaternary ammonium surfactants with the alkyl spacers* [3, 7-17, 26, 27] (Fig. 2. 2A, 3A, 4A) – they are usually denoted as m-s-$(m$-$s)_x$-m, where m represents number of carbon atom in hydrophobic chain, s number of the carbon atoms in the spacer and $x=j-2$, where j is the degree of oligomerization. Up to now, these surfactants have been the most investigated oligomeric surfactants, due to the relative ease of their synthesis and possibility to tailor surfactant properties by changing spacer and chain length. Majority of these surfactants are bromide salts, chlorides were synthesized in lesser extent.
  - *oligomeric quaternary ammonium surfactants prepared by epichlorohydrin* [28, 29] (Fig. 2. 2B, 3B) – these surfactants have short polar spacers containing –OH groups.
  - *oligomeric quaternary ammonium chlorides with trans-1,4-buten-2-ylene, m-xylylene and p-xylylene spacers* (Fig. 2. 2C, 3C, 4B) – the spacers groups differ in both nature and length, and are all rigid. Chloride was chosen as counterion to increase solubility in water and to provide lower Krafft temperatures [30, 31].
  - *polyoxyethylene ether trimeric quaternary ammonium surfactants* (Fig. 2. 3I) – these surfactants contain chains consisting of both polyoxethylene and dodecyl alkyl groups [32].
  - *star-shaped* trimeric quaternary ammonium bromides with different chain length [33] (Fig. 2. 3D).
  - *star-shaped trimeric, tetrameric and hexameric quaternatry ammonium salts with amide groups* (Fig. 2. 3E, 3F, 4C, 6A) – amide groups were chosen to increase solubility of these surfactants. The spacers are rigid. Non symmetric and symmetric trimeric surfactants were prepared with a slight difference in the spacer [22, 34-36].
  - Tris[2-hydroxy-3-(alkyldimethyammonio)-propoxymethyl]ethane]-these surfactants were designed in order to obtain surfactants with enhanced antimicrobial properties [37] (Fig. 2. 3G, 3H).
- **anionic**
  - *ring-type trimeric* surfactants synthesized by introducing three hydrocarbon chains to cyanuric chloride (Fig. 3. 3J) –they can dissolve in water only at pH around 13, which renders their applicability [38].

- *triple chain* surfactants with three hydrocarbon chains and two or three carboxylate headgroups (Fig. 3. 3K, 3L)– they exhibit same problem with solubility as ring type trimeric surfactants [39].

- *tetrameric surfactants with multiple-ring spacers* based on dioxane rings with different flexibility (flexible, semi-flexible and rigid) of spacers [40] (Fig. 3. 4D-G).

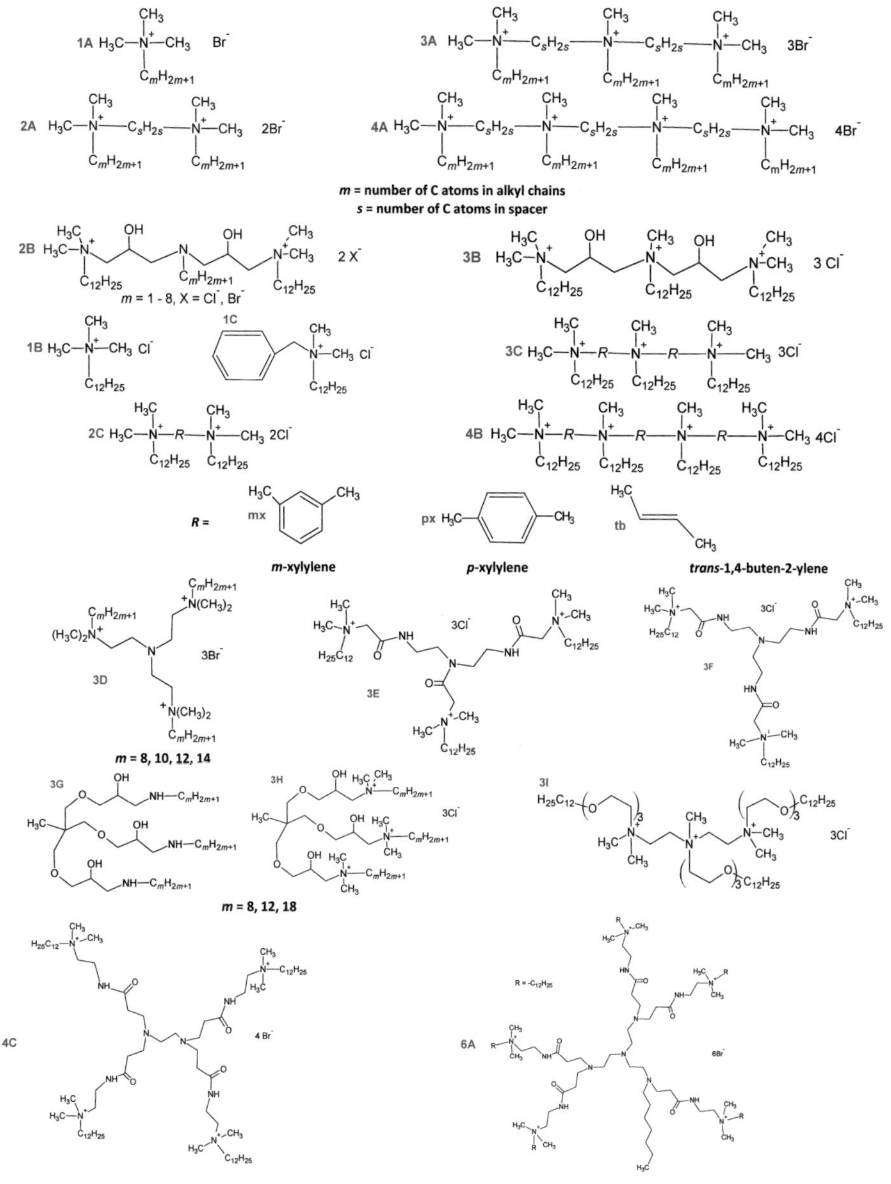

Figure 2. Molecular structure of cationic oligomeric surfactants-quaternary alkyl ammonium salts.

- **nonionic**
  - *tyloxapol* (Fig. 4. 7A) – repeating unit is close to Triton X-100 and maximum degree of polymerization is about 7 [21]. This surfactant is commercially available.
  - *n-alkylphenol polyoxyethylene trimeric surfactants* with different lengths of hydrophilic group oxyethylene chains and hydrophobic group methylene chains [41] (Fig. 4. 3M).
  - trimeric surfactants derived from tris(2-aminoethy)amine (Fig. 4. 3N) – these surfactants offer possibility to change hydrophilic/lipophilic balance while keeping molecular skeleton the same [42].

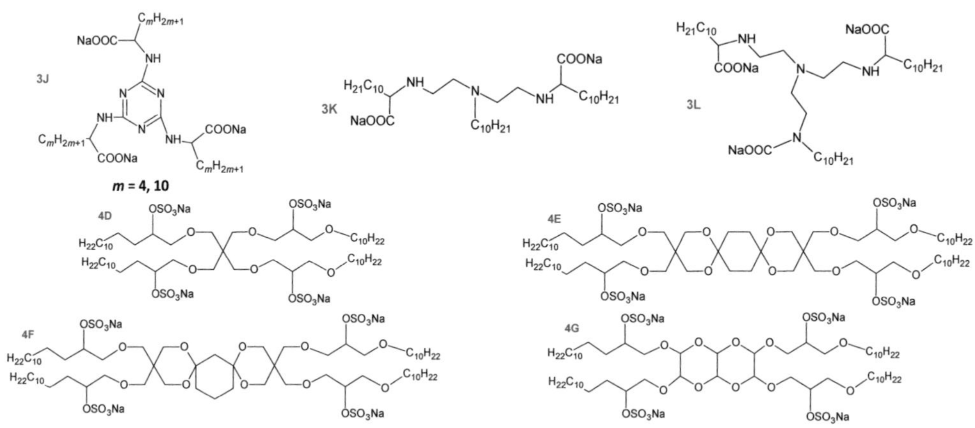

Figure 3. Molecular structure of anionic oligomeric surfactants.

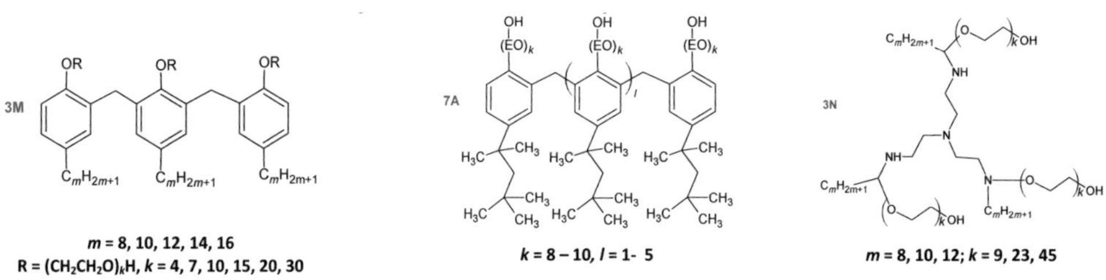

Figure 4. Molecular structure of nonionic oligomeric surfactants.

http://dx.doi.org/10.5772/57655

3. Oligomeric surfactants in solution

Unique properties that surfactants exhibit in aqueous solution and in the solid state are consequence of their amphiphilic nature. When present in low concentrations in aqueous solution, surfactants tend to concentrate at the available interfaces and in that way reduce the free energy of the system. At higher concentration, when all the interfaces are saturated the reduction of the system energy can be achieved, depending on the experimental conditions, by crystallization of the surfactant from the solution or by the formation of supramolecular aggregates (micelles, vesicles, liquid crystals etc., Fig. 5). The concentration above which micelle are formed is called critical micellization concentration (cmc). Micelles are thermodynamically stable dispersed species in equilibrium with surfactant monomers [1, 2].

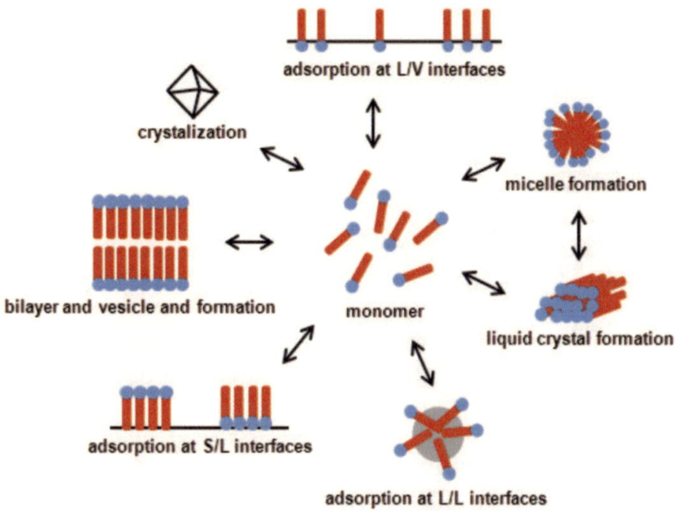

Figure 5. Modes of surfactant reduction of surface and interfacial energies. After ref [1].

3.1. Solubility of surfactants

Overall solubility of many ionic compounds increases as temperature increases. This effect is the result of the physical characteristics of the solid phase, namely crystal lattice energy and heat of hydratation of the material being dissolved.

In the case of ionic surfactants, it is often observed that the solubility undergoes a sharp, discontinuous increase at some characteristic temperature, named the Krafft temperature (T_K) (Fig. 6). Below the Krafft temperature solubility of the surfactant is determined by the solid state properties, while above it the surfactant solubility increases due to formation of micelles, which are thermodynamically favored form [1, 2, 43].

The Kraft temperature varies with alkyl chain length and structure, as well as with counterion. Lowering of the Krafft temperature can be achieved by introducing chain branching, multiple

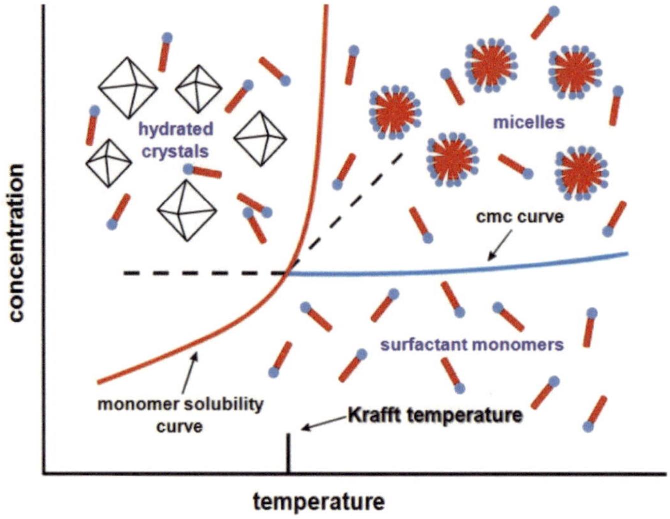

Figure 6. Schematical representation of the solubility curve for the ionic surfactants. The Krafft temperature (T_K) is the temperature at which surfactant solubility equals the cmc. Above T_K surfactant molecules form a dispersed phase; below T_K hydrated crystals are formed. After ref. [43].

bonds in the alkyl chain or bulkier hydrophilic groups in the surfactant molecules. In this way intermolecular reactions that promote crystallization are reduced [1, 2, 43].

The Krafft temperature is usually determined either by measuring the change of electrical conductivity with temperature or visually observing the change of turbidity of supersaturated surfactant solution (usually 1 wt %).

Knowledge of the Krafft temperature is crucial in many applications since below T_K the surfactant will clearly not perform efficiently; hence typical characteristics such as maximum surface tension lowering and micelle formation cannot be achieved.

Ionic dimeric *m-s-m* surfactants with $m \leq 12$ are generally highly soluble in water. The Krafft temperatures below 0 °C have been reported for many series of anionic dimeric surfactants with hydrophobic or hydrophilic spacers [3].

Only the Krafft temperatures of cationic higher oligomeric surfactants were reported, to the best of our knowledge. Majority of the reported values are below 0 °C, which is important for their possible applications in cold water. Besides relative ease of their synthesis, low Krafft temperature is one of the main reasons why oligomeric quaternary ammonium surfactants have received much of the attention.

While trimeric 12-2-12-2-12 surfactant has Krafft temperature in the vicinity of 0 °C, T_K for corresponding tetramer12-2-12-2-12-2-12 is 32 °C [27]. It is interesting to note that the Krafft temperature does not regularly change with the degree of oligomerization in this series of oligomeric dodecyl quaternary ammonium surfactants since the Krafft temperature of monomeric DTAB (1A) is below 0 °C and that of dimeric 12-2-12 is 15 °C [3].

http://dx.doi.org/10.5772/57655

Trimeric quaternary ammonium surfactants with polar spacers (3B) have also shown good water solubility and their Krafft temperatures are all below 0 °C [28].

In the series of quaternary ammonium surfactants with *trans*-1,4-butenylene, *m*-and *p*-xylylene spacers (2C, 3C, 4B) it was shown that the Krafft temperatures are reduced by a higher degree of oligomerization [30]. Interestingly, T_K of the trimeric and tetrameric surfactants with the *p*-xylylene spacer (3Cpx and 4Bpx) are below 0 °C, whereas the analogous dimer 2Cpx surfactant has a Krafft temperature of 23 °C [30].

The Krafft temperature of trimeric surfactants 3E and 3F [34], and polyoxyethylene ether trimeric quaternary ammonium surfactant 3I is also found to be below 0 °C [32]. The T_K of star-shaped trimeric surfactants 3D were found to be lower than 5 °C [33].

Despite the large number of hydrophobic alkyl chains, the Krafft temperature of hexameric surfactant 6A is also found to be below 0 °C [22].

3.2. Adsorption at the air/water interface

Many surfactant applications are based on their ability to absorb at various interfaces in an oriented fashion. The difference in the surface activity of different surfactants is a consequence of the difference in their packing density at the air/water interface. The packing density is reflected in the values of surfactant surface excess concentration (Γ_{max}) and surface area occupied by a surfactant molecule (a_{min}). The higher the value of surface excess concentration, consequently the lower the value of surface area, the more efficient surfactant is. a_{min} is not only measure of efficacy of adsorption, but it is also the first information about the orientation and packing of the surfactant at the interface [1, 2, 43].

Maximum surface excess concentration of a surfactant (Γ_{max}), can be calculated from the surface tension (γ) measurements, i.e. from the maximal slope (dγ/dlog*c*) in the γ *vs.* log *c* before cmc (Fig. 7):

$$\Gamma_{max} = -\frac{1}{2.303nRT}\left(\frac{\partial \gamma}{\partial \log c}\right)_T \tag{1}$$

where *c* denotes concentration, *R* is the gas constant (8.314 Jmol^{-1} K^{-1}), *T* is absolute temperature and *n* is the number of solute species whose concentration at the interface changes with change in the value of surfactant bulk concentration (*c*) [2].

From the surface excess concentration, the area per molecule at the interface, a_{min}, in square nms is calculated from the relation:

$$a_{min} = \frac{10^{14}}{N_A \Gamma_{max}} \tag{2}$$

where N_A is Avogardo's number and Γ_{max} is expressed in mol/m^2 [2].

In addition to Γ_{max} and a_{min} following quantities can be used to assess surfactant performance in lowering surface tension and its preference for adsorption in comparison to micellization [1, 2, 43]:

- the maximum reduction of surface tension (γ_{cmc}) which can be attained in the solution of certain surfactant regardless of its concentration. Lower γ_{cmc} means more surface active surfactant.
- the concentration required to produce a surface tension reduction of 20 mN m^{-1} (C_{20}), usually expressed as negative logarithm of such concentration, pC_{20}. The larger pC_{20} the more efficiently the surfactant is adsorbed at the interface and the more efficiently it reduces surface tension.
- the cmc/C_{20}, ratio is a convenient measure of the relative effects of structural factors on the micellization and adsorption processes; the larger the values of the cmc/C_{20} ratio, the greater the tendency of the surfactant to adsorb at the interface, relative to its tendency to form micelles.

Surface activity of higher oligomeric surfactants was assessed based on Γ_{max}, a_{min}, γ_{cmc} and pC_{20} values. Data reported in available literature are summarized in Table 1.

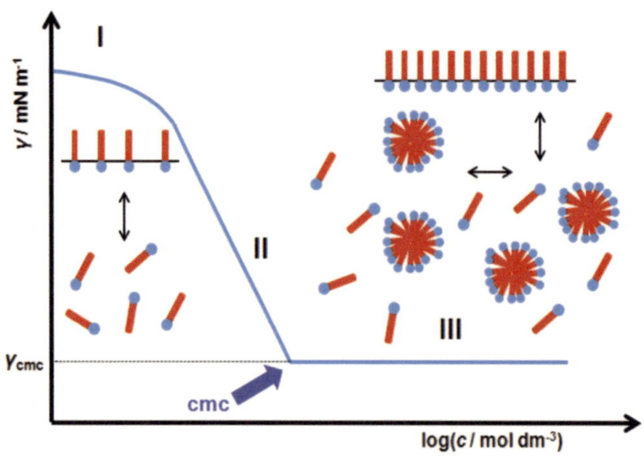

Figure 7. Plot of surface tension versus log of the bulk phase concentration for an aqueous solution of a surfactant. Three distinctive parts of the curve represent [45]:

i. At low surfactant concentrations surfactant monomers are forming monolayer at the air/water interface. Surface tension is decreasing with the surfactant bulk concentration due to the increasing surfactant surface concentration at the air/water interface.

ii. At concentrations below, but close to cmc slope of the curve is constant because surface concentration reached its maximum value.

iii. At concentrations above cmc, surface tension remains almost constant, due to constant monomer concentration.

Surfactant	$10^6 \cdot \Gamma_{max}$ [mol molecule/ m^2]	$10^6 \cdot \Gamma_{max}$ [mol alkyl chain / m^2]	a_{min} [nm^2 /molecule]	a_{min} [nm^2 / alkyl chain]	n	γ_{cmc} [mN / m^{-1}]	pC_{20}	T [°C]	Ref.
Cationic									
DeTAB*						40		25	[26]
1A	3.5	3.5		0.59	2	38.6		25	[15, 16]
	2.7		0.62			38.9	2.3	30	[27]
10-2-10						31.8		25	[44]
12-2-12	2.7	5.4	0.72		3		31.4	25	[15,16]
	1.7		1.04			30.3	3.8	30	[27]
12-3-12	2.3	4.6		0.48	3			25	[15]
12-6-12	1.35	2.7		0.72	3			25	[7,15]
8-2-8-2-8	1.07		1.55		4	35.1		25	[26]
10-2-10-2-10	1.38		1.21		4	25.8		25	[26]
12-2-12-2-12	1.11		1.49		4	36.4		25	[26]
	1.3		1.27			36.0	3.7	30	[27]
12-3-12-3-12	1.75	5.25		0.49	4			25	[15]
12-6-12-6-12	0.7	2.1		0.83	4			25	[15]
12-2-12-2-12-2-12	0.9		1.84		5	29.7	4.0	40	[27]
12-3-12-4-12-3-12	1.3	5.2						25	[15]
1B						40.5		23	[30]
BQADC**						37		20	[28]
2B-12 Cl						35		20	[28]
3B -12 Cl						32		20	[28]
1C						39		23	[30]
2Ctb						41.5		23	[30]
2C mx						43		23	[30]
2Cpx						45.0		23	[30]
3Ctb						41		23	[30]
3Cmx						40		23	[30]
3Cpx						40		23	[30]
4Btb						41		23	[30]
4Bmx						38.5		23	[30]
4Bpx						35		23	[30]
3I						38.9		25	[32]
3H - 8	2.8		0.593			34.78	4.21	20	[37]
3H - 10	2.88		0.575			34.03	4.70	20	[37]
3H - 12	2.94		0.565			33.02	5.39	20	[37]

*decyltrimethylammonium bromide, ** bisquaternary ammonium dichloride with dodecyl chain $(R(CH3)_2N^+CH_2CH\text{-}(OH)CH_2{}^+N(CH_3)_2R \cdot 2Cl^-$

Surfactant	$10^6 \cdot \Gamma_{max}$		a_{min}		n	γ_{cmc}	pC_{20}	T	Ref.
	[mol molecule/ m^2]	[mol alkyl chain / m^2]	[nm^2 /molecule]	[nm^2 / alkyl chain]		[mN / m^{-1}]		[°C]	
Cationic									
3D-8	0.875		1.90		4			25	[33]
3D-10	0.883		1.88		4	33.4		25	[33]
3D-12	0.820		2.03		4	32.3		25	[33]
3D-14	0.970		1.70		4	32.1		25	[33]
4C						47.6		25	[36]
Anionic									
$C_{11}H_{23}COONa$	2.43		0.69			37.5	2.3	20	[38]
$C_{12}H_{25}OSO_3Na$	3.16		0.53			32.5	2.5	25	[38]
3J-4	3.54		0.47		1	32.9	3.8	25	[38]
3J-10	4.84		0.34		1	28.6	5.0	25	[38]
$C_{10}H_{21}CH(NH)_2COONa$	3.37		0.49		1	46.0	3.3	25	[39]
3K	9.13		0.18		1	29.3	5.9	25	[39]
3L	2.57		0.65		1	33.2	6.1	25	[39]
4D			0.95-1.18		4-5	30	5.13	20	[40]
4E			0.83-1.04		4-5	28	4.50	20	[40]
4F			1.02-1.28		4-5	32	4.54	20	[40]
Nonionic									
3N 8-PEG400	1.82		0.91		1	35.4		25	[42]
3N 8- PEG1000	2.21		0.89		1	33.2		25	[42]
3N 8- PEG2000	2.57		0.83		1	31.1		25	[42]
3N 10- PEG400	1.84		0.93		1	36		25	[42]
3N 10- PEG1000	3.33		0.90		1	35.8		25	[42]
3N 10- PEG2000	3.78		0.85		1	33.3		25	[42]
3N 12- PEG400	1.93		0.94		1	36.8		25	[42]
3N 12- PEG1000	2.45		0.91		1	36		25	[42]
3N 12- PEG2000	4.02		0.85		1	34.3		25	[42]

Table 1. The surface excess concentration ($_{max}$), the minimum area per molecule at the air/solution interface (a_{min}), the number of solute species whose concentration at the interface changes with change in the value of surfactant bulk concentration (n), the maximum reduction of surface tension ($_{cmc}$) and the negative logarithm of concentration required to produce a surface tension reduction of 20 mN m^{-1} (pC_{20}) obtained at given temperature (T) for higher oligomeric and corresponding monomeric and dimeric surfactants. Abbreviated names of listed surfactants are given according to the figures 2-4.

The comparison of the experimentally obtained Γ_{max} and a_{min} values for single chain surfactants is straight forward. However, this is far from being so for the oligomeric surfactants. The problem lies in the n in the expression for the surface excess concentration (eq. 1). The prefactor n represents the number of species at the interface, the concentration of which changes with the bulk surfactant concentration. This value depends on the degree of dissociation of ionic surfactants, which is not known exactly for all surfactants. There are two extreme cases, one assuming no dissociation, thus treating the surfactants as one particle, or the other assuming complete dissociation. In the case of single-chained surfactants situation is simple, this value is 1 for the nonionic, which are not ionized, and 2 for the ionic surfactants, which are considered to be fully ionized, when both ions are univalent. However introducing additional structural elements to the surfactant molecule complicates the situation. Not only that for many oligomeric surfactants degree of dissociation is unknown, but it can vary in the same series of surfactants either with changing the degree of oligomerization or the spacer length.

This problem was first encountered with dimeric *m-s-m* surfactants. The reported values were obtained using n=2 assuming that one of the headgroup is neutralized by a counterion [46, 47], or n=3 assuming that surfactant is fully ionized [7]. In order to be able to compare different series of the surfactants some studies reported values obtained using both values. Not even studies in which surface was determined directly by small angle neutron scattering manage to solve the problem, because it was shown that degree of dissociation depends on the nature of the spacer [48]. Nowadays it is commonly assumed that for series of linear quaternary ammonium surfactants with dodecyl chains and short ethylen space, n=3 for dimer, 4 for trimer and 5 for tetramer [15, 27].

In addition, when comparing Γ_{max} and a_{min} obtained in different studies, one should be aware that they can be expressed either per molecule or per alkyl chain. The later is more convenient to determine spacer influence in the series of surfactants with the same degree of oligomerization.

Based only on the structure of the surfactant molecule it could be expected that a_{min} will increase with the degree of oligomerization due to the increased number of headgroups in the molecule. However, since higher oligomers are more surface active, in some cases, their molecules are more closely packed at the air/water interface. As a result, sometimes a_{min} doesn't vary with the degree of oligomerization and therefore it may be the same or even lesser than for corresponding monomer.

Most of the pioneering work in investigating oligomeric surfactant adsorption at the air/water interface has been done by Zana and collaborators [3,7], who have laid the basis for the understanding of oligomeric surfactant behavior at air/water interface.

The influence of the oligomerization degree, spacer and alkyl chain length of *m-s-(m-s)x-m* (1A-4A) surfactants on surface activity was investigated by different groups [15-17, 26, 27]. The dependence of surface area occupied by a surfactant molecule (expressed per alkyl chain) on oligomerization degree depends on spacer length for the surfactants with dodecyl chains. For spacer length s=2 and 3 surface area a_{min} remains nearly constant going from monomer to trimer [15, 26, 27], while for *s*=6 a slight increase going from dimer to trimer has been observed [15].

For *s*=2 values of both maximum surface excess concentration and the minimum area per molecule adsorbed at the air/solution indicate vertical orientation of hydrophobic tails towards the air of monomeric and oligomeric surfactants with the spacer of the latter located at the interface [27].

The γ_{cmc} value for dimeric (2A; *m*=12, *s*=2) and tetrameric (4A; *m*=12, *s*=2) surfactants are much lower than those for their monomeric counterpart, 1A. However, the γ_{cmc} value of the trimer (3A; *m*=12, *s*=2) is higher than those of the dimer and tetramer, but still lower compared to that of monomer 1A. Both surface efficiency and effectiveness of adsorption at the air/solution interface are lower for the trimer than these parameters for the dimer and tetramer. The differences found between surface activity within the oligomeric series may be attributed to different packing densities at the air/solution interface. The pC_{20} values of these oligomeric surfactants are much higher than for monomer 1A. Within the oligomeric series, however, the degree of oligomerization does not significantly affect the pC_{20} value [27].

In the series of trimeric 12-*s*-12-*s*-12 surfactants a_{min} increases as *s* is increased, as is the case for the same series of surfactant dimers [15]. Among trimeric surfactants of quaternary ammonium bromide *m-2-m-2-m* type, where *m*=8, 10, or 12, 10-2-10-2-10 surfactant has highest efficiency, it occupies lowest surface at the air/water interface and has the lowest γ_{cmc} [26].

The reliable values of surface tension could not be obtained for 12-3-12-4-12-3-12 tetramer [15].

It can be concluded that in the series *m-s-(m-s)x-m* (1A-4A) higher oligomeric surfactants exhibits somewhat better adsorption properties in comparison with monomeric and dimeric surfactants. However, the difference in the surface activity is highest between monomer and dimer than between dimer and higher oligomers.

Kim [28] and Chelebicki [29] synthesized oligomeric quaternary ammonium salts using epichlorohydrin and epibromohydrine. In that way alkyl spacer containing hydroxy group was introduced (2B, 3B). The γ_{cmc} of compounds without central alkyl chain (gemini surfactant; 2B, *m*=1) were higher than that of coresponding lower analogs; bisquaternary ammonium dichloride with dodecyl chain ($R(CH3)_2N^+CH_2CH\text{-}(OH)CH_2{}^+N(CH_3)_2R\ 2Cl^-$) and monomeric 1B. Contrary to that, compounds having three dodecyl chains (timeric surfactant, 3B) show lower γ_{cmc} than both monomer and dimer. In addition, positive charge in the center of the molecule contributes to the lowering of γ_{cmc} [28]. Chelebicki et al. [29] varied counterions (Cl^-, Br^-) and alkyl chain length at central nitrogen atom (2B; *m*=2 – 8). Obtained Γ_{max} and a_{min} values indicated that 2B surfactants with longer alkyl chain length are packed more tightly at the air–water interface. pC_{20} values of all 2B (X=Br^-) surfactants were higher than monomer 1B and analogous dichloride bis-ammonium salts 2B (X=Cl^-). According to obtained γ_{cmc} values, investigated bromide salts reduce surface tension of water more than gemini surfactants and analogous dichloride bis-ammonium salts [29].

In conclusion, introducing additional hydrophobic chain results in the better surface activity of 2B and 3B surfactants. Surface activity is also influenced by counterion, bromides are more surface active than chlorides.

Laschewsky et al. [30] prepared large number of oligomeric quaternary ammonium surfactants with different degree of oligomerisation and various spacers. In the three investigated series (2C, 3C, 4B) the spacer length was chosen to be in the range between C4 and C6, because the most pronounced changes in the properties of dimeric surfactants were reported for rather short spacer groups. The spacer groups employed, namely *trans*-1,4-buten-2-ylene, *m*-xylylene, and *p*-xylylene, can be considered as rigid, thus fixing chemically the distance between the cationic groups within the same molecule. Γ_{max} i a_{min} values were not reported in order to avoid controversy which n value should be used (see section 3.2). Therefore surface activity was assessed based only on γ_{cmc}. The surface tension at cmc decreases with the degree of oligomerization, the effect being more pronounced for the longer spacers. The nature of the spacer has important influence on packing of the molecules in monolayer. Based on the γ_{cmc} obtained for the three series with different spacers suggest that the packing density of the dodecyl chains in the adsorbed monolayer stays approximately the same for the series with the *trans*-butenylene spacer, increases somewhat for the series with *m*-xylylene spacer, and improves much for the series with *p*-xylylene spacer with increasing degree of oligomerization [30].

Greater possibility of adjusting surfactant physico-chemical properties through changing molecular structure and conformation has motivated synthesis and characterization of several star-shaped surfactants.

Yoshimura et al. [33] investigated star-shaped trimeric surfactants consisting of three quaternary ammonium surfactants linked to a tris(2-aminoethyl)amine core (3D). Each ammonium group had two methyls and a straight alkyl chain of 8, 10, 12, or 14 carbons. In comparison with corresponding monomeric and gemini surfactants with dodecyl chains, Γ_{max} and a_{min} of 3D are smaller and much larger, respectively. However, a_{min} calculated per hydrocarbon chain are close to that of a monomeric surfactant and a_{min} per molecule slightly larger than that of the linear-type trimeric surfactants. This indicates that 3D adsorbed to the air/water interface in an orientation which cause high surface activity, by interactions of the multiple hydrocarbon chains despite the strong electrostatic repulsion between multiple quaternary ammonium headgroups. These results were also supported by pC_{20} and the cmc/C_{20} ratio [33].

Trimeric surfactant 3D exhibited γ_{cmc} values smaller than those of the corresponding monomeric surfactants, and almost the same γ_{cmc} values as those of the dimeric surfactants. Their γ_{cmc} were also lower in comparison with afore mentioned linear-type cationic trimeric surfactants with three dodecyl chains and two spacers such as ethylene (3A; m=12, s=2), *trans*-1,4-buten-2-ylene (3Ctb), *m*-xylylene (3Cmx), or *p*-xylylene (3Cpx). Although the adsorption kinetics of trimeric strar–shape surfactants to the air/water interface was slow, they have strongly adsorbed and oriented themselves at the interface, indicating that they efficiently lower the surface tension of water [33].

Wang et al. [22, 34-36] synthesized diversity of oligomeric star-shape surfactants: trimeric (3E, 3F), tetrameric (4C) and hexameric (6A). However, because of their unusual aggregation behavior and properties surface activity was investigated only for tetrameric molecule (4C). The obtained γ_{cmc} was higher than the most of the reported cationic gemini and higher oligomeric surfactants. In addition, the surface tension continues to decrease significantly after cmc in large concentration range. This was attributed to the formation of premicellar aggregates before cmc through hydrophobic interaction among hydrophobic chains of different molecules [36].

Neutral and cationic series of trimeric β-hydroxy amino ammonium surfactants with different alkyl chain length (3G, 3H; m=8, 12, 18) was synthesized by Grau et al. [37]. Poor solubility of neutral trimers (3G) prevented a reliable determination of their surface active properties. On the other hand cationic compounds (3H) displayed a sharp break in the surface tension *vs.* concentration (on log scale) curves and a final plateau indicating a well-defined cmc. The variation of Γ_{max} and a_{min} with alkyl chain length is almost negligible. The values of a_{min} decrease somewhat with the increase in the alkyl chain length, which was attributed to the flexibility of the spacing group and stronger intermolecular van der Waals forces at increasing chain lengths [37].

All three cationic compounds (3H) have good surface activity as indicated by pC_{20} values. The cmc/C_{20} ratios indicate that compound with dodecyl alkyl chain has a slightly greater preference for adsorption than for micellization [37].

Contrary to the relatively large number of data for oligomeric cationic surfactants, data for anionic and nonionic surfactants are scarce.

Yoshimura and Esumi [38, 39] have synthesized diverse anionic surfactants. Physico-chemical properties of all surfactants were investigated in alkaline solution at pH 13 (3J-L).

Surface activity of two trimeric surfactants (3K, 3L) was compared to the activity of corresponding single-chain sodium 2-aminododecanoate surfactant. All parameters, Γ_{max}, a_{min} and γ_{cmc} show that investigated trimeric surfactants provide greater efficiency in lowering the surface tension than the single-chain surfactant. Surface area occupied by 3L molecule is somewhat larger than that of single-chained surfactant. On the other hand, a_{min} for 3K is very small. This can be attributed to the smaller electrostatic repulsion between the chains in the molecule of 3K because central chain bears no charged groups unlike the 3L. This enables closer packing of the chains. Among two triple-chain surfactants, 3K shows lower surface tension than 3L. It is considered that the orientation of the latter derived from tris(2-aminoethyl)amine is less effective at the air/water interface, due to the bulky structure compared to the former from 3-aza-1,5-pentanediamine [39].

The obtained cmc/C_{20} values for both trimeric surfactants 3K and 3L indicates their preference to adsorb at air/water interface due to the difficulty of packing three hydrocarbon chains into the micelles. The cmc/C_{20} of 3L is also much larger than that of 3K, suggesting that it is easier for the former to adsorb at the interface than for the latter [39].

Although it is expected that the areas occupied per molecule of the ring-type trimeric surfactants are large (3J), because they possess three hydrocarbon chains and a bulky triazine ring, they turned out to be small. The a_{min} of the surfactant with shorter chain was comparable to the ones of single-chained surfactants. Increase in hydrocarbon chain length resulted in even smaller a_{min}. It is suggested that the ring-type trimeric surfactant molecules pack densely at the air/water interface and therefore are highly surface active, probably due to the hydrophobic interactions between multi-hydrocarbon chains [38].

Ring-typed trimeric surfactants showed higher efficiency at reducing the surface tension in the alkali solution in comparison with single-chain sodium dodecanoate and sodium dodecyl sulfate (SDS). Increasing the length of the side chains results in lowering γ_{cmc}. The obtained values are similar to those obtained for 3K and 3L anionic trimeric surfactants [38].

The pC_{20} values of the ring-type trimeric surfactants were larger than those of the monomeric anionic surfactants (sodium dodecanoate and SDS) and increased when the hydrophobic chain length increased, indicating that a long hydrophobic chain in the molecule facilitates a more close packed arrangement at the air/water interface and a more efficient adsorption [38]. In comparison with 3K and 3L, based on cmc/C_{20} values, ring like surfactants are more likely to form micelles in the bulk solutions.

Grau et al. [40] synthesized tetrameric anionic surfactants with multiple ring spacers which are flexible (4E), semi-flexible (4D, 4F) or rigid (4G). To the best of our knowledge this is the only study of the influence of spacer nature on surface activity of anionic oligomeric surfactants. Dioxane groups in the spacer confer wet-ability of synthesized surfactants. It was shown that surfactant 4G with the most rigid spacer displayed a minor surface activity at 20 °C, but all four surfactants are surface active at 40 °C.

Obtained a_{min} values are less than four times the value for the single-chain surfactant sodium 1-decanesulfonate ($C_{10}H_{21}SO_3Na$) indicating that these tetrameric surfactants are somewhat more closely packed at the air–solution interface than the single chain reference compound [40].

The γ_{cmc} values are smaller than those of single-chain surfactants, but similar to those of double- and triple-decyl chain surfactants with two sulfonate groups. The γ_{cmc} values of the investigated surfactants are not meaningfully different [40]. The pC_{20} value of the surfactant 4D is larger than those of the corresponding 4 E-G surfactants, which are quite similar. This reveals that the pC_{20} values decrease with an increase in the number of dioxane rings, in accordance with the fact that the shortest spacing group provides the maximum efficiency [40].

Although nonionic surfactants are widely applied, especially in the emulsion formulations and in drug delivery, there are only few papers dealing with higher oligomeric surfactants. One of the reasons can be that different polymeric nonionic surfactants are commercially available.

Surface activity of nonionic oligomeric 3M surfactants was compared with oligomeric trimeric nonylphenol polyoxyethylene surfactants (TNP) and monomeric nonylphenol polyoxyethylene ether surfactants (NP) [41]. γ_{cmc} and the Γ_{max} values of the TNP surfactants are lower than those of corresponding monomer NP surfactants. For both TNP and NP surfactants a_{min} decreases greatly as the length of oxyethylene chain number is increased. In general, low a_{min} values suggest close packing of the surfactants at the air/water interface with almost perpendicular orientation of the surfactant molecule. The area per surfactant molecule of TNP is three times smaller than that of NP, which indicates that the trimeric surfactant molecules of TNP are not arranged side by side at the air–water interface, but staggered three-dimensional arrangement. This could be explained with increased hydrophobic interactions due to the increase in number and length of the chains. Hence, TNP surfactants exhibit much better surface activities, including strong adsorption at the surface and wetting ability [41].

Surface activity of 3M surfactants is not improved as much compared to monomeric surfactants. a_{min} increases with increasing hydrophobic spacer length. This may be explained as follows: with the number of carbon atoms in the hydrophobic spacer of the 3M increasing, the hydrophobic property is stronger. Therefore, it decreases the amount of the saturation of adsorption for 3M at the air–water interface, resulting in the increase of the surface tension at cmc, γ_{cmc}.

Mohamed et al. [42] prepared a series of trimeric nonionic surfactants based on tris(2-aminoethyl)amine with varying alkyl and poly(ethylene glycol) chain length, 3N. It was found that a_{min} for the prepared trimeric surfactants increases by increasing the alkyl chain length because the surfactant molecules adsorb at the air/water interface to orient themselves so that the hydrophobes are directed away from water. But it is obvious that a_{min} decreases by increasing the hydrophilic chain length within the group. This behavior was previously mentioned for nonionic surfactants and is explained by the increase in polyethylene glycol chain leading to coiling the chains in order to minimize any probable interactions between them [42].

In terms of minimum surface tension octyl-3N surfactants proved to be the most efficient in lowering the surface tension of aqueous solutions. The increase in the hydrocarbon chains from octyl to decyl to dodecyl caused an increase of the γ_{cmc}, whereas the increase in the poly(ethylene glycol) chain length within the same group leads to a decrease in the γ_{cmc} [42].

In conclusion, described higher oligomeric surfactants do poses better surface activity compared with monomeric surfactants. However, the change is, in most cases, lesser than for going from monomer to dimer. It was found that the influence of the nature and length of the spacer, as well as alkyl chain length, on adsorption is not the same for different surfactants series.

3.3. Micellization

One of the main characteristics of surfactants is that physico-chemical properties of surfactant solutions abruptly change over small concentration range (see for example Fig. 7). This is a consequence of a significant change in the nature of a solute species, i.e. formation of supramolecular aggregates called micelles. The surfactant concentration at which the change occurs is called the critical micelle concentration (cmc). cmc can also be defined as a minimum surfactant concentration at which micelles are formed and remain in dynamical equilibrium with free monomers [1, 2, 43].

The main driving force for the micelle formation in aqueous solution is the effective interaction between the hydrophobic parts of the surfactant molecules. Interactions opposing micellization may include electrostatic repulsive interactions between charged head groups of ionic surfactants, repulsive osmotic interactions between chainlike polar head groups such as oligo(ethylene oxide) chains, or steric interactions between bulky head groups [49].

Cmc values are characteristic for given surfactant. Factors influencing cmc value are alkyl chain length, type of hydrophilic group and/or counterions, ionic strength, pH, pressure, temperature, etc.

Micelles size and shape is another property of crucial importance for the application of surfactants. Micelles can vary in size and shape (e.g. spherical, cylindrical, disklike, wormlike), depending on the structure of the molecule and experimental conditions (e.g. surfactant concentration, presence of the electrolytes, temperature, etc.). Aggregation number of the micelles, i.e. number of the surfactant molecules present, ranges between 50 and 200, and can be determined by static light scattering (SLS) or small angle neutron scattering (SANS).

The change of free energy of micellization (ΔG_{mic}) tells us whether it is a spontaneous process ($\Delta G_{mic}< 0$) or not ($\Delta G_{mic}> 0$) and the magnitude of its driving force. Expression for the free energy of micellization of oligomeric surfactants has been derived by Zana [12]:

$$\Delta G_{\mathrm{mic}} = \left(\frac{1}{j} + \beta\right) RT \ln \mathrm{cmc} - \left(\frac{RT}{j}\right) \ln j \tag{3}$$

where β is the degree of counterion association to the micelle/solution interface and j is the number of alkyl chains connected by some spacer groups. Micelle ionization degree, α, is defined as α=1-β.

The fact that dimeric and higher oligomeric surfactants have several degrees of the magnitude lower cmcs than corresponding monomeric surfactans was among main reason for the investigation of these surfactants. Lower cmcs means that less surfactant is needed, which has both financial and environmental benefit. Observed general trend is decrease of cmc with increasing oligomerization degree which is mainly attributed to thermodynamical reasons [3, 30] as basically the entropic loss resulting from micellization of the surfactants becomes smaller. However, the difference in cmcs in the series of the same surfactants with the different degree of oligomerization is largest going from monomer to dimer, and gets smaller in going from one to the other higher oligomeric molecule. Micellisation properties of higher oligomeric surfactants were assessed based on cmc and ΔG_{mic} values. Available data are summarized in Table 2.

For the dimeric, trimeric and tetrameric dodecyl quaternary ammonium bromides (1A-4B) the general trend of decreasing cmc with degree of the oligomerization was observed [15, 16, 27]. The cmcs values increase with the spacer length for trimeric surfactants, as it is the case for corresponding dimeric surfactants. In addition, the values of ΔG_{mic} (per mole of dodecyl chain) are all around-20 kJ/mol, irrespective of the values of degree of oligomerization and/or spacer length as reported by In et al. [15]. Recent study showed that on a per chain basis for 1A – 4A surfactants (m=12, s=2) ΔG_{mic} is more negative for the monomer than for oligomers [27]. This may be attributed to steric hindrance of the short ethylene spacer from becoming part of the micelle core. The ΔG_{mic} values within the oligomeric series become less negative with increasing the degree of oligomerization as a consequence of lower driving force for micellization. Both ΔG_{ads} and ΔG_{mic} are negative, showing spontaneous adsorption and micellization.

Differences in their magnitudes reveal that these surfactants have greater preference toward adsorption in comparison with micellization. In the series of trimeric *m*-2-*m*-2-*m* surfactants (3A; *m*=8, 10, 12) cmc linearly decreases with increasing alkyl chain length, like for monomeric surfactants [26]. However, the effect of hydrocarbon chain length on the cmc is smaller for trimeric than for monomeric surfactants.

Kim et al. [28] observed that the cmc values of the quaternary ammonium compounds with two dodecyl chains (2B; *m*=1) are 2 orders of magnitude lower, and those of the compounds with three dodecyl chains (3B; *m*=12) are 4 orders of magnitude lower compared with the conventional dodecyltrimethylammonium chloride (1B), regardless of the number of hydrophilic ammonio groups in the molecule. The cmc values of compounds having a dimethylammonio group in the center of the molecule are of the same order as that of compounds in which it is not present. This means that the additional charge in the center of the molecule has only a small effect on the cmc of the quaternary ammonium salts. In addition, there is a linear relationship between the total carbon number in the hydrophobic group and the cmcs on the semilogarithmic scale for all investigated quaternary ammonium salts, interestingly irrespective of the number of hydrophilic ammonio groups in the molecule [28].

Increase in alkyl chain length from 2 to 8 resulted in linear decrease of the cmcs values of 2A surfactants with both bromide and chloride counterions [29]. However, the type of counterion influences the cmc values, the values for the chlorides were higher than for bromides. Obtained values were 2 orders of the magnitude lower than those of corresponding monomeric 1A and 1B. They were also lower than that of dimeric *m*-*s*-*m* surfactants with hydrophobic flexible spacers (2A; *s*=3, 4, 6). The authors explained this with the formation of H-bonds between two OH groups in the spacer and water molecules, which facilitate the bending of the spacer toward the aqueous phase forming the convex micellar surface [29].

In conclusion, Kim and Chlebicki showed that introducing additional hydrophobic chain results in lower cmc values. Type of counterion also influences micellization process, bromide salts aggregates at lower concentrations than chloride. Obtained results are in accordance with generally observed trend for oligomeric surfactants.

The cmc values of oligomeric quaternary ammonium surfactants with various rigid spacers 2C, 3C, 4B are much lower than those of the structurally closely related surfactant monomers, 1B and 1C [30]. The decrease in cmc is more pronounced going from monomers to dimers, than going from trimers to tetramers. It was shown that the chemical nature of the spacer has an influence on the cmc values among the surfactants of the same degree of oligomerization. Comparing the isomeric spacers *m*-xylylene and *p*-xylylene the cmc slightly increases with increasing spacer length, similar to the behavior of oligomeric surfactants with flexible alkyl spacers. The oligomers with *m*-xylylene spacer have lower cmcs than those with butenylen, despite their increased length, due to the higher hydrophobicity [30]. Exception to this behaviour was 4Bpx surfactant. The reason lies in the formation of premicellar aggregates at very low concentration, resulting in shifting cmc to values higher than expected. Premicellar aggregation can occur in solutions of conventional surfactants that are sufficiently hydrophobic (at least 14 carbon atoms) and in those of dimeric *m*-*s*-*m* surfactants (for 12-*s*-12 with $m \geq 12$, for *m*-8-*m* with $m \geq 14$, and for 16-*p*-xylylen-16 [11].

Surfactant	cmc_γ [mmol / dm^3]	cmc_κ [mmol / dm^3]	α	$-\Delta G_{mic}$ [kJ / mol alkyl chain]	N_{agg}	T [°C]	Ref
Cationic							
DeTAB*	67					25	[26]
1A	14	15.1	0.25	18.3		25	[15, 16]
	14	15	0.29	18.1		30	[27]
10-2-10	5.0					25	[44]
12-2-12	0.9					25	[16]
	0.8	1	0.27	23.2		30	[27]
12-3-12	0.96		0.22	20.8		25	[15, 50]
12-6-12	1.03		0.33	18.8		25	[15, 50]
8-2-8-2-8	14					25	[26]
10-2-10-2-10	0.95					25	[26]
12-2-12-2-12	0.065					25	[26]
	0.6	0.8	0.26	19.3		30	[27]
12-3-12-3-12	0.14		0.19	21.5		25	[15]
12-6-12-6-12	0.28		0.30	19		25	[15]
12-2-12-2-12-2-12	0.7	0.8	0.35	16.7		40	[27]
12-3-12-4-12-3-12	0.06		0.20	22.6		25	[15]
1B	18.3	22	0.34	21.94	34.3	23	[29, 30, 51]
BQADC**	0.78					20	[28]
2B-12 Cl	0.0062					20	[28]
3B – 12 Cl	0.0096					20	[28]
1C	7.0				27.3	23	[30, 51]
2Ctb	2.0				15.5	23	[30, 31, 51]
2C mx	1.5				11.3	23	[30, 31, 51]
2Cpx	2.1				10.5	23	[30, 31, 51]
3Ctb	0.36				5.0	23	[30, 51]
3Cmx	0.28				5.4	23	[30, 51]
3Cpx	0.29				3.5	23	[30, 51]
4Btb	0.12				3.8	23	[30, 51]
4Bmx	0.09					23	[30]
4Bpx	1.3				3.5	23	[30, 51]
3I	1.0	1.0				25	[32]
3H - 8	0.622					20	[37]
3H - 10	0.223					20	[37]
3H- 12	0.049					20	[37]

*decyltrimethylammonium bromide, ** bisquaternary ammonium dichloride with dodecyl chain ($R(CH3)_2N^+CH_2CH\text{-}(OH)CH_2{}^+N(CH_3)_2R{\cdot}2Cl^-$

Surfactant	cmc$_\gamma$ [mmol / dm^3]	cmc$_\kappa$ [mmol / dm^3]	α	-ΔG_{mic} [kJ / mol alkyl chain]	N_{agg}	T [°C]	Ref
Cationic							
3D-10	1.17	1.60		30.0		25	[33]
3D-12	0.139	0.177		36.3		25	[33]
3D-14	0.00647	0.384		33.7		25	[33]
3E	0.20^a	0.32^a	0.45			25	[34]
3F	0.33^a	0.39^a	0.50			25	[34]
4C	0.08	0.12	0.73			25	[36]
6A		0.11^a	0.29			25	[22]
	1.31^a	1.13^a	0.24				
Anionic							
$C_{11}H_{23}COONa$	20					20	[38]
$C_{12}H_{25}OSO_3Na$	8.2					25	[38]
3J-4	0.19				32	25	[38]
3J-10	0.011				580±20	25	[38]
$C_{10}H_{21}CH(NH)_2COONa$	1			34.2		25	[39]
3K	0.0063			10.7		25	[39]
3L	0.0167			9.9		25	[39]
4D	0.0468^a					20	[40]
4E	0.178^a					20	[40]
4F	0.105^a					20	[40]
Nonionic							
3N 8-PEG400	$8.23 \cdot 10^{-4}$					25	[42]
3N 8- PEG1000	$7.57 \cdot 10^{-4}$					25	[42]
3N 8- PEG2000	$7.1 \cdot 10^{-4}$					25	[42]
3N 10- PEG400	$7.92 \cdot 10^{-4}$					25	[42]
3N 10- PEG1000	$7.34 \cdot 10^{-4}$					25	[42]
3N 10- PEG2000	$6.91 \cdot 10^{-4}$					25	[42]
3N 12- PEG400	$7.50 \cdot 10^{-4}$					25	[42]
3N 12- PEG1000	$7.15 \cdot 10^{-4}$					25	[42]
3N 12- PEG2000	$6.63 \cdot 10^{-4}$					25	[42]

[a] the critical aggregation concentration is reported

Table 2. The critical micellization concentration determined by surface tension (cmc) and electrical conductivity measurement (cmc), the degree of counterion dissociation (), aggregation number (N_{agg}) and the free energy of micellization (G_{mic}) obtained at given temperature (T) of higher oligomeric and corresponding monomeric and dimeric surfactants. Abbreviated names of listed surfactants are given according to the figures 2-4.

The cmc of the cationic trimeric surfactants 3H linearly decreases with the number of carbon atoms in the alkyl chain. In other words, cmcs values decreases with alkyl length even for very hydrophobic molecule with tree octadecyl chains. Calculated cmc/C_{20} values point that trimeric surfactant with 16 C atoms in alkyl chains has slightly greater tendency to adsorb at air/water interface in comparison with tendency to form micelles [37].

In the 3D series [33] the cmc values decrease with increasing chain length from 10 to 14. The cmc of 3D-8 was not possible to determine because micelles did not form in solution even at the highest concentration studied, due to the short chains, and the solution simply became turbid. The cmc value of 3D-12 surfactant is lower than that of corresponding monomeric and gemini surfactants. The cmc in that series of surfactants decreases for an order of magnitude for each additional surfactant moiety. Compared with linear 12-2-12 surfactant, cmc of 3D-12 was slightly higher, in line with authors assumption that tris(2-amionoethyl)amine would have greater hydrophobicity. 3D-12 also exhibited cmc lower than that of the linear cationic trimeric surfactants with spacers such as *trans*-1,4-buten-2-ylene (3Ctb), *m*-xylylene (3Cmx), and *p*-xylylene (3Cpx), respectively [33]. The relationship between the logarithm of cmc and hydrocarbon chain lengths (as for monomeric and gemini surfactants) or chain number was linear [33]. This means that the longer the chain length and the higher the chain number of surfactants, the lower the cmc will be.

The values of pC_{20} and cmc/C_{20} ratio in 3D series were much larger than those of most ionic monomeric surfactants. In addition, the absolute values of ΔG_{ads} are significantly greater than those of ΔG_{mic} for all the hydrocarbon chain lengths. Both these facts suggest that in comparison to micellization, the adsorption of star-type trimeric surfactants is preferred [33].

Trimeric surfactants 3E and 3F display unusual aggregation behavior in aqueous solution [34]. Both trimeric molecules form vesicles just above critical aggregation concentration (cac) and then vesicles gradually transform to the micelles with the increase of surfactant concentration. Normally, surfactants form small aggregates first, and then the aggregates may become large with an increase of the surfactant concentration. The reported cac value of 3E is lower than that of the 3F. Both values are slightly lower from cmc of cationic gemini surfactants. The enthalpy changes for aggregation have large negative value for both surfactants indicating that they have similar aggregation behavior and ability, and that their aggregation is dominated by hydrophobic interaction. This is expected because investigated trimeric surfactants have similar molecular structure. 3F is a completely symmetric molecule, whereas 3E is not symmetric with spacer slightly different from 3F [34].

Tetrameric (4C) and hexameric (6A) surfactants synthesized by the same group of authors [22, 36] also display interesting aggregation behavior which will be address in more detail in the next part of the chapter. Reported cmc value of tetrameric 4C (0.08 mmol dm^{-3}) is at least an order of magnitude smaller than those of cationic gemini surfactants [36]. This is in accordance with the fact that increasing number of alkyl chains in the surfactant molecule results in decreasing cmc values.

Hexameric 6A (as 3E and 3F) above cac forms vesicles not micelles. It is reported that 6A displays two cac (C1 and C2) as a consequence of aggregate transformation caused by the changes of the surfactant configuration through hydrophobic interaction among the hydrocarbon chains [22].

The enthalpy change for the aggregation of 6A exhibits a very large negative value, much larger than those of other surfactants with closely similar cationic ammonium amphiphilic moiety. This enthalpy change should concern the entire aggregation process including both the first and the second aggregation processes at C1 and C2 [22].

Enthalpy change of aggregation for 6A is much more negative than for 1A, 2A (*m*=12, *s*=6), 3F and 4C, due to the much stronger inter-and intramolecular hydrophobic interactions between the alkyl chains. That is to say, the cooperative hydrophobic interaction becomes stronger with the increase of the number of the hydrophobic chains in a surfactant molecule. Of course, hydrogen bonding between amide groups can also increase the enthalpy change per amphiphilic moiety for 3F, 4C, and 6A. However, each amphiphilic moiety of these molecules has one amide group, and the significantly enhanced enthalpy change per amphiphilic moiety for 6A confirms that the contribution of each hydrocarbon chain to inter-and intramolecular hydrophobic interaction in the 6A aggregation becomes much stronger than that for other surfactants [22].

The cmc values of ring type anionic trimeric surfactant 3J decrease with increasing the alkyl chain length from 4 to 10 [38]. Obtained cmc/C_{20} ratios are very small compared with the single chained surfactants suggesting that trimeric anionic ring type surfactants are more likely to form micelles in the bulk solution due to the interactions between multi-hydrocarbon chains.

Anionic triple chain surfactants [39] 3K and 3L have cmc values one-two order of magnitude lower than corresponding single chained sodium 2-aminododecanoate. This indicates that both surfactants have excellent micelle-forming ability at low concentration. The cmc of 3K is also lower than that of 3L, showing the effect in the number of hydrophilic groups on the cmc. On the contrary, absolute values of ΔG_{mic} per hydrocarbon chain of 3K and 3L are much smaller than that of the single chain surfactant. It indicates that the steric hindrance of closely connected hydrocarbon chains makes it difficult for the triple-chain surfactants to form micelles. This result is supported by the large cmc/C_{20} ratio [39]. The opposite result for cmc/C_{20} ratio was obtained for ring type anionic trimeric surfactant 3J [38].

The cmc values of 4D-4G tetrameric surfactants depend on the architecture of the spacer [40]. Increasing the number of dioxane rings in the spacer increases the cmc value, surfactant 4D has the smallest cmc value among investigated surfactants. The dioxane rings appear to act as hydrophilic units, which contribute to increasing cmc values. Furthermore, a comparison between homologues compounds, as surfactants 4E and 4F, reveals that, as expected, the higher the flexibility of the spacer the higher the cmc. This behavior is more evident at higher temperature (40 °C) [40].

Cmc/C_{20} values indicate that 4D-4G surfactants have higher preference to aggregate than gemini surfactants. Authors proposed that this might be because the four hydrophobic groups are more suitably oriented to accommodate themselves in the internal part of the aggregates than the two hydrophobic groups of the gemini surfactants. The preference for the aggregation, relative to the adsorption, increases with the number of dioxane rings and rigidity of the spacer [40].

As a rule cmc values of nonionic surfactants are lower in comparison with ionic surfactants due to the lesser electrostatic repulsion of the hydrophilic groups at the micelle/water interface. Although, the data for nonionic higher oligomeric surfactants are scarce, this trend has been observed in investigated higher oligomeric nonionic surfactants. The cmc value of Tyloxapol (7A) determined by time-resolved fluorescence quenching (TRFQ) is in the micromolar range, i.e., about a hundred times lower than for the "monomer" TX100 [21].

The cmc values of trimeric nonylphenol polyoxyethylene and monomeric nonylphenol polyoxyethylene ether surfactants 3M increase with the number of the oxyethylene groups in the spacer due to the greater hydrophilic characteristics. On the other hand, increasing the hydrophobic chain length results in lower cmc. Short hydrophobic chains are stretched and in contact with water. Therefore they need more free energy to form micelle and as a results the cmc of these surfactants are higher. Long hydrophobic chains, longer than the equilibrium distance of electrostatic repulsion of the head groups, pack inside the micelle hydrophobic core, which may decrease the free energy and consequently lower the cmc value [41].

The cmc values of trimeric nonionic 3N surfactants are small, which suggest that they easily form aggregates in solution. As expected, the cmc values decrease with increasing alkyl chain length. But they also decrease with increasing poly(ethylene glycol) chain length [42].

In conclusion, the cmc values of higher oligomeric surfactants are smaller in comparison with corresponding monomeric and dimeric surfactants. However, the changes are becoming less significant with increase of degree of oligomerization. Reported cmc/C_{20} ratios indicate that, as expected, structural factors have dominant role in determining higher oligomeric surfactants preference toward adsorption or micellization. This is collaborated with wide range of obtained ΔG_{mic} values for different type of oligomeric surfactants (Table 2).

3.4. Properties of higher oligomeric surfactants aggregates

The most striking feature of the dimeric surfactants with short spacers, in comparison with monomeric, is their tendency to form elongated micelles already at relatively low concentrations, without added electrolyte. Zana have pointed out that the origin of the different aggregation behavior of monomeric and dimeric surfactants lies in the different distribution of the head group distances at the micelle/water interface in these two classes of surfactants [3, 10]. For the monomeric surfactants, the head groups are randomly distributed on the surface separating the aqueous phase and the micelle hydrophobic core. The distribution of distances between head groups has a maximum at a thermodynamic equilibrium distance (d_T) determined by the equilibrium of the opposite forces involved in micelle formation. In the case of dimeric surfactants, the distribution becomes bimodal. One maximum corresponds to the thermodynamic distance, the other one, narrower, at a distance d_S that corresponds to the length of the spacer. The length of the spacer is determined not only by the number of atoms in the spacer but also by its conformation. The distance d_S can be adjusted to be smaller, equal or larger than d_T by changing structure of the spacer opening a possibility to create variety of structure [3]. The bimodal distribution of head group distances and the presence of the chemical link between head groups strongly affect the curvature of surfactant layers, and thus the micelle shape. Additional alkyl chains and hydrophilic or hydrophobic spacers in molecules of higher oligomeric surfactants complicate aggregation processes even more from this point of view. For higher oligomeric surfactants to date most data about aggregation behavior is given for linear and star-shaped dodecyl quaternary ammonium surfactants. Experimental techniques used for determining the aggregation number and structure of oligomeric surfactant micellar solutions are SANS, TRFQ, SLS, dynamic light scattering (DLS) and transmission electron microscopy at cryogenic temperatures (cryo-TEM).

In et al. [15] have shown, using cryo-TEM, that the sequence of the aggregates morphology of linear dodecyl quaternary ammonium surfactants 1A-4A, (m=12, s=3) with increasing degree of oligomerization from 1 to 4 is:

spherical micelles (monomer) → linear wormlike micelles (dimer) → branched wormlike micelles (trimer) → closedloop (ring) micelles (tetramer).

These results were confirmed with molecular modeling and molecular dynamic simulations [19, 20]. The changes of micelle shape can have a strong impact on rheology of the oligomeric surfactants solution, as will be discussed in following part.

In 12-s-(12-s)$_x$-12 series with a long spacer, s=6, there is practically no change of micelle shape with the increasing degree of oligomerization. It was shown that aggregation numbers for these oligomeric surfactants are similar to the aggregation number of DTAB (1A) spherical micelles [15].

The size distributions for the micelles of the trimeric surfactants m-2-m-2-m (3A; m=8, 10, 12) were obtained by DLS at concentrations 2-to 16-fold cmc [26]. For all three surfactants bimodal size distribution was obtained. The smaller hydrodynamic diameter was determined to be 3.8–5.2, 3.8–5.4 and 4.8–6.6 nm for 8-2-8-2-8, 10-2-10-2-10, and 12-2-12-2-12, respectively. The hydrodynamic diameter of larger particles was 30–50 nm. The aggregation numbers at cmc, determined by SLS, increases with increasing alkyl chain length from 7±3 for 8-2-8-2-8, to 8±2 for 10-2-10-2-10 and 18±1 for 12-2-12-2-12. The aggregation number of 12-2-12-2-12 corresponds to 54 (3x18) alkyl chains in the micelle, which is almost identical to the aggregation number of DTAB (1A) and twofold for 12-2-12 [26]. The small aggregation numbers of 8-2-8-2-8 and 10-2-10-2-10 were attributed to shorter hydrocarbon chains [26].

The particle size measurements at a concentration 2-fold of the cmc revealed that the apparent hydrodynamic diameter of monomeric, dimeric trimeric and tetrameric dodecyl ammonium bromides with ethylene spacer is around 3.6 nm in all cases, indicating the formation of spherical micelles [27]. This is in accordance with the theoretical considerations [19] and experimentally obtained data for oligomeric surfactants, showing that their micelles are spherical at the cmc [15]. It is shown that both DTAB (1A) and 12-2-12 form spherical micelles, DTAB even at fairly high concentrations and the 12-2-12 up to 1.3 wt% [50, 51]. The larger particles observed in the Yoshimura's study [26] of trimer were not detected in this case.

Chlebicki et al. [29] reported the coexistence of small spherical micelles and large nearly spherical aggregates when the length of the alkyl chain at the central nitrogen is short (2B).

Furthermore, when this alkyl chain is longer, only larger aggregates are observed. In addition, it was also found that the aggregate size increased with an increase in the surfactant concentration. It can be concluded that the size distribution of investigated 2B surfactants depends on the alkyl chain length and concentration but not on the counterion, Br^-or Cl^-.

DLS revealed the existence of only small aggregates (2 nm or smaller) in the solution of 2C, 3C and 4B (regardless of the spacer) surfactants up to 1% wt of surfactant [51]. TRFQ measurements were used to determine aggregation numbers (N_{agg}). The relatively low $N_{agg} < 40$ for solutions of ca. 3 wt %, was ascribed to the chloride counterion which are less bound to the cationic head groups (higher ionization degree) than bromide, resulting in a stronger electrostatic repulsion of the hydrophilic heads as well as a steric hindrance due to higher hydration of the latter. The nature of the spacer has a major influence on the aggregation number of the oligomeric surfactants. In general, the shorter the spacer the larger aggregation number is. For the trimers aggregation number decreases in series N_{agg} (tb) > N_{agg} (mx) > N_{agg} (px) in accordance with the behavior of dimeric surfactants. Contrary to *m-s-m* type of surfactants, the aggregation number of 2C, 3C and 4B surfactants decrease with the degree of oligomerization. In this case a decrease of N_{agg} with increasing degree of oligomerization, could be ascribed to the fact that further addition of long rigid spacers between the head groups reduces the overall flexibility of the structures and, hence, makes it difficult for the higher oligomers to pack tightly [51].

The structure of the micelles formed in the solution of star-shaped trimeric surfactants 3D was significantly influenced by alkyl chain length, for m=10, ellipsoidal micelle formed, for m=12, the ellipsoidal micelle transformed to the threadlike micelles with increasing concentration, and for m=14, threadlike micelles were formed at low concentration and no transitions were observed as concentration increased [33].

Unusual aggregation behavior of star-shaped trimeric 3E and 3F, tetrameric 4C and hexameric 6A surfactants was explained by the dominant role of hydrophobic interactions which enables the configuration of the molecules to change [22, 34-36].

Due to the rigid spacer and the intramolecular electrostatic repulsion among the quaternary ammonium headgroups, the hydrophobic chains of 3E and 3F pack loosely, and the 4C molecule presents a stretched configuration at low concentrations [34-36].

Trimeric surfactants (3E and 3F) form vesicles just above the cac. With increasing concentration hydrophobic interaction becomes strong enough to pack the hydrophobic tails tightly and turn the molecular conformation into a pyramid-like shape, which results in the gradual transformation of vesicles in spherical micelles.

For tetrameric 4C these interactions are strong enough already below cac so 4C form large network-like premicellar aggregates that changes to small spherical micelles at high surfactant concentration due to the same reason as for trimeric surfactants [34-36].

Star-shaped hexameric (6A) surfactant also form network-like premicellar aggregates well below the cac and transfer to small spherical micelles at high concentration. Its aggregation behavior is more complex since two cacs are observed. Between two cacs the hydrophobic interaction becomes stronger so that the 6A may transfer to a claw-like configuration. Above second cac the hydrophobic interaction continues to strengthen, and to cause the molecular configuration to convert into a pyramid-like shape, which generates the transition of the large spherical aggregates to small spherical micelles as for 3E, 3F and 4C surfactants [22]. This research also proves that introducing more alkyl chains in molecules results with more complex aggregation process of oligomeric surfactants.

Yosimura and Esumi [38, 39] investigated size and aggregation number of anionic trimeric surfactants (3K, 3L, 3J) aggregates in solution by DLS and SLS.

The aggregation number of the ring-type trimeric surfactants 3J-4 and 3J-10 were 32 and 580±20, respectively as determined by SLS [38]. The N_{agg} of 3J-10 is very large, probably due to the strong cohesion derived from the interactions between longer three hydrocarbon chains in the ring-type trimeric surfactants. The size of the micelles was determined by DLS at the concentration 4-6-fold of the cmc of 3J. In the solution of both surfactants smaller and larger aggregates were detected. Both kinds of aggregates increased in size with increasing length of alkyl chains [38].

The large difference in aggregation numbers of 3K (N_{agg}=1304±8) and 3L (N_{agg}=39±2) was explained by the strong attractive interactions between hydrocarbon chains as well as the decrease of electrostatic repulsion due to less hydrophilic groups of 3K in comparison with 3L, which has usual micelle size [39]. DLS measurements revealed that in solution of 3K the aggregates 20–30 nm in size are formed, while in the solution of 3L aggregates of sizes 2–10 and 15–40 nm coexist. In the case of 3L, it is considered that the smaller ones correspond to the micelles and the larger ones correspond to the large aggregates similar to 3K [39].

The aggregation numbers of Tyloxapol (7A) and of TX100 micelles were determined by TRFQ. The Tyloxapol micelles were found to be smaller than the TX100 micelles. This behavior is opposite to that found for ionic surfactant oligomers (2A-4A) with respect to their corresponding monomers. Cryo-TEM showed that the Tyloxapol micelles remain spheroidal up to a concentration of about 10 wt%. At 15 wt%, some regions of ordered elongated micelles were also observed, which could be the precursors of the hexagonal phase known to occur at about 35 wt% [21].

3.5. Viscosity

The peculiar aggregation behavior of oligomeric *m-s-*$(m\text{-}s)_x$*-m* surfactants significantly influences the rheological behavior of their solutions. The change of the micelle shape, in the series of surfactants with short spacers, from spherical to wormlike or threadlike micelles with increasing degree of oligomerization affects viscosity of the surfactant solution. The phenomenon that attracts special attention, in this sense, is viscoelastic behavior that is observed in solution of long wormlike micelles. For example dimeric surfactants with a short spacer, such as 12-2-12, give rise to worm-like micelles at fairly low surfactant concentrations, even in the absence of added salt [3, 9-11, 14, 15, 52].

No such changes were observed in the solution of the surfactants with longer spacers, like in going from the monomeric [53] to the dimeric (12-6-12) [10] and trimeric surfactant (12-6-12-6-12) analogues [15]. This indicates that the large, rodlike aggregates were not formed in the solution of these surfactants. The same absence of the effect was observed by Laschewsky et al. [30] who concluded that despite general theoretical predictions made for oligomeric surfactants [19, 20] the remarkable thickening power of certain oligomeric surfactants is apparently restricted to molecular structures with very short spacer groups (namely s=2 or 3).

Rheological behavior of star-shaped trimeric surfactants strongly depends on alkyl chain length [33].The shear-rate dependence of viscosity of the star-shaped 3D-10 solutions was the same as that for water. On the basis of previous findings it was concluded that 3D-10 forms the spherical or ellipsoidal micelles [33].

At lower concentrations the viscosity of 3D-12 solution was also the same as that for the water. But at higher concentrations viscosity decreased with increasing shear rate the behavior known as shear thinning typical for the rod-or wormlike micelles and chainlike polymers. The results point out that with increasing concentration of 3D-12 surfactant a transition of the spherical micelle-to-rodlike micelle occurs, which was confirmed with SANS and cryo-TEM techniques. When concentration was further increased, growth of wormlike micelles was observed, accompanying the extrusion of the water from the micelles. In order to elucidate this behavior the surface charge per unit length and the end-cap energy of monomeric 1A, dimeric 12-2-12 and trimeric 3D-12 were obtained from the analysis of the volume fraction dependence of the zero-shear viscosity. Results have shown that molecular structure (linear or star-shaped) and degree of oligomerization (j=1, 2 or 3) have no influence on the surface charge per unit length. However, the end-cap energy of wormlike micelles decreased in order 3D-12 > 12-2-12 > 1A, indicating that wormlike micelles form more easily in the solutions of trimeric and dimerics surfactants even at lower concentrations. The crucial difference in the molecular structure between 3D-12 and 12-2-12 is the number of spacer chains. It seems that it is the spacer responsible for the increase of end-cap energy, one of the reasons being limitation of intramolecular motion by increasing the number of spacer groups in the molecule. The shape of the molecule also affects the end-cap energy. Trimeric star-shaped 3D surfactant is more round shaped than the linear one which results in lower end-cap energy of its wormlike micelles [54].

Viscosity of 3D-14 solutions was higher than that of water and it didn't change with increasing shear rate which was attributed to presence of threadlike micelles in solution [33].

Already in these several studies the complex rheological behavior of higher oligomeric surfactants was observed. Results have shown that number and length of spacer in surfactant molecule have key role in their aggregation and thus rheological behavior. It seems that the interplay of contribution of spacer and alkyl chains in resulting behavior is even more subtle than for dimeric surfactants. Considering the unusual aggregation behavior of tetrameric 4C and hexameric 6A it would be interesting to see how it reflects to their rheological behavior.

4. Applications

Synthetic surfactants are nowadays present in many everyday products and are utilized in many industrial processes. Although, dimeric and higher oligomeric surfactants have shown better properties compared to conventional ones, the difficulties in their synthesis in sufficient quantities are rendering their commercial application. In addition, in order for new surfactant to be used in certain application, its toxicity and harmfulness should be assessed, which requires both time and money. Considering their improved physico-chemical properties it would be worthwhile to perform such tests. Possibility to use higher oligomeric surfactants for several different applications has been investigated.

4.1. Solubilization

Solubilization of poorly water-soluble or insoluble compounds is among most frequent surfactant application.

Laschewsky at el. [30] investigated solubilization capacity of a 2C, 3C and 4B for *p*-xylene and 2,3-dimethylbut-2-ene. The results indicated that the solubilization depends on both the chemical nature of the spacer and the couple surfactant-solubilizate used. Degree of oligomerization, within a given series of oligomers, didn't have significant influence on the solubilization capacity.

4.2. Foaming

Foams are encountered in many important technological areas.

It was shown that the trimeric surfactant 3B has almost the same foaming ability and the foam stability as corresponding bisquaternary ammonium dichlorides which indicates that balance of hydrophobicity and hydrophilicity strongly affects the foaming properties [28]. Simple test of bubbling the air through aqueous solution of a fixed concentration until a given height of foam was produced was used for determining foaming ability of 2C, 3C and 4B [30]. For both trimeric and tetrameric surfactants it was observed that the short spacer favor foam formation and stability. Trimeric surfactants stabilize foam significantly more than corresponding dimeric, but the difference between corresponding trimers and tetramers is marginal. This effect cannot be explained by the difference in viscosity.

4.3. Emulsification

The oil/water emulsions are fine dispersions, encountered in many household and industrial products. The role of surfactants in emulsions is to stabilize them by adsorbing at the water/oil interface preventing phase separation into oil and water phases.

Ability of hexameric quaternary ammonium salt 6A to emulsify heptanes, dodecane, toluene, xylene was investigated. As described in section 3.4, hexameric 6A forms network-like premicellar aggregates at very low concentration, which have compeled the authors to study emulsion forming efficiency at the concentrations far below first cac. It was found that oil/water emulsion form quickly after vigorous shaking. The surfactant 6A can emulsify heptanes and dodecane, emulsions being stable at the level of 60 – 70 %, but can't emulsify toluene and xylene. The reason was proposed to be the greater compatibility of 6A with linear fatty acids due to its long hydrophobic alkyl chains. Due to the hexameric structure, in comparison with monomeric and dimeric surfactants, the emulsification ability and the strength of interfacial films are greatly enhanced [22].

Abdul-Raouf et al. have studied factors affecting stability of oil-in-water emulsions prepared by shearing together known amounts of Land Belayim crude oil and aqueous solutions of nonionic 3N surfactants. It was observed that equilibrium interfacial tension of nonionic 3N surfactants decreases with the length of ethylene oxide chain for the same alkyl chain length. After 24 h coalescence process has started only in a few emulsions, indicating that the emulsions were mostly stable. The stability of the emulsions depended on oil percentage and oil/water ratio. The emulsions with higher oil percentage were found to be more stable. The effectiveness of the emulsifiers decreases with increasing alkyl chain length for 3N surfactants. The emulsions stability increased with increasing surfactants concentration from 300 to 400 ppm, but decreased by further increasing the surfactants concentration to 500 ppm. These findings led to the conclusion that there is an optimum concentration at which the emulsion droplets are fully encapsulated, preventing agglomeration and coalescence which occur at lower concentrations. At higher concentrations surfactant molecules interact with each other and cause the disorder of the arrangement at the interface which facilitates the coalescence [42].

4.4. Lubricants

Surfactants that can self-assemble in smooth bilayers are promising water-based lubricants. However, the cohesion of adsorbed layers is not always satisfactory. Lagleize et al. [55] have investigated the possibility of combining surface-adsorbing surfactants and coadsorbing polymer in order to obtain dense and cohesive lubricant films by self-assembly on mica surfaces. The three studied quarternary ammonium surfactants, monomeric cetyl triethylammonium chloride, dimeric 12-3-12 and trimeric 12-3-12-3-12 form flat bilayers at the negatively charged mica surface at the concentrations above cmc [3]. It was shown that the degree by which coadsorption of the anionic–neutral poly(acrylic acid)–polyacrylamide diblock copolymer reinforces the adsorbed layers, as well as the nature and the characteristic times of shear-induced dynamic transitions between states of low-and high-friction forces depend on the degree of oligomerization. Behaviour of systems containing dimeric and trimeric surfactants under shear and compression are qualitatively simillar, while systems with monomeric surfactant have shown different response indicating lower cohesion of the layers [55].

4.5. Adsorption at mineral surfaces

Many practical applications of the surfactants are based on their ability to adsorb at different surfaces. Understanding the interactions between surfactants and mineral surfaces are of special interest for agriculture and oil recovery and consequently for the environment.

Esumi et al. [17] investigated adsorption of 2A and 3A (m=12, s=2) surfactants, on silica as well as adsolubilization of 2-naphtol. Adsorption of dimeric and trimeric surfactant differs from the monomeric as revealed by adsorption isotherms. Density of adsorbed surfactants decreases with oligomerization degree. The ratio of maximum amount of 2-naphtol adsolubilized to the adsorbed amount of surfactant on silica increases with the oligomerization degree. From a two-step process of adsorption-adsolubilization it was concluded that oligomers are adsorbed at silica much more strongly than the monomeric surfactant keeping 2-naphtol in their adsorbed layers.

The same group investigated competitive adsorption of pesticide paraquat and 2A and 3A (m=12, s=2) surfactants, on clay [56]. The results indicate competitive adsorption between paraquat and the surfactants. Oligomerization degree didn't have a significant influence on replacement of paraquat.

In et al. [15] investigated adsorption of 12-3-12-3-12 and 12-6-12-6-12 trimers and 12-3-12-4-12-3-12 tetramer at silica and compared it with behavior of corresponding dimers. The values of surface surface excess concentration per mol of adsorbed chain indicate that 12-3-12-3-12 and 12-3-12-4-12-3-12 form bilayer on silica, while value for 12-6-12-6-12 indicates that it forms cylindrical micelles, similar to 12-6-12.

The efficiency of oil recovery process is affected by the wettability of oil reservoir rocks. It was shown that trimeric 12-2-12-2-12 surfactant can alter wettabilty of both water-wet and oil-wet mica surfaces more efficiently then monomeric or dimeric cationic surfactants. The change in wettabillity is a consequence of surfactant adsorption in the form of monolayer at the mica surface. The orientation of the surfactant molecules in the monolayer depends on the type of the surface (oil or water wet) [57].

Atomic force microscopy (AFM), X-ray photoelectron spectroscopy (XPS) and PC model calculation showed that trimeric 3F forms highly-ordered bilayers on the mica surface [35]. Such an ordered structure is induced by match of 3 charged surfactant head groups with negatively charged sites on mica surface. In addition, the formation of the bilayer is promoted by intermolecular bonding and hydrophobic interactions. It was concluded that structural features of an oligomeric surfactant can greatly affect its manner of adsorption, which can help in design of self-assembling molecules for the fabrication of surface patterns.

4.6. Synthesis of advanced materials

So far higher oligomeric surfactants have been used in synthesis of two classes of materials – mesoporous silica materials and ZnO quantum dots.

Preparation of the materials with pores of controlled size, shape and connectivity is of great importance for the practical applications where the shape of the molecules should be recognized. Examples are catalysis, molecular sieves, selective adsorption, sensing, etc. Different applications put different demands for the material properties. Therefore the new routs of synthesis or modifications of existing ones are constantly thought. Surfactants, in general, are frequently used as templates in the synthesis of inorganic materials. Among different meso- and microporous materials, silica based materials take special place.

Although different gemini surfactants of bis-quarternay ammonium type has been used in preparation of different cubic, hexagonal, lamellar mesoporous silica materials [3], only 14-2-14-2-14 [58] and 10-2-10-2-10 [59] have been used as structure directing agent. By using 14-2-14-2-14 surfactant it was possible to synthesize, high quality, ordered two-dimensional hexagonal mesoporous silica under mild conditions. In the presence of 10-2-10-2-10 surfactant highly ordered supermicroporous silica with the pore structure belonging to the two-dimensional hexagonal structure and pore size from 1.92 to 2.16 nm was obtained.

Quantum dots are inorganic nanoparticles with photoluminiscent properties. They can be prepared either in water or in organic solvents. Synthesis in organic solvents do offer better shape control and achieving higher crystallinity of the product, since organic ligands (most often alkyl amine) are used to control crystal growth. As a result particles coated with hydrophobic layer are obtained. This layer should be removed if nanoparticles are to be used in the water, without affecting their optical properties. Strategy based on encapsulation of nanoparticles in amphiphilic molecules aggregates is proving to be successful in this sense.

Dazzazi et al. [60] investigated the phase transfer of highly monodispers ZnO nanocrystals using monomer (DTAB, 1A), dimer (12-3-12 $2Br^-$), trimer (12-3-12-3-12 $3Br^-$) and polymer alkyl ammonium surfactants. It was shown that 60 % transfer yields could be obtained with the oligomers and polymer, but no measurable transfer was observed with monomeric surfactant. Trimer and polymer were more efficient than dimer. The results were explained by more quantitative molecular aggregation of surfactants at nanocrystals surface with increasing degree of oligomerization. In addition, the dynamics of molecular exchange between the bulk and double layer coating decreases with increasing degree of oligomerization. Obtained nanocrystals exhibited strong photoluminiscence in the water, as well as long term chemical and photo-chemical stability.

4.7. Antistatic properties

It was shown that the antielectrostatic effect of oligomeric quaternary ammonium derived from epichlorohydrin 2B strongly depends on the counterion, i.e. bromide or chloride. However, both chlorides and bromides have shown very good antistatic properties, similar to commercially available antistatic agent Catanac. No influence of increasing the central chain length on the antielectrostatic properties was observed [29].

4.8. Biomedical applications

Surfactants are frequently used in pharmacy in preparation of drug carriers or systems for targeted drug delivery. In addition, many drug molecules are amphiphilic and therefore surface active.

Motivated by the fact that some dimeric surfactants have antimicrobial activity, neutral and cationic trimeric surfactants 3G and 3H were tested against fungi, Gram positive and Gram-negative bacteria. Both types of trimeric surfactants were the most efficient against Gram positive bacteria. Cationic trimers were more efficient then neutral. Compounds with 12 C atoms in alkyl chain were found to be most active, while those with 8 and 18 C atoms were found to be almost inactive [37].

A proposed approach in the prevention or therapy of Alzheimer's disease is decreasing or eliminating the neuritic plaques composed of fibrillar β-amyloid (Aβ). The strong tendency of 4C to self-assemble even below cmc prompted authors to study disassembly of amyloide fibrils in its presence. It was shown that both 4C premicellar and micellar aggregates can effectively disassemble matutre Aβ(1-40) fibrils in aqueous solution. Unlike 4C, 12-6-12 loses its efficiency with decreasing the concentration, indicating that 4C significant self-aggregation ability below cmc could be the key factor in the fibril disassembly. Authors have proposed two key features of the fibril disassembly. First is binding of positively charged 4C surfactant with negatively charged fibriles through electrostatic interactions. Second is self-assembly of the bound 4C molecules which lead to fibrils disaggregation and formation of mixed surfactant/ Aβ(1-40) molecules aggregates [61].

5. Conclusion

Current investigations in surfactant science are driven by the requirements to design surfactants that possess enhanced physico-chemical properties, new surfactant utilization in complex systems and specific applications in modern technologies. Hence, investigations of the structure–property relationship in surfactant systems are very important in order to be able to design new surfactants and their supramolecules for specific applications [62]. Although, in theory, higher oligomeric surfactants are text book example how the properties could be tailored by changing structural elements of the surfactants molecule, in the practice investigation of higher oligomeric surfactants are still more driven by the feasibility of the synthesis than the intended application. Additional reason is that the process of their aggregation is still largely unpredictable.

So far mostly cationic oligomeric quaternary ammonium surfactants have been synthesized and investigated, due to the relative ease of their synthesis, low Krafft temperatures and interesting rheological behavior. Oligomeric quaternary ammonium surfactants with different molecular architecture, linear [15-17, 26, 27], dissymmetric [62-65] and star-shaped molecules [22, 33, 34, 36], have been synthesized. Anionic and nonionic oligomeric surfactants have been studied to much lesser extent.

Common conclusions that can be drawn for these different classes of higher oligomeric surfactants are:

- most frequently the dodecyl chain was chosen as hydrophobic building block because it is long enough to confer good surfactant properties to amphiphiles, while it is still short enough that good water solubility for higher oligomeric or polymeric derivatives could be expected. Also, in the most cases, within certain series of surfactants the best properties have surfactants with dodecyl chains,
- the increase of the number of alkyl chains, i.e. degree of oligomerization, within oligomeric surfactant series:
    - enhances the surface activity,
    - shifts the critical micelle concentration (cmc) to lower concentrations, although the changes are becoming less significant with increase of degree of oligomerization above 2,

http://dx.doi.org/10.5772/57655

- linear relationship between the total carbon number in the hydrophobic group and the cmcs on the semilogarithmic scale has been shown for different oligomeric quaternary ammonium surfactants series, as for monomers and dimers

- quaternary ammonium oligomeric surfactants with short spacers (m=12, s=2 or 3) and star-shaped topology exhibit peculiar aggregation behavior, which significantly influences the rheological behavior of their solutions. Such a behavior was not observed for the oligomeric surfactants with longer spacers.

- the length and nature of the spacer within the series of surfactants with the same alkyl chain length and hydrophilic group are most dominant factors in determining the overall surfactant behavior. However, influence of the nature and length of the spacer on adsorption and micellization is not the same for different surfactants series, i.e. it depends on entire molecular architecture.

Although solid state properties of surfactant, in general, attract considerable attention due to their polymorphism and mesomorphism, to the best of our knowledge only one study of higher oligomeric surfactants has been reported [67].

Despite the obstacles, the results of a number studies which have shown the potential of oligomeric surfactants for different applications give additional motivation for the future research.

Acknowledgements

To our dear mentor dr.sc.Nada Filipović-Vinceković for introducing us to the wonderful world of surfactans.

Author details

D. Jurašin and M. Dutour Sikirić

Laboratory for Synthesis and Processes of Self-Assembling of Organic Molecules, Division of Physical Chemistry, Ruđer Bošković Institute, Zagreb, Croatia

References

[1] Meyers D. Surfactants Science and Technology. 3rd Ed. Hoboken, New Jersey: John Wiley & Sons Inc.; 2006.

[2] Rosen M. J. Surfactants and Interfacial Phenomena, 3rd Ed, Hoboken, New Jersey: Wiley Interscience; 2004.

[3] Zana R., Dimeric and Oligomeric Surfactants. Behavior at Interfaces and in Aqueous Solution a Review. Advances in Colloid and Interface Science 2002;97 205-253.

[4] Bunton CA, Robinson L, Schaak J, Stam MF. Catalysis of Nucleophilic Substitutions by Micelles of Dicationic Detergents. Journal of Organic Chemistry 1971;36 2346 – 2350.

[5] Devinsky F, Masarova L, Lacko I. Surface Activity and Micelle Formation of Some New Bisquaternary Ammonium Salts. Journal of Colloid and Interface Science 1985;105 235-239.

[6] Zhu Y-P, Masuyama A, Okahara M. Preparation and Surface Active Properties of Amphipathic Compounds With Two Sulfate Groups and Two Lipophilic Alkyl Chains Journal of American Oil Chemist Society 1990;67 459-463.

[7] Alami E, Beinert G, Marie P, Zana R. Alkanediyl-α,ω-bis(dimethylalkylammonium bromide) surfactants. 3. Behavior at the air-water interface Langmuir 1993;9 1465-1467.

[8] Zana R, Levy H, Papoutsi D, Beinert G. Micellization of Two Triquaternary Ammonium Surfactants in Aqueous Solution. Langmuir 1995;11 3694-3698.

[9] Danino D, Talmon Y, Levy H, Beinert G, Zana R. Branched Threadlike Micelles in an Aqueous Solution of a Trimeric Surfactant. Science 1995;269 1420-1421.

[10] Danino D, Talmon Y, Zana R. Alkanediyl-α,ω-bis(dimethylalkylammonium bromide) Surfactants (Dimeric Surfactants). 5. Aggregation and Microstructure in Aqueous Solutions. Langmuir 1995;11 1448-1456.

[11] Zana R. Dimeric (Gemini) Surfactants: Effect of the Spacer Group on the Association Behavior in Aqueous Solution. Journal of Colloid and Interface Science 2002;248 203–220.

[12] Zana R. Critical Micellization Concentration of Surfactants in Aqueous Solution and Free Energy of Micellization. Langmuir 1996; 12 1208-1211.

[13] Zana R, In M, Levy H, Duportail G. Alkanediyl-α,ω-bis(dimethylalkylammonium bromide). 7. Fluorescence Probing Studies of Micelle Micropolarity and Microviscosity. Langmuir 1997;13 5552-5557.

[14] In M, Warr GG, Zana R. Dynamics of Branched Threadlike Micelles. Physical Review Letters 1998;83 2278-2281.

[15] In M, Bec V, Aguerre-Chariol O, Zana R. Quaternary Ammonium Bromide Surfactant Oligomers in Aqueous Solution: Self-Association and Microstructure. Langmuir 2000;16 141-148.

[16] Esumi K, Taguma K, Koide Y. Aqueous Properties of Multichain Quaternary Cationic Surfactants. Langmuir 1996;12 4039-4041.

[17] Esumi K, Goino M., Koide Y. Adsorption and Adsolubilization by Monomeric, Dimeric or Trimeric Quarternary Ammonium Surfactant at Silica/water Interface. Journal of Colloid and Interface Science 1996;183 539-545.

[18] Hait SK, Moulik SP. Gemini surfactants: A distinct class of self-assembling molecules. Current Science 2002;82 1101-1111.

[19] Maiti PK, Lansac Y, Glaser MA, Clark NA, Rouault Y. Self-Assembly in Surfactant Oligomers: A Coarse-Grained Description through Molecular Dynamics Simulations. Langmuir 2002;18 1908–1918.

[20] Wu H, Xu J, He X, Zhao J, Wen H. Mesoscopic Simulation of Self Assembly in Surfactant Oligomers by Dissipative Particle Dynamics. Colloids and Surfaces A: Physicochemical and Engineering Aspects 2006;290 239–246.

[21] Regev O, Zana R. Aggregation Behavior of Tyloxapol, a Nonionic Surfactant Oligomer in Aqueous Solution. Journal of Colloid and Interface Science 1999;210 8–17.

[22] Fan Y, Hou Y, Xiang J, Yu D, Wu C, Tian M, Han Y, Wang Y. Synthesis and Aggregation Behavior of a Hexameric Quaternary Ammonium Surfactant. Langmuir 2011;27 10570–10579.

[23] White KA, Warr GG, Linkage by elimination/addition: A simple synthesis for a family of oligomeric alkylpyridinium surfactants. Journal of Colloid and Interface Science 2009;337 304-306

[24] Su X, Feng Y, Wang B, Lu Z,Wei L. Oligomericcationic surfactants prepared from surfmersvia ATRP: synthesis and surface activities. Colloid and Polymer Science 2011, 289, 101–110.

[25] Su X, Chu Z, Shuai Y, Guo Z, Feng Y. Oligomeric alkylpyridinium surfactants prepared via ATRP. *e-Polymers,* 2011, no. 046

[26] Yoshimura T, Yoshida H, Ohno A, Esumi K. Physicochemical Properties of Quaternary Ammonium Bromide-type Trimeric Surfactants. Journal of Colloid and Interface Science 2003; 267 167–172.

[27] Jurašin D, Habuš I, Filipović-Vinceković N. Role of the Alkyl Chain Number and Head Groups Location on Surfactants self-assembly in Aqueous Solutions. Colloids and Surfaces A: Physicochemical and Engineering Aspects 2010;368 119–128.

[28] Kim T-S, Kida T, Nakatsuji Y, Ikeda I. Preparation and Properties of Multiple Ammonium Salts Quaternized by Epichlorohydrin. Langmuir 1996;12 6304-6308.

[29] Chlebicki J, Węgrzyńska J, Wilk K. A. Surface-active, Micellar, and Antielectrostatic Properties of Bis-ammonium salts. *Journal of Colloid and Interface Science* 2008;323 372–378.

[30] Laschewsky A, Wattebled L, Arotcaréna M, Habib–Jiwan JL, Rakotoaly RH. Synthesis and Properties of Cationic Oligomeric Surfactants. Langmuir 2005;21 7170–7179.

[31] Laschewsky A, Lunkenheimer K, Rakotoaly RH, Wattebled L. Spacer Effects in Dimeric Cationic Surfactants. Colloid and Polymer Science 2005;283 469-479.

[32] Zhang J, Zheng Y, Yu P, He L, Wang H, Wang R. Synthesis, Characterization and Surface-Activity of a Polyoxyethylene Ether Trimeric Quaternary Ammonium Surfactant. Journal of Surfactants and Detergents 2010;13 155–158.

[33] Yoshimura T, Kusano T, Iwase H, Shibayama M, Ogawa T, Kurata H. Star-Shaped Trimeric Quaternary Ammonium Bromide Surfactants:Adsorption and Aggregation Properties. Langmuir 2012;28 9322–9331.

[34] Wu C, Hou Y, Deng M, Huang X, Yu D,Xiang Y, Liu Y, Li Z, Wang Y. Molecular Conformation-Controlled Vesicle/Micelle Transition of Cationic Trimeric Surfactants in Aqueous Solution. Langmuir 2010;26 7922–7927.

[35] Hou Y, Cao M, Deng M, Wang Y. Highly-Ordered Selective Self-Assembly of a Trimeric Cationic Surfactant on a Mica Surface. Langmuir 2008;24 10572-10574.

[36] Hou Y, Han Y, Deng M, Xiang J, Wang Y. Aggregation Behavior of a Tetrameric Cationic Surfactant in Aqueous Solution. Langmuir 2009;26 28–33.

[37] Murguia MC, Cristaldi MD, Porto A, Di Conza J, Grau RJ. Synthesis, Surface-Active Properties, and Antimicrobial Activities of New Neutral and Cationic Trimeric. Surfactants. Journal of Surfactants and Detergents 2008;11 41–48.

[38] Yoshimura T, Esumi K. Physicochemical Properties of Ring-Type Trimeric Surfactants from Cyanuric Chloride. Langmuir 2003;19 3535-3538.

[39] Yoshimura T, Esumi K. Physicochemical properties of anionic triple-chain surfactants in alkaline solutions. Journal of Colloid and Interface Science 2004;276 450–455.

[40] Murguia MC, Cabrera MI, Guastavino JF, Grau RJ. New oligomeric surfactants with multiple-ring spacers: Synthesis and tensioactive properties. Colloids and Surfaces A: Physicochemical and Engineering Aspects 2005;262 1–7.

[41] Yang F, Li G, Qi J, Zhang SM, Liu R. Synthesis and surface activity properties of alkylphenol polyoxyethylene nonionic trimeric surfactants. Applied Surface Science 2010;257 312–318.

[42] El-Sayed Abdul-Raouf M, Mahmoud Abdul-Raheim AR, El-Sayed Maysour N, Mohamed H. Synthesis, Surface-Active Properties, and Emulsification Efficiency of Trimeric-Type Nonionic Surfactants Derived From Tris(2-aminoethyl)amine. Journal of Surfactants and Detergents 2011;14 185–193.

[43] Eastoe J. Surfactant Aggregation and Adsorption at Interfaces. In: Cosgrove T. (ed.) Colloid Science Principles, Methods and Applications, 2nd Edition. Chichester, UK: John Wiley & Sons; 2010. 61-89.

[44] Yoshimura T, Nagata Y, Esumi K. Interactions of quaternary ammonium salt-type gemini surfactants with sodium poly(styrene sulfonate). Journal of Colloid and Interface Science 2004;275 618-622.

[45] Ottewill RH. Introduction. In: Tadros ThF. (ed) Surfactants. London; Academic Press, Inc.: 1984. 1-18.

[46] Pinazo, A, Diz M., Solans C, Pes MA, Erra P, Infante, MR. Synthesis and Properties of Cationic Surfactants Containing a Disulfide Bond Journal of American Oil Chemist Society 1993;70 37-42.

[47] Devínsky F, Lacko I, Bitterová F, Tomečková L. Relationship between Structure, Surface Activity and Micelle Formation of Some new Bisquaternary Isosteres of 1,5-Pentanediammonium Dibromides. Journal of Colloid and Interface Science 1986;114 314-322.

[48] Li ZX, Dong CC, Thomas RK. Neutron Reflectivity Studies of the Surface Excess of Gemini Surfactants at the Air–Water Interface Langmuir 1999;15 4392-4396.

[49] Malmsten M. Surfactants and Polymers in Drug Delivery. New York; Marcel Dekker, Inc.: 2002.

[50] Zana R, Benrraou Rueff, Alkanediyl-α,ω-bis(dimethylalkylammonium bromide) Surfactants. 1. Effect of the Spacer Chain Length on the Critical Micelle Concentration and Micelle Ionization Degree Langmuir 1991;7 1072-1075.

[51] Wattebled L., Laschewsky A., Moussa A., Habib–Jiwan J.–L. Aggregation Numbers of Cationic Oligomeric Surfactants: A Time-Resolved Fluorescence Quenching Study. Langmuir 2006;22 2551–2557.

[52] Bernheim-Groswasser A, Zana R, Talmon Y. Sphere-to-cylinder Transition in Aqueous Micellar Solution of a Dimeric (Gemini) Surfactant. Journal of Physical Chemistry B 2000;104 4005–4009.

[53] Lianos P, Lang J, Zana R. Fluorescence Probe Study of the Effect of Concentration on the State of Aggregation of Dodecylalkyldimethylammonium Bromides and Dialkyldimethylammonium Chlorides in Aqueous Solution *Journal of Colloid and Interface Science* 1983;91 276-279.

[54] Kusano T, Iwase H, Yoshimura T, Shibayama M. Structural and Rheological Studies on Growth of Salt-Free Wormlike Micelles Formed by Star-Type Trimeric Surfactants. Langmuir 2012; 28 16798-16806.

[55] Lagleize J.-M, Richetti P, Drummond C. Effect of Surfactant Oligomerization Degree on Lubricant Properties of Mixed Surfactant-Diblock Copolymer Films. Tribology Letters 2010;39 31–38.

[56] Esumi K, Takeda Y, Koide Y. Competitive adsorption of cationic surfactant and pesticide on laponite, Colloids and Surfaces A: Physicochemical and Engineering Aspects 1998;135 59-62.

[57] Zhang R, Qin N, Peng L, Tang K, Ye Z. Wettability alteration by trimeric cationic surfactant at water-wet/oil-wet mice mineral surfaces. Applied Surface Science 2012;258 7943-7949.

[58] Yan X, Han S, Hou W, Yu X, Zeng C, Zhao X, Che H. Synthesis of highely ordered mesoporous silica using cationic trimeric surfactant as structure-directing agent, Colloids and Surfaces A: Physicochemical and Engineering Aspects 2007;303 219-225.

[59] Sun Y, Han s, Yu X, Che H, Liu A, Wang S, Synthesis of highly ordered supermicroporous silica using short-chain cationic trimeric surfactant as structure-directing agent. Journal of porous Materials 2010;17:597-603.

[60] Dazzazi A, Coppel Y, In M, Chassenieux C, Mascalchi P, Salome L, Bouhaouss A, Kahn ML, Gauffre F. Oligomeric and polymeric surfactants for the transfer of luminiscent ZnO nanocrystals to water. Journal of Materials Chemistry C 2013;11 2158-2165.

[61] He C, Hou Y, Han Y, Wang Y. Disassembly of Amyloid Fibrils by Premicellar and Micellar Aggregates of a Tetrameric Cationic Surfactant in Aqueous Solution. Langmuir 2011;27 4551-4556.

[62] Svenson S. Controlling surfactant self-assembly. Current Opinion in Colloid and Interface Science 2004;25 201–212.

[63] Oda R, Huc I, Candau SJ. Gemini surfactants, the effect of hydrophobic chain length and dissymmetry. Chemical Communications 1997;2105-2106.

[64] Oda R, Huc I, Homo JC, Heinrich B, Schmutz M, Candau SJ. Elongated Aggregates Formed by Cationic Gemini Surfactants. Langmuir 1999;15 2384-2390.

[65] Sikirić M, Primožič I, Talmon Y, Filipović-Vinceković N. Effect of the spacer length on the association and adsorption behavior of dissymmetric gemini surfactants. Journal of Colloid and Interface Science 2005;281 473-481.

[66] Sikirić M, Šmit I, Tušek-Božić Lj, Tomašić V, Pucić I, Primožič I, Filipović-Vinceković N. Effect of the spacer length on the solid phase transitions of dissymmetric gemini surfactants. Langmuir 2003;19 10044-10053.

[67] Jurašin D, Pustak A, Habuš I, Šmit I, Filipović-Vinceković N. Polymorphism and Mesomorphism of Oligomeric Surfactants: Effect of the Degree of Oligomerization. Langmuir 2011;27 14118-14130.

Chapter 6

Oligomerization of Nucleic Acids and Peptides under the Primitive Earth Conditions

Kunio Kawamura

Additional information is available at the end of the chapter

http://dx.doi.org/10.5772/58222

1. Introduction

1.1. Why are nucleic acids and peptides so important for the emergence of life-like systems?

Question how life originated on the primitive earth is still a frontier of science. Nowadays, primitive life-like system is mostly considered to have emerged by chemical evolution on the Earth although some scientists are evaluating the possibility of panspermia hypothesis [1-3]. External and internal energies from the earth, such as cosmic rays, ultra violet radiation, meteorite impacts, volcanic eruption, submarine hydrothermal vent system, etc. resulted in organic molecules from inorganic materials, such as primitive atmospheric gas, minerals, materials solved in the ocean. Organic molecules were polymerized and finally resulted in several biological functions. Complex mixtures of the organic molecules should have evolved to primitive life-like systems. By the Miller-Urey experiment at 1953, a scenario from simple molecules to organic molecules in a simulated primitive atmosphere – ocean system was evaluated for the first time [4]. After the Miller-Urey experiment, the origin of life problem became as a scientific subject and a number of simulation experiments were carried out to elucidate how life-like system was originated on this planet [5-7].

The knowledge regarding the primitive environments of the Earth and surrounding universe in relation to chemical evolution has been accumulated as well as the simulation experiments. For instance, an environment, which was assumed at the time of the Miller-Urey experiment, is not already acceptable nowadays although it does not mean that the Miller-Urey experiment was devalued; the knowledge on the primitive earth has continuously accumulated to improve the reliability of simulation experiments. The primitive earth environments are difficult to estimate since the evidences regarding oldest rocks and fossil organisms are merely capable to trace back to near 3.9 billion years ago, where almost oldest evidence of life-like system has been obtained [8, 9], although the best current evidences are closer to 3.5 billion years ago [10].

DNA, RNA, and proteins are realized as most important biological materials, which play central roles in the present life-like systems. These molecules possess a number of functions, such as genetic information storage, transcription and translation of genetic information, enzymatic activities, which are based on their unique three-dimensional structures of oligo- and polynucleotides and proteins. A certain length of polymerized nucleic acid and peptide chains is the essential to display systemized biological functions although small molecules also play important roles in living organisms. Thus, the formation and accumulation of a certain length of polymers of nucleic acids and peptides are essential process toward the emergence of life-like system.

Metabolism, amplification (replication), and evolution are generally considered as essential requisites, for which a life-like system is regarded as alive [11]. The requisites are strongly related to a train of machinery, which maintains the genetic coding and translation system in cell consisting of DNA, RNA, and proteins as shown in Figure. 1. It was generally considered that the phenotype molecules of cell-type of organism are proteins while the genotype molecules are DNA molecules until the discovery of ribozymes. The function of DNA is storage of genetic information. A DNA sequence is copied to mRNA and translated to amino acid sequence of polypeptide on ribosome using tRNA (Figure 1 left). The peptide becomes a protein by final processing and folding of three-dimensional structure. In the modern organisms, DNA is regarded as blueprint and protein is regarded as building block created on the basis of the blueprint. The information flow from DNA and protein is universal in all the present organisms [12]. On the other hand, one of the important functions of proteins is enzymatic function. It is considered that protein enzymes control the formation of other biologically important molecules, such as sugar, lipid, vitamin, coenzyme; these functional molecules are constructed indirectly by the catalytic actions of protein enzymes. In addition, protein enzymes maintain the several reactions regarding the DNA replication, the RNA formation, and the protein metabolism itself.

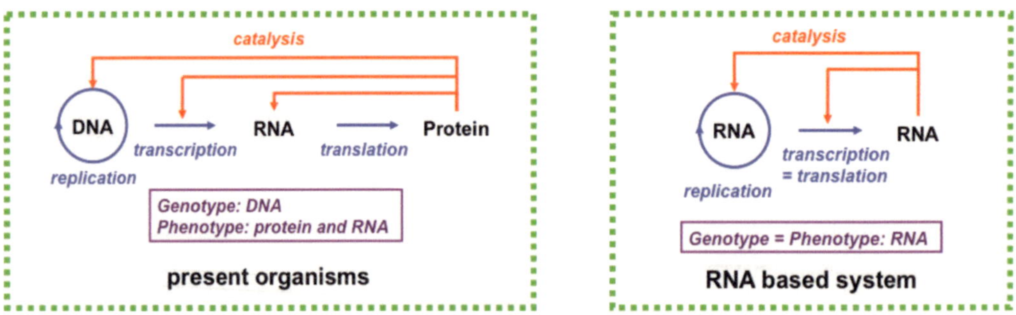

Figure 1. Information flow in the present organisms (left) and in an RNA-based life-like system (right)

A schematic pathway for chemical evolution of RNA and protein-like molecules is illustrated in Figure 2. According to the modern functions of RNA and proteins, it is assumed that proteins could not have possessed capability for storage of information during the chemical evolution in contrast to RNA molecules. However, the interactions between RNA and protein-like

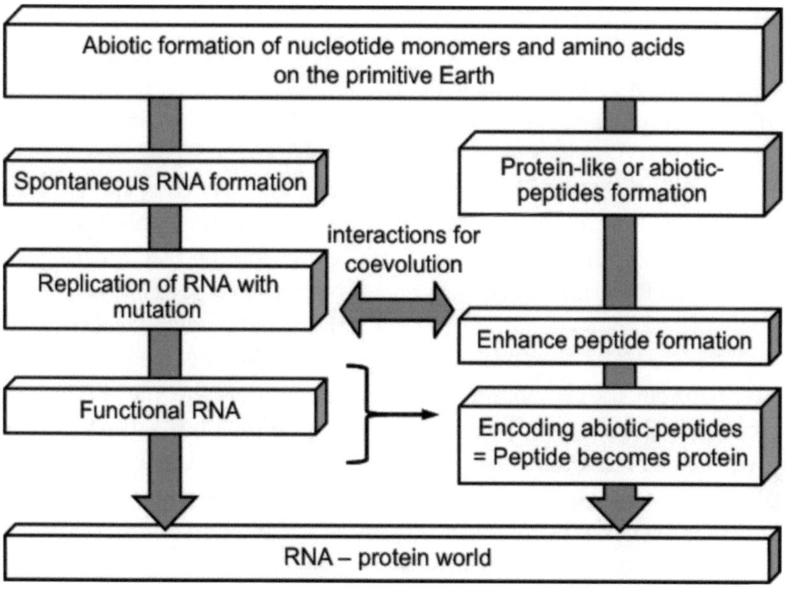

Figure 2. Chemical evolution of RNA and protein-like molecules towards RNA-protein world

molecules would have facilitated chemical evolution each other. Such cooperative chemical evolution would have resulted in an RNA – protein world after the RNA world system.

Here, the term "protein" and "peptide" should be reconsidered in the present chapter. Normally, a protein is ambiguously defined just as a very long peptide, which has a biochemical function. On the other hand, peptides are frequently regarded as just smaller molecules than proteins. However, an important characteristic from the viewpoint of the origin of life, that is, the fact that proteins are synthesized on the basis of genetic translation system in cell should be pointed out. So, if a primitive long peptide were synthesized on the basis of genetic information, which might be constructed from RNA molecules, the long peptide is regarded as a protein. Thus, the assignment method must be elucidated simultaneously during the evaluation of the origin of protein. Thus, in the present chapter the term "a protein" is defined as a certain length of peptide molecule, which is created using an assignment method in a life-like system. Thus, the term "abiotic-peptide" or "protein-like" will be used as describing oligomerized molecules made from amino acids including peptide bonding. On the other hand, the length of abiotic-peptides, which could have gained biological functions would be also important to consider the oligomerization of amino acids. Biological functions of proteins and peptides are based on the three-dimensional structures. Thus, a minimum length of oligopeptides would be regarded as a candidate functional molecule. The minimum length of a peptide possessing a biological function might be down to 10-mer oligopeptides since a 10-mer peptide could form a rigid three-dimensional structure [13].

Conclusively, DNA, RNA, and proteins are regarded as most important molecules, which play important roles regarding the genetic information flow. Then, the origin of the genetic information is important issue in relation to the origin of life. The relationship between DNA and protein has been an issue for many years, which is so-called as "egg and chicken problem". DNA replication, the transformation of DNA to RNA, the translation to proteins are maintained by protein enzymes and the genetic information of such proteins is coded in DNA sequences. Thus, the relationship between DNA and protein resembles the relationship between egg and chicken. However, discovery of RNA enzyme [14], that is ribozyme, suggested that RNA molecules should have played a central role in ancient life-like systems, where RNA preserves genetic information, but also enzymatic functions. This is so-called the RNA world hypothesis [15-17] and extensively investigated for the last 2 decays. The genetic information flow in an RNA-based life-like system is shown in Figure 1 (right). RNA molecules play both the functions, that are, storage of genetic information and expression of enzymatic activities. Evidences, such as, important functions of mRNA, tRNA, rRNA during the translation of DNA to protein, the enzymatic functions of ribozymes, the role of ATP as high energy phosphate, the roles of coenzyme possessing nucleotide moieties, support the RNA world hypothesis. Nowadays, it is understood that the DNA sequences as genotype molecules are assigned to the amino acid sequences and RNA sequences as phenotype molecules in cell. The presence of an assignment method between genotype and phenotype is added as an important requisite for life-like system [18, 19]. In an RNA based life-like system, RNA molecules should have played both the roles of genotype and phenotype.

2. Difficulties of dehydration in aqueous solution

Both the formations of nucleic acids and proteins *in vivo* are the dehydration process in aqueous medium. In other words, the hydrolytic degradation of both nucleic acids and proteins spontaneously proceeds in aqueous medium from the thermodynamic viewpoint. Thus, the formation of these biopolymers is not advantageous from the viewpoint of thermodynamics unless using a condensation reagent and/or an activation technique of monomer. In the present organisms, triphosphate nucleotide monomers are primarily used to form nucleic acids as monomer unit, and the activated amino acids by forming aminoacyl-tRNA (Figure 3) are adopted to form peptide bonding on ribosome. Similarly, the dehydration using condensation agent and/or activation of the monomers in organic solvent are a standard strategy for the organic chemical synthesis of nucleic acids and proteins.

On the other hand, RNA oligomerization proceeds by RNA polymerases on a DNA template from nucleotide 5′-triphosphates (Figure 4), which are the activated nucleotide monomers *in vivo*. A polypeptide forms on a ribosome in the presence of mRNA molecules and aminoacyl-tRNA molecules, which are formed from amino acids and tRNA molecules by aminoacyl tRNA synthetase molecules. On the other hand, modern DNA templates, ribosomes, and enzymes were not present on the primitive earth. Thus, the processes of RNA formation without these organic materials should be identified to clarify the origin of primitive life-like system.

Furthermore, an important factor should be pointed out for the investigation of prebiotic oligomerization; life-like system is a thermodynamically open system, of which energy and materials inflow and outflow from and to the environment. Thus, the formation of dissipative

http://dx.doi.org/10.5772/58222

structures could be formed in an open system, which could not be formed under the equilibrium state. Thus, it should be considered that oligonucleotides and abiotic-peptides could have accumulated under conditions, which are far from equilibrium conditions.

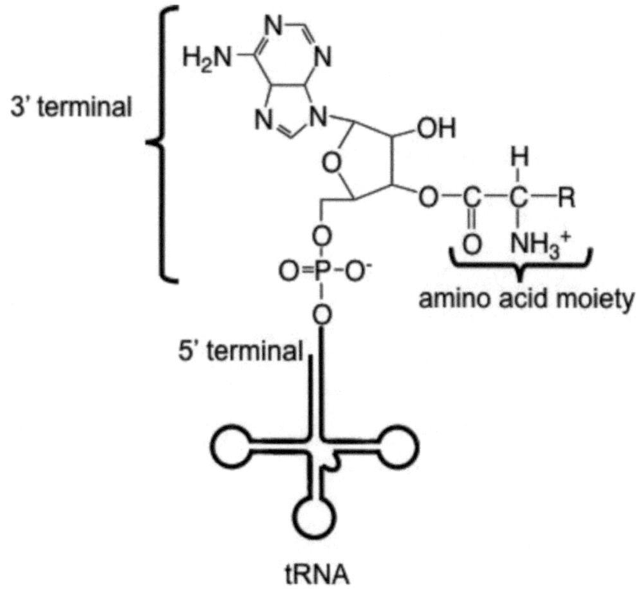

Figure 3. Activation of amino acid in aminoacyl-tRNA

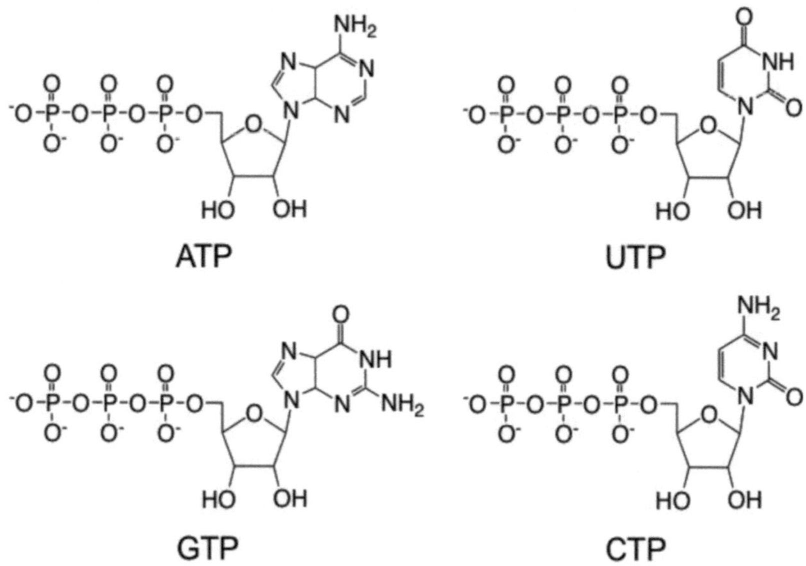

Figure 4. Structures of nucleoside 5'-triphosphate

3. Primitive earth environments: So extreme as comparing to the present earth environments

The time that a most primitive life emerged on the primitive earth is considered between from 4.6 to 3.9 billion years ago; the earth formed 4.6 billion years ago and the oldest evidence in 3.8 billion year old rock has been pointed out [8]. Thus, since physical evidences between 4.6 – 3.9 billion years ago are hardly obtained the theoretical estimation on the basis of astrophysics and earth science regarding the environments of the earth between 4.6 – 3.9 billion years ago has been carried out. At the early stage of the earth, the surface of the earth is covered with magma-ocean. Presumably, a large number of giant meteorite impacts, celestial bodies, and planetesimals attacked the primitive earth until 3.8 billion years ago [20]. Thus, the giant meteorite impact could have evaporated totally the water of the ocean while the ocean would have formed until 4.2 billion years ago [21, 22]; life-like system, which might have emerged before 3.8 years ago, could have been destroyed under such extreme conditions. Furthermore, the moon was much closer than today which caused strong tidal actions. These factors should have strongly affected the surface temperature of the earth. On the other hand, some scientists assumed that the primitive earth was frozen since the solar luminosity was relatively less than at present [23, 24]. Thus, the temperature of the primitive ocean, in which life originated, remains speculative [25-27] although the temperature would have been higher than the present.

At the same time, it should be also considered that the ancient environments of the earth should be heterogeneous as the present environments are. It would be more difficult to elucidate the local heterogeneity of the ancient environments because of the limitation by theoretical estimation. The presence of the ocean would suggest the presence of mantle convection on the ancient earth before 4.2 billion years ago [28] so that hydrothermal vent systems would have been present in the primitive ocean. In addition, it should be considered that large continents were not formed on the primitive earth. On the other hand, the primitive atmospheric conditions should be taken into account since the primitive atmosphere is nowadays considered including CO_2, CO, N_2, H_2O mainly as such strong reductive one [4] dispite the relatively strong reductive atmosphere at the time of Miller-Urey experiment was assumed to include CH_4, NH_3, H_2O [27].

Conclusively, the primitive earth environments should have been extreme and heterogeneous. Nevertheless, the simulation experiments of chemical evolution of nucleic acids and abiotic-peptides have been normally carried out under mild conditions. One reason is that the ancient environments of the earth should have been so extreme that these simulation experiments could not be easily carried out. Thus, these extreme conditions must be considered as possible earth environments for the simulation experiments of chemical evolution.

4. Successful examples of the formation of nucleic acids

4.1. RNA formation using the present high-energy nucleotide monomers

Discovery of ribozyme suggested that RNA molecules played important roles in the emergence of life. If the RNA world hypothesis is correct then RNA or RNA-like molecules should have accumulated on the primitive earth. It is now getting known that diversity of RNA molecules in the present organisms is very large including mRNA, tRNA, rRNA, ribozymes, small non-coding RNA [29]. The biological functions of tRNA, rRNA, and ribozymes are caused by the formation of three-dimensional structures. Presumably, a certain length of RNA oligomers would be necessary to display such functions.

The importance of RNA molecules to the origin of life had been speculated many years ago [30, 31] before the time when the RNA world hypothesis was proposed [15]. Thus, the simulation experiments of chemical evolution of RNA molecules were extensively carried out before the proposal of the RNA world hypothesis. RNA molecules should have evolved without enzymes and template DNA molecules. On the other hand, there are some enzymes in modern organisms and viruses (Table 1) that catalyze the formation of RNA molecules with and without using a template nucleic acid. RNA polymerase catalyzes the RNA formation from nucleotide 5′-triphosphate (NTP) in the presence of DNA template [32], and Qβ replicase from a virus catalyzes the RNA formation in the presence of an RNA template [33]. Besides, polynucleotide phosphorylase (PNPase) catalyzes the formation of RNA from nucleotide 5′-diphosphate (NDP) without a template nucleic acid [34]; it is considered that the role of PNPase is primarily to form NDP from polynucleotides, but the enzyme also catalyzes the formation of polynucleotide from NTP. Thus, the enzyme is used for the preparation of polynucleotide in the laboratory. The presence of different enzymes in the formation of RNA indicates that oligonucleotides would have formed under the primitive earth conditions through multiple pathways.

The pathways for the formation of oligonucleotides without enzymes were extensively investigated. The potential of the present high-energy nucleotide phosphate as monomer unit, that is, NDP and NTP, was examined in the absence of the enzyme and a template nucleic acid. However, the efficient formation of oligonucleotides from neither NDP nor NTP has been observed [35]. The high-energy nucleotide phosphates possess the sufficient Gibbs free energy to form oligonucleotides so that the reason of the difficulty of oligomerization of NDP and NTP is probably due to the relatively small rate constants of formation of oligonucleotide without enzyme as comparing to the degradation of these monomers or the hydrolysis of monomers. Thus, condensation agents were used for the oligomeriation of nucleotide monomers on a template nucleotide polymer although these reactions are normally not so efficient [36-40]. These condensation agents are not considered as prebiotic so cyclic 2′, 3′-phosphate were also verified [41, 42]. Recently, the acceptable prebiotic pathway of 2′, 3′-cyclic pyrimidine nucleotide monomers was proposed, which might have been an activated nucleotide monomer for oligoucleotides [43]. Very recently, nucleoside 3′, 5′-cyclicmonophosphate was used to produce oligonucleotides at medium temperatures, where oligonucleotides were detected using gel electrophoresis [44].

Enzyme	Monomer	Template
Ribonuclease	Nucleoside 5′-triphosphate	DNA
Qβ replicase	Nucleoside 5′-triphosphate	RNA
Polynucleotide phosphorylase	Nucleoside 5′-diphosphate	No template

Table 1. Formation of RNA in the presence of different types of enzymes.

It was not successful for an efficient template-directed formation of oligonucleotides from nucleoside 5′-monophosphate monomers [36-40, 45] or using an activated nucleotide intermediate [46] in the presence of the condensation agent. Finally, it was discovered an acceptable primitive activated nucleotide monomer, that is, nucleotide 5′-monophosphorimidazolide (ImpN) (Figure 5), which is activated with N-P bound on the phosphate group of the nucleotide monomer [47-49]. For instance, adenosine 5′-monophophorimidazolide (ImpA) forms oligoadenylate (oligoA) in the presence of Zn^{2+}without a template [50] and guanosine 5′-monophosphorimidazolide (ImpG) forms efficiently oligoguanylate (oligoG) up to 40 nucleotide units in the presence of a polycytidylic acid template (polyC) without using enzyme in the presence of Pb^{2+}or Zn^{2+}[51, 52]. The prebiotic RNA formation reactions in detail will be summarized in the following sections.

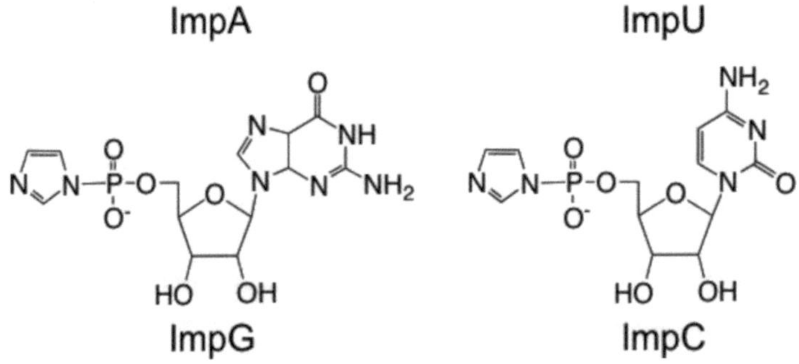

Figure 5. Activated nucleotide monomers

The details and variation of the reaction using activated nucleotide monomers have been extensively investigated in the absence and presence of template, in the presence of metal ion catalysts, clay mineral catalysts, using different types of activated nucleotide monomers, using non-ribose based nucleotide analogues. As a variation of ImpN, nucleoside 5′-monophospho-2-methylimidazolde (2MeImpN) has been frequently used [53].

4.2. Prebiotic formation and organic synthesis of the activated nucleotide monomer

It was successfully demonstrated that oligonucleotides form from the activated nucleotide monomers without enzyme. However, the formation of the activated nucleotide under the primitive earth conditions is also an important issue. Possible pathways for the spontaneous formation of the activated nucleotide had been barely elucidated [48, 54, 55]. However, the accumulation of the activated nucleotide monomers would not be so likely since the moisture controlled N-P bond formation and the formation of nucleotide phosphate under dry conditions are included [54]. Furthermore, the activated nucleotide monomers are substantially hydrolyzed to nucleoside 5′-monophosphate [56]. The yield of the activated nucleotide monomers formed under these simulation conditions is not so high that the simulation experiments of oligonucleotide formation from the activated nucleotide monomers under the primitive earth are normally carried out using a sufficient amount of the activated nucleotide monomers, such as ImpN, which are prepared using an organic synthetic technique. Organic synthesis of ImpN was established and the yields of ImpN and the analogues are normally over 95 % by coupling method using a sulfide compound [57].

4.3. Spontaneous formation of oligonucleotide using metal catalyst (Metal-catalyzed reaction)

Spontaneous oligomerization of the activated nucleotide monomers should be a next step for the chemical evolution of oligonucleotide once these formed under the primitive earth conditions. A spontaneous formation pathway in the presence of metal ion catalysts was found for the first time (Figure 6) [50, 58]. The maximum length of the oligonucleotide formed by the Metal-catalyzed reaction of RNA reaches to ca. 10-mer nucleotide units [59-61]. The efficiency of Metal-catalyzed reaction is less dependent on the type of nucleotide bases. Catalytic metal ions, such as Zn^{2+}, Pb^{2+}, and UO_2^{2+}, are normally active at fairly high concentrations in the simulation experiments as comparing to the concentration in the present ocean. Regioselective formation of 3′, 5′-or 2′, 5′-linked oligonucleotides has been observed that the formation of 2′, 5′-linked oligonucleotide is preferable using Pb^{2+}or $UO_2^{2+}$and that of 3′, 5′-linked oligonucleotide is preferable using Zn^{2+}[50, 59-61]. Although the reason 3′, 5′-linked RNA oligonucelotide was selected by the chemical evolution of RNA is not yet established, these experimental data would be related to the reason 3′, 5′-linked olignucelotide was selected. In addition, it is pointed out that the 2′, 5′-linked oligonucelotide formation was applied to a practical synthetic method of 2′, 5′-linked oligonucleotide [59]. Mechanistic analysis of Metal-catalyzed reactions suggested that the acceleration by metal ions is due to the formation of the metal ions with phosphate group and/or bases, which enhance the association between two monomers or that between a monomer and an elongating oligomer prior to the phosphodiester bond formation [62].

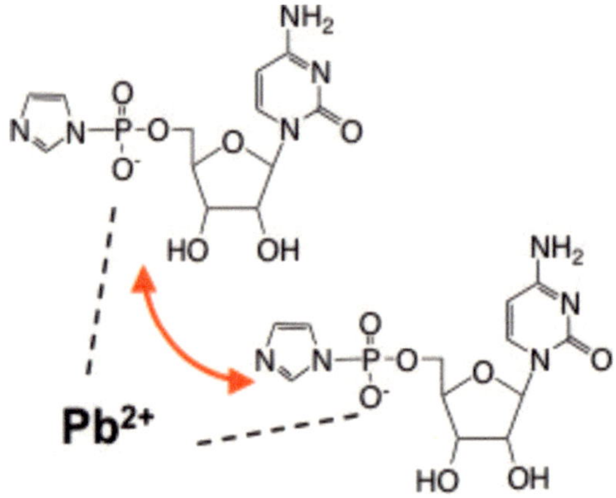

Figure 6. Metal-catalyzed reaction

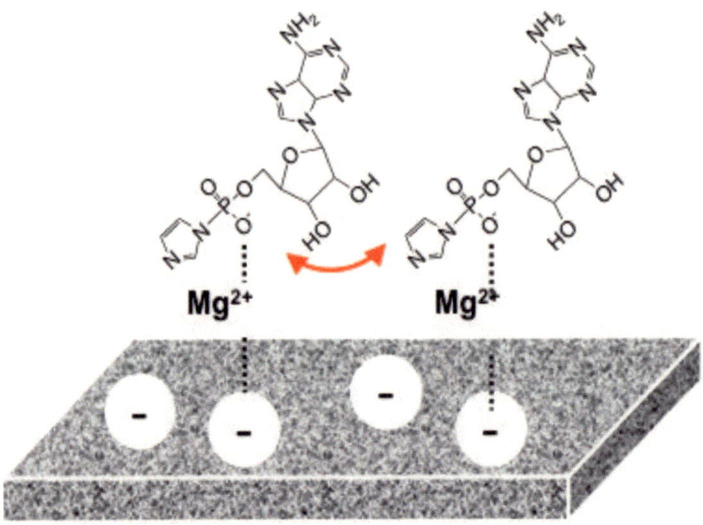

Figure 7. RNA formation of montmorillonite clay

4.4. Spontaneous formation of oligonucleotide using clay catalyst (Clay-catalyzed reaction)

The importance of minerals for the chemical evolution of biomolecules and biopolymers was frequently assumed from old days [63]. Thus, the simulation experiments of oligonucleotide formation have been extensively carried out [64-66]. An efficient formation of oligonucleotide

http://dx.doi.org/10.5772/58222

on a clay mineral, naturally-occurring montmorillonite, was discovered for the first time at 1992 and the reaction was extensively investigated [67]. The oligonucleotides up to ca. 10-mer nucleotide monomer units are formed spontaneously in one pot reaction in the presence of montmorillonite catalyst. The efficiency of Clay-catalyzed reaction is somewhat dependent on the type of nucleotide bases; the formation of oligoguanylate (oligoA) is somewhat less effective as comparing to the others [68-71]. By these reactions, both the 2′, 5′-linked and 3′, 5′-linked oligonucleotides form and pyrophosphate-linked isomers were observed. There is a trend that the oligonucleotides formed from ImpN consisting of purine base involve mainly 3′, 5′-linked isomers and those formed from ImpN consisting of pyrimidine base involve mainly 2′, 5′-linked isomers. The percentage of pyrophosphate-linked isomers is normally small.

Furthermore, it is noted that a fairly large amount of cyclic oligonucleotides, such as cyclic-2-mer, cyclic-3-mer, and cyclic-4-mer, are generally observed. The ratio of the cyclic isomers is also dependent on the type of nucleotide bases [70]. For instance, the 2-mer fraction on an anion-exchange HPLC column of the products using ImpU and ImpI involves 60 – 96% of cyclic-3-mers. Longer cyclic oligomers are also observed in higher length fractions. The cyclization of nucleotide 6-mer and higher oligomers, which possess phosphate group at 5′-terminal and 2′-OH at 3′-terminal, was observed in the presence of condensation agent [72, 73]; this reaction was designed for the competition reaction between elongation and cyclization. The cyclization of oligonucleotides is considered as termination reaction for the chemical evolution of RNA.

The oligonucleotide formation proceeds using different sources of montmorillonite, but the oligonucleotide formation does not proceed at all using a clay from Otay [69]. The binding of ImpN on different clay mineral is observed in these reaction systems, and there is a strong correlation between the yield of oligonucleotide and the binding constant of the activated nucleotide monomer on the clay. Thus, binding is necessary step for the acceleration of oligonucleotide formation on the clay. The activated nucleotide monomer is bound to the surface of montmorillonite ca. 80%, where negative charges are present, and bound to the edge of the montmorillonite 20% [74]. The activated nucleotide monomer is bound through Mg^{2+}ion bridge, which forms complex with phosphorimidazolide group (Figure 7) [69]. Low efficiency of oligonucleotide formation in the absence of Mg^{2+}supports the Mg^{2+}bridge binding model, and Ca^{2+}ion can also form the bridge [69].

On the other hand, the binding of the activated nucleotide monomers is fairly dependent on the types of nucleotide bases [69, 70]. The strength of binding of ImpN consisting of purine base is larger than that consisting of pyrimidine base. Although the apparent binding ratio of ImpC and ImpU on the clay is very low, the yields of oligonucleotide are not much different from those of ImpA and ImpG. This fact suggests that the apparent binding of ImpN includes an effective and non-effective binding on the clay mineral for the oligonucleotide formation. Non-effective binding reflects that the binding is dependent on the hydrophobicity and staking by nucleotide bases.

By using Clay-catalyzed reaction, it was elucidated that oligonucleotides up to ca. 50-mer nucleotide units form by a continuous feeding of ImpN to the montmorillonite-aqueous solution system [75]. Furthermore, a ImpN analogue replaced the imidazole moiety with purine derivative forms oligonucleotides with ca. 50-mer nucleotide units in one-step reaction [76, 77]. The modifications of Clay-catalyzed reaction indicate that oligonucleotides could have accumulated under the primitive earth conditions if the conditions were mild as present.

The reaction mechanism of Clay-catalyzed reaction was extensively investigated on the basis of reaction kinetics and binding thermodynamics. The binding of ImpN or analogues is the first step for the formation of oligonucleotides. Mg^{2+}ions are necessary for the binding of ImpN through the bridge with Mg^{2+}, which attaches to phosphate group of ImpN on the negatively charged clay surface [69, 74]. Kinetic analysis showed that the formation rate constant of 2-mer is much smaller than that of 3-mer and longer oligonucleotides. This fact would suggest that the association of two ImpN molecules to form 2-mer is much weaker than that of ImpN with elongating olignucleotide. A similar trend was observed in the cases of Metal-catalyzed reaction [62]. The regioselectivity of 2′, 5′-or 3′, 5′-linked isomers using pyrimidine or purine ImpN is probably due to the different binding conformation of the activated nucleotide monomer on clay surface. During the oligomerization, the hydrolysis of ImpN simultaneously proceeds, but the formation of oligonucleotide is much faster than that of ImpN hydrolysis.

4.5. Template-directed formation of oligonucleotide (Template-directed reaction)

According to the scenario of chemical evolution of RNA, the replication of oligonucleotides should be a next step after the spontaneous formation of oligonucelotides by Metal-catalyzed and Clay-catalyzed reactions since the replication of RNA would have resulted in the replication of genetic information. Orgel and coworkers showed that template-directed formation of oligoG from ImpG or 2MeImpG on a polyC without using any enzyme (Figure 8) [51-53]. These reactions produce oligonucleotides up to ca. 40-mer. The efficiency of Template-directed reaction using ImpG is dependent on metal ions, where Pb^{2+}ion catalyzes the 2′, 5′-linked oligoG and Zn^{2+}ion catalyzes the 3′, 5′-linked oligoG. It is interesting that $Zn^{2+}$is present in the active center of modern RNA polymerase (Figure 9) [78]. Besides, 2MeImpG forms 3′, 5′-linked oligoG without using any metal ions. Formation of cyclic isomers is less in Template-directed reaction while the cyclic 2-mer, 3-mer, and higher oligomers are frequently observed for Clay-catalyzed reactions. The Template-directed reaction is dependent on pH and temperature. Normally, the efficiency of the reaction is highest at around pH 8 [53]. This is due to the activity of the activity of the activated nucleotide monomers.

It is noted that Template-directed reactions using different ImpN or 2MeImpN with different nucleotide bases, that is, adenine, uracil, and cytosine do efficiently not proceed to form oligonucleotides; oligoA forms with very low efficient from ImpA on polyU template [49] and oligoU and oligoC are not formed at all from ImpU and ImpC on a complementary template. The low efficiency using activated nucleotide monomers is probably due to that the association between two ImpN molecules or that between ImpN and an elongating oligonucleotide on a complementary template is very weak for ImpN consisting of adenine, uridine, or cytidine base.

A possible pathway was demonstrated that Template-directed reaction from a mixture of 4 types of ImpN with different bases on a template including complementary bases enable to

http://dx.doi.org/10.5772/58222

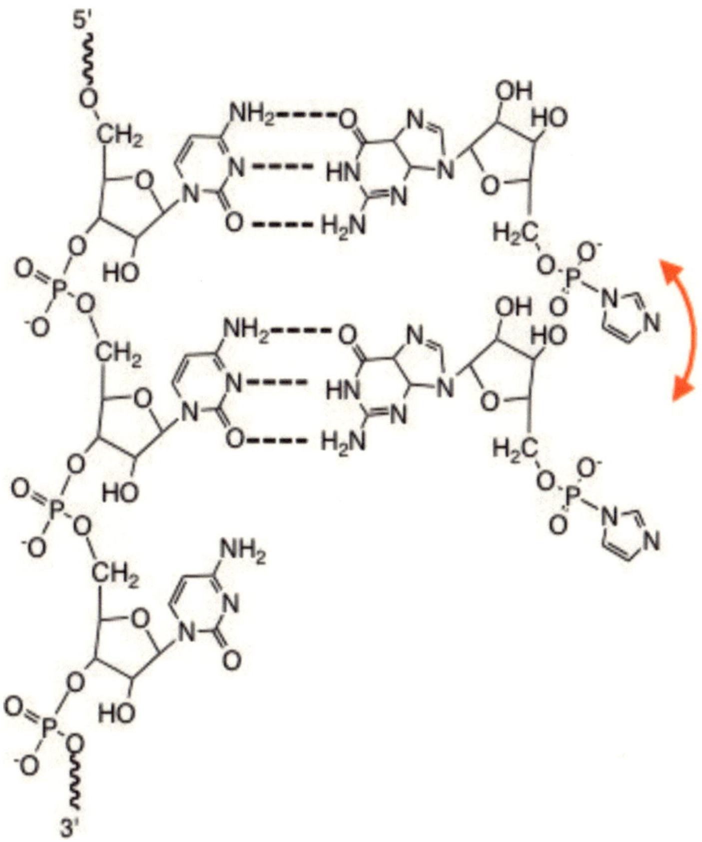

Figure 8. Template-directed formation of RNA

partially incorporate different bases in the oligonucleotides other than oligoG [57, 79]. The efficiency of the Template-directed reaction increases with increasing of the ratio of cytidine of the template. In addition, it was confirmed that Template-directed reaction using ImpA, ImpU, and ImpC would partially proceeds using hairpin oligonucleotides [80]. Furthermore, Template-directed reaction proceeds if the activated oligoU is used as starting material on a polyA template, of which the association of oligoU would be stronger than that of ImpU monomers [81]. Although these results suggest that Template-directed reaction partially proceeds in mixed oligonucleotides, thus the fidelity of Template-directed reaction is not high. The limitaition of Template-directed reaction as prebiotic formation of RNA has been pointed out [82]. In other words, these facts would suggest that there is still a drawback how the replication of oligonucleotide could have proceeded in the absence of enzyme. In addition, protocell and vesicles could have enhanced the chemical evolution of RNA molecules [83, 84]. Conclusively, a universal pathway for the replication using activated nucleotide monomers is not yet identified.

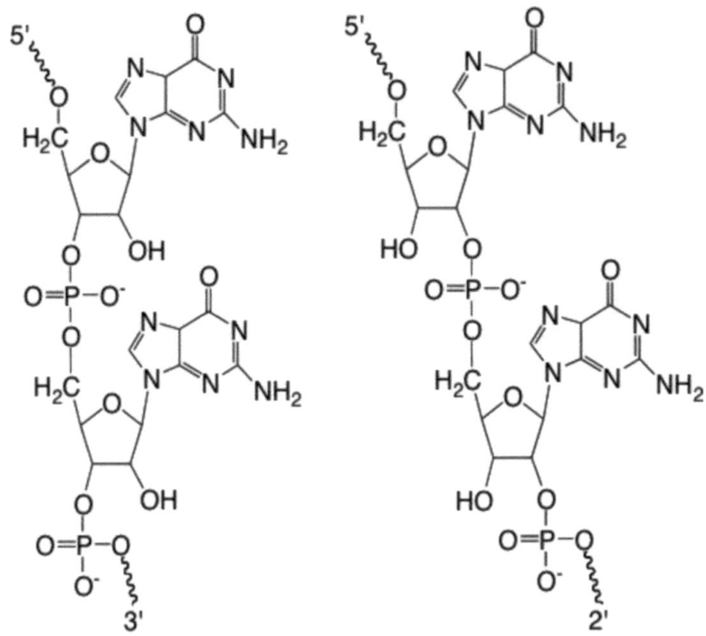

Figure 9. 3′, 5′-linked (left) and 2′, 5′-linked (right) RNA

4.6. Kinetic analysis of prebiotic oligonucleotide formation

First, the reaction mechanism of Template-directed reaction has been extensively investigated on the basis of kinetic analysis [85]. The rate constants for the formation of 2-mer, 3-mer and longer increase in the order 2-mer << 3-mer < 4-mer and higher. The hydrolysis of the activated nucleotide proceeds simultaneously to the oligomerization (Figure 10) [56, 85]. The rate constants of oligomerization are greater than that of hydrolysis in the presence of template. The association between two activated monomers and that between activated monomer and the elongating oligoG is important for the oligonucleotide formation. A similar trend was observed for Clay-catalyzed reaction [69, 70]. This was evaluated in detail and kinetic analysis in details showed a trend that two or three activated nucleotide monomers align on the polyC template prior to the phosphodiester bond formation [85]. For instance, the association between two ImpG molecules would be stronger than that between two ImpC molecules, which is determined by the strength of stacking of nucleotide bases. This is compatible to the fact that the efficiency of Template-directed reaction with ImpA, ImpU, and/or ImpC is much lower than Template-directed reaction with ImpG.

On the other hand, kinetic analysis of the reaction mechanism of Clay-catalyzed reaction showed that the rate constants for the formation of 2-mer, 3-mer, and 4-mer and higher oligonucleotide increase in the order of 2-mer < 3-mer < 4-mer and higher; this trend is consistent with Template-directed reaction [69, 70]. Thus, the hydrolysis of the activated nucleotide monomer is somewhat competitive reaction to the oligomerization in the presence of clay. The kinetics suggested that the association between two activation monomers and that

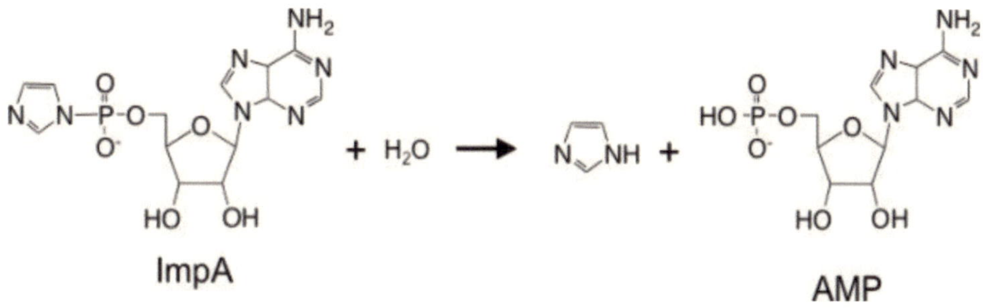

Figure 10. Hydrolytic degradation of the activated nucleotide

between activated monomer and elongating oligomer are important to determine the rate constants of 2-mer, 3-mer, and 4-mer formations, which is similar to Template-directed reaction. The association is mainly due to the stacking through nucleotide bases, which increases with length of oligonucleotide up to 4-mer and remains constant. Kinetic analysis of Metal-catalyzed reaction also implies a similar mechanism to that for Template-directed and Clay-catalyzed reactions (Figure 11). Conclusively, the kinetic investigations summarize that the association of an activated nucleotide monomer with another one activated nucleotide monomer or an elongating oligomer is important to determine the efficiency of oligomerization in the presence of any additives, such as a polynucleotide template, clay surface, and metal ions.

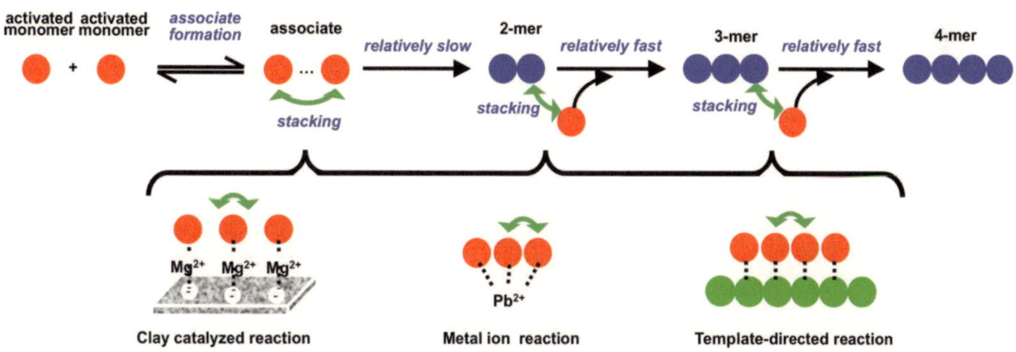

Figure 11. Importance of associate formation for oligonucleotide elongation

4.7. Variations of the activated nucleotide monomers

The activated nucleotide monomers with N-P bond were normally tested using ImpN or 2MeImpN, where imidazole and 2-methylimidazole moieties are the leaving group from N-P bound with phosphate group. Although the methyl group does not seem to be so effective, this causes fairly large influence for the formation of long oligonucleotides [86]. For instance, the efficiency of Template-directed reaction using 2MImpG in the absence of metal catalyst is higher than that using ImpG. This is probably due to the enhancement of association between monomers or that between monomer and elongating oligomer by stacking of 2-methyl-imidazole. The solubility of 2MeImpN is lower than that of ImpN in aqueous solutions. Both imidazole and 2-methy-limidazole are potentially prebiotic compounds. Analogues of ImpN with different leaving group were examined. It was found that the efficiency of the oligonucleotide formation using different leaving group with N-P bond is correlated to the acidity of the leading group. Oligonucleotide formation indeed proceeds efficiently using such ImpN analogues for Clay-catalyzed reactions, where higher oligonucleotides were observed in some cases (Figure 12) [87, 88]. These studies suggest that a variation of pathways should have been possible for the formation of RNA under the primitive earth conditions.

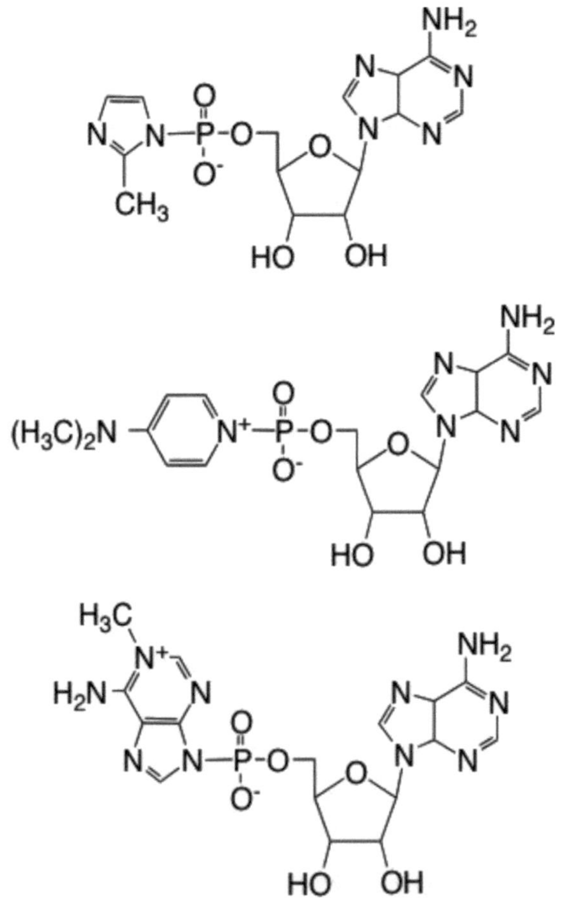

Figure 12. Variations of the activated nucleotide

On the other hand, it is claimed that RNA would not be suitable as an initial material to preserve genetic information since ribose as an RNA moiety is considered normally difficult to form under the primitive earth conditions. Ribose does not form efficiently under prebiotic conditions although many types of sugars are readily formed by formose reaction [89, 90]; it is

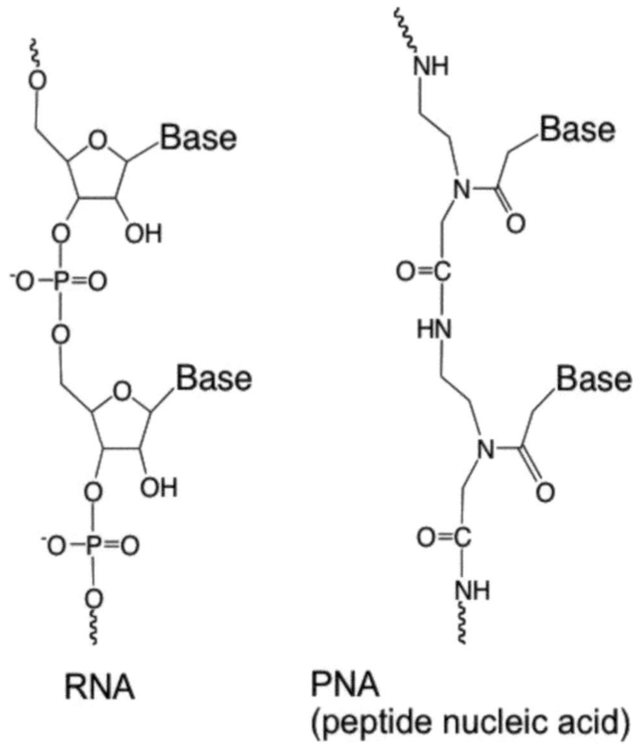

Figure 13. RBA and peptide nucleic acid

considered as a major pathway for the formation of sugars. In addition, RNA is considered unstable as comparing to hexose. Naturally, the nucleotide analogues of RNA instead of ribose could be formed under prebiotic conditions. Furthermore, the backbone of phosphodiester bond could be replaced with peptide bonding backbone [91-93] (Figure 13). These possibilities were also experimentally verified. Conclusively, the reactions of the activated nucleotide analogues using hexose backbone also proceed to form oligonucleotide analogues using Template-directed reaction [94, 95]. These examples suggest that template direction could have played important roles in a variety of analogical materials under the prebiotic conditions.

Finally, as mentioned above, it is known that ImpN monomers consisting of deoxyribose do not form oligonucleotides since the reactivity of ImpN with deoxyribose is very low. This is due to the less reactivity of 2′-H group of ribose. The formation of DNA from such activated nucleotide monomers is not yet deeply investigated since DNA molecules are not considered as initial genetic material on the primitive earth.

4.8. Chiral selection of oligonucleotide using the prebiotic RNA formation models

The RNA polymerization model reactions were extensively studied. Normally, these studies were normally carried out using homochiral materials because of the difficulty of preparation of starting materials. The selection of a single chirality, which is L-ribose for the present nucleic acid, is a great issue in the field of the origin of life study. Although this question is not yet solved, some attempts were carried out for the evaluation of the efficiency of Template-directed, Clay-catalyzed, and Metal-catalyzed reactions by using heterochiral materials [96-99]. The preparation of the activated nucleotide monomers with L-ribose was succeeded although it involves very complicated organic synthetic procedures. These reactions showed that the homochiral oligomerization is preferable instead of the heterochiral oligomerization for both the Clay-catalyzed and Template-directed reactions. The results would imply the reason how homochiral biochemistry was selected during the chemical evolution of nucleotide and abiotic-peptide (Figure 14).

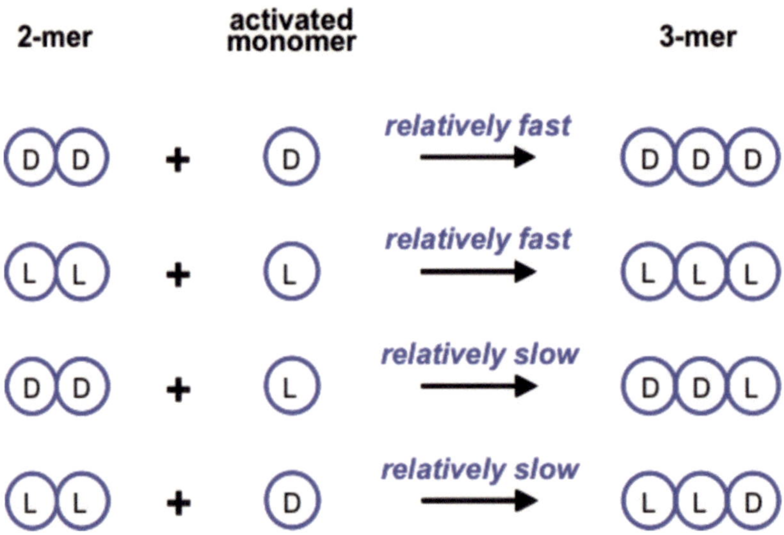

Figure 14. Homochiral selectivity for prebiotic RNA formation

4.9. Temperature dependence of the pebiotic oligomerization at medium to high temperature

According to the evidences of geoscience, the temperature of the primitive earth would be much higher than that of present. In addition, submarine hydrothermal vent system would have been present and played important roles for the emergence of life and the last universal common ancestor, which is deduced from phylogenetic tree analysis, could be close to hyperthermophiles [100, 101]. Naturally, the heterogeneity of the primitive earth environment should be considered so the previous studies on the formation of RNA are regarded to bias very mild conditions. As mentioned above, the oligomeriation of the activated nucleotide monomers does not proceed in acidic and alkaline solutions because of the hydrolysis of the activated nucleotide. Besides, the temperature dependence of these reactions was extensively investigated in relation to the hydrothermal origin of life hypothesis.

http://dx.doi.org/10.5772/58222

In general, the efficiency of oligonucleotide formation by Metal-catalyzed, Clay-catalyzed, and Template-directed reactions decreases with increasing temperature. Kinetic analyses clarified successfully the reason of the low efficiency at high temperatures [102-104]. The investigations provided the rate constants regarding these reaction systems, that is, the formation of oligomers 2-mer (k_2), 3-mer (k_3), 4-mer (k_4), the hydrolysis of the activated nucleotide (k_{hy}), the formation of pyrophosphate-linkage (k_{py}), and the hydrolysis of oligonucleotide formed by these prebiotic oligomerizations. Comparison of the temperature dependence of k_2, k_3, k_4 shows that k_2 has a smaller activation energy than that for k_3 and k_4. In addition, the relative magnitude of k_2 becomes small as comparing to k_{hy} with increasing temperature although that of k_3 and k_4 does not; the activation energy for k_2 is smaller thant that for k_{hy}, but that for k_3 and k_4 is comparable to that for k_{hy}. The main reason of low efficiency of oligonucleotide formation at high temperature is that the formation of 2-mer becomes relatively slow as comparing to the hydrolysis of the activated nucleotide at high temperature.

The similar trend was observed for Metal-catalyzed and Clay-catalyzed reactions on the basis of kinetic analysis [103, 104]. In addition, the association does not only affect the efficiency of these reactions but also the regioselectivity, and effective to the formation of cyclic-isomers. The fact that the formation of DNA double-helix is primarily stabilized by the base-stacking rather than hydrogen bonding is minor factor [105] is consistent with the trend in the primitive RNA formation reactions. These analyses suggest that the accumulation of the activated nucleotide monomer would be difficult at high temperatures since the hydrolysis of the activation nucleotide becomes fast unless the activated nucleotide monomers are continuously supplied by a plausible pathway. Nevertheless, such conformation of DNA and RNA must be held at high temperatures since a thermophilic organism could survive at least 100-120 °C [106, 107].

5. Successful examples of the formation of abiotic-peptides

5.1. Introduction

Proteins are regarded the practical entity for maintaining life-like system while DNA in the present organisms or the primitive RNA in RNA worlds is regarded as blueprint of life-like systems. The question, which was the first material between nucleic acids and proteins for the emergence of life-like system on the primitive earth, has been controversy for many years although this might be solved by the proposal of the RNA world hypothesis. However, there are a number of drawbacks regarding the RNA world hypothesis so the question is continuously discussed. As mentioned above, the term "protein" in the present chapter involves the meaning that protein must be a functional material, which is synthesized on the basis of genetic information. If a protein-like molecule is merely consisting of peptide, but the sequence is not encoded in a genetic coding system, the material is called as just abiotic-peptide. As mentioned, such peptides are called as abiotic-peptides to distinguish from the peptides possessing biological functions *in vivo*. Naturally, for instance, this is not inconsistent with a hypothesis that both the RNA and protein-like molecules formed independently and became interrelated only after the accumulations of these molecules.

In the first place, abiotic-peptide formation from amino acids is also the dehydration so this is not preferable in aqueous medium unless using the activation of amino acid or a condensation agent. Peptides are normally regarded more stable against the hydrolysis as comparing to nucleic acids although peptides in aqueous medium are normally exposed under the pressure to hydrolytic degradation. In the present organisms, peptides form on ribosomes from aminoacyl-tRNA, which is regarded as activation of amino acid activation. Furthermore, condensation agent and activation techniques are used for modern organic synthesis of peptides. Thus, it should be noted that abiotic-peptides do not form spontaneously from the viewpoint of thermodynamics even if amino acids are mixed in aqueous medium. This is the same situation as the oligomerization of DNA and RNA.

5.2. Thermal condensation of amino acid mixture

Spontaneous condensation of amino acid mixtures does not or less proceeds in aqueous solutions. This is due to that dehydration is disadvantageous in aqueous medium. It was found that thermal condensation products form from amino acid mixtures, such as glutamic acid and aspartic acid, under dry conditions at temperatures 180-200 °C [108-110] (Figure. 15). Microspheres can be formed by treatment of the condensation products in boiling water; the microspheres were named as proteinoid. It is surprising that these condensation products readily form with normally molecular weight over 10000 Da involving peptide bonding, but the peptide bonds are not controlled; probably branched sequences and non-peptide bonding would be involved. Lacton formed by heating glutamic acid etc. is melted at high temperatures, where other amino acids are dissolved and then result in polymerization. The biochemical functions of proteinoids have been extensively investigated [111-113].

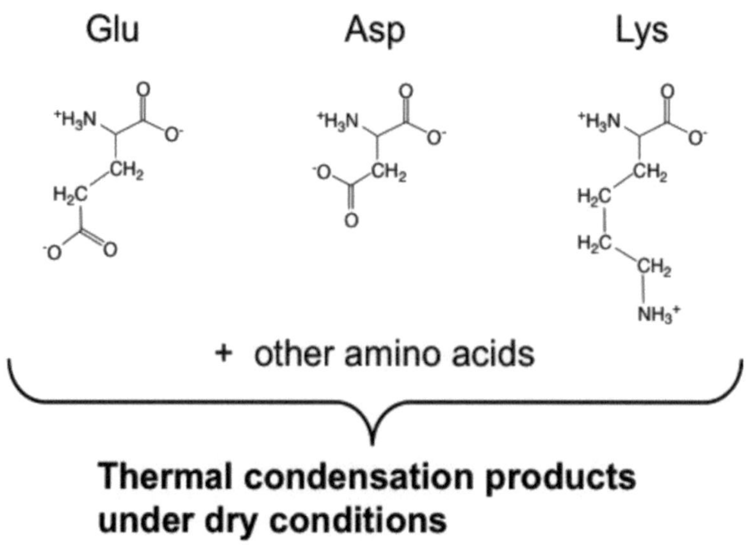

Figure 15. Thermal condensation of amino acids

Although the proteinoids involve partially peptide bonding, proteinoids might not reflect a plausible ancient accumulation pathway of protein-like molecules since large continents were not present on the primitive earth. The question how the proteinoids containing peptide bonding incompletely had evolved to the modern proteins is a great drawback. In addition, as mentioned above these condensation products are not categorized as proteins since these are not synthesized by the direction of genetic information. Normally, it is realized that protein does not replicate alone. Furthermore, these condensation products do not seem to be useful in an organic synthetic method. Thus, the research regarding the thermal condensation of amino acids does not progress so far.

5.3. Formation of protein-like and abiotic-peptide molecules in aqueous medium

The condensation from amino acids to form peptide bonding is disadvantageous in aqueous medium [114]. Activation of amino acids or usage of a condensation agent is necessary for peptide bond formation [115-120]. Thus, the pathways for oligopeptides have been investigated as simulation experiments under the primitive earth environments. Several types of minerals and metal ions were evaluated to see whether these materials enhance the formation of long oligopeptides. Nevertheless, the plausible oligomerization has met with limited success [119].

5.4. Hydrothermal synthesis of oligpeptides

Presumably, the temperature of the primitive earth surface should have been much higher than present. Furthermore, the phylogenetic analysis of present organisms suggested that last universal common ancestor (LUCA), which is located to the branches between Bacteria and Archaea, would have been a hyperthermophilic organism [100, 101] although this assumption is still strongly disputed [121, 122]. The importance of such hydrothermal systems has been extensively investigated [123]. The hypothesis is named as "hydrothermal origin of life hypothesis". Although LUCA is not the origin of life, it would reflect old organisms and the environments, where life had emerged. Recent discoveries of ecosystem present near submarine hydrothermal vent systems in deep-ocean support the hydrothermal origin of life hypothesis [124]. Thus, simulation experiments have been carried out for the formation of oligopeptides under the hydrothermal conditions using batch reactors at the beginning of such simulation experiments. Condensation products, which involve peptide bonding and bonding between silicate and organic materials, were observed in the products [125].

Recently, flow reactors were developed and behaviors of biomolecules in hydrothermal systems have been extensively investigated to simulate the submarine vent system in deep-ocean [126]. The simulation experiment showed that oligopeptides are formed from glycine monomers at temperature over 250 °C, where oligoglycine up to 6 amino acid units was detected. The efficiency of the direct oligopeptide formation from amino acid is normally low; the maximum yield is normally 0.1 – 1 % at most [126, 127]. The reason is due to primarily the fact that dehydration in aqueous solution is not advantageous so that the oligopeptides once produced are always exposed under the degradation pressure. In addition, the formation of diketopierazine (DKP) inhibits the further elongation of oligopeptides (Figure 16); DKP is cyclic dipeptide and considered as very stable. Rapid accumulation of DKP from glycine is also observed in the simulation reaction of submarine vent system.

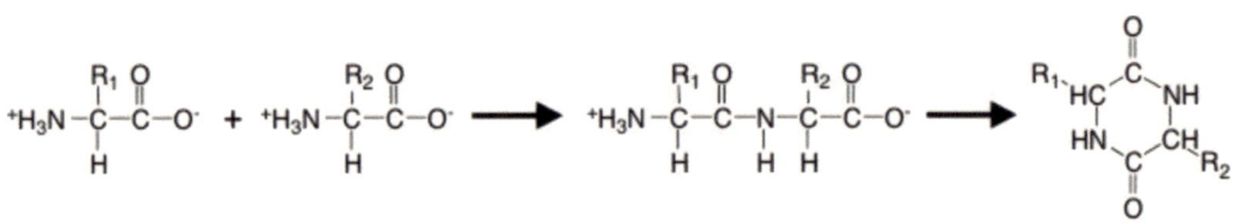

Figure 16. DKP formation from amino acids

On the other hand, in our group, a hydrothermal micro flow reactor using fused-silica capillary has been developed (Figure 17) [128-130]. This enables to monitor hydrothermal reactions at temperature up to 400 °C at time scale between 2 ms – 200 s. By using the flow reactor, a more efficient elongation of 4-mer and 5-mer of alanine to 5-mer and 6-mer was discovered at 250 – 310 °C within 5 – 20 s [131]. Furthermore, it was found that oligopeptides including 20 amino acid units form within 180 s at temperatures 270 – 310 °C [132]. These reactions were designed to avoid the formation of DKP during the direct oligomerization of monomeric amino acids; the formation of DKP is regarded as a problem for the organic synthesis of peptide. DKP readily forms from dipeptide at high temperatures, for instance DKP forms from alanine dipeptide within 10 s at 275 °C (Figure 18). On the contrary, DKP formation is relatively slow if the elongation starts from 4-mer and 5-mer of alanine. In these reactions, 4-mer and 5-mer are converted to DKP finally, but the DKP formation is relatively slow. Thus the elongation of these peptides is observed during the conversion of the 4-mer and 5-me to DKP. The reaction scheme is shown in Figure 18. The elongation yield reaches to 10 %, which is ca. 100-fold greater than that of direct formation from monomeric amino acids. This is surprising since neither condensation reagent nor catalyst is necessary. The kinetic analysis implies that peptide bond within 4-mer would compensate the elongation of peptide bonding to form 5-mer at very high temperatures.

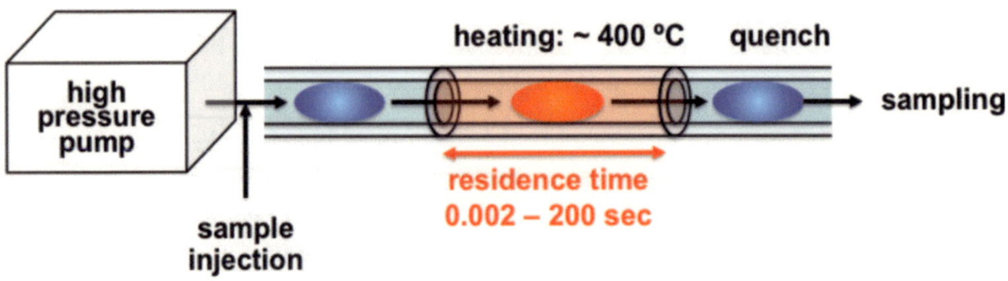

Figure 17. Principle of hydrothermal flow reactor

To investigate the role of minerals under hydrothermal conditions, a new type of flow reactor technique was established in our group [133]. The system consisting a mineral-packed column

http://dx.doi.org/10.5772/58222

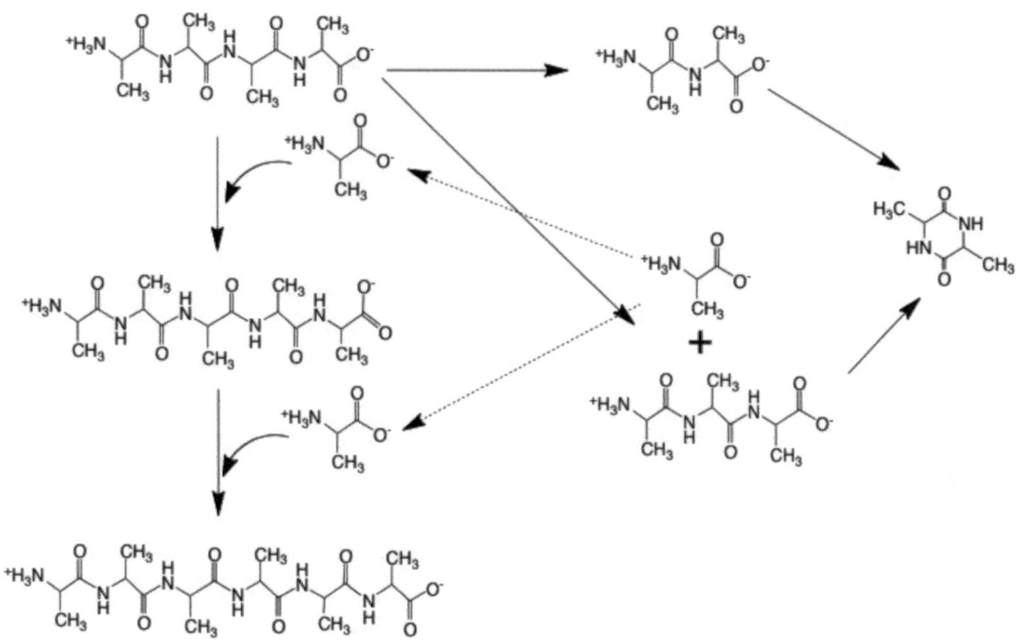

Figure 18. Pathways of elongation and degradation of oligopeptides

reactor is settled in a high temperature heating system. This system enables to monitor hydrothermal reactions at temperatures up to 300 °C at the time scale 2 – 200 s in the presence of mineral particles. By using the method, it was discovered that the oligopeptide elongation reaction from alanine 4-mer to 5-mer within 5 – 20 s is notably enhanced by naturally-occurring carbonate minerals, such as calcite and dolomite.

6. Unsolved drawbacks regarding the origin of building blocks for a primitive life-like system

6.1. General difficulties of oligomerization of nucleic acids and abiotic-peptides

First, the dehydration in aqueous solution is disadvantageous from the viewpoint of thermodynamics. As we evaluated the difficulties of oligonucleotides and oligopeptides in a former section, activation method is a key technique to form long oligomers. For the case of nucleotide, the activated nucleotide monomer, that is, imidazolide of nucleoside 5′-monophosphate is useful material. However, the prebiotic activation method, which is regarded as compatible to the primitive earth environments, is not identified for the peptide formation. Aminoacyl-tRNA is universally used in the present organisms and some minerals would have played roles for activation or condensation agents for the oligopeptide formation. However, these possibilities are not yet extensively evaluated.

Second, it is generally true that the formation of short cyclic-oligonucleotides [70, 72, 73, 134] and DKP [131] inhibits the further elongation of oligomers (Figure 19). The cyclization of oligonucleotides is completely controlled by RNA polymerase and peptide synthetic method on ribosomes *in vivo*. Nucleic acid templates, mineral catalysts, and metal ion catalysts tend to inhibit the cyclization reaction. The additives probably enhance suitable steric alignment of monomers to inhibit cyclization although the general rules for inhibition of cyclization are not elucidated. Actually as a primitive oligomerization, there is less cyclic-dinucleotide for the formation of oligoG in Template-directed reaction. This is probably due to Watson-Crick type hydrogen bonding and base stacking. On the contrary, cyclic-2-mer, 3-mer, and 4-mer somewhat readily form in Clay-catalyzed reactions. On the other hand, the elongation from 4mer or higher is effective for inhibition of DKP formation. In the present organisms, the alignment of amino acid monomers is supported by the machinery of ribosome as well as mRNA and tRNA. However, if the hydrothermal hypothesis is correct, then the oligopeptide formation would have occurred even under such extreme conditions involving the submarine hydrothermal vent system. In such a system, it is assumed that the alignment of amino acids would be so difficult since weak interactions among biomolecules would not act at high temperatures.

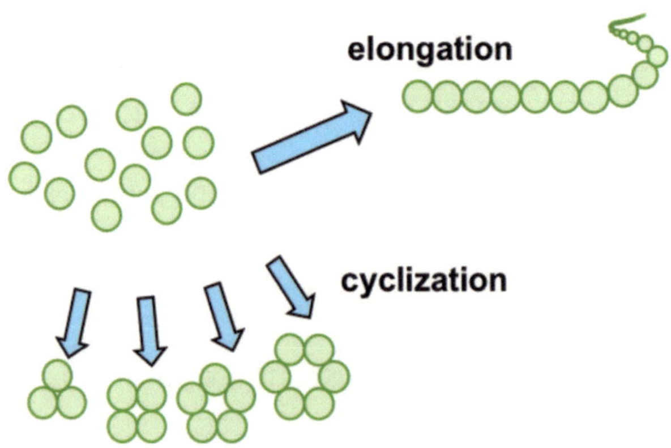

Figure 19. Cyclization and elongation of prebiotic oligomers

Third, the estimation of the primitive earth environments is not yet succeeded. Furthermore, the heterogeneity of earth conditions should be considered. Improvement of knowledge about the primitive environment is essential and a variety of conditions of simulation experiments would be helpful. Most of simulation reactions for the oligonucleotide formation have been carried out under the fairly mild temperatures and pressures. Beside, abiotic-peptide formations were carried out under a variety conditions from mild conditions to hydrothermal conditions. Wider range of simulation experiments should be carried out unless the exact earth environments are readily identified. General rules would be found on the basis of such a variety of simulation experiments in future.

6.2. Formation of simple oligomers to functional entities

The genetic information flow in present organisms consists of DNA replication, transformation of DNA sequence to mRNA, translation of mRNA sequence to protein. According to the RNA world hypothesis, the roles of DNA and protein should have been covered by RNA molecules so the replication of RNA molecules is the first step in an RNA based life-like systems [135, 136]. Here, the possibility of the prebiotic RNA replication is briefly discussed.

In a former section, the difficulty if Template-directed reaction for the cases of A, U, C as nucleotide bases on the complementary template was described. This would be due to the weak stacking between the activated monomers and elongating oligomer. Thus, the enhancement of the association has been attempted. It has been shown that intercalator somewhat enhances the efficiency of oligonucleotide by enhancement of the association [137]. However, the practical replication conditions using primitive activate nucleotide monomers are not identified [82, 135].

The biological functions of oligonucleotide and oliopeptide would be displayed with a certain length of oligomers, such as 30-mer, 50-mer, and 100-mer. Transfer-RNA (tRNA) would be one of smallest functional RNA consisting of ca. 80-mer. On the other hand, recently it has been extensively found that small non-coding RNA and so-called microRNA possess important roles *in vivo*. In addition, it is also known that short peptides possess several important roles while the average length of proteins is ca. 100-mer. Oligonucleotides with 40 – 50-mer can form from Template-directed and Clay-catalyzed reactions. This fact suggests that different types of functional RNA molecules should have been included in a randomly formed large RNA pool on the primitive earth. Several functional RNA sequences would have amplified by replication machinery, that is ribosome, of which some main parts are made of RNA molecules. As mentioned, Template-directed reaction is only efficient for the case of the formation of oligoG on a polyC template. However, the Template-directed reaction could proceed using the activated oligonucleotide as a monomer for the case of oligoU formation of a polyA template. If this type of Template-directed reaction is universal for the different types of combination of monomer and template, a replication of RNA molecules could have evolved under the primitive earth conditions. Nevertheless, it was confirmed that a heterogeneous oligoC template spontaneously formed by Clay-catalyzed reaction containing 2′, 5′-and 3′-5′-linked isomers could have preserved as template for the Template-directed reaction of oligoG formation [138]. On the other hand, the oligopeptide formation from amino acid mixtures is not yet elucidated although some thermal condensation products of amino acids, such as proteinoids, were observed. Actually, it is considered that the formation of amino acids seems to be easier than that of nucleotide monomers since amino acids are readily found in gas reaction products with different types of energy sources. However, it could not straightforwardly mean that the oligopeptide formation using prebiotic materials is easier than the oligonucleotide formation.

In vitro selection technique has been extensively investigated to create several types of artificial functional RNA molecules [139, 140]. This technique elucidated that a RNA replication system catalyzed by RNA molecule could have evolved if amplification and selection processes were present. Naturally, the molecular biological tools and materials are necessary for the *in vitro* selection technique. However, if the amplification on the basis of Template-directed reaction under the primitive earth conditions, amplified RNA molecules are exposed under different types of selection pressure. This would cause the accumulation of functional RNA molecules, such as RNA replicase function, under the primitive earth conditions.

6.3. Relationship between RNA and protein-like molecules

Cooperation of RNA and proteins is definitely important in modern organisms. Thus, it should be evaluated that the formation of RNA and that of peptides could be enhanced by each other. According to previous studies, Template-directed reaction is not enhanced in the presence of proteinoids [141]. On the other hand, possible roles of nucleotides and oligonucleotides for the formation of oligopeptides are not yet well evaluated. Presumably, nucleotides and oligonucleotides would not be helpful for the abiotic-peptide synthesis under hydrothermal conditions since nucleotides and oligonucleotides are much more unstable than amino acids and peptides [129, 135, 136, 142]. This fact implies that chemical evolution of RNA would have started after the accumulation of oligopeptides.

6.4. Accumulation of RNA molecules and abiotic-peptide under the primitive earth conditions

According to the former sections, RNA could have formed from the activated nucleotide in the absence of enzyme. Abiotic-peptides also form without using the *in vitro* translation system on ribosome. However, the accumulation of oligonucleotides and oligopeptides would not have been easy under the primitive earth conditions since these are spontaneously hydrolyzed in aqueous medium although the rates of hydrolysis of these oligomers are normally slower than those of formation at low temperatures. The hydrolysis rate increases with increasing temperature.

The accumulation of oligomers should be considered from the viewpoint that the life-like system is considered as thermodynamically open system. That is, the accumulation of oligomers is determined by both the rates of formation and degradation of oligomers. Thus, although the rate of degradation of oligomers at high temperatures is large, the accumulation would be possible if the formation rate is much lager than that of degradation. The formation rate would be possibly enhanced by several prebiotic catalysts. Experimental simulations of these conditions are not yet well succeeded while most of simulation experiments were carried out under static conditions. Some experiments, of which activated monomers were fed to batch reaction, successfully showed the possible formation of long oligonucleotides and oligopeptides [75]. Flow reactor would be useful to design where the activated monomers are fed to a chemical evolutionary system.

http://dx.doi.org/10.5772/58222

7. Conclusive remarks: Possible future applications by learning the primitive oligomerization processes

These studies support the fact that oligomer formation in aqueous solution in the absence of enzyme is possible. The reaction processes are different from normal organic synthetic techniques. Therefore, the improvement of the formation reactions of oligomers, which possess biological functions, should be achieved by learning chemistry of living organisms. Recently, organic chemistry is expected to be a technology compatible with the global environmental protection. Obviously, organic synthesis requires a large amount of organic solvent, which is not suitable to the environmental protection. On the contrary, nowadays, creation of a environmental harmless system for organic synthetic methods is expected by learning the living organisms. However, the organic synthesis is generally difficult in aqueous medium. According to the examples shown here, oligonucleotides and oligopeptides could be synthesized in aqueous medium. The fundamental researches will facilitate such demand of environment harmless organic synthesis.

Acknowledgements

This research was supported by Grant-in-Aid for Exploratory Research in Japan, 24657166 from JSPS of Japan.

Author details

Kunio Kawamura*

Address all correspondence to: kawamura@shudo-u.ac.jp

Department of Human Environmental Studies, Hiroshima Shudo University, Hiroshima, Japan

References

[1] Arrhenius S. Die Verbreitung des Lebens im Weltenraum. Die Umschau 1903; 7: 481–485.

[2] Crick FHC., Orgel LE. Directed Panspermia. Icarus 1973; 19(3): 341-346.

[3] Horneck G., Rettberg P., Reitz G., Wehner J., Eschweiler U., Strauch K., Panitz C., Starke V., Baumstark-Khan C. Protection of Bacterial Spores in Space, A Contribution to the Discussion on Panspermia. Origins Life Evol Biospheres 2001; 31: 527-547.

[4] Miller SL. A Production of Amino Acids under Possible Primitive Earth Conditions. Science 1953; 528-529.

[5] Ferris JP., Joshi PC., Edelson EH., Lawless JG. HCN-Plausible Source Of Purines, Pyrimidines And Amino-Acids on Primitive Earth. J Mol Evol 1978; 11: 293-311.

[6] Schlesinger G., Miller, SL. Prebiotic Synthesis in Atmospheres Containing CH_4, CO, and CO_2. J Mol Evol 1983; 19: 376-382.

[7] Cleaves HJ. Chalmers JH, Lazcano A. Miller SL, Bada JL. A Reassessment of Prebiotic Organic Synthesis in Neutral Planetary Atmospheres. Origins Life Evol Biospheres 2008; 38(2): 105-115.

[8] Mojzsis SJ., Arrhenius G. McKeegan KD., Harrison TM., Nutman AP., Friend CRL. chopfbefore 3, 800 Million Years Ago. Nature 1996; 384(7): 55-59.

[9] Van Kranendonk, MJ. Volcanic Degassing, Hydrothermal Circulation and the Flourishing of Early Life on Earth: A Review of the Evidence from c. 3490-3240 Ma Rocks of the Pilbara Supergroup, Pilbara Craton, Western Australia. Earth Sci Rev 2006; 74(3-4): 197-240.

[10] Kuppers, BO. Molecular Theory of Evolution. Outline of a Physico-chemical Theory of the Origin of Life. Berlin: Springer-Verlag; 1985.

[11] Schopf, JW. Microfossils of the Early Archean Apex Chert: New Evidence of the Antiquity of Life. Science 1993; 260: 640-646.

[12] Crick F. Central Dogma of Moecular Biology. Nature 1970; 227: 561-563.

[13] Honda S., Yamasaki K., Sawada Y., Morii H. 10 Residue Folded Peptide Designed by Segment Statistics. Structure 2004; 12: 1507-1518.

[14] Cech TR., Zaung AJ., Grabowski PJ. In Vitro Splicing of the Ribosomal RNA Precousor of Tetrahymena: Involvment of a Guanosine Nucleotide in the Excision of the Intervening Sequence. Cell 1981; 27: 487-496.

[15] Gilbert W. Origin of life: the RNA world. Nature 1986; 319: 618–618.

[16] Cech TR. A model for the RNA-catalyzed replication of RNA. Proc Natl Acad Sci USA 1986; 83: 4360–4363.

[17] Lazcano A. The Biochemical Roots of the RNA World: from Zymonucleic Acid to Ribozymes. Hist. Phil. Life Sci. 2012; 34(3): 407-423.

[18] Nemoto N., Husimi Y. A Model of the Virus-type Strategy in the Early Stage of Encoded Molecular Evolution. J. Theor Biol 1995; 176: 67-77.

[19] Kawamura K. Civilization as a Biosystem Examined by the Comparative Analysis of Biosystems. BioSystems 2007; 90: 139-150.

[20] Gomes R., Levison HF., Tsiganis K., Morbidelli A. Origin of the cataclysmic Late Heavy Bombardment period of the terrestrial planets. Nature 2005; 435(7041): 466-469.

[21] Maher KA., Stevenson, DJ. Impact Frustration of the Origin of Life. Nature 1988; 331(6157): 612-614.

[22] Sleep NH., Zahnle KJ., Kasting IF., Morowitz HJ., Annihilation of Ecosystems by Large Asteroid Impacts on the Early Earth. Nature 1989; 342(6246): 139-142.

[23] Sagan C., Mullen G. Earth and Mars: Evolution of Atmospheres and Surface Temperatures. 1972; 177(4043): 52-56.

[24] Newman MJ., Rood RT. Implications of Solar Evolution for the Earth's Early Atmosphere. Science 1977; 198(4321): 1035-1037.

[25] Walker JCG. Carbon-dioxide on the Early Earth. Origins Life Evol Biosphere 1985; 16(2): 117-127.

[26] Kasting JF., Ackerman TP. Climatic Consequences of Very High-carbon Dioxide Levels in the Earths Early Atmosphere. Science 1986; 234(4782): 1383-1385.

[27] Kasting JF. Earth's early atmosphere. Science 1993; 259(5097): 920-926.

[28] Mojzsis SJ., Harrison TM., Pidgeon, RT. Oxygen-isotope Evidence from Ancient Zircons for Liquid Water at Earth's Surface 4, 300 Myr Ago. Nature 2001; 409: 178-181.

[29] Olivas WM., Muhlrad D., Parker R. Analysis of the Yeast genome: Identification of New Non-coding and Small ORF-containing RNAs. Nucl Acid Res 1997; 25(22): 4619-4625.

[30] Eigen M. Selforganization of Matter and the Evolution of Biological Nacromolecules. Naturwissenschaften 1971; 58: 465-523.

[31] Orgel LE., Crick FHC. Anticipating an RNA World Some Past Speculations on the Origin of Life: Where are They Today? FASEB J 1993;7: 238-239.

[32] Chamberlin N., berg P. Deoxyrinonucleic Acid-directed Synthesis of Ribonucleic Acid by an Enzyme from Escherichia Coli. Proc Natl Acad Sci USA 1962; 48(2): 81-94.

[33] Haruna I., Spiegelman S. An RNA "Replicase" Induced by and Selective for a Viral RNA: Isolation and Properties. Proc Natl Acad Sci USA 1963; 50(2): 905-911.

[34] Grunberg-Manago M. Enzymatic Synthesis of Nucleic Acids. Ann Rev Biochem 1962; 31: 301-332.

[35] Kawamura K., Kuranoue K., Umehara M. Chemical Evolution of RNA and RNA Polymerases: Implications from Search for the Prebiotic Pathway of formation of RNA

from Adenosine 5′-Triphosphate in the Presence of Thermal Condensation Products of Amino Acids as Primitive Enzymes. Viva Origino 2002; 30(3): 123-134.

[36] Sulston J., Lohrmann R., Orgel LE., Todd Miles H. Nonenzymatic Synthesis of Oligoadenylates on a Polyuridylic Acid Template. Proc Natl Acad Sci USA 1968; 59: 726-733.

[37] Sulston J., Lohrmann R., Orgel LE., Todd Miles H. Specificity of Oligonucleotide Synthesis Directed by Polynucleic Acid. Proc Natl Acad Sci USA 1968; 60: 409-415.

[38] Lohrmann R., Orgel LE. Prebiotic Synthesis: Phosphorylation in Aqueous Solution. Science 1968; 161(3836): 64-66.

[39] Sulston J., Lohrmann R., Orgel LE., Schneider-Bernloehr H., Todd Miles H. Non-enzymatic Synthesis of Deoxyadenylate Oligonucleotides on a Polyuridylate Template. J Mol Biol 1969; 37: 151-155.

[40] Sulston J., Lohrmann R., Orgel LE., Schneider-Bernloehr H., Todd Miles H. Non-enzymatic Oligonucleotide Synthesis on a Polycytidylate Template. J Mol Biol 1969; 40: 227-234.

[41] Renz M., Lorhmann R., Orge LE. Catalyst for the Polymerization of Adenosine Cyclic 2′, 3′-Phosphate on a Poly(U) Template. Biochim Biophys Acta 1971; 240: 463-471.

[42] Verlander MS., Lorhmann R., Orge LE. Catalysts for the Self-Polymerization of Adenosine Cyclic 2′, 3′-Phosphate. J Mol Evol 1973; 2: 303-316.

[43] Powner MW., Gerland B., Sutherland JD. Synthesis of Activated Pyrimidine Ribonucleotides in Prebiotically Plausible Conditions. Nature 2009; 459: 239-242.

[44] Costanzo G., Pino S., Ciciriello F., Di Mauro E. Generation of Long RNA Chains in Water. J Biol Chem 2012; 284(48): 33206-33216.

[45] Ibanez JD., Kimball AP., Oró J. Possible Prebiotic Condensation of Mononucleotide by Cyanamide. Science 1971; 173: 444-445.

[46] Schneider-Bernloehr H., Lorhmann R., Sulston JE., Orge LE., Todd Miles H. Specific of Template-directed Synthesis with Adenine Nucleotides. J Mol Biol 1970; 47: 257-260.

[47] Weimann BJ., Lorhmann R., Orge LE., Schneider-Bernloehr H., Sulston JE. Template-Directed Synthesis with Adenosine-5'-phosphorimidazolide. Science 1968; 161(3839): 387-387.

[48] Orgel LE., Lohrmann R. Prebiotic Chemistry and Nucleic Aicd Replication. Acc Chem Res 1974; 7: 368-377.

[49] Lohrmann R., Orgel LE. Studies of Oligoadenylate Formation on a Poly(U) template. J Mol Evol 1979; 12: 237-257.

[50] Sawai H., Orgel LE. Oligonucleotide Synthesis Catalyzed by the Zn^{2+}ion. J Am Chem Soc 1975; 97(12): 3532-3533.

[51] Lohrmann R., Orgel LE. Efficient catalysis of polycytidylic acid-directed oligoguanylate formation by Pb^{2+}. J Mol Biol 1980; 142(4, 5): 555-567.

[52] Bridson P.K., Orgel LE. Catalysis of accurate poly(C)-directed synthesis of 3′-5′-linked oligoguanylates by Zn^{2+}. J Mol Biol 1980; 144(4): 567-577.

[53] Inoue T., Orgel LE. Oligomerization of (Guanosine 5′-Phosphor)-2-methylimidazolide on Poly(C): An RNA Polymerase Model. J Mol Biol 1982(1); 201-207.

[54] Lohrmann R. Formation of Ncleoside 5′-Phosphamidates under Potentially Prebiological Conditions. J Mol Evol 1977; 10: 137-154.

[55] Lohrmann R., Orgel LE. Prebiotic Activation Processes. Nature 1973; 244: 418-420.

[56] Kanavarioti A., Bernasconi CF., Doodokyan DL., Alberas DJ. Magnesium-ion Catalyzed P-N Bond Hydrolysis in Imidazolide-Activated Nucleotides-Relevance to Template-directed Synthesis of Polynucleotides. J Am Chem Soc 1989; 111(18): 7247-7257.

[57] Joyce GF., Inoue T., Orgel LE. Non-enzymatic template-directed synthesis on RNA random copolymers: Poly(C, U) templates. J Mol Biol 1984; 176(2): 279-306.

[58] Sawai H. Oligonucleotide Formation Catalyzed by Divalent Metal Ions. The Uniqueness of the Ribosyl System. J Mol Evol 1988; 27(3): 181-186.

[59] Sawai H., Shibata T., Ohno M. Preparation of Oligoadenylates with 2'-5' Linkage Using Pb^{2+}Ion Catalyst. Tetrahedron 1981; 37(3): 481-485.

[60] Sawai H., Kuroda K., Hojo T. Uranyl Ion as a Highly Effective Catalyst for Internucleotide Bond Formation. Bull Chem Soc Jpn 1989; 62(2): 2018-2023.

[61] Sawai H., Higa K., Kuroda K. Synthesis of Cyclic and Acyclic Oligocytidylates by Uranyl Ion Catalyst in Aqueous Solution. J Chem Soc Perkin Trans 1 1992; (4): 505-508.

[62] Kawamura K., Maeda J. Kinetic Analysis of Oligo(C) Formation from the 5′-Monophosphorimidazolide of Cytidine with Pb(II) Ion Catalyst at 10-75 °C. Origins Life Evol Biospheres 2007; 37: 153-165.

[63] Bernal JD. The Physical Basis of Life. Proc R Soc Lond A 1949; 62: 537-558.

[64] Ponnamperuma C., Shimoyama A., Friebele E. Clay and the Origin of Life. Origins Life 1982; 12: 9-40.

[65] Ferris JP., Huang CH., Hagan WJ. Montmorillonite: A Multifunctional Mineral Catalyst for the Prebiological Formation of Phosphate Esters. Origins Life Evol Biosphere 1988; 18(1-2): 121-133.

[66] Ferris JP., Ertem G., Agarwak VK. The Adsorption of Nucleotides and Polynucleotides on Montmorillonite Clay. Origins Life Evol Biosphere 1989; 19: 153-164.

[67] Ferris JP., Ertem G. Oligomerization of Ribonucleotides on Montmorillonite: Reaction of the 5'-Phosphorimidazolide of Adenosine. Science 1992; 257(5075): 1387-1389.

[68] Ferris JP., Ertem G. Montmorillonite Catalysis of RNA Oligomer Formation in Aqueous Solution. A model for the Prebiotic Formation of RNA. J Am Chem Soc 1993; 115(26): 12270-12275.

[69] Kawamura K., Ferris JP. Kinetics and Mechanistic Analysis of Dinucleotide and Oligonucleotide Formation from the 5'-Phosphorimidazolide of Adenosine on Na^+-Montmorillonite. J Am Chem Soc 1994; 116(17): 7564-7572.

[70] Kawamura K., Ferris JP. Clay Catalysis of Oligonucleotide Formation: Kinetics of the Reaction of the 5'-Phosphorimidazolides of Nucleotides with the Non-basic Heterocycles Uracil and Hypoxanthine. Origins Life Evol Biosphere 1999; 29(6): 563-591.

[71] Ertem G., Ferris JP. Template-directed Synthesis Using the Heterogeneous Templates Produced by Montmorillonite Catalysis. A Possible Bridge between the Prebiotic and RNA Worlds. J Am Chem Soc 1997; 119(31): 7197-7201.

[72] Kawamura K., Nakahara N., Okamoto F., Okuda N. Temperature Dependence of the Cyclization of Guanine and Cytosine Mix Hexanucleotides with Water-soluble Carbodiimide at 0-75 °C. Viva Origino 2003; 31(4): 221-232.

[73] Kawamura K., Okamoto F., Okuda N. Influence of Template Oligonucleotides on the Condensation of Oligonucleotides in the Presence of Water-soluble Carbodiimide: Search for Model Reactions of the Formation of RNA Which Could Be Effective at High Temperatures. Viva Origino 2004; 32: 68-80.

[74] Ertem G., Ferris JP. Formation of RNA Oligomers on Montmorillonite: Site of Catalysis. Origins Life Evol Biosphere 1998; 28: 485-499.

[75] Ferris JP., Hill AR., Liu RH., Orgel LE. Synthesis of Long Prebiotic Oligomers on Mineral Surfaces. Nature 1996; 381: 59-61.

[76] Huang WH., Ferris JP. Synthesis of 35-40 Mers of RNA Oligomers from Unblocked Monomers. A Simple Approach to the RNA World. Chem Comm 2003; 12: 1458-1459.

[77] Huang WH., Ferris JP. One-step, Regioselective Synthesis of up to 50-mers of RNA Oligomers by Montmorillonite Catalysis. J Am Chem Soc 2006; 128(27): 8914-8919.

[78] Lorhman R., Bridson PK., Orgel LE. Efficient Metal-ion Catalyzed Template-Directed Oligonucleotide Synthesis. Science 1980; 208(4451): 1464-1465.

[79] Inoue T., Orgel LE. A nonenzymatic RNA polymerase model. Science 1983; 219(4586): 859-862.

[80] Wu TF., Orgel LE. Nonenzymatic Template-Directed Synthesis on Oligodeoxycytidylate Sequences in Hairpin Oligonucleotides. J Am Chem Soc 1992; 114(1): 317-322.

[81] Sawai H., Wada K. Nonenzymatic Template-Directed Condensation of Short-chained Oligouridylates on a Poly(A) Template. Origins Life Evol Biosphere 2000; 30: 503-511.

[82] Hill Jr AR, Orgel LE., Wu T. The Limits of Template-directed Synthesis with Nucleoside-5′-Phosphoro(2-Methyl)Imidazolides. Origins Life Evol Biosphere 1993; 23(5-6): 285-290.

[83] Szostak JW., Bartel DP., Luisi PL. Synthesizing Life. Nature 2001; 409: 387-390.

[84] Mansy SS., Schrum JP., Krishnamurthy M., Tobe S., Treco DA., Szostak JW. Template-directed Synthesis of a Genetic Polymer in a Model Protocell. Nature 2008; 454: 122-126.

[85] Kanavarioti A., Bernasconi CF., Alberas DJ., Baird, EE. Kinetic Dissection of Individual Steps on the Poly(C)-Directed Oligoguanylate Synthesis from Guanosine 5'-Monophosphate 2-Methylimidazolide. J Am Chem Soc 1993; 115(19): 8537-8546.

[86] Inoue T., Orgel LE. Substituent Control of the Poly(C)-Directed Oligomerization of Guanosine 5′-Phosphorimidazolide. J Am Chem Soc 1981; 103: 7666-7667.

[87] Prabahar KJ., Cole TD., Ferris JP. Effect of Phosphate Activating Group on Oligonucleotide Formation on Montmorillonite-the Regioselective Formation of 3', 5'-Linked Oligoadenylates. J Am Chem Soc 1994; 116(24): 10914-10920.

[88] Prabahar KJ., Ferris JP. Adenine Derivatives as Phosphate-activating Groups for the Regioselective Formation of 3', 5'-Linked Oligoadenylates on Montmorillonite: Possible Phosphate-activating Groups for the Prebiotic Synthesis of RNA. J Am Chem Soc 1997; 119(19): 4330-4337.

[89] Oscar L. Ueber Formaldehyd und Dessen Condensation. J Prak Chem 1886; 321: 321-351.

[90] Breslow R. On the Mechanism of the Formose Reaction. Tetrahedron Lett 1959; 21: 22-26.

[91] Eschenmoser A., Loewenthal E. Chemistry of Potentially Prebiological Natural-products. Chem Soc Rev 1992; 21(1): 1-16.

[92] Dueholm KL., Egholm M., Behrens C., Christensen L., Hansen HF., Vulpius T., Petersen KH., Berg RH., Nielsen PE., Buchardt O. Synthesis of Peptide Nucleic-acid Monomers Containing the 4 Natural Nucleobases-Thymine, Cytosine, Adenine, and Guanine and their Oligomerization. J Org Chem 1994; 59(19): 5767-5773.

[93] Joyce GF. The Antiquity of RNA-based Evolution. Nature 2002; 418: 214-221.

[94] Pitsch S., Krishnamurthy R., Bolli M., Wendeborn S., Holzner A., Minton M., Lesueur C., Schlonvogt I., Jaun B., Eschenmoser A. Pyranosyl-RNA (P-RNA)-Base-Pairing Selectivity and Potential to Replicate-Preliminary Communication. Helv Chim Acta 1995; 78(7): 1621-1635.

[95] Bolli M., Micura R., Eschenmoser, A. Pyranosyl-RNA: Chiroselective self-assembly of base sequences by ligative oligomerization of tetranucleotide-2', 3'-cyclophosphates (with a commentary concerning the origin of biomolecular homochirality). Chem & Biol 1997; 4(4): 309-320.

[96] Joyce GF., Schwartz AW., Miller SL., Orgel LE. The Case for an Ancestral Genetic System Involving Simple Analogs of the Nucleotides. Proc Natl Acad Sci USA 1987; 84: 4398-4402.

[97] Joshi PC., Pitsch S., Ferris JP. Homochiral Selection in the Montmorillonite-Catalyzed and Uncatalyzed Prebiotic Synthesis of RNA. Chem Commun 2000; 2497–2498.

[98] Urata H., Ando C., Ohmoto N., Shimamoto Y., Kobayashi Y., Akagi M. Efficient and Homochiral Selective Oligomerization of Racemic Ribonucleotides on Mineral Surface. Chem Lett 2001; 324–325.

[99] Osawa K., Urata H., Sawai H. Chiral Selection in Oligoadenylate Formation in the Presence of a Metal Ion Catalyst or Poly(U) Template. Origins Life Evol Biosphere 2005; 35: 213-223.

[100] Nisbet EG. Origin of life: RNA and Hot-water Springs. Nature 1986; 322: 206–206.

[101] Pace NR. Origin of Life—Facing up to the Physical Setting. Cell 1991; 65: 531–533.

[102] Kawamura K., Umehara M. Kinetic Analysis of the Temperature Dependence of the Template-directed Formation of Oligoguanylate from the 5'-Phosphorimidazolide of Guanosine on a Poly(C) Template with Zn^{2+}. Bull Chem Soc Jpn 2001; 74(5): 927-935.

[103] Kawamura K., Maeda J. Kinetic Analysis of Oligo(C) Formation from the 5′-Monophosphorimidazolide of Cytidine with Pb(II) Ion Catalyst at 10–75°C. Origin Life Evol Biospheres 2007; 37: 153–165.

[104] Kawamura K., Maeda J. Kinetics and Activation Parameter Analysis for the Prebiotic Oligocytidylate Formation on Na^{+}-Montmorillonite at 0–100°C. J Phys Chem A 2008; 112: 8015–8023.

[105] Yakovchuk P., Protozanova E., Frank-Kamenetskii MD. Base-stacking and Base-pairing Contributions into Thermal Stability of the DNA Double Helix. Nucl Acid Res 2006; 34(2): 564-574.

[106] Blochl E., Rachel R., Burggraf S., Hafenbradl D., Jannasch HW., Stetter KO., Pyrolobus Fumarii, Gen. and Sp. Nov., Represents a Novel Group of Archaea, Extending the Upper Temperature Limit for life to 113 °C. Extremophiles 1997; 1: 14-21.

[107] Cowan DA. The Upper Temperature for Life-Where Do We Draw the Line? Trends Micobiol 2004; 12: 58-60.

[108] Fox SW., Harada K. Thermal Copolymerization of Amino Acids to a Product Resembling Protein. Science 1958; 128(3333): 1214-1215.

[109] Fox SW., Harada K. The Thermal Copolymerization of Amino Acids Common to Protein. J Am Chem Soc 1960; 82: 3745–3751.

[110] Harada K., Fox SW. The Thermal Copolymerization of Aspartic Acid and Glutamic Acid. Arch Biochem Biophys 1960; 86(2): 274-280.

[111] Fox SW., Krampitz G. Catalytic Decomposition of Glucose in Aqueous Solution by Thermal Proteinoids. Nature 1964; 203: 1362-1364.

[112] Nakashima T., Fox SW. Selective Condensation of Aminoacyl Adenylates by Necleoproteinoid Microparticles. Proc Natl Acad Sci USA 1972; 69(1):106-108.

[113] Fox SW. Metabolic Microspheres Origins and Evolution. Naturwissenschaften 1980; 67: 378-383.

[114] Shock EL. Stability of Peptides in High-temperature Aqueous-solutions. Geochim Cosmochim Acta 1992; 56(9): 3481-3491.

[115] Rode BM., Schwendinger MG. Copper-catalyzed Amino Acid Condensation in Water — A Simple Possible Way of Prebiotic Peptide Formation. Origins Life Evol Biosphere 1990; 20(5): 401-410.

[116] Bujdak J., Rode BM. Silica, Alumina, and Clay-Catalyzed Alanine Peptide Bond Formation. J Mol Evol 1997; 45: 457-466.

[117] Rode BM. Peptides and the Origin of Life. Peptides 1999; 20(6): 773-786.

[118] Meng M., Stievano L., Lambert JF. Adsorption and Thermal Condensation Mechanisms of Amino Acids on Oxide Supports. 1. Glycine on Silica. Langmuir 2004; 20(3): 914-923.

[119] Leman L., Orgel LE., Ghadiri MR. Carbonyl Sulfide-mediated Prebiotic Formation of Peptides. Science 2004; 306(5694): 283-286.

[120] Lambert J-F., Adsorption and Polymerization of Amino Acids on Mineral Surfaces: A Review. Origins Life Evol Biospheres 2008; 38: 211-242.

[121] Forterre P., A Hot Topic: the Origin of Hyperthermophiles. Cell 1996; 85: 789–792.

[122] Galtier N., Tourasse N., Gouy M. A Nonhyperthermophilic Common Ancestor to Extant Life Forms. Science 1999; 283: 220–221.

[123] Holm NG., Marine Hydrothermal Systems and the Origin of Life. Origin Life Evol Biosphere 1992; 22: 1–242.

[124] Baross JA., Hoffman SE. Submarine Hydrothermal Vents and Associated Gradient Environments as Sites for the Origin and Evolution of Life. Origin Life Evol Biosphere 1985; 15(4): 327-345.

[125] Yanagawa H., Kojima K. Thermophilic Microspheres of Peptide-Like Polymers and Silicates Formed at 250°C. J Biochem 1985; 97: 1521-1524.

[126] Imai E., Honda H., Hatori K., Brack A., Matsuno K. Elongation of Oligopeptides in a Simulated Submarine Hydrothermal System. Science 1999; 283: 831-833.

[127] Imai E., Honda H., Hatori K., Matsuno K. Autocatalytic Synthesis of Oligoglycine in a Simulated Submarine Hydrothermal System. Origin Life Evol Biosphere 2000; 29: 249-259.

[128] Kawamura K. Monitoring of Hydrothermal Reactions In 3 ms Using Fused-silica Capillary Tubing. Chem Lett 1999; 125-126.

[129] Kawamura K. Monitoring Hydrothermal Reactions on the Millisecond Time Scale Using a Micro-tube Flow Reactor and Kinetics of ATP Hydrolysis for the RNA World Hypothesis. Bull Chem Soc Jpn 2000; 73: 1805-1811.

[130] Kawamura K. Development of Micro-flow Hydrothermal Monitoring Systems and their Applications to the Origin of Life Study on Earth. Anal Sci 2011; 27(7): 675-683.

[131] Kawamura K., Nishi T., Sakiyama T. Consecutive Elongation of Alanine Oligopeptides at the Second Time Range Under Hydrothermal Condition Using a Micro Flow Reactor System. J Am Chem Soc 2005; 127: 522-523.

[132] Kawamura K., Shimahashi M. One-step Formation of Oligopeptide-like Molecules from Glu and Asp in Hydrothermal Environments. Naturwissenschaften 2008; 95: 449-454.

[133] Kawamura K., Takeya H., Kushibe T., Koizumi Y. Mineral-enhanced Hydrothermal Oligopeptide Formation at the Second Time Scale. Astrobiology 2011; 11: 461-469.

[134] Horowitza ED., Engelharta AE., Chena MC., Quarlesa KA., Smitha MW., Lynna DG., Hud NV. Intercalation as a Means to Suppress Cyclizationand Promote Polymerization of Base-pairing Oligonucleotides in a Prebiotic World. Proc Natl Acad Sci USA 2010; 107(12): 5288-5293.

[135] Kawamura K. Drawbacks of the Ancient RNA-based Life-like System Under Primitive Earth Conditions. Biochimie 2012; 94(7): 1441-1450.

[136] Kawamura K. Reality of the Emergence of Life-like Systems from Simple Prebiotic Polymers on Primitive Earth. In: Seckbach J., Gordon R. (ed.) Genesis-in the Beginning: Precursors of Life, Chemical Models and Early Biological Evolution. Springer; 2012. p123-144.

[137] Hud NV., Jaina SS., Lia X., Lynn DG. Addressing the Problems of Base Pairing and Strand Cyclization in Template-Directed Synthesis A Case for the Utility and Necessity of 'Molecular Midwives' and Reversible Backbone Linkages for the Origin of proto-RNA. Chem & Biodiversity 2007; 4: 768-783.

[138] Ertem G., Ferris JP. Synthesis of RNA Oligomers on Heterogeneous Templates. Nature 1996; 379: 238-240.

[139] Tuerk C., Gold L. Systematic Evolution of Ligands by Exponential Enrichment: RNA Ligands to Bacteriophage T4 DNA Polymerase. Science 1990; 249: 505-510.

[140] Ellington AD., Szostak JW. In Vitro Selection of RNA Molecules That Bind Specific Ligands. Nature 1990; 346: 818-822.

[141] Kawamura K., Kuranoue K., Nagahama M. Prebiotic Inhibitory Activity of Protein-like Molecules to the Template-directed Formation of Oligoguanylate from Guanosine 5′-Monophosphate 2-Methylimidazolide on a Polycytidylic Acid Template. Bull Chem Soc Jpn 2004; 77(7): 1367-1375.

[142] Kawamura K., Yukioka M. Kinetics of the Racemization of Amino Acids at 225-275 °C Using a Real-time Monitoring Method of Hydrothermal Reactions. Thermochimica Acta 2001; 375(1-2): 9-16.

Section 2

Biological Oligomers

Chapter 7

Oligomerization of Biomacromolecules – Example of RNA Binding Sm/LSm Proteins

Bozidarka L. Zaric

Additional information is available at the end of the chapter

http://dx.doi.org/10.5772/57592

1. Introduction

1.1. A multitude of RNAs

The central dogma of molecular biology stated that genetic information only flows in one direction, from DNA to proteins via an intermediate called messenger ribonucleic acid (mRNA) [1,2,3]. Originally, ribonucleic acid (RNA) was thought to have roles in information transfer and structure maintenance. Today, we know that RNA performs a remarkable range of functions in the living cell, (control of gene expression, chromosome –end maintenance, housekeeping activities, sorting of proteins in the cell and defines metazoan development) [3].Although, proteins have enzymatic activities mostly, in the early 1980s has been shown that RNA molecules can catalyze a chemical reaction and RNAs with catalytic activity are called **ribozymes**. The discovery of ribozymes led to the hypothesis that RNA could have been the original molecule of life on earth about four billion years ago; a biopolymer with the ability to self-replicate and that could both store information and catalyze chemical reactions. RNA would have been self-sufficient as the original molecule of life [4]. Discovery of the unexpectedly wide variety of functions carried out by RNA was accompanied by the identification of a multitude of further types of small, non-coding RNAs (small nuclear RNA, small nucleolar RNAs, small interfering RNAs, micro RNAs) highlighting the versatility of RNA as a biochemical tool for the cell [2].

rRNAs represent structural and catalytic elements of the ribosome. In the nucleolus of eukaryotic cells, more than 100 tandemly repeated units of rRNA genes are transcribed into long precursor transcripts [7,8]. Following transcription, pre-rRNA is subsequently cleaved to form mature rRNAs and with approximately 80 proteins to form the large and small ribosomal subunits prior to their export to the cytoplasm. **SnoRNAs (Small nucleolar**

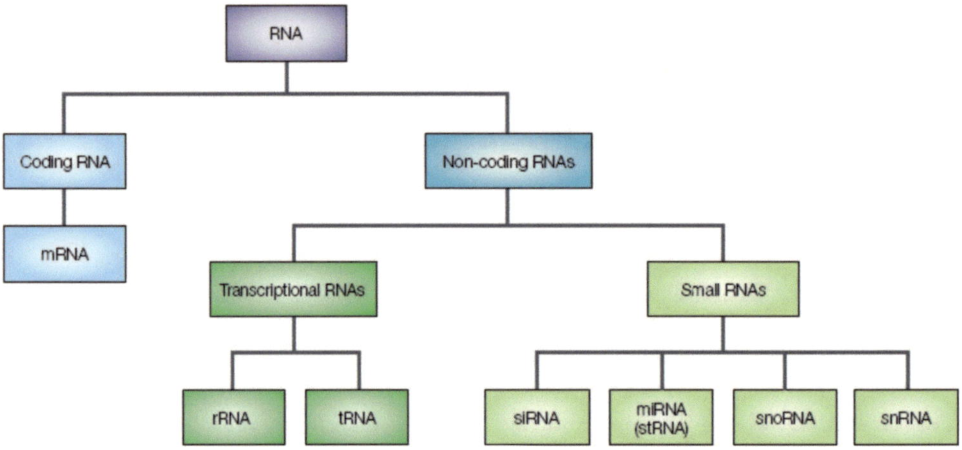

Figure 1. The RNA family (Reprinted from reference 2)

RNAs) participate in both the modification and cleavage events that occur during ribosome biogenesis [5]. A second group of small RNAs are the **tRNAs,** which are essential in translation [8,10]. **Micro RNAs** are non-coding RNAs of 22-24 nucleotides in length. They down regulate gene expression by attaching themselves to mRNA, thereby preventing them from being translated into protein. Another type of non-coding RNA is the **small interfering RNA**. These small RNAs mediate life time of RNA by interacting with mRNAs and labeling it for destruction [2, 4]. **PiRNA** is the large class of small non-coding RNAs which acts with piwi proteins. These piRNA have been linked to both epigenetic and posttranslational gene silencing of retrotransposons and other genetic elements in germline cells [11,12]. Telomerase is complex of proteins and RNA,and is responsible for maintaining of natural end of chromosomes. Telomerase acts as reverse transcriptase because its mechanism of action is copying RNA template into DNA [10].

Small nuclear RNAs are components of the macromolecular machinery (spliceosome) that has a role in the maturation of mRNA. They are termed **U snRNAs** (stands for uridyl rich small nuclear RNA). U1, U2, U4, U5, U7, U11, and U12 are synthesized in the nucleus by RNA polymerase II. After that they are transported to the cytoplasm where association with the U snRNP proteins (proteins which associate with uridine rich small nuclear RNAs, generating UsnRNP, uridyl rich small nuclear ribonuclear particles) occurs, followed by re-import into the nucleus [1, 13].U6 and U8 snRNA belong as well to class of small nuclear RNA but their synthesis, biogenesis and function differs from mentioned UsnRNAs. U3 snRNA shares common denomination but as well this small RNA is found in the nucleolus, and has role in pre-rRNA processing and has C/D box motif, which technically make it a member of the C/D class of snoRNAs [14].

1.2. Structure of RNA and association with proteins

DNA and RNA have similar covalent structures, the only difference being the change from a 2`-deoxyribose sugar to a ribose sugar and from a methyl group in thymine to a hydrogen in uracil. RNA has much wider biological activities and adopts a wider range of structures. DNA double helices preferentially assume the B-form structure in solution and RNA double helices are found in the A-form. The **RNA A-form double helix** has a narrow and deep major groove, which prevents proteins to recognize RNA in a manner analogous to the way they recognize DNA. An RNA molecule can locally adopt several types of secondary structure (bulges, hairpins, internal loops) [15, 16].Eukaryotic mRNAs are almost always associated with RNA-binding proteins. RNA-binding proteins generally have a modular structure and contain RNA-binding domains of 70–150 amino acids that mediate RNA recognition. Three major classes of eukaryotic RNA-binding protein domains are known: the RNA-recognition motif (RRM), the double stranded RNA binding domain (dsRBD) and the K-homology (KH) domain [17].

1.3. All mRNA processing steps are coupled

Eukaryotic gene expression is a complex, stepwise process that begins with transcription (synthesis of pre-mRNA) [18]. Mature mRNAs are produced in the cell nucleus from primary transcripts of coding genes (pre-mRNAs) by a series of processing events which include capping, splicing, and 3` end polyadenylation. Mature mRNAs are transported to the cytoplasm. All modification steps are coupled and influence each other. RNA polymerase II is a key molecular coordinator of these processing events, and phosphorylation of it has regulatory role [19, 20,21,].

1.4. Removal of introns and the splicing reaction

In 1977, a number of research groups discovered that the genes of higher organisms are often made up of a sequence of coding (called exons) and non-coding base sequences (introns). During transcription, all parts of the gene are copied to form a strand of pre-mRNA. Introns are removed and the exons stitched together so that the now continuous exons can be translated to produce a protein. This splicing of the pre-mRNA is a multistage process, carried out by complex macromolecular machinery known as the spliceosome, which is among the most complex macromolecular machineries in the cell [22].

Splicing of precursors to mRNAs occurs in two steps, both involving a single transesterification reaction [23].Assembly and function of the spliceosome requires approximately 300 polypeptides and five snRNAs, not considering gene-specific RNA-binding factors [23]. There are two distinct types of spliceosome in most cells. The major class U2-type spliceosome is universal in eukaryotes, whereas the minor class or U12-type spliceosome is not present in some organisms. The evolutionary relation between these two spliceosomes is uncertain.

1.5. Types of introns

The pre-mRNA contains conserved elements at its intron/ exon boundaries that determine the proper sites for the splicing reaction (Figure 3.). The 5′ splice site contains a conserved consensus sequence, which is AG/GURAGU (R=purine, / denotes the exon/ intron boundary). The branch site lies between 100 and 18 bases upstream of the 3′ splice site and has the consensus: CUR**A**Y for vertebrates (**A** branching nucleotide, Y is pyrimidine). In higher

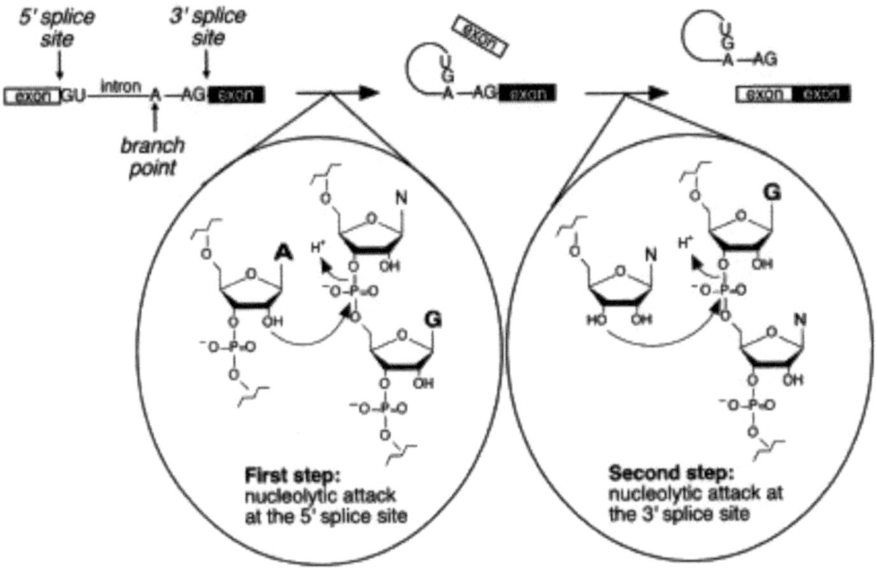

Figure 2. The two chemical steps of splicing (Reprinted from reference 23)

eukaryotes, a polypyrimidine tract variable in length is often located between the branch site and the 3′ splice site. The 3′ splice site has the consensus:YAG/R for mammals. This class of introns is spliced by U2 spliceosome. The U12 type introns have different consensus sequences and are spliced by the U12 spliceosome [24]. The number of known U12 introns is still very small. U12 type introns are present in many vertebrates, nematodes, insect, and plant species.

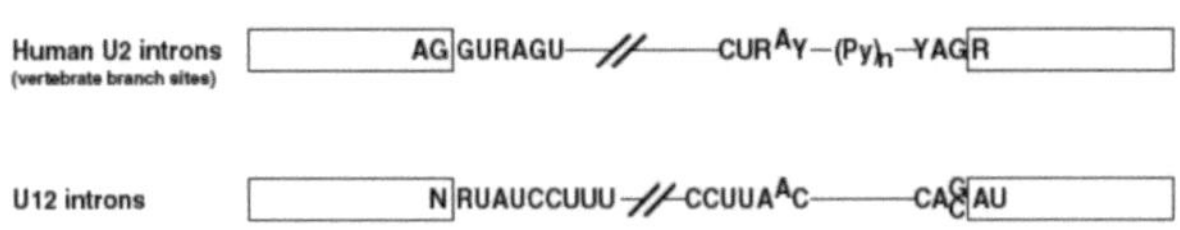

Figure 3. Splice site consensus sequences. Comparison of splice site consensus sequences for human U2 dependent and U12 dependent introns. The most conserved regions, 5' splice site (5' SS), branch point (BP), 3'splice site (3' SS), are shown with their consensus sequences (R=purine, Y=pyrimidine). The polypyrimidine tract often present in metazoan U2 dependent introns is indicated as (Py). (Reprinted from reference 25)

1.6. Small nuclear ribonucleoproteins, snRNPs

Small ribonucleoproteins (RNPs) are tight complexes of one or more proteins with a short RNA molecule (usually 60-300 nucleotides). RNPs inhabit nuclear and cytoplasmatical compartments of the eukaryotic cell [25]. Those that reside in the nucleus, the small nuclear ribonucleoproteins (snRNPs) can themselves be divided into two families. There are **snRNPs** of the nucleoplasm, whose function lies in preparing messenger RNA for export into the cytoplasm. A different set of snRNPs, called **snoRNPs**, reside in the nucleolus [26]. The **nucleolus** is a structure in the nucleus composed of proteins and nucleic acids. Its function is to transcribe ribosomal RNA (rRNA) and combine it with proteins to form ribosomes. There are about 200 distinct kinds of snRNPs (they differ according to the RNA or protein components) with abundances between 10^4 (for snoRNPs) to over 10^6 copies per cell (for snRNPs in major spliceosomes). They generally play a role in gene expression. One exception is the telomerase snRNP, essential for genome maintenance, which is present only in a few copies per cell. Spliceosomes are formed around the pre-mRNA substrate by the successive assembly of five small nuclear ribonucleoprotein-particles (snRNPs): U1, U2, U4, U5, and U6. These particles are composed each of a small nuclear RNA (snRNA), seven Sm core proteins common to all snRNPs (except for the U6 snRNP, which contains a related set of seven proteins, the Sm-like proteins) and several snRNP-specific proteins. The snRNPs play a central role in the process of splicing. They are responsible for the recognition of splice sites, definition of exon/ intron boundaries. These interactions are partially mediated through base pairing and are dynamic so that the spliceosome complex changes during the process of splicing.

1.7. U snRNP biogenesis

Subsequent to transcription by RNA polymerase II and capping, pre-U1 snRNA assembles with several factors including cap-binding proteins (CBP), a phosphorylated adaptor for RNA export (PHAX), Crm1, and Ran-GTP, which all together mediate export of U1 snRNA to the cytoplasm. After export, Sm proteins interact with the U snRNAs to form the snRNP Sm core. This step is facilitated by the SMN complex (survival of motor neurons complex). The SMN complex is composed from SMN protein and the other proteins called Gemins (Gemins 2-7). Nuclear re-import is mediated by snurportin-1 (SPN1), which binds to the snRNAs m3G cap structure. After import, these factors dissociate. The U1 specific proteins are imported independently into the nucleus, where assembly into mature U1 snRNP occurs [27]. This is a pathway shared with U1, U2, U4 and U5 snRNPs.

1.8. Spliceosome assembly

Assembly of a spliceosome for excision of an intron requires recognition of sequences at the 5' splice site as well as the branch site and nearby 3' splice site. U1 snRNA binds to the 5` end of the intron using sequence complementarities. There are reports which show that the U1 snRNA recognizes the 5` splice site in a preassembled penta-snRNP complex [28]. U2 snRNP complex associates with the branch region. Early snRNP/pre-mRNA complexes are preferentially committed to splicing as compared to free RNA and thus are called **commitment complexes** (CCs). The process of U2 snRNP association is ATP-dependent and four proteins are critical for recognition [29]. Subsequent to the binding of U2 snRNP complex, a tri-snRNP complex containing U4/U6 snRNP and U5 snRNP associates in an ATP-dependent manner to form complex A2-1. It is likely that the U1 snRNA/pre-mRNA duplex dissociates at this stage. The 5' splice site sequence is probably paired on the intron side to U6 snRNA and on the exon side to U5 snRNA. The transition between complex A2-1 and A1 requires destabilization of at least U4/U6 di-snRNA. As only three snRNAs U2, U6 and U5 are associated with the spliceosome at the moment of catalysis, and as U5 snRNA pairing with exon sequences is not essential, the

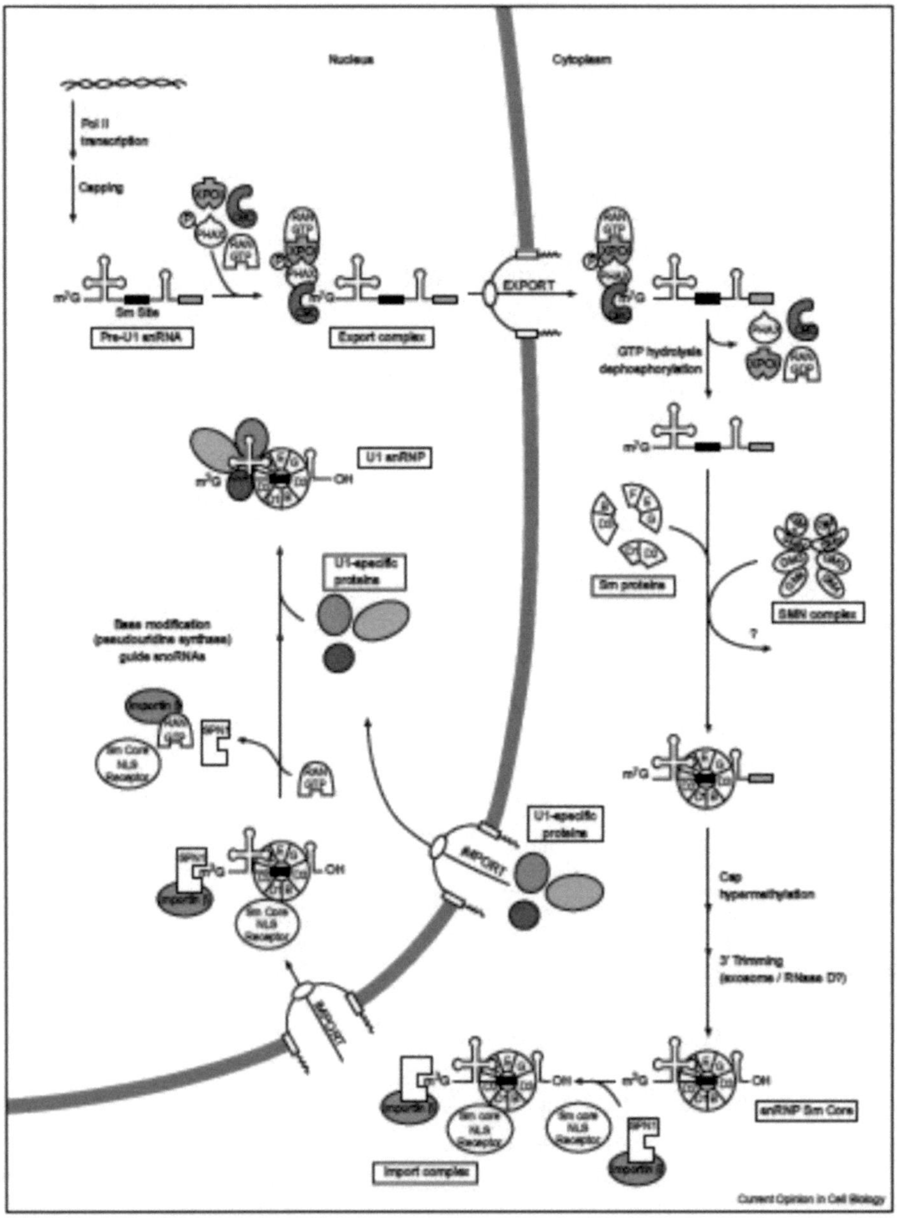

Figure 4. The U snRNP biogenesis pathway (Reprinted from reference 27)

catalytic site is either created by U6 snRNA by U2 snRNA or both. The action of certain proteins is required for the transition to the second step in splicing. The catalytic site for the second step is created by either U6 snRNA, U2 snRNA, or associated proteins. The reannealing of released U4 and U6 snRNP and association with U5 forms the U4/U6 U5 tri-snRNP complex which is then ready to reassemble on another commitment complex.

http://dx.doi.org/10.5772/57592

The classical view of spliceosome assembly has been challenged by Stevens et al [30]. This group isolated from yeast a penta-snRNP complex which when supplied with soluble components, does splice pre-mRNA.

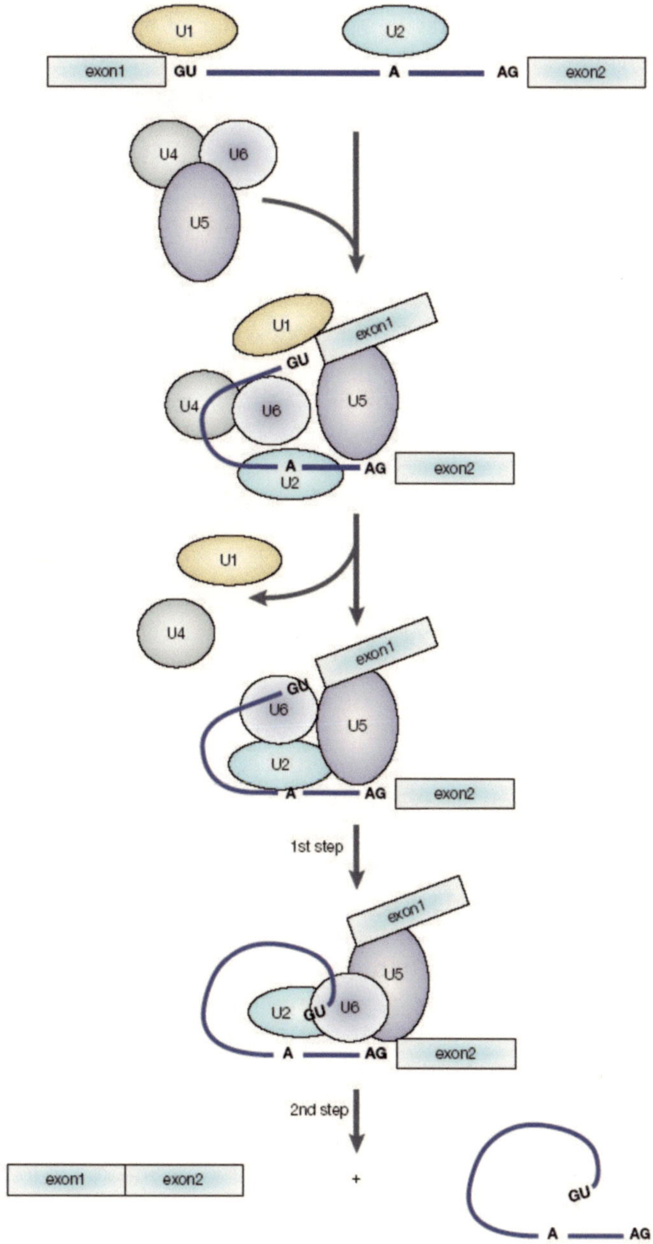

Figure 5. A simplified view of the splicing process (Reprinted from reference 5)

1.9. mRNA stabilization, degradation

Regulation of mRNA decay rates is an important control point in determining the abundance of cellular transcripts. Some mRNA has half-lives that are 100 times shorter than cellular generation times and some mRNA have half-lives spanning several cell cycles [21]. The poly (A) tail is important in stabilization of mRNA. It interacts with the poly (A) binding protein (PABP), which makes direct contact with a specific region of the translation-initiation factor (eIF4E). Translation initiation factor (eIF4) interacts with the cap binding proteins. In this way, a ternary (PABP-translation initiation,-cap binding protein, poly (A) tail) complex is formed which circularizes mRNA *in vitro,* promoting translation and stabilization of mRNAs [21]. Several sequence elements can regulate the mRNA turnover rate, either by its promotion (destabilizer elements) or by its inhibition (stabilizer elements). Important elements are A+U-rich elements (ARE), located in the 3` untranslated regions (UTR) of mRNAs [31]. At least four different ways of mRNA degradation have been reported in eukaryotic cells [32]. In most cases, degradation of the transcript begins with the shortening of the poly (A) tail at the 3` end of mRNA. After shortening of the poly (A) tail follows the removal of the 5` cap structure (decapping), thereby exposing the transcript to digestion by a 5` to 3` exonuclease. Family of LSm proteins is involved in degradation of mRNAs, in the deccaping step Transcripts can be degraded in the 3`-5` direction after deadenylation. This process is catalyzed by the exosome [33]. One mRNA degradation pathway is the **nonsense mediated decay (NMD)**, which provides strongest evidence for a link between translation and turnover [34].

2. Family of Sm-LSm proteins

2.1. Sm proteins, assembly of U1, U2, U4, U5 snRNPs

The Sm proteins were first discovered as antigens targeted by so-called **Anti-Sm antibodies** in a patient with a form of Systemic lupus erythematosus (SLE), a debilitating autoimmune disease. They were named Sm proteins in honor of Stephanie Smith, a patient who suffered from SLE. Other proteins with very similar structures were subsequently discovered and named LSm proteins. The common proteins for U1, U2, U4 and U5 snRNPs are named Sm proteins due to their recognition by anti-Sm autoantibodies (isolated from the serum of patients with autoimmune diseases [35, 36].Eight proteins: B`, B, D1-D3, E, F and G have been characterized in human cells. All of the Sm core proteins are encoded by separate genes [37], with the exception of B and B`. The B and B` that result from an alternative splicing of gene 6628 located on chromosome 20, locus 20p13, only differs in 11 amino acids at the C-terminus [38]. In neural tissues, SmN replaces SmB and SmB` [39]. Two sequence motifs, named Sm1 and Sm2, are found in all known Sm proteins, what is reason that they are called Sm proteins [40]. The N terminal Sm1 motif is composed of 32 amino acids. The Sm2 motif, located closer to the C terminus, is shorter spanning only 14 amino acids [35]. Sm motif 1 and Sm motif 2 are separated by a linker of variable length. The alignment of the sequences of human Sm proteins reveals a striking conservation of the two motifs. Majority of the Sm proteins have amino or carboxy-terminal extensions.

http://dx.doi.org/10.5772/57592

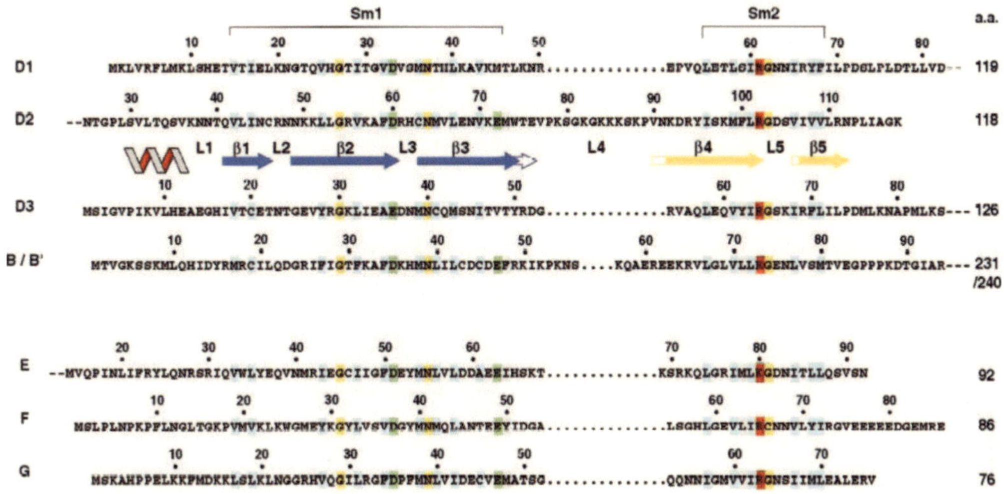

Figure 6. Primary and secondary structure of Sm proteins (Reprinted from reference 41). Amino acid sequence alignment of the human Sm (D1, D2, D3, B/B`, E, F, and G) proteins with secondary structure elements. Wavy line, helix; arrows, β strands. The β strands within the Sm1 and Sm2 motifs are colored blue and yellow, respectively. The β strands and interconnecting loops are numbered consecutively from the N terminus. The conserved Sm1 and Sm2 motifs are indicated and the conserved residues within these motifs are highlighted in blue (hydrophobic), grey (hydrophobic, less well conserved), orange (neutral polar), red (basic) and green (acidic). (Reprinted from reference 41)

Solved structures of this protein family members (pdb codes: 1d3b,1b34,1hk9, 1h64,1i8f,1i4k, 1kq1,3bw1,1th7) show that the fold is highly conserved. It is defined by an N-terminal helix, followed by a five-stranded anti-parallel β sheet. Strands β1, β2, and β3 are part of the Sm1 motif, whereas the Sm2 motif forms strands β4 and β5. The five stranded β sheet is strongly bent in the middle and the conserved hydrophobic residues form a hydrophobic core [41].

The Sm proteins bind to the Sm site of U snRNAs [42]. The Sm site consensus sequence (PuAU4-6GPu) has a central, uridine rich tract and flanking purines. *In vitro* studies reveal that the single–stranded U rich region and the 5` adenosine of the Sm site play critical roles in Sm protein assembly. The uridine bases and the 2` hydroxyl groups collectively provide binding determinants [43]. In the absence of U snRNA, the seven Sm proteins form three stable subcomplexes (D3B, or D3B′, D1D2, and EFG). These sub complexes then form a heptameric ring around the snRNA Sm site, and as such the complex is termed the **Sm core**. SnRNP core assembly is an ordered pathway that involves formation of a sub-core particle followed by formation of the full Sm core, which promotes cap hypermethylation and pre-snRNP import [44]. The Sm fold is necessary and sufficient for the formation of specific inter-subunit interactions. Biochemical results indicate that there is one copy of each Sm protein in the snRNP core domain and therefore support the heptameric ring model of the snRNP core domain [45]. None of the single Sm proteins has a known RNA recognition motif, so another type of interactions with RNA must be involved. Crosslinking studies indicate that Sm motif 1 is responsible for interactions with RNA, and Sm motif 2 for protein –protein interactions [43].

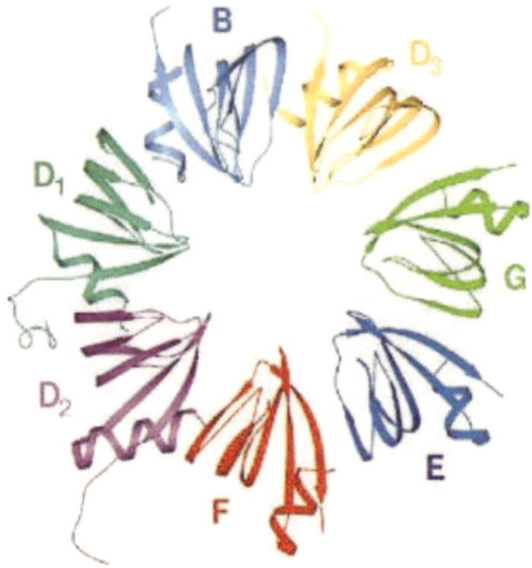

Figure 7. Proposed Higher-Order Assembly of the Human Core snRNP Proteins. The seven core Sm proteins (B/B`, D1, D2, D3, E, F, and G) are arranged within the seven-membered ring based on the crystal structures of the D1D2(1b34) and D3B (d3b)complexes and pairwise interactions deduced from biochemical and genetic experiments. (Reprinted from reference 41)

Basic residues of human and yeast SmB, SmD1 and SmD3 are reported to be responsible for import of the Sm core particle [45]. *In vitro*, the snRNP core domain can be assembled from purified components [46]. Assembly of the spliceosomal class of snRNPs *in vivo* is an active process that is mediated by several factors, including the product of the SMN gene (survival of motor neurons gene). Mutations of SMN gene are responsible for spinal muscular atrophy disease (SMA). Spinal muscular atrophy is an autosomal recessive disorder correlated with loss of motor neurons, as a result of a mutation on the SMN gene [47]. The SMN protein is ubiquitously expressed in all tissues of metazoan organisms reflecting the fact that it provides a fundamental activity required by all cells. The SMN protein is predominantly cytosolic but it is also found in the nucleus, namely in a few spherical nuclear domains that overlap with the so-called Cajal bodies (where snRNPs and snoRNPs are localized). These spherical domains have been called Gemini of Cajal bodies (Gems). Proteins associated with the SMN protein are called Gemins. The SMN complex interacts *in vitro* with Sm and LSm proteins which contain symmetrically methylated RG (arginine –glycine) repeats [48]. Symmetrically methylated RG repeats of SmD1, SmD3 and LSm4 are generated by action of the so-called methylosome [48,49].The SMN complex binds to the human hypermethylase which suggests that SMN may have a role in formation of the snRNA m^3G cap structure. It has been proposed that after binding of SMN to the Sm core proteins, SMN promotes engagement of TGS1 to the m^7G-capped snRNP particle. According to that model, SMN dissociates from the C terminal part of the B/B` Sm proteins followed by association of TGS1 and Sm core. This step allows formation of the m^3G-cap. The association of Snurportin 1 with the m^3G-cap can promote release of TGS1 and generate import-competent snRNP [50]. According to these data, the SMN complex interacts with protein components of U snRNPs, but there are reports [51] on sequence-specific interactions between U1 snRNA and the SMN complex. The nuclear

http://dx.doi.org/10.5772/57592

localization signal of U snRNP is composed of the U snRNAs 2,2,7 tri-methylguanosine cap and Sm core domain. Snurportin 1 binds to the m^3G cap but not to the Sm core. Snurportin1 has an N-terminal importin beta binding domain and a carboxy terminal m^3G-cap binding domain [52]. The Importin beta binding domain allows for snRNPs cargo to be imported in a Ran independent fashion. After import of snRNPs into the nucleus, Snurportin1 dissociates from its cargo and is exported back into the cytoplasm using Crm1, a receptor for leucine-rich nuclear export signals [53]. The SMN complex not only mediates snRNP core assembly but is an integral complex component during the entire snRNP core biogenesis in the cytoplasm.It is not excluded that SMN is actually the long-sought Sm core nuclear localization signal receptor [54].

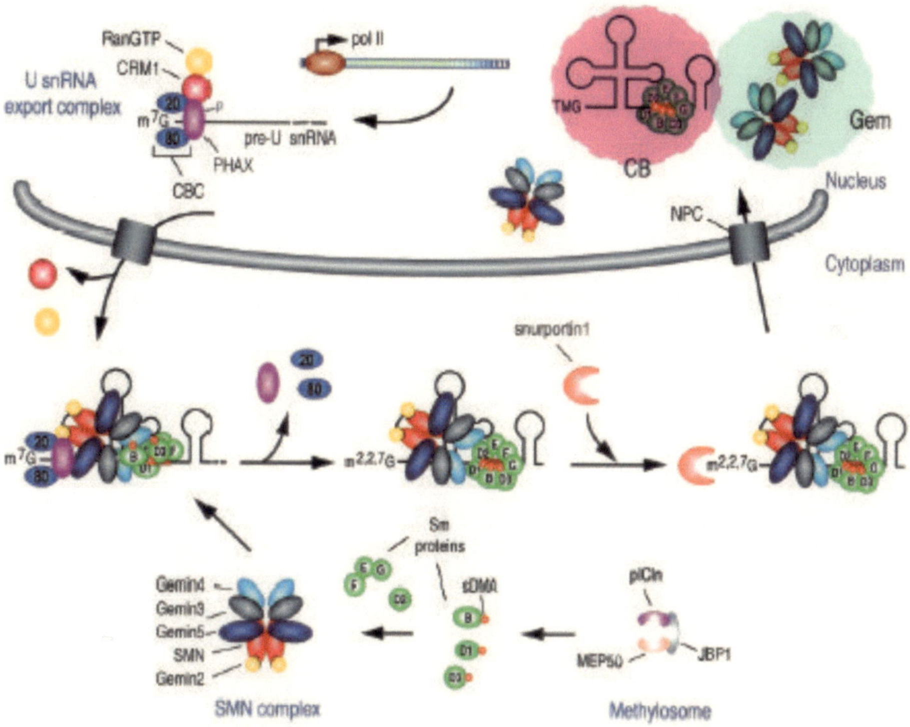

Figure 8. Schematic model of the role of SMN in snRNP core biogenesis in the cytoplasm. (Reprinted from reference 54)

2.2. LSm proteins

Sm and Sm-like proteins are found in all kingdoms of life: *eukarya, archaea and bacteria.* These proteins were found even in *Archaea*. Because Archaea have been proposed to be related to the ancestor of the eukaryotic nuclear genome, this fact suggests that an LSm protein gene was present in the last common ancestor. *Archaebacteria* harbour between one and two genes wich encode for Sm motif containing proteins. The *in vivo* functions of archaeal Sm proteins remain unknown (in constrast to the eukaryotic and bacterial homologs, and fact that high resolution structure from archaeal systems is known (pdb code 1ljo) [55]. LSm proteins have been identified in plants as well [56]. Eukaryotic genomes have more than 20 Sm/LSm genes each, corresponding to the LSm and Sm proteins which are components of Sm and LSm complexes. Database searches in the yeast genome, revealed 16 Sm motif containing proteins. Some Sm-like proteins were found to interact weakly with some Sm proteins, most probably via non-specific Sm domain interactions [57], but some of the LSm proteins interact with Sm proteins as part of U7 snRNP. In yeast there are nine LSm proteins, in humans more than eight. Each of the human LSm proteins has one orthologue in yeast. Yeast LSm2p-LSm7p share sequence identity with human LSm2-LSm7 ranging from 41-62%. LSm9p appears to be present only in yeast. Yeast LSm8p aligns best with human LSm8 (26% identity). In addition, LSm proteins are highly conserved throughout all eukaryotic kingdoms, as the homologues in insect, nematode and plant database share between 50 and 75 % identity with their human counterparts. Each of the LSm proteins in humans can clearly be best aligned with one of the canonical Sm proteins. However, their sequence identities are not high enough to allow the conclusion that LSm proteins undergo the same protein–protein interactions like Sm proteins [58].

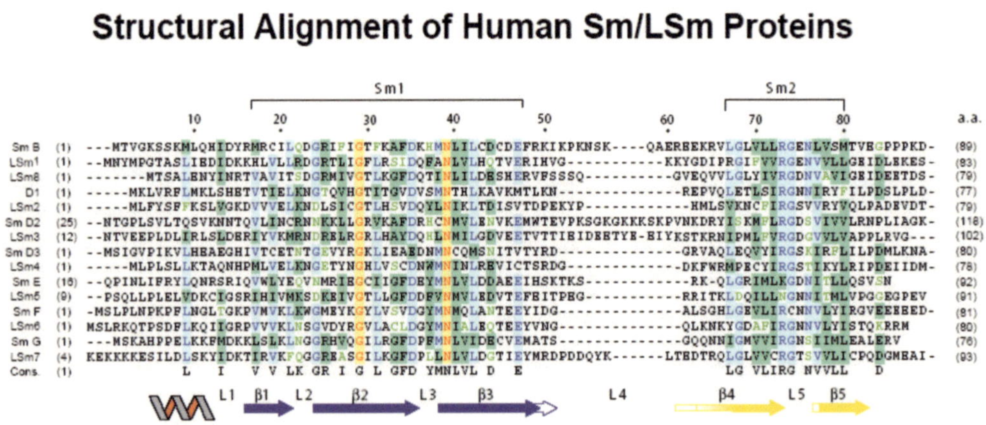

Figure 9. Structural Alignment of Human Sm/LSm proteins

Similar to canonical Sm proteins, the LSm proteins are recognized by antibodies from patients suffering from systemic lupus erythematosus (SLE) [59]. Sm/LSm proteins always appear as homomeric (in the case of prokaryotes) or heteromeric (in eukaryotes) ringlike multimers. These ring-shaped complexes, generally containing either six or seven subunits, are the functional LSm protein unit. All canonical Sm proteins are essential for vegetative growth of yeast. LSm proteins have variable effects after depletion in yeast. In mice embryos, LSm4-null zygotes survived to the blastocyst stage, but died shortly after [60].

2.3. Role of LSm 2-8 oligomers in U6snRNP assembly

The LSm2-8 complex was isolated from Hela cells nuclear extract in an RNA free form. Electron micrographs revealed a doughnut–shaped heterooligomer, similar to the Sm core snRNPs [58]. LSm proteins have a high affinity for single-stranded oligo-U, but they do not recognize the canonical Sm binding site. In yeast and humans, LSm2-8 forms a heteroheptameric ring around the 3` end of U6 snRNA, consisting of a U rich tract. The Sm core RNP is extremely salt stable; however, LSm-U6 snRNA dissociates at salt concentrations higher than 0.5M, or in the presence of competitor RNA, suggesting that the LSm-U6 complex is less stable [58]. U6 snRNA has no conserved Sm site and does not associate with Sm proteins. Its biogenesis pathway differs in many respects from the U1, U2, U4 and U5snRNP pathways; it is transcribed by RNA polymerase III and capped by γ-monomethyltriphosphate. The 3` end of pre-U6 snRNA is elongated during maturation and subsequently trimmed leaving in most organisms a 2`-3` cyclic phosphate. The enzymes involved in this process are specific for U6 snRNA, and U6 snRNA does not leave the nucleus [61]. Mature U6 snRNA shows nucleoplasmic localization [62]. Experimental evidence suggests that U6 snRNA is present in the cytoplasmic compartment of mouse fibroblast cells [63]. This result suggests that the LSm2-8 complex may act as a nuclear localization signal, but the cytoplasmic localization of the U6 snRNP is highly questionable. The actual function of the LSm 2-8 complex associated with U6 snRNA appears to be connected to U6 snRNP assembly and function. Mutants with decreased levels of LSm2-8 show splicing defects correlating with a reduced level of U6 snRNA. How the LSm2-8 complex affects U6 snRNP remains unclear. One possibility is that LSm proteins facilitate conformational rearrangements during the splicing cycle, U4/U6 annealing and formation of U4/U6/U5 tri-snRNP [64].

2.4. Role of LSm proteins in protecting mRNA 3` end termini from degradation

LSm proteins have additional roles apart from splicing. Yeast strains which lack LSm1-7p fail to grow at higher temperatures, and accumulate mRNA shortened at the 3` end by 20-30 nucleotides. The simplest model proposes that LSm1-7 complex binds to the mRNA and sterically inhibits endo and exo-nucleases. Nuclear LSm2-8 binds to the U6 snRNA 3` end, suggesting, that LSm2-8 protects the 3` end of U6 snRNA from degradation [65].

2.5. Role of LSm oligomer proteins in U8 snoRNP organization

U8 snoRNP is required for processing of 5.8S and 28S rRNAs, which together with the 5S rRNA build up the large ribosomal subunit. In Xenopus extract, LSm2, 3, 4, 6, 7, and 8 are bound as hetero hexamer to U8 snoRNA on the conserved third stem-loop sequences [66].

2.6. LSm oligomers as part of U7 snRNP

Maturation of the non-polyadenylated histone mRNAs 3' ends occurs by endonucleolytic cleavage mediated by U7 snRNP [67]. U7 snRNA contains a non-canonical Sm site. Purified U7 snRNP lacks D1 and D2 proteins but has LSm10 (14kDa) and LSm 11 (50kDa) instead [68].

2.7. LSm protein oligomers in mRNA degradation

Yeast two hybrid assays reveal multiple interactions between the eight LSm proteins, suggesting the existence of more than one LSm protein complex. Each human LSm protein is capable of interacting with multiple other LSm proteins and splicing factors, like prp24, prp4, and SmD1 [69]. Coprecipitation experiments demonstrated that LSm1p (LSmXp, is the nomenclature for yeast LSm proteins) together with LSm2p-LSm7p forms a new seven-subunit complex [70, 71]. The LSm complex LSm1-7 plays a role in mRNA degradation [72], and LSm2-8 has a role in the stabilization of U6 snRNP. These two protein complexes thus have very different functions. LSm1p mutants accumulate full length capped transcripts, but mutations on LSm1p do not stabilize mRNA containing premature stop codons, suggesting that the LSm1-7 complex is not involved in NMD [71]. The function of the LSm1-7 complex is most likely to interact with the mRNA substrate and accelerate decapping. Decapping is mediated by a decapping enzyme that is consisting of Dcp1a, Dcp1b, and the catalytic subunit Dcp2. The LSm1-7 proteins are localized in discrete cytoplasmic foci. The foci contain key decapping factors required for 5`-3` mRNA degradation. Coexpression of LSm proteins increases the number of foci. The cytoplasmic foci contain LSm1-7 proteins [73]. LSm1 and LSm8 are closely related to each other, and to the SmB protein. The 33 C terminal amino acids of LSm1 are necessary but not sufficient for proper cellular localization of hLSm1 [73]. Finally it has been demonstrated [73] that the foci are actual degradation centers, where mRNA degradation occurs. This suggests that the cytoplasm of cells is more organized than previously thought. Bacterial Hfq protein (pdb 1hk9) is able to chaperone RNA-RNA interactions similarly like LSm proteins ability to chaperone RNA/protein interactions and protect the 3' end of a transcript from exonucleolytic decay while encouraging degradation through other pathways [74].

2.8. LSm proteins in the processing of pre-tRNAs

It has been reported that depletion of LSm proteins in yeast leads to strong accumulation of unspliced tRNA species. The absence of LSm proteins most probably alters the pattern of processing intermediate [75].

2.9. LSm 2-7 complex associated with snR5

An LSm2-7 hexameric complex is found to be associated with snR5 in *Saccharomyces cerevisiae*. This RNA is a member of the class of snoRNAs that function in pseudouridylation of rRNA. The SnR5 associated LSm complex may be a hexamer, but it is not excluded that one LSm protein in this complex is present in more than one copy, or an as yet unidentified yeast protein associates with LSm2-7, thereby closing the heptameric ring [76]. LSm2-8 interacts with external 3` sequences on U6 snRNA. The LSm1-7 complex interacts with 3` UTR mRNA but there could well be secondary structure elements between the LSm1-7 binding site and the mRNA 3′ end. U8snoRNA and snR5 bind LSm proteins via internal RNA sequences, suggesting that LSm rings can assemble onto the RNA. LSm proteins have a role in the biogenesis and function of at least a subset of nucleolar RNAs. One possibility is that LSm proteins assist base

http://dx.doi.org/10.5772/57592

pairing of snoRNAs with their rRNA targets, to conduct pseudouridylation and ribose modifications.

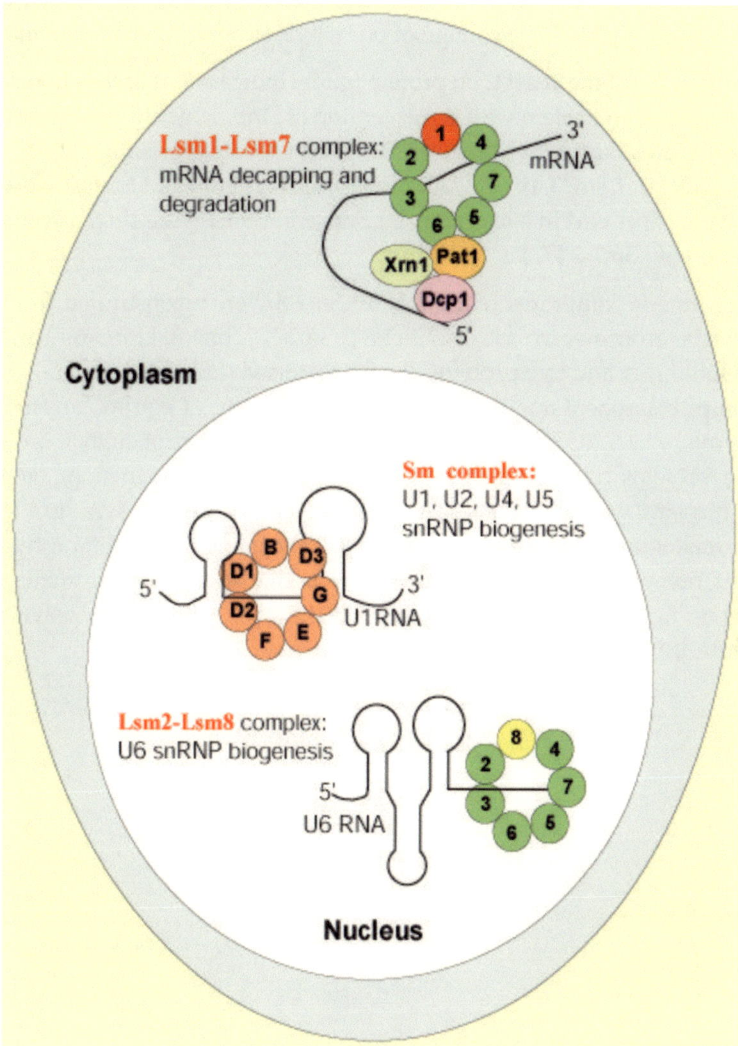

Figure 10. Three different heptameric complexes contain Sm or LSm proteins, reprinted from [71]

A particularly interesting example of forming higher order complexes-oligomers is the Sm/LSm protein family (whose various complexes are described above), whose members are engaged in a variety of RNA processing events, forming complexes which differ sometimes only by one out of seven subunits. Another important aspect of the Sm/LSm protein family is that these proteins never occur in isolation; for proper functioning they require complex formation. Hence, the way to better understand Sm/LSm protein function is to study Sm/LSm complexes. It is difficult to determine the connection between the oligomeric state of a given protein and its function *in vivo*. Reconstitution in vitro of two human LSm complexes with seven subunits each, LSm1-7 and LSm 2-8, has been described [77, 78, 79]. The LSm2-8 complex binds to the 3′-end of U6 snRNA in the cell nucleus. The closely related cytoplasmic LSm1-7complex binds to the 3′UTR of mRNAs destined for degradation. Remarkably, LSm1-7 differs from LSm2-8 only by the exchange of one single subunit, LSm1 for LSm8.

Sequence comparisons of the yeast LSm protein family indicate that each canonical Sm protein has a corresponding LSm protein with the exception of SmB, which aligns almost equally well with LSm1 and LSm8. Based on sequence comparisons co expression vectors encoding the homologs of SmD1D2, LSm23, of SmD3B, LSm48, and of SmEFG, LSm567 were constructed and proteins were expressed in bacteria [77]. LSm4 and LSm1 were singly over expressed for the reconstitution of LSm1-7 [77].

Two heteroheptameric complexes LSm1-7 and LSm2-8 were reconstituted from two heterodimers and one heterotrimer in case of LSm2-8 (LSm2-3, LSm4-8, LSm5-6-7) and one heterotrimer, one heterodimer and two proteins singly expressed (LSm2-3, LSm5-6-7, LSm1, LSm8). Reconstitution of heteroheptamers was achieved by mixing of equimolar amounts of each appropriate protein at 37°C adding 4 M urea in order to disrupt higher order structures, because those proteins have tendency to oligomerize. After incubation, mixture of pure recombinant proteins was dialyzed against native buffer. Mixture was applied on to size exclusion chromatography, followed by the anion exchange chromatography. Last step in purification of homogenous heteroheptamers was size exclusion chromatography (peak profile shown on figure 11., and respective fractions were analyzed on polyacrylamide gel (shown of figure 12).

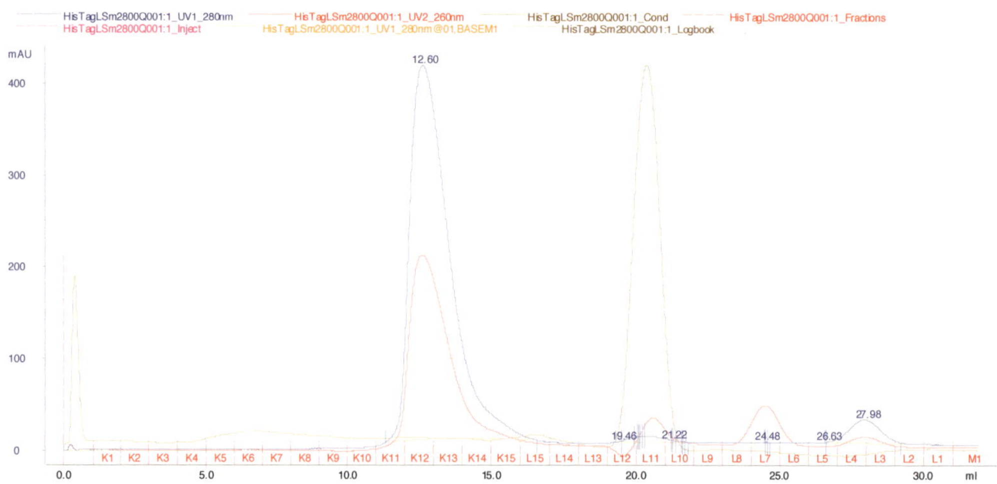

Figure 11. Second size exclusion chromatography step

Negative stain electron micrographs show that reconstituted LSm2-8 has a ring-like architecture with a diameter of about 8 nm. The overall dimensions are similar to those previously observed for the native LSm2-8 complex isolated from HeLa cell nuclear extract (8 nm) and

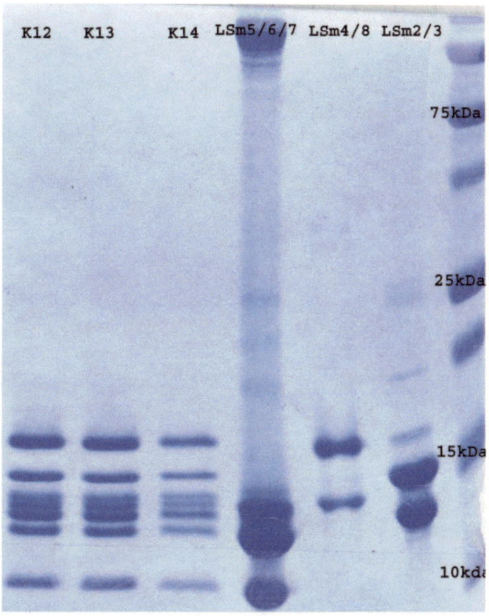

Figure 12. SDS PAGE gel

core snRNP domain [58]. The central cavity observed for the recombinant LSm2-8 complex is larger than in the native LSm2-8 complexes (3 vs. 2 nm, respectively). The LSm1-7 rings appear to be slightly smaller, measuring ~ 7 nm across with a pore diameter of less than 1.5 nm. Thus, recombinant LSm1-7 and LSm2-8 complexes are similar to one another and to the native Sm/LSm complexes at this level. In all LSm co-crystal structures solved with RNA oligonucleotides, the RNA molecules mainly wrap around the rim of the pore.

One of the methods which can be used for the identification and characterization of the RNA binding proteins is the electrophoretic mobility shift assay (EMSA). The basis of this method is the change in the electrophoretic mobility of a nucleic acid molecule upon binding to a protein or another molecule. Initially a labeled RNA, which contains the binding sequence, is incubated with a sample containing the RNA binding proteins and the mixture is then analyzed on a non-denaturing gel. The unbound RNA will have a characteristic electrophoretic mobility. Functionality of reconstituted LSm2-8 and LSm1-7 complexes has been demonstrated using this essay in vitro [77]. That oligomer complexes are functional in vivo has been shown [77], by injecting fluorescently labeled complexes into cytoplasm of living cells. They localized in expected cellular compartment, namely LSm 2-8 took nuclear localization and LSm1-7 complex remained in the cytoplasm. The structure-function relationships within the Sm/LSm protein family reflect three major interconnected features which illustrate why it is so important to solve the structures of Sm/LSm hetero-oligomeric complexes: First, Sm/LSm protein function is in general strictly dependent on complex formation. This holds for RNA binding, Sm/LSm-protein containing RNP biogenesis, interaction with non-Sm protein effector proteins, and RNA processing activity. The required interaction interfaces are apparently always three dimensional structural sites generated from several Sm/LSm subunits. High resolution

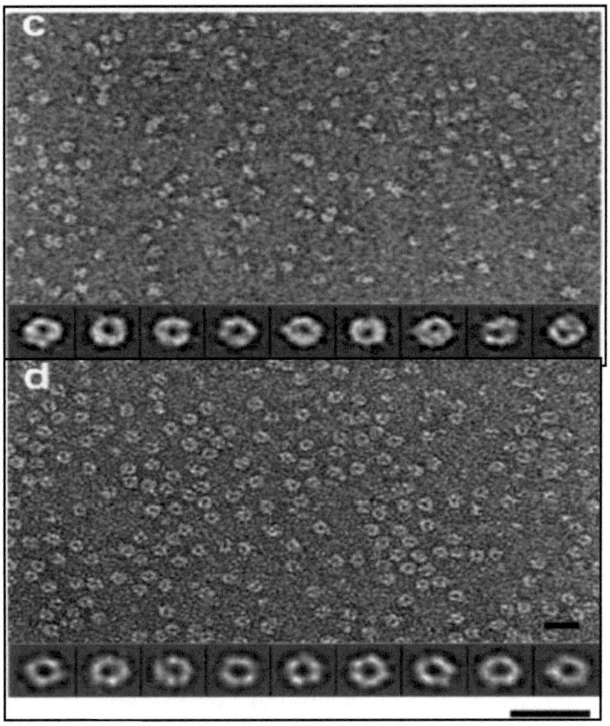

Figure 13. Electron micrographs of reconstituted complex LSm1-7 (c) and LSm2-8 (d) (reprinted from reference 77).

structural information is clearly required to explain the molecular basis for this phenomenon. Second, exchange of only one or two subunits from one to another heterooligomeric (mostly heptameric) Sm/LSm complex changes its whole biology (see above). How such subtle structural changes can have these very large functional effects can only be addressed by solving the crystal structures of the respective complexes. Lastly, the ability of individual Sm/LSm proteins to assemble with different homologous binding partners to form architecturally very similar, yet functionally diverse complexes argues for a very fine balance between flexibility and specificity for the respective Sm-Sm interactions. Clearly, in order to understand the "molecular recognition code" governing the specificity balance mentioned above, more structural information on such interactions is indispensable. Recently crystal structure of Saccharomyces cerevisiae LSm2-8 complex bound to U6 snRNA had been determined (pdb code 4M7D) [80].

Acknowledgements

This work was supported by the grant No. 172001 from the Ministry of Science and Education, Republic of Serbia.

Author details

Bozidarka L. Zaric

ICTM – Centre of Chemistry, University of Belgrade, Belgrade, Serbia

References

[1] Mark G. Caprara and Timothy W. Nilsen (2000) RNA: Versatility in form and function, *Nature Structural Biology*, (7): 831-835

[2] Steven Buckingham (2003) The Major World of microRNAs, Horizon symposia, 1-4, http://www. nature. com/horizon/rna/background/understanding. html

[3] Molecular Biology of the Cell (2001) Bruce Alberts, Dennis Bray, Julian Lewis, Martin Raff, Keith Roberts, James D. Watson, Chapter 6, Third edition, Garland Publishing, Inc. NewYork and London.

[4] Pietzsch J. (2003) Understanding the RNAissance, Horizon Symposia, 1-4, http:// www. nature. com/horizon/rna/background/understanding. html

[5] Hartman E. and Hartmann R. (2003) The enigma of ribonuclease P evolution, *Trends in Genetics*, 19(10): 561-568.

[6] Altman S., Ribonuclease P, Chapter 14 in *The RNA World* (Second edition), Cold Sping Harbor, New York.

[7] Tollervey D. and Kiss T. (1997) Function and synthesis of small nucleolar RNAs. *Current Opinion Cell Biology*, (3): 337-342.

[8] Gerbi S. A. and Lange T. S. (2002) All Small Nuclear RNAs (snRNAs) of U4/U6U5 tri-snRNP Localize to Nucleoli, Identification of the Nucleolar Localization Element of U6 snRNA., *Molecular Biology of the Cell*, (13): 3123-3137.

[9] Trotta C. R., and Abelson J., (2000) tRNA splicing: An RNA World Add-on or an Ancient Reaction? *RNA World*, second edition, Cold Spring Harbor Laboratory Press.

[10] Wong JM, Collins K. Telomere maintenanace and disease (2003) *Lancet*;362(9388): 983-8.

[11] Bortvin A. PIWI-interacting RNAs (piRNAs)-a mouse testis perspective. *Biochemistry (Mosc).* (2013):78(6):592-602.

[12] Burroughs AM, Ando Y, Aravind L. (2013) New perspectives on the diversification of the RNA interference system: insights from comparative genomics and small RNA sequencing. (2013) Wiley Interdiscip Rev RNA Dec 5. doi: 10. 1002/wrna. 1210.

[13] Steitz J. A. (1998) ``Snurps``, *Scientific American*, 36-41.

[14] Pendrak ML, Roberts DD. (2011) Ribosomal RNA processing in candida albicans *RNA.* (2011) 17(12):2235-48.

[15] Varani G. and Pardi A. (1994) Structure of RNA, Chapter 1, in *Protein –RNA interactions,* Nagai and Mattaj, IRL press, Oxford

[16] Varani G. and Kiyoshi N. (1998) RNA Recognition by RNP proteins during RNA processing, *Annu. Rev. Biophys. Biomol. Struct.* 27: 402-45.

[17] Pérez-Cañadillas J. M. and Varani G. (2001) Recent advances in RNA–protein recognition, *Current Opinion Structural Biology,* 11, 53-58.

[18] Maniatis T and Reed R. (2002) An extensive network of coupling among gene expression machines, *Nature,* 416, 499-504

[19] Proudfoot NJ, Furger A, Dye MJ.,(2002) Integrating mRNA processing with transcription, *Cell,* 108 (4), 501-512.

[20] Kornblihtt AR, de la Mata M, Fededa JP, Munoz MJ, Nogues G. (2004) Multiple links between transcription and splicing, *RNA,* 10, 1489-1498.

[21] Cougot N, van Dijk E, Babajko S, Seraphin B. (2004) Cap-Tabolism, *Trends in Biochemical Sciences,* 29(8): 436-44.

[22] Jurica Melissa and Melissa J. Moore: Capturing splicing complexes to study structure and mechanism (2002) *Methods,* 28: 336-345.

[23] Nilsen Timothy: The spliceosome: The most complex macromolecular machine in the cell? (2003) *Bioassays,* 25: 1147-1149

[24] Will CL, Schneider C, Hossbach M, Urlaub H, Rauhut R, Elbashir S, Tuschl T, Luhrmann R (2004) The human 18S U11/U12 snRNP contains a set of novel proteins not found in the U2-dependent spliceosome. *RNA,* 10(6):929-41.

[25] Yi-Tao Yu,. Scharl E.,Smith C. M., and A. Steiz J. A. The growing world of small nuclear ribonucleoproteins, *RNA WORLD,* second edition, Cold Spring Harbor Laboratory Press

[26] Balakin AG, Smith L, Fournier MJ. (1996) The RNA world in the nucleolus: Two major families of small RNAs defined by different box elements with related function, *Cell,* 86: 823-834.

[27] Will C. L and Lührmann R. (2001) Spliceosomal U snRNP biogenesis, structure and function, *Current Opinion in Cell Bilogy,* (13): 290-301.

[28] Malca H, Shomron N, Ast G (2003) The U1 snRNP Base Pairs with the 5` Splice Site within a Penta-snRNP complex, *Molecular and Cellular Biology,* 23(10): 3442-3455.

[29] Pascolo E. and Seraphin B. (1997) The Branchpoint Residue Is Recognized during Commitment Complex Formation before Being Bulged out of the U2snRNA-pre-mRNA Duplex, *Molecular and Cellular Biology,* 17(7): 3469-3476.

[30] Stevens SW, Ryan DE, Ge HY, Moore RE, Young MK, Lee TD, Abelson J (2002) Composition and functional characterization of the yeast spliceosomal penta-snRNP, *Molecular Cell*, 9: 31-44.

[31] Chen Ching-Yi, Roberto Gherzi, Shao-En Ong, Edvard L. Chan, Reinout Raijmakers, Ger J. M. Pruijin, Georg Stoecklin, Christof Moroni, Matthias Mann and Mitchael Karin (2001) AU binding proteins recruits the exosome to degrade ARE-containing mRNAs. *Cell*, 107(4) :451-64.

[32] Tucker M. and Parker R. (2000) Mechanisms and Control of mRNA decapping in Saccharomyces cerevisiae, *Annual Rev. Bioch.*, 69: 571-95.

[33] Lejeune Fabrice, Xiaojie Li and Lynne E. Maquat (2003) Nonsense-Mediated mRNA Decay in Mammalian Cells Involved Deccaping, Deadenylating and Exonucleolytic Activities, *Molecular Cell*,. 12: 675-687.

[34] Miles F. Wilkinson and Ann-Bin Shyu (2002) RNA Surveillance by nuclear scanning? *Nature Cell Biology*, 4: 144-147.

[35] Luhrmann, R., Kastner, B. and Bach, M. (1990) Structure of spliceosomal snRNPs and their role in pre-mRNA splicing, *Biochim. Biophys. Acta*, 1087: 265-292.

[36] Lerner, E. A., Lerner, M. R., Hardin, J. A., Janeway, C. A and Steitz, J. A. (1981) Monoclonal antibodies to nucleic acid containing cellular constituents: probes for molecular biology and autoimmune disease. *PNAS*, 78: 2737-2741.

[37] Hermann H, et al., (1995) snRNP Sm proteins share two evolutionary conserved sequence motifs which are involved in Sm protein –protein interaction, *EMBO Journal.*, 14(9): 2076-2088.

[38] van Dam A, Winkel I, Zijlstra-Baalbergen J, Smeenk R, Cuypers HT. (1989) Cloned human snRNP proteins B and B' differ only in their carboxy-terminal part. *The EMBO Journal*. 8(12): 3853-60.

[39] McAllister, G., Amara, S. G and Lerner, M. R. (1988) Tissue-specific expression and cDNA cloning of small nuclear ribonucleoprotein-associated polypeptide N. *PNAS, USA*, 85(14): 5296-300.

[40] Salgado-Garrido,J. ; Bragado-Nilsson,E. ; Kandels-Lewis,S. ; Seraphin,B. (1999) Sm and Sm-like proteins assemble in two related complexes of deep evolutionary origin, *The EMBO Journal*, 8(12), 3451-3462.

[41] Kambach,C. ; Walke,S. ; Young,R. ; Avis,J. M. ; de la,Fortelle E. ; Raker,V. A. ; Luhrmann,R. ; Li,J. ; Nagai,K. (1999) Crystal Structures of Two Sm protein Complexes and their Implications for the Assembly of the Spliceosomal snRNPs, *Cell*, 96:375-387.

[42] Raker, V. A., Plessel Gabriele and Lurmann Reinhard (1996) The snRNP core assembly pathaway: identification of stable core protein heteromeric complexes and an snRNP subcore particle in vitro, *The EMBO Journal*, 15(9): 2256-2269.

[43] Raker,V. A. ; Hartmuth,K. ; Kastner,B. ; Luhrmann,R. (1999) Spliceosomal U snRNPcore assembly: Sm proteins assemble onto an Sm site RNA Nonanucleotide in a specific and Thermodynamically Stable Manner, *Molecular and Cell Biology*, 19(10): 6554-6565.

[44] Walke,S. ; Bragado-Nilsson,E. ; Seraphin,B. ; Nagai,K. (2001) Stoichiometry of the Sm proteins in yeast spliceosomal snRNPs supports the heptamer ring model of the core domain, *J. Mol. Biol.*, 308:49-58.

[45] R. Bordonne (2001) Functional characterization of nuclear localization signals in yeast Sm proteins, *Molecular Cell. Biology*, 20(21):7943-54.

[46] Stark H, Dube P, Luhrmann R, Kastner B. (2001) Arrangement of RNA and proteins in the spliceosomal U1 small nuclear ribonucleoprotein particle, *Nature,* (409): 539-542.

[47] SchrankB., GötzR., Gunnersen J. M., UreJ. M.,, Toyka K. V., Smith A., and Sendtner M. (1997) Inactivation of the survival motor neuron gene, a candidate for human spinal muscular atrophy, leads to massive cell death in early mouse embryos, *PNAS*. 94: 9920-9925.

[48] Brahms,H. ; Meheus,L. ; de,Brabandere,V; Fischer,U. ; Luhrmann,R. (2001) Symmetrical dimethylation of arginine residues in spliceosomal Sm protein B/B and the Sm-like protein LSm4, and their interaction with the SMN protein, *RNA*, 7: 1531-1542.

[49] Friesen,W. J. ; Wyce,A. ; Paushkin,S. ; Abel,L. ; Rappsilber,J. ; Mann,M. ; Dreyfuss,G. (2002) A Novel WD repeat protein Component of the Methylosome Binds Sm proteins, *Journal of Biol. Chemistry*, 277: 8243-8247.

[50] Gubitz AK, Feng W, Dreyfuss G. (2004) The SMN complex, *Experimental Cell Research* 296:, 51-56.

[51] Yong J, Pellizzoni L, Dreyfuss G. (2002) Sequence –specific interaction of U1 snRNA with the SMN complex, *The EMBO Journal*, 21(5):1188-1196.

[52] Narayanan U, Ospina JK, Frey MR, Hebert MD, Matera AG. (2002) SMN, the spinal muscular atrophy protein forms a pre-import snRNP complex with snurportin1 and importin β, *Human Molecular Genetics*, 11(15): 1785-1795.

[53] Paraskeva E, Izaurralde E, Bischoff FR, Huber J, Kutay U, Hartmann E, Luhrmann R, Gorlich D. (1999) CRM-1 mediated Recycling of Snurportin1 to the Cytoplasm, *The Journal of Cell Biology*, 145(2): 255-264.

[54] Massenet,S. ; Pellizzoni,L. ; Paushkin,S. ; Mattaj,I. W. ; Dreyfuss,G. et al (2002) The SMN Complex is Associated with snRNPs throughout their Cytoplasmic Assembly Pathway, *Molecular and Cellular Biology*, 22(18):6533-6541.

[55] Cameron Mura,Peter S. Randolph, Jennifer Patterson and Aaron e. Cozen (2013) Archaeal and eukaryotic homologs of Hfq A structural and evolutionary perspective on Sm function. *RNA Biology* 10:4, 636–651.

[56] Anna Golisz, Pawel J. Sikorski, Katarzyna Kruszka and Joanna Kufel (2013) Arabidopsis thaliana LSM proteins function in mRNA splicing and degradation *Nucleic Acid Research* 41:6232-6249.

[57] Camasses,A. ; Bragado-Nilsson,E. ; Martin,R. ; Seraphin,B. ; Bordonne,R. (1998) Interactions within yeast Sm core complex: from proteins to amino acids., *Mol. Cell. Biol.* 18: 1956-1966.

[58] Achsel,T. ; Brahms,H. ; Kastner,B. ; Bachi,A. ; Wilm,M. ; Luhrmann,R. (1999) A doughnut-shaped heteromer of human Sm-like proteins binds to the 3` end of U6 snRNA, thereby facilitating U4/U6 duplex formation in vitro, *The EMBO Journal,* 18(20):5789-5802.

[59] Eystathioy T, Peebles CL, Hamel JC, Vaughn JH, Chan EK. (2002) Autoantibody to the hLSm4 and the Heptameric LSm complex in Anti-Sm Sera, *Arthritis and Rheumatism,* 46(3), 726-734.

[60] Hirsch,E. ; Oohashi,T. ; Ahmad,M. ; Stamm,S. ; Fassler,R. (2000) Peri-Implatantation Lethality in Mice Lacking the Sm Motif-containing protein LSm4, *Molecular and Cellular Biology,* 20 (3): 1055-1062.

[61] Vankan,P. ; McGuigan,C. ; Mattaj,I. W. (1990) Domains of U4 and U6 snRNAs required for snRNP assembly and splicing complementation,*The EMBO Journal.*,9: 3397-3404.

[62] Lange TS, Gerbi SA. (2000) Transient Nucleolar Localization of U6 snRNA in Xenopus Laevis Oocytes, *Molecular Biology of the Cell,* 11:2419-2428.

[63] Fury, M. G. and Zieve, G. W. (1996) U6 snRNA maturation and stability, *Experimental Cell Research,* 228: 63-69.

[64] Mayes,A. E. ; Verdone,L. ; Legrain,P. ; Beggs,J. D. (1999) Characterization of Sm-like proteins in yeast and their association with U6 snRNA, *The EMBO Journal,* 18 (15): 4321-4331.

[65] Weihai He and Parker R. (2001) The Yeast Cytoplasmic LSmI/pat1p complex protects mRNA 3` termini from partial degradation, *Genetics,* 158:1445-1455.

[66] Tomasevic,N. ; Peculis,B. A. (2002) Xenopus LSm proteins Bind U8 snoRNA via an Internal Evolutionarily Conserved Octamer Sequence, *Molecular and Cellular Biology,*. 22(12): 4101-4112.

[67] Muller B. and Schumperli D. (1997) The U7 snRNP and the hairpin binding protein: Key players in histone mRNA metabolism, *Seminars in cell and developmental biology,* 8: 567-577.

[68] Pillai RS, Grimmler M, Meister G, Will CL, Luhrmann R, Fischer U, Schumperli D. (2003) Unique Sm core structure of U7 snRNPs: assembly by a specialized SMN complex and the role of a new component, LSm11 in histone RNA processing, *Genes and Development* 17: 2321-2333.

[69] Fromont-Racine M, Mayes AE, Brunet-Simon A, Rain JC, Colley A, Dix I, Decourty L, Joly N, Ricard F, Beggs JD, Legrain P.. (2000) Genome-wide protein interaction screens reveal functional networks involving Sm-like proteins, *Yeast,* 95-110.

[70] Tharun,S. ; He,W. ; Mayes,A. E. ; Lennertz,P. ; Beggs,J. D. ; Parker,R. (2000) Yeast Sm-like proteins function in mRNA deccaping and decay, *Nature,* 404: 515-518.

[71] Bouveret,E. ; Rigaut,G. ; Shevchenko,A. ; Wilm,M. ; Seraphin,B. 1 (2000) A Sm-like protein complex that participates in mRNA degradation, *The EMBO Journal,* 19 (7): 1661-1671.

[72] Panone B. K. and WolinS. L. (2000) RNA degradation: Sm-like proteins WRING the neck of mRNA, *Current Biology,* 10: 478-482.

[73] Ingelfinger D, Arndt-Jovin DJ, Luhrmann R, Achsel T. (2002) The human LSm1-7 proteins colocalize with the mRNA-degrading enzymes Dcp1/2 and Xrn1 in distinct cytoplasmic foci, *RNA,* 8: 1489-1501.

[74] Cougot N, van Dijk E, Babajko S, Seraphin B. (2004) Cytoplasmic foci are sites of mRNA decay in human cells, *The Journal of Cell Biology,* 165(1): 31-40.

[75] Carol J. Wilusz and Jeffrey Wilusz (2013) Lsm proteins and Hfq Life at the 3' end". *RNA Biology* 10:4, 592–601.

[76] Kufel J, Allmang C, Verdone L, Beggs JD, Tollervey D. (2002) LSm proteins Are Required for Normal Processing of the Pre-tRNA and Their Efficient Association with La-Homologous Protein Lhp1p, *Molecular and Cellular Biology,* 22(14):5248-56.

[77] Fernandez CF, Pannone BK, Chen X, Fuchs G, Wolin SL. (2004) An LSm2-7 Complex in Saccharomyces cerevisiae Associates with the Small Nucleolar RNA snR5, Molecular Biology of the *Cell,* (15):2842-2852.

[78] Zaric B., Chami M., Remigy H., Engel A., Ballmer-Hofer K., Winkler FK and Kambach C. (2005) Reconstitution of two Recombinant LSm protein complexes reveals aspects of their architecture, assembly and function, *Journal of Biological Chemistry,* 280: 16066-16075.

[79] Karen C. M. Moraes, Naomi Bergman, Bozidarka Zaric, Christian Kambach, Carol J. Wilusz& Jeffrey Wilusz,(2007)Lsm proteins bind to and stabilize RNAs containing; poly(A) tracts *Nature Structural Molecular Bioogy.* ;14(9):824-31.

[80] Bozidarka Zaric and Christian Kambach (2008)"Reconstitution of recombinant human LSm complex for biochemical, biophysical and cell biological studies". *Methods in Enzymology* 448:57-74.

[81] Zhou L, Hang J, Zhou Y, Wan R, Lu G, Yin P, Yan C, Shi Y(2013) Crystal structures of the Lsm complex bound to the 3' end sequence of U6 small nuclear RNA. Nature Nov 17. doi: 10. 1038/nature12803.

Chapter 8

Protein Oligomerization

Giovanni Gotte and Massimo Libonati

Additional information is available at the end of the chapter

http://dx.doi.org/10.5772/57489

1. Introduction

Protein oligomerization is a wide and fascinating topic concerning the behavior of proteins that can form supramolecular structures, either naturally or artificially. Proteins can homo- or hetero-oligomerize through a covalent, almost always irreversible stabilization, or through often reversible associations mediated by electrostatic and hydrophobic interactions, or H-bonds. The structural and functional aspects of protein oligomerization have acquired increasing importance especially in the last two decades. The improvement of the X-ray analyses quality, and NMR potential, as well as the incoming of dynamic light scattering (DLS) or surface Plasmon resonance (SPR) techniques allowed to understand features unknown before or to correct notions that were wrongly believed true. Protein oligomerization is often a phenomenon crucial in triggering various physiological pathways. On the contrary, in different compartments other protein oligomers can be the first deleterious seed driving to protein fibrillization, an event implicated in several devastating neurodegenerative diseases. In the latter case, the isolation and analysis of the oligomeric species, considered as the real toxic agents, remained elusive for a long time. Only very recently new techniques, such as solid-state NMR, Cryo-transmission electron microscopy (Cryo-TEM), High-Resolution Atomic-Force Spectroscopy, Molecular Modeling, allowed to discover structural and functional data that can clarify the determinants of a very complicated pathway.

Under the light of these recent discoveries, the chapter aims at exploring the most important structural and functional aspects of natural or artificial oligomers.

In the first paragraph after this we introduce a tentative rationalization of the terms related to the wide world of protein oligomerization; in the second we unveil the structural and mechanistic features of the different protein oligomers that can natively or artificially form; in the third we analyze the stability of protein oligomers and the factors influencing or affecting it; in the fourth we describe the functional (benign) versus the aberrant interactions determined by protein oligomerization; in the fifth paragraph, finally, some hints related to possible industrial applications of protein oligomers will be mentioned.

Numerous cases and literature reports related to different protein will be described, especially the ones whose oligomers have been extensively studied. In this context, ribonucleases (RNases), and ribonuclease A in particular, which have become useful and interesting models, will be considered.

2. Natural and/or artificial protein polymers: Oligomers, multimers, aggregates, fibrils; tentative definitions and classifications

Before describing and analyzing the protein oligomerization events in detail, some definitions are necessary.

First of all, proteins can self- or cross-associate either naturally or artificially. The artificial phenomenon can occur when the environmental conditions of a protein solution are changed, or cross-linking chemical reaction(s) are introduced.

If a monomeric protein, i.e., lacking quaternary structure, is considered as the starting species, we can obtain the first polymeric seed when a dimer, or different dimers, form. Then, polymerization can continue towards trimer(s), tetramer(s), up to decamer(s) and so on. All these protein species are polymers that in the literature are termed, in turn, oligomers, multimers, or large aggregates, protofibrils and fibrils. The smallest subset of different subunits forming an oligomer is the structural unit of an oligomeric protein, and can also be called protomer. A promoter can be a protein subunit or several different subunits that assemble(s) in a defined stoichiometry to form an oligomer. For example, hemoglobin consists of four subunits, two α-chains and two β-chains. The oligomer stoichiometry is thus $\alpha_2\beta_2$. Hemoglobin is a hetero-tetramer, but it is also a dimer of two $\alpha\beta$-protomers. Several authors use the terms protein oligomers, multimers or aggregates, to define protein species that range between a dimer and a proto-fibril. Consequently, the reader cannot withdraw the right meaning of what is reported unless from the context in which these terms are used. These different definitions are due to the lack of a clear-cut terminology assignment to the various supra-molecular protein products forming.

Although the elegant classification of homomers recently reported by Levy & Teichmann [1] is based on the symmetry and size of the structures, i.e., the number of subunits, it does not distinguish between homo-oligomers and large homo-polymers. Thus, we introduce here a tentative rationalization of the terminologies. In addition, rules that may help to assign the right terminology are lacking also because protein oligomers can be distinguished in two different ways: a) on the basis of the number of subunits forming the polymer (i.e., on the basis of the polymerization degree) or b) on the basis of the molecular weight (M.W.) of the polymer.

These two different approaches represent the principal controversy that may generate confusion. In fact, if an oligomer can be formed by two to eight-ten protein subunits, a protein of 50 kDa can be considered oligomeric also when its M.W. reaches 400-500 kDa. Instead, a 10 kDa protein is oligomeric when it reaches a weight of 80-100 kDa. On the contrary, if the oligomeric limit were a MW of, hypothetically, 100 kDa, the former protein would be larger than an oligomer already at a trimeric status, while the latter would be an oligomer also when composed by ten subunits, thus being a decamer. Protein oligomerization is heavily involved also in the formation of amyloidogenic (or not) fibrils, and in this scenario it is difficult to find the most appropriate terms able to indicate oligomers or other supramolecular structures preceding fibrillization [2] because of the transient nature of these structures. Moreover, it is almost impossible to determine the exact number of subunits forming a fibril, although the main criterion to distinguish an oligomer from a fibril is that the former is soluble, while the latter is not.

Consequently, the M.W. classification (low-molecular/high-molecular weight oligomers) becomes prevalent in the 'fibrillization' context [3, 4], while the classification based on the polymerization degree [5, 6] can be useful in the context of natural or artificial, limited or controlled, self- or hetero-protein(s) association.

To make the story short and to tangibly classify the protein polymerization (oligomerization) events we can enter two paths, one following controlled oligomerization, the other trapping protein oligomers before undergoing fibrillization and using Aβ-amyloid peptide of 40-42 residues as the monomeric unit [7]. From these two different, although possibly overlapping terminologies, although referring to different structures and/or structural elements, the following classification could be adopted:

1. Limited protein polymerization (not resulting in fibers)
   a. Oligomers: from dimer(s) to pentamer(s)
   b. Multimers or higher-order oligomers: from hexamer(s) to pentadecamers
   c. Very large oligomers or large multimers: from pentadecamers up to more than 20-30mers.
2. Uncontrolled and extensive oligomerization driving to fibrillization:
   a. (Aβ) Pre-Fibrillar Oligomers: from a tetramer of 18 kDa to structures of about 75 kDa [7].
   b. (Aβ) Fibrillar Oligomers: from a dimer or a tetramer of 9-18 kDa to structures of about 500 kDa [7].
   c. Proto-fibrils: structures lacking order and periodic symmetry of mature fibrils; shorter and less linear than fibrils [8], and often deriving from 'globulomers' [9] or 'annular oligomers' [10], already called 'ring-like shaped' annular aggregates [11].
   d. Fibrils: ordered, symmetrical, long, varying in length, extending towards several micrometers [12], stacked filaments of unbranched cross-β-spine [13] pairs, ranging from about 60 up to 200 Å in diameter or width [8, 14].

Thus, different types of oligomers deriving from the same protein or domain can certainly overlap in terms of size and M.W. Consequently, it has to be also taken into account that the same protein can be considered as to be fibrillogenic or not fibrillogenic not only on the basis of its dimensions or M.W., but also because of its morphology, toxicity, pathway of formation, or method of artificial formation of its oligomers [2, 7, 8]. Furthermore, these supramolecular species can be toxic or harmless provided they satisfy, or not, some requisites [2, 15] that will be discussed in the following paragraphs.

Based on what it has been reported, the terms oligomers or others related to supramolecular protein structures may be still somehow misleading and equivocal. Anyway, the tentative classification here proposed and based on two different contexts, i.e., fibrillogenesis or not, will be used in the following paragraphs.

3. Protein oligomers

3.1. Homo/hetero-dimeric or oligomeric proteins

Several studies have been focused, especially in the last two decades, on the features of homo-polymeric or, more precisely, homo-oligomeric proteins, i.e., supramolecular structures formed by self-associating proteins. In this context, a great amount of structural, mechanistic, physicochemical, and functional elements have been discovered and elucidated as important determinants which tune and control protein self-association.

Instead, except the very well known heterotetrameric hemoglobin, less is known about hetero-oligomeric protein complexes. These hetero-structures refer to chains of different sequences, which undergo association pathway in a way less statistically favorable and easily controllable, either qualitatively or quantitatively, than protein self-association. Nevertheless, protein hetero-oligomerization represents a very important phenomenon in the formation of molecular machines, like for motor proteins (kinesin, microtubules,), or, alternatively, can be obtained artificially by the use of asymmetric bifunctional reagents, while new discoveries of natural hetero-protein association events lighted up again the interest over this topic.

3.2. Covalently linked oligomers

Although not very frequently, protein cross-linking can occur naturally forming covalently linked species that display quaternary structures, or active covalent complexes starting from inactive monomeric precursors. Post-transduction modifications, photochemical event(s), or co-enzyme binding (i.e., going from apo-to holo-forms), can also induce protein self-or cross-linking.

Natural cross-linking can sometimes occur through free cysteines of two different subunits that can couple to form intermolecular disulfides depending on the redox state of the environment. This is the case, for example, of bovine seminal RNase (BS-RNase), the unique member of the large pancreatic-type secretory ribonuclease super-family which is dimeric in nature [16]. Other proteins that can form covalent supramolecular structures are structural proteins such as collagen or elastin, the latter forming or disrupting in tissues or vessels desmosine bridges forming between lysine residues after elastic stress or relaxation [17].

Alternatively, artificial cross-linking(s) can produce oligomers after the reaction of a protein with dehydrating molecules, such as EDC or carbodihymides in general, or through the use of several bifunctional reagents, such as dialdehydes or diimidoesters [18, 19]. These two latter chemicals display two terminal reactive groups separated by a variable number of unreactive spacers, such as methylenes. They are often symmetric, but sometimes asymmetric. The alternative use of them becomes advantageous depending on the goal to be reached, as for example for a homo-or hetero-protein cross-linking, respectively.

Some of these different reagents were extensively used already in the late '50s and '60s to produce protein oligomers that have been useful, after limited proteolysis, to study protein primary and tertiary structures, and, thus, protein conformations. Later, the formation of covalently linked oligomers allowed sometimes the production of protein derivatives with increased activity, and also higher stability against proteases, and so on.

One of the first class of reagents used in protein cross-linking were dialdehydes, such as glutaraldehyde (Figure 1 a) or diimidoesters [19], like dimethyladipimidate, dimethylpimelimidate and dimethylsuberimidate (Figure 1 b).

Both classes of these cross-linkers react mainly with lysine residues, which act as nucleophyles towards the aldehydic or imidic carbon of the cross-linker (Fig. 1 c) under slightly basic environmental conditions (pH ranging from 7.5 to 8.5-9). By this reaction a yield of more than 20% of dimers can be obtained, as well as a decreasing amount of trimers, tetramers and traces of higher-order oligomers.

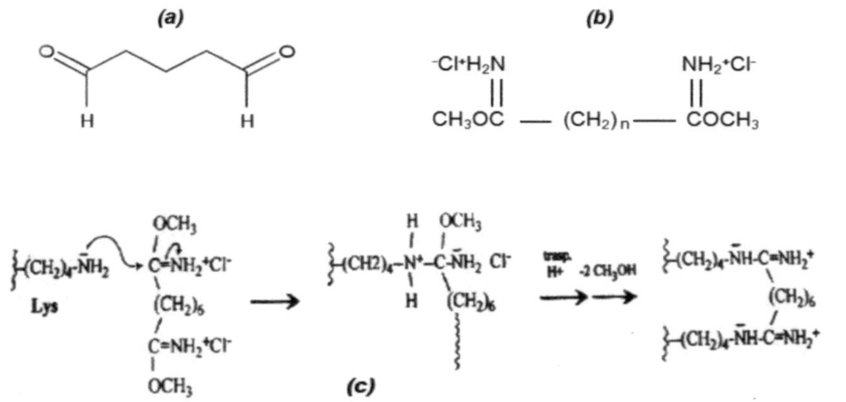

Figure 1. (a) Glutaraldehyde. **(b)** Diimidoesters: n=4, dimethyl-adipimidate; n=5,-pimelimidate; n=6,-suberimidate. **(c)** Mechanism of reaction of a protein with an imidoester (dimethylsuberimidate).

The advantage represented by diimidoesters with respect to dialdehydes, like glutaraldehyde, is that the latters are toxic, generally more reactive than diimidoesters and consequently often able to introduce unwanted changes in a protein. Moreover, sometimes they strongly drive towards side intramolecular reactions with a possible inactivation of the protein. Diimidoesters reactivity, on the contrary, can be better controlled although protein lysines may be modified, and allows to maintain the overall charge of the protein unmodified. Consequently, despite the oligomeric yields obtained with diimidoesters are lower than those of dialdehydes, the products obtained are more specific and active than the ones obtained with the latters. Finally, the longer the spacer, the higher the intermolecular yield and the less rigid oligomeric product obtainable [19]. Thus, dimethylsuberimidate is more useful to artificially oligomerize a protein than the other shorter diimidoesters mentioned above.

Other reagents that can be useful to obtain protein oligomers are the bifunctional N-substituted maleimide derivatives [20], such as *N,N'*1,2-or 1,3-or 1,4phenyldimaleimide (orto-meta-para-PDM) (Figure 2 a-c), or derivatives displaying spacers of varying length between the two reactive maleimides (Figure 2 d-f). The reaction occurs between the maleimide and the free sulfhydryl group of a cysteine of the protein as an addiction which saturates the double link of the maleimide conjugated with two carbonyls (Figure 2 g) [20].

The cross-linking event can be irreversible or reversible. Reversibility can be favored, for example, by the use of spacers containing disulfide bonds (Figure 2 g). These "redox" spacers can be applied to all bifunctional reagents displaying spacers if the reversibility of the reaction is desired, and are very tricky for asymmetric bifunctional linkers.

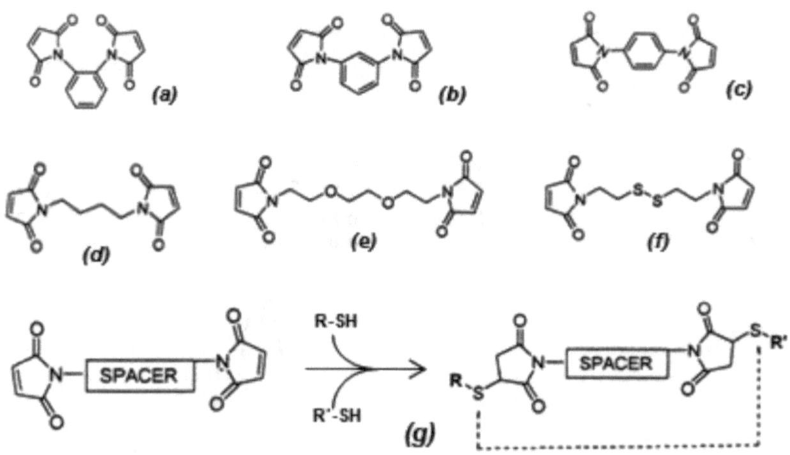

Figure 2. Some bifuctional maleimides.(a), (b), (c) orto-meta-para N, N'phenylenemaleimmide (o,m,pPDM); **(d)** 1,4-bismaleimidobutane (BMB); **(e)** 1,8-bismaleimidotriethyleneglycol (BM[PEO]$_3$); **(f)** dithio-bis-maleimidoethane (DMTE); **(g)** Reaction of a bis-maleimide derivative with sulfhydryl groups (free, reduced cysteines). Figure adapted from [20].

This can allow, for example, to obtain species that need to be driven inside cells as heterodimers and then to be released as dissociated monomers exploiting the reducing environment of the cytosol. Thus, spacers containing disulfides can be applied to all bifunctional reagents if the reversibility of the reaction is desired.

Asymmetric bifunctional reagents can be very useful to covalently link antibodies or part of them (light/heavy chains) with proteins, protein domains, or toxins, or whatever of biological interest. They can be a combination of maleimides or succinimides displaying or not dithioderivatives in the spacers, and coupled, in the second terminus, with imidoesters, diones, thiones, 2-iminothiolane, etc. Almost all of these bifunctional cross-links foresee a two-step reaction in which one of the partners is firstly modified and activated (for example, with 2-iminothiolane) in order to become able to react with the appropriate partner and form the new hetero-dimeric adduct. Some of the most important reactants used for artificial heterodimerziation are 1-(3-((2,5-Dioxopyrrolidinyl)oxy-carbonyl)phenyl)-1H-pyrrole-2,5-dione (MBS) N-succinimidyl-3-(2-pyridyldithio)propionate (SPDP), S-4-succinimidyloxycarbonyl-α-methyl benzyl thiosulfate (SMBT), or 4-succinimidyloxycarbonyl-α-methyl-α(2pyridyldithio)toluene (SMPT). The details of these hetero-oligomerization reactions are beyond the aim of this chapter but are clearly described in the review of Fracasso and colleagues [21].

Again, also another dione, 4-phenyl, 1,2,4-triazoline-3,5-dione (PTD), structurally similar to maleimides but containing nitrogens in the pentagon ring, can be useful to covalently cross-link proteins through amino-groups [22].

Other important bifunctional reagents are carbodiimides (R_1-N=C=C-R_2^+), such as EDC, which exploits its dehydrating power to create new isopeptide bonds between the side chains of Lys and Glu or Asp residues of proteins [23], thus producing a "zero-length" cross-linking. This reagent can be very useful to covalently fix previously formed oligomeric protein aggregates [24] and allows to avoid unwanted insertions of chemicals or net charge modifications in the protein complex.

Divinyl sulfone (DVS) [25] and dinitrodifluorobenzene (DFDNB) [26] are two bifunctional reagents that deserve to be mentioned (Figure 3). They lack spacers, and to stabilize preformed structures without affecting oligomers' conformations [25, 27-29].

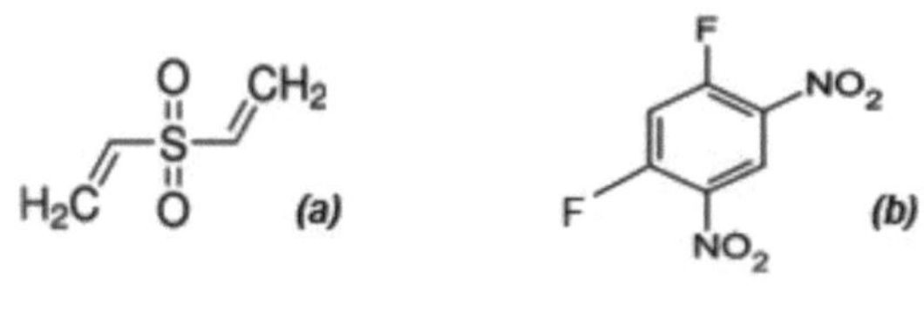

Figure 3. (a) Divinyl sulfone (DVS). **(b)** Dinitrodifluorobenzene (DFDNB)

DVS is specific for histidines, while DFDNB reacts with lysines. The limited dimensions of both molecules allow cross-linking only between residues that are very close to each other. For this reason subsequent chromatographic or electrophoretic analyses can highlight important conformational features of the protein or its oligomers. The cross-linking can drive towards intra-or inter-molecular adducts, thus revealing if the protein was monomeric or already oligomeric before the covalent stabilization [25].

Finally, covalent protein "zero-length" oligomerization can be obtained, without chemicals, by sealing a lyophilized protein under vacuum at high temperature, up to 85 °C, for 24-96 hours [30]. This process permits to obtain dimers, trimers, and traces of tetramers of ribonuclease A (RNase A) and lysozyme, without the introduction of chemical groups which could totally or partially inactivate the protein residues involved in the reaction. In fact, the heat-vacuum treatment of proteins induces the dehydration of some of the Lys and Asp or Glu side-chains, thus producing newly formed intermolecular isopeptide bonds. However, this reaction, firstly considered to be specific of a single couple of Lys and Glu of RNase A [31], was later found to involve more than one of these acid or basic residues, thus producing a mixture of heterogeneous products [32].

Last, but not least, cross-linked oligomers can be produced also through UV photochemistry. The UV-treatment of a monomer or pre-formed non-covalent oligomers [33] produces covalently stabilized oligomers. If the photochemical treatment inactivates the native protein or its possibly pre-formed oligomers, structural important information can be anyway withdrawn from these artificial modifications [34, 35].

3.3. Non-covalent protein oligomers

Protein association can occur very often naturally, and without covalent modification(s), through a homo-or hetero-association mediated by a weak-bond-network, formed by electrostatic or hydrophobic interactions, and/or specific H-bond(s). If the interface between monomers or protomers is large almost all types of interactions can occur and are more frequently conserved. Small interfaces have been probably acquired recently in evolution [36]. These interactions can be crucial for the active forms of several class of proteins, such as enzymes and transporters. These interaction(s)/association(s) may occur naturally because of the sequence and structural features of the subunits which build the oligomeric complex(es). Otherwise, they can form because of environmental changes, like pH or ionic strength, or as a consequence of the increase of the monomers local concentration.

These events allow the protein to overpass its dimerization dissociation constant(s) $K_{D1/2}$ (equation 1 and Figure 4, red and orange).

$$\mathbf{K_D} = [\mathbf{M}]^2 / [\mathbf{D}] \qquad \textit{with } M = \textit{monomer},\ D = \textit{dimer} \tag{1}$$

A further increase of the concentration can augment the degree of polymerization with the formation of trimers (Figure 4, green pictures), tetramers (Figure 4 magenta, pink, violet), and larger oligomers or multimers if the entropy cost is balanced by favorable interface interaction(s).

The free energy of multimerization is reported in equation 2 [37].

$$\Delta \mathbf{G} = \Delta \mathbf{G^0} + \mathbf{RT\ ln}\left([\mathbf{nP}] / [\mathbf{P}]^{\mathbf{n}} \right) \tag{2}$$

in which [P] is the concentration of protein segment(s) exposed, or protein interfaces [37], and prone to interact with other protomers.

http://dx.doi.org/10.5772/57489

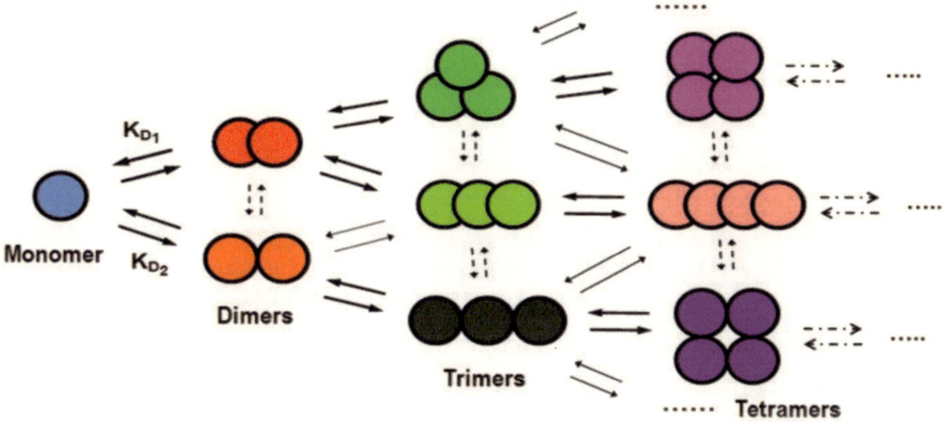

Figure 4. Schematic view of protein oligomerization. A native monomer (blue) can become a dimer if K_D of dimerization (equation 1) is overpassed. Two possible dimeric conformers with different interfaces or interface areas are shown to form with K_{D1} and K_{D2} respectively. These events can lead to the formation of different trimers (three green conformers are shown), tetramers (magenta, pink, violet) and so on. Trimers and tetramers unlikely form from isolated monomers because of entropic reasons, but can grow from the pre-existing oligomer(s) (or also from two dimers). Oligomers can follow different pathways (dotted lines) to form higher-order oligomers or multimers, and can undergo conformational rearrangements which can compensate the entropy cost to be payed to associate the subunits.

An example of natural oligomeric protein is represented by the oxygen transporter hemoglobin (Hb), which is functionally active only as an $\alpha_2\beta_2$ tetramer (or, in the fetus, $\alpha_2\gamma_2$, endowed with a greater affinity to oxygen) with its subunits being associated through salt bridges and other weak interactions. The loss of these interactions drives Hb to switch its conformation from the deoxygenated (tense, T) to the oxygenated (relaxed, R) form, and this modification is due to allosteric interactions triggered by the first oxygen bound (Figure 5). Hb, unless being tetrameric, cannot be active, while myoglobin, the oxygen collector in tissues, is active as a monomer.

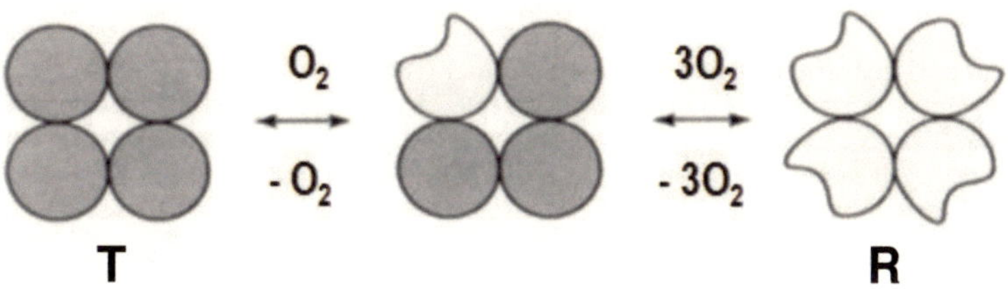

Figure 5. Allosteric control of haemoglobin (Hb) induced by oxygen binding.

Another important oligomeric protein transporter is transthyretin (TTR), formerly called pre-albumin, one of the transporters of the hormone thyroxine and of the lipocalin retinol-binding protein (RBP), the specific carrier of A-vitamin [38]. TTR is natively a 55-kDa dimer of dimers, or homotetramer, mainly composed of β-sheets [39]. Its peculiarity consists of the pathologic pathway it follows if destabilized by malignant mutations: if so, monomers dissociate from tetrameric assembly and undergo uncontrolled aggregation and fibrillization through the formation of intermediate annular oligomers [10]. These findings will be further discussed below.

Other cases of non-covalent association between proteins can be represented, for example, by 5'pyridoxal-phosphate (PLP) enzymes, such as the aspartic aminotransferase (AAT), alanine-glyoxylate-aminotransferase (AGT) (Figure 6a), dopa-decarboxylase (DDC) (Figure 6 b), and cystalisin (a lyase, like DDC). Inactive as monomers, these enzymes are active only when they are in form of dimers, although a detailed analysis of their dimerization pathways has been performed only with *Treponema Denticola* cystalisin mutants [40].

The family of the PLP-dependent enzymes is very large, forming five different fold types, and some members of the family are active as tetramers or hexamers. AGT [41] and DDC [42], shown in Figure 6, are dimers belonging to Fold Type I, as well as cystalysin and AAT. Belonging to Fold Type I means that each subunit of the holo-form host a PLP molecule, but the active site is composite, i.e., formed by residues of both subunits. Instead, enzymes belonging to Fold Type II are active as dimers or oligomers binding one PLP each at the same time, but they evolved to form one active site per subunit, and are often accompanied by the presence of allosteric regulation domains.

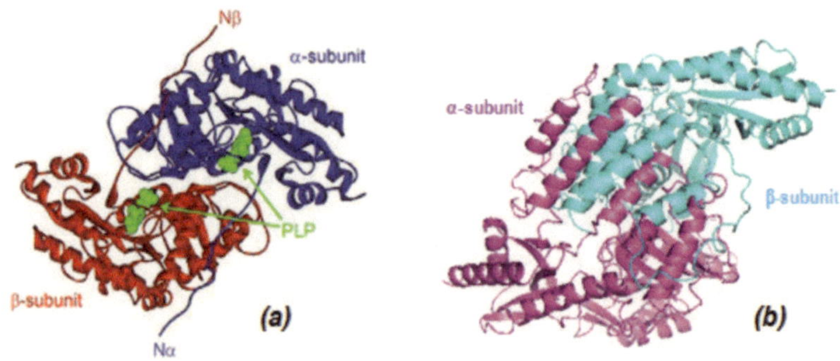

Figure 6. Three dimensional structure of holo natively dimeric(a) human liver alanine-glioxylate aminotransferase (hAGT) [41] **(b)** pig kidney dopa decarboxylase (pkDDC) [42] (Burkhard P. *et al* 2001, *NSB* **8** (11) 963-7). The two α/β subunits are shown with different colours and for AGT the location of the two PLP molecules is shown in green. AGT is peculiar for the wrapping of each terminal arm (Nα/Nβ) into the region occupied by the complementary subunit.

This evolution pathway allowed them to build domains which are active as allosteric regulators. The association of the subunits of these PLP-enzymes is often mediated mainly by hydrophobic interactions because the monomeric forms are readily prone to uncontrolled extensive aggregation [35]. Anyway, the inter-subunit surfaces being often very large, several types of contacts occur, such as electrostatic interactions or H-bonds.

Another interesting example of oligomeric protein is represented by mammalian Phenylalanine Hydroxylase (PAH), which is a homo-tetrameric enzyme made of four 50 kDa subunits whose overall 3D structure has not been solved yet. Its activity depends on its tetrameric structure and the presence of tetrahydrobiopterin (BH_4), and is allosterically regulated by the substrate itself (Phe). Some mutations become pathogenic because they destabilize and inactivate the tetramer, and consequently drive the organism towards the incoming of Phenylketonuria (PKU) [43].

A natively oligomeric protein can also switch towards higher-order oligomers. Indeed, the natively homotetrameric L-rhamnulose-1-phosphatase aldolase becomes an octamer only upon a single A88F mutation [44]. The introduction of a single residue displaying large non-polar side-chains (Phe, Trp) can be sufficient to drive the native oligomer towards larger multimeric complex(es).

It has to be mentioned here that from a total of about 450 well characterized enzymes, only about 140 of them are monomeric. Of the other 310, 200 are homo-oligomers/multimers and in detail: 125 homodimers, 50 homotetramers and 25 are structures larger than tetramers. Finally, the remaining about 110 are hetero-oligomers/multimers [45].

Protein oligomerization can also be a non-native event: indeed, natively monomeric proteins can naturally and non-covalently undergo oligomerization as a sort of post translational event which can become a switch between active and non active products. This is true, for example, for several trans-membrane receptors, which often display kinase activity. Upon ligand binding, the intracellular domain dimerizes, this event triggering (auto)phosphorilation of the intracellular domain which undergoes conformational changes and is able to activate a signal transduction cascade that induces or tunes important physiological phenomena. Examples of families of this type of receptors are: growth ormone, interferon, cytokine and Tyr-kinase, G-protein-coupled receptors families (GPCRs) [46]. The components of the latter family were initially thought to act as monomers, but several pharmacological, biochemical and biophysical data indicate that GPCRs function as cooperatively controlled dimers [47].

Another very interesting example is represented by the Caspase-3,-7, and-9, a family of proteins involved in apoptosis. Under physiological conditions Caspase-9 exists as an inactive monomer forming an 1:1 complex with the Apaf-1 cofactor in the presence of Cytochrome C and ATP to produce a heteromultimer. This complex co-localizes with a multiple array of Caspase-9 molecules, which consequently increase their concentration above the Caspase-9 homodimer dissociation constant K_D. This event allows the homodimer to be formed through the exposure of an activation loop, and the active dimer provides the catalytic activity necessary to activate Caspase-3 and-7 [48].

Finally, other interesting examples of oligomeric natural proteins are the membrane channel-forming tetrameric complexes that allow specific ions (Na+ or K+) or water to permeate cells, such as aquaporins or aquaglyceroporins [49].

Proteins can also form large pathogenic oligomers or multimers that can evolve towards pathogenic supramolecular structures. Important examples of these malignant events are the uncontrolled aggregation of the Glu6Val Hb mutant of Hb (E6V-Hb or HbS) in sickle cell anaemia or the formation of amyloid or amyloid-like fibrils, as it occurs with several proteins related to severe neurodegenerative diseases. These latter products, overpassing the oligomeric status, will be discussed later in a greater detail, within the physio-pathological functional consequences of protein oligomerization.

3.4. Self/cross-association through three dimensional (3D) domain swapping

A peculiar, interesting way to form protein dimers, oligomers or large multimers can occur naturally or artificially through the reciprocal exchange of small or large regions (peptide(s) or entire domain(s)) of the monomeric subunits. Monomers exploit short flexible hinge-loops present in their sequence to address a definite domain (or more than one) into the corresponding partner subunit that will reciprocally swap an identical domain with the former (Figure 7). This mechanism was called three dimensional domain swapping (3D-DS) by Eisenberg and co-workers when they discovered that diphtheria toxin (DT) can form a dimer intertwining an entire domain [50]. Beyond dimers, this mechanism, known for DT as well as for other proteins, can also lead to the formation of larger oligomers [51], exploiting a small flexible loop (Figure 7) that is able to adopt different conformations within various different environmental conditions.

The domain-swapped oligomer reconstitutes the native contacts present in the monomer (closed interface [52], green in Figure 7) except the hinge loop, while a new interface (open interface [52], magenta in Figure 7) forms in the oligomer only, stabilizing it.

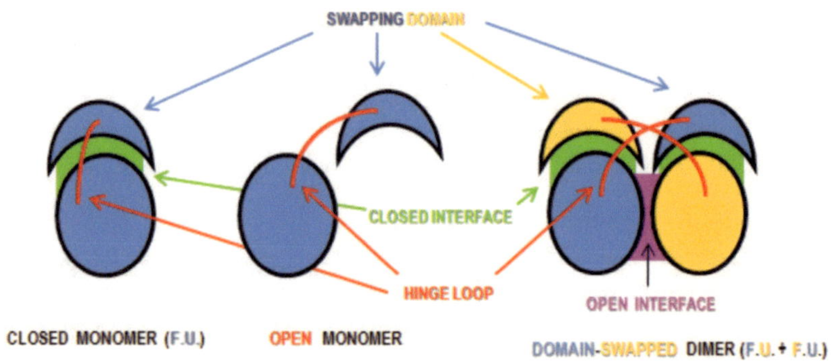

Figure 7. Schematic view of the 3D domain swapping (3D-DS) mechanism. The movement opening of the loop-present in the 'starting' monomer (blue) allows the formation of the dimer by recreating the intramolecular interdomain interface present in the monomer (closed interface, green [52]) in an intermolecular dimeric interface instead. This subsequently drives the formation of a new dimeric interface (open interface, magenta [52]), absent in the monomer. Through this mechanism a protein can form an active dimer still maintaining functional units (F.U.). The same mechanism can drive to oligomers of higher stoichiometry (number of associated chains) than dimers. (Modified from [53]).

3D domain swapping (3D-DS) was hypothesized about fifty years ago [54] for RNase A, and was then confirmed by several brilliant crystallographic results obtained in the '90s, the first with BS-RNase [55]. In the last two decades, domain swapping has been discovered to involve more than sixty proteins [56], and about 300 domain-swapped structures have been solved in crystals or solution, while even a higher number of models of oligomeric swapped proteins have been proposed. Furthermore, 3D-DS is almost uniquely related to protein self-association, although rare cases of hetero-association exist.

Eisenberg and co-workers, who named 3D-DS this mechanism of reciprocal exchange of domains between proteins, immediately underlined the double-face nature, malignant and/or benign, of this protein-protein interaction, defining it as "entangling alliances between proteins" [50]. The term "entanglement" can be forced to be considered as expressing a negative fate of a protein without any possibility to evade the interaction although this 'jailing' can not be known 'a priori'. This is the case of amyloidogenic proteins which form, through this mechanism, fibrils involved in neurodegenerative diseases. Conversely, the term "alliances" clearly indicates the benign face of 3D-DS. In fact, a protein can self-associate to acquire novel activities absent in the monomer, or could also enforce pre-existing ones [57, 58], or control them, for example allosterically, as it occurs for swapped dimeric RNases [59, 60].

The mechanism is made possible thanks to the presence of a hinge-loop (red in Figure 7) located between two different protein structured domains. The flexible loop changes its conformation depending on the environment, and can address the swappable domain(s) into the complementary subunit(s). The parts of a protein to be exchanged can be elements of secondary structure, such as α-helices or β-sheets (RNase A, BS-RNase), or entire domains (DT). The loop is usually composed of few residues which can be different from one protein to another. Interestingly, a single point mutation can induce dramatic changes in the loop flexibility and switch the protein towards or against self association through 3D-DS [61-64]. Other important factors that govern the 3D-DS event(s) are obviously protein concentration and inter-molecular interactions [65, 66].

Some proteins can be constitutively domain-swapped, as it is for the member of the cyclin-dependent kinase p13suc1, which is natively a mixture of a monomer and a domain-swapped dimer [67]. The two monomeric/dimeric states are in equilibrium and the domain-swapped dimer is favored by the presence of proline residues in the hinge loop [68].

Another very interesting natively domain-swapped protein is one of the two conformers of BS-RNase (Figure 8). This protein is the unique natural dimer of the large pancreatic-type RNases superfamily, whose proto-type is RNase A. Furthermore, BS-RNase is a mixture of two isoforms [60], which are both covalently dimeric because of the presence of two antiparallel disulfide bonds, i.e., occurring between Cys-31 and-32 of one subunit and Cys-32 and-31, respectively, of the other [69]. About 70% of the molecules spontaneously swap their N-terminal helices to form the conformer called MxM shown in the right panel of Figure 8. This swapping event implies interesting functional consequences that will be discussed later. Thus, About 30% of BS-RNase is dimeric only thanks to the cited disulfides and called M=M [60, 70] (Figure 8, left panel), while about 70% of the molecules additionally swap their N-terminal helices to form the second BS-RNase conformer, called MxM (Figure 8, right panel) [55, 60]. This swapping event implies interesting functional consequences that will be discussed later.

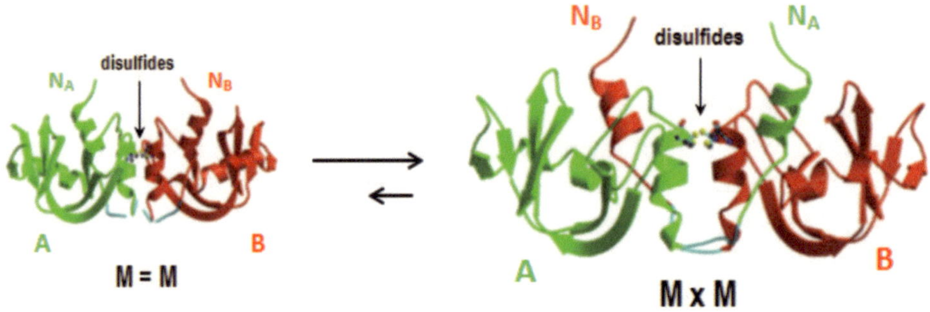

Figure 8. Structure of the two dimeric conformers of BS-RNase: left panel, the unswapped (M=M) conformer [70]; right, the swapped isoform (MxM) [55]. The two subunits, A and B, and their N-termini, are highlighted, as well as the disulfide bonds that covalently link the two subunits in both isoforms [60].

The BS-RNase hinge loop and open interface have been exstensively studied [64] and, while Pro 19 and Leu 28 are key-residues for the stability of the MxM isoform [65], the entire 16-21 loop mutation together with a R80S mutation dramatically inverted the swapping tendency of the protein [63].

Other proteins, being native monomers, can be induced to form domain-swapped oligomers either naturally or artificially. This is the case of the 13.7 kDa RNase A (Figure 9 A). This enzyme can form various non-covalent oligomers when is lyophilized from 40-50% acetic acid solutions, and was the first protein for which 3D-DS mechanism was hypothesized to occur through the swapping of its N-terminal ends [54]. In 1998 this idea was confirmed by the analysis of the crystal structure of the N-term-swapped dimer of RNase A [71], now called RNase A N-dimer, or N_D [5] (Figure 9 B). Three years later, the crystallographic structure of another dimeric conformer of RNase A was solved, discovering that the protein also swaps its C-terminus to form a second dimeric conformer [72], then called the C-dimer or C_D [5] (Figure 9 C).

Several RNase A oligomers larger than dimers have also been found to form [6, 29, 73, 74], Among them, three trimers (Figure 9 D-F), and six tetrameric different conformers have been found and extensively or partially characterized (Figure 9, panels G-N) as well as several other larger multimers [6, 27, 29, 73-75].

These findings indicate that the folds of RNase A are highly versatile, despite its overall known stability. Its dimers are not exclusively artificial, given that traces of N_D are present in a native mixture [77, 78], and that C_D has been detected to form and be subsequently degraded during protein expression in cells [79]. The capability of this enzyme to swap both termini definitely increases the number of possible structures it can form: in fact, the linear, or quasi-linear but not cyclically-closed oligomeric structures reported in Figure 9 (panels D and G-M), can form thanks to the contemporary swapping of N-and C-termini [6, 27, 28, 75, 76].

The capability to swap multiple domains highly increases the 'swapping capacity' (SC), which is defined as the upper limit of subunits with which a protein may interact [80]. In fact, if a

http://dx.doi.org/10.5772/57489

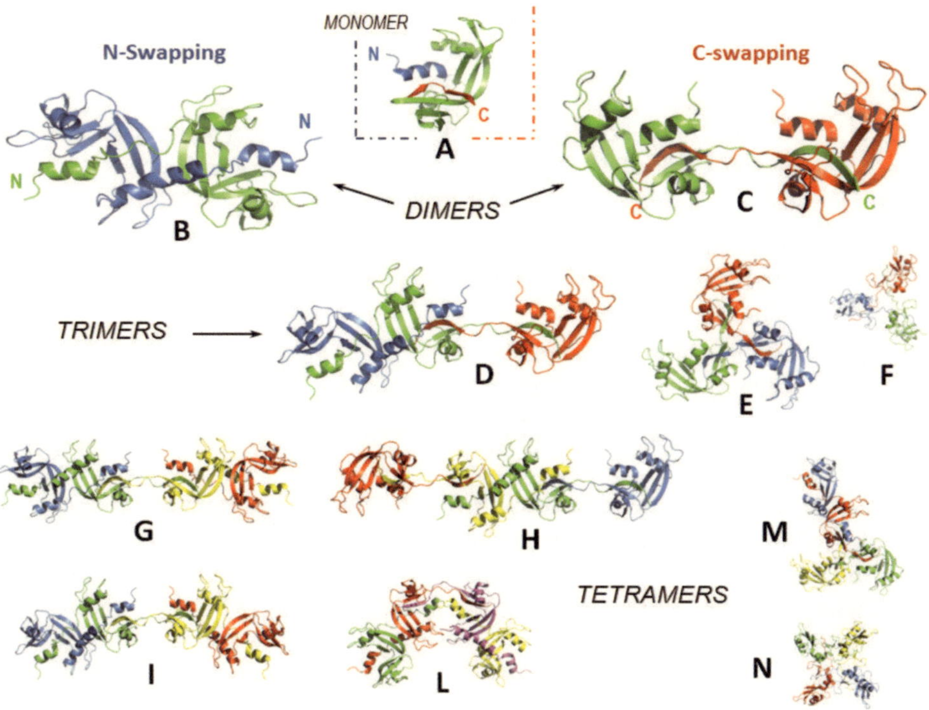

Figure 9. RNase A 3D domain-swapped oligomers.A, native RNase A monomer; **B**, N-dimer, N_D [71]; **C**, C-dimer, C_D [72]; **D,** N+C-swapped trimeric model [27, 75]; **E**, cyclic C-swapped-only, trimer, [75]; **F**, cyclic C-swapped-only, trimeric model [75]; **G-N**: tetramers, all N+C-swapped [6, 27, 28, 76], except the cyclic C-swapped-only model of panel **N** [76]. The dimensions of the oligomers here reported are tentatively representing the relative proportions withdrawable after the lyophilization from 40% acetic acid solutions. RNase A also forms pentamers, hexamers [6, 29, 73], and larger multimers, up to tetradecamers [6, 74], not shown here.

protein displays N swappable domains, SC is defined to be equal to N^2+N [80]. Thus, if N=1, then SC=2 (1+1), while if N=2 SC speeds up to 6 (4+2), and so on. RNase A was thought to be unique in its capability to swap more than one domain, but recently also BS-RNase [81], cyanovirin-N and the mammalian DUF59-Fam96a protein displayed a similar multiple DS behavior [82, 83]. In particular, the natively N-swapped dimeric BS-RNase, known to be able to self-associate since 1969 [57], was found to form either N-and C-swapped tetramers and multimers [81, 84]. In addition, it has recently been reported that a monomerized BS-RNase [85] can be induced to form a C-swapped dimer similar to the one of RNase A [86].

Furthermore, RNase A was firstly thought to be induced to oligomerize through an initial only partial denaturation [72] via 40% aqueous acetic acid treatment [54], while later it was shown that the protein undergoes almost complete denaturation (except its four disulfides) under acidic conditions [87]. Then, when lyophilization is followed by a re-dissolution of the powder in 'benign' buffers [87], i.e., buffers that can slow the regression towards monomer, as phosphate does for RNase oligomers, the protein can re-gain its native monomeric form for about 70% of the initial amount, and the remaining 30% forms various domain-swapped oligomers. The debate between the theories based on a partial [88], or total [87] denaturation pathway that a generic protein must follow to oligomerize through 3D-DS is still open, and what has been established for RNase A [87] is not possible to be absolutely stated for all proteins.

RNase A and BS-RNase can oligomerize following the same 3D-DS mechanism also if very highly concentrated water-alcohol solutions of the enzyme are heated up to 60 °C, and then stabilized by phosphate buffers avoiding the lyophilization step [78, 81]. In this way, the absolute and relative amounts of the various N- or C-swapped oligomers change depending on the environmental conditions. Point mutations which modulated the hydrophilic/hydrophobic nature of the N-/C-swappable domains of RNase A confirmed this tendency [89].

Also human RNases like human pancreatic RNase (hp-RNase) can spontaneously dimerize through 3D-DS when some point mutations are introduced [90-92], although no oligomers larger than dimers have been detected among its aggregation products.

The 3D-DS mechanism can let a protein to overpass the dimeric status and go towards larger oligomerization, multimerization and possibly fibrillization through multiple swapping and/or other alternative ways to stabilize the aggregates. For instance, the formation of disulfide bonds, or the repeat for several times of the same type of swapping with the formation of open-ended structures through a propagative or runaway 3D-DS [93, 94] (Figure 10) are proper alternative ways.

Figure 10. DS in not-cyclic oligomers larger than dimers. (modified from [94]). The first two models display open-ended edging subunits. The increase in stability of an open-ended structure is proportional to the number of subunits that are present between the two edges. The three models reported are an evolution of the ones reported in [93].

Beyond RNases, several other proteins involved in important biologic processes show to form domain-swapped oligomers. One domain swapping-prone protein which in the last years has been discovered to be highly structurally versatile is cyanovirin-N, an 11 kDa protein that inhibits HIV [95]. It can be active either as a monomer or as a metastable domain-swapped dimer [96]. Interestingly, some mutants that become active only as domain-swapped dimers were recently found to form two different relatively stable 3D-DS dimeric conformers, one 3D-DS trimer, and two 3D-DS tetramers [82].

Other important examples of proteins able to form domain-swapped structures are the following: i) Cytochrome C, which was known to polymerize since 1962 [97], but only recently showed to form these inactive supramolecular structures protomers via a runaway 3D-DS of its C-termini. In particular, domain-swapped dimers, trimers, tetramers and polymers up to ~ 40-mers have been characterized [98]; ii) BCL-X_L, an anti-apoptotic protein belonging to the BCL-2 family which can form active C-term-swapped dimers when highly concentrated [99] or alternatively when heated up to 50 °C [100]. iii) Cadherins, which are cell adhesion proteins, dimerize through β-strand swapping to mediate the adhesion itself [101]. Incidentally, it has to be mentioned that several protein cell receptors are known to dimerize to become active (see above), but less is known about the mechanism responsible for the dimerization. Thus, possibly some of them could undergo 3D-DS. iv) Finally, also histones are known to fold through 3D-DS in their evolutionary pathway [102].

3D-DS can also be favored, or hindered, by point or multiple mutations, as is well known for hp-RNase, or by the conditions under which the crystallization process occurs. This is true for barnase, a 12kDa RNase from *Bacillus amyloliquefaciens*, which forms a DS cyclic trimer under not too harsh conditions [103], and for the DS dimer formed in crystals by Grb2-SH2 domains [104].

Last, but not least, some amyloidogenic proteins form fibrils through the initial formation of domain-swapped dimers and oligomers, which are the starting point of their massive self-association [80]. The possibility to overlap 3D-DS with the mechanism of formation of amyloid fibrils was firstly hypothesized by Eisenberg and colleagues. They explained it as a compatibility existing between 3D-DS and the polyglutammine(polyQ)-cross-β steric zippers [72]. This idea was supported by the structural similarity existing between the Asn-based open interface of RNase A C_D [72] and the fibrillogenic nature of poly-Q expansions [105] which are structured as cross-β-spines [13]. The validity of this theory was also enforced by the discovery that either prion protein or cystatin-C, two amyloidogenic cross-β-spine-prone proteins, form 3D-DS dimers [106, 107] (Figure 11, panels A and B). Later, numerous different experimental evidences confirmed this hypothesis. In fact, after discovering that the prion protein (PrP), which is associated with the lethal neurodegenerative Creutzfeld-Jacobs Disease (CJD) and Scrapie, dimerizes through 3D-DS [106], the dimerization event was shown to be the rate-limiting step in the conversion towards the infectious fibrillogenic form of PrP [108]. Then, conversion to fibrils is promoted by an unlocking of the globular domain combined to a redox process, both triggered by 3D-DS [109, 110]. These events drive to the formation of domain-swapped oligomers and multimers stabilized by intermolecular newly formed disulfide bonds [110], but do not affect the overall tertiary structure of the globular main domain of PrP [109] (Figure 11 A).

Furthermore, cystatins, a class of proteins which comprises also stefins and that inhibit cysteine proteases, can also dimerize through 3D-DS [113]. In particular, the L68Q human cystatin-C (hCC) mutant in particular, can dimerize through 3D-DS [107], then inducing severe massive amyloidosis in brain arteries and lethal cerebral hemorrhages. The 13.3 kDa hCC forms fibers through a preliminary domain-swapped dimer+dimer tetrameric rearrangement [114] and a

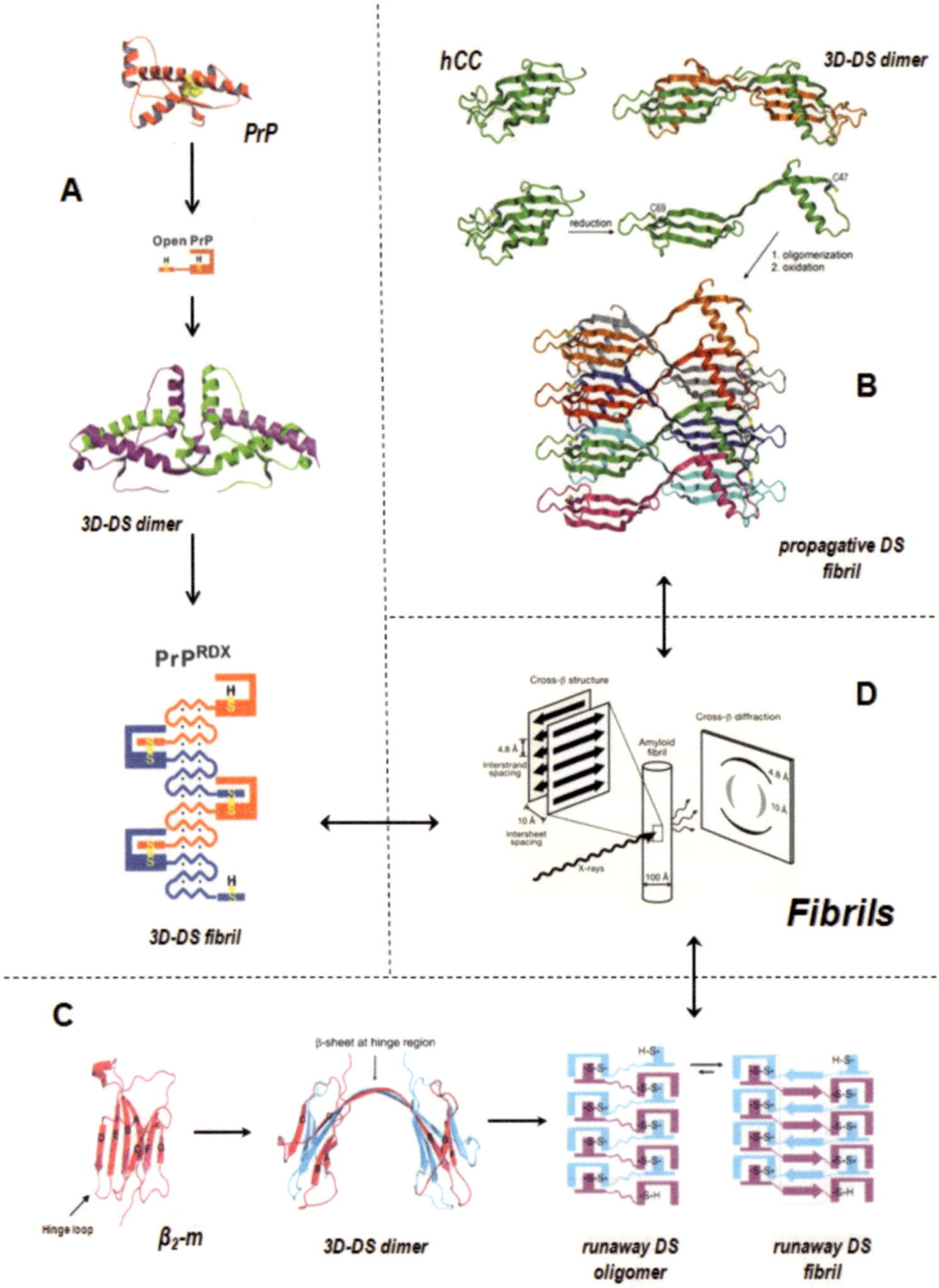

Figure 11. DS Amyloid fibrilsof(A) human prion protein (hPrP), **(B)** human cystatin-C, hCC, and **(C)** β2-microglobulin (β2-m). In all cases, fibrillogenesis is promoted by the preliminary formation of a domain-swapped dimer which evolves towards fibrils through a redox pathway. All the three fibers display the features of a generic amyloid fiber shown in panel **D**. The figure summarizes pictures reported in [106, 107, 110-112].

subsequent propagation of 3D-DS [111]. The finding that prevention of 3D-DS inhibits cystatin-C dimerization and multimerization [115] and studies on the hinge-loop governing the 3D-DS event [116] confirm that 3D-DS plays a key-role in cystatin-C fibrillogenesis [117] (Figure 11 B).

Another important amyloidogenic protein able to self-associate through 3D-DS is β_2-microglobulin (β_2-m), the 10.9 kDa light chain of type-I histocompatibility complex, which can seed as amyloid fibrils during long-term hemodyalisis treatments. Like PrP, β_2-m dimerizes through 3D-DS [118] and forms propagated domain-swapped amyloid fibrils stabilized by disulfide bonds [112] (Figure 11 C).

Again, other amyloidogenic proteins dimerize and massively aggregate through 3D-DS. These are: i) the immunoglobulin G-binding B1 domain, which forms 3D-DS conformationally different dimers [119] and tetramers [120] induced by core-domain mutations before forming fibrils [121]; ii) T7-endonuclease I which forms runaway domain-swapped fibrils stabilized by core-domain intermolecular disulfides [122]. iii) Cell cycle protein Cks1, which fibrillize through the preliminary formation of a domain-swapped dimer [123]. Conversely, for another important amyloidogenic protein, TTR, 3D-DS is to date only hypothesized [124], while the direct stacking model interaction between subunits [125] is the one still nowadays accepted.

Before leaving the "3D-DS toward amyloidosis" topic, it has to be mentioned that wt RNase A, despite its high structural versatility [5], and its high SC, and although displaying some amyloid-prone short sequences [126, 127], is not able to form amyloid or amyloid-like (i.e., *in vitro*) fibers. Its core domain, in fact, is stabilized by four disulfide bonds, and 'self-chaperones' the whole protein from falling towards fibrillization [128]. Incidentally, the only pancreatic-type RNase known to date to form fibrils is the eosinophil cationic protein ECP [129]. Again, Eisenberg and co-workers showed that, contrary to wt RNase A, some mutants, such as poly Q- and poly G-RNase A, spontaneously form "native-like" amyloid fibrils through C-terminal and N-terminal 3D-DS, respectively [128, 130]. The term "native-like" is referred to the evidence that the core-domains of each protomer forming the fiber remain conserved and natively structured in it [128, 130]. These findings confirm once again that 3D-DS and cross-β-zipper spines are events that can overlap [72].

Finally, the 3D-DS protein dimers or oligomers mentioned above are homo-polymers, and almost all domain-swapped proteins nowadays known are indeed homo-oligomers. Domain-swapped hetero-oligomers are extremely rare, but one to be mentioned is the IX/X-bp anticoagulant complex. This is a domain-swapped dimer forming between two homologous subunits which cross-associate through an intermolecular disulfide bond, but also intertwining a flexible loop located in the central part of each subunit [131]. Contrary to what it could be expected, the dimeric hybrid obtained by associating RNase A with a monomerized BS-RNase [85] did not show a domain-swapped nature. It consists, instead, of two different conformers associated through hydrophobic and electrostatic interactions [132].

4. Stability of the protein oligomers

Protein oligomers are supramolecular structures which are sometimes 'chosen'by nature *ab-initio*, or often built up as a response to natural or non-natural events. Anyway, in both cases oligomers can follow different fates. They can be highly stable, mainly when formed by irreversible phenomena, or can represent metastable or even transient events, thus undergoing fast or slow dissociation. Therefore, there are clearly several cases to be analyzed.

First of all, covalently linked oligomers are obviously the most stable supra-molecular protein structures, except if they interact with proteases, or of course if they have disulfide bonds. Disulfides can be affected by slight redox changes of the environment and reduced to free cysteines or cysteine-like adducts, thus unchaining the two (or more) covalently linked protein molecules. This can be the case of hetero-dimers formed by immunotoxins or other artificial conjugates that can be released in the cell by the reducing power of cytosol [21]. A similar but natural event is what concerns BS-RNase, whose two intermolecular disulfide bridges can be reduced in the cytosol with the formation of two monomers derived from the M=M isoform, while a non-covalent dimer (NCD) [133] survives from MxM [55] thanks to the 3D-DS of its N-termini.

The majority of protein oligomers forms through non-covalent weak associations which can often lead to metastable dimers or oligomers. Their lability is essentially related to the nature of the interaction(s) between the subunits, and to the extent of interface area. H-bonds are weaker than electrostatic interactions, but they can be crucial in anchoring a domain into a specific orientation that can be further stabilized by hydrophobic or electrostatic interactions. Thus, the balance of enthalpy and entropy contributions is decisive in driving a protein oligomer to survive or not, taking into account that entropy is against protein-protein association, a phenomenon which is instead favored by a high concentration of the protomers [37, 50, 77], as was reported above in equations 1 and 2.

For example, human AGT (see Figure 6 a) is active only in the dimeric status, while the enzyme loses its activity if dimerization is hindered. This occurs because the two composite active sites are incomplete despite a PLP co-enzyme is present in each subunit [41]. It is now well known that mutations destabilizing the large interface located between the two subunits can dramatically lower the activity of AGT and possibly drive the enzyme towards monomerization [35]. This latter event induces an unwanted and uncontrolled aggregation which traps monomers, blocking the natural dimerization of AGT [35]. The pathologic effects due to the loss of dimeric, or in general oligomeric, native structures is a general feature of PLP-dependent enzymes.

In general, mutations can affect the overall conformation of protein dimers and/or oligomers and induce monomerization or vice versa uncontrolled aggregation, and even fibrillization. This is the case of TTR (see above), whose native homotetramers are destabilized by point mutations, such as V30M or L55P. The tetrameric assembly in these variants is weakened and the protein easily monomerizes, then undergoing fibrillization through dimeric and octameric annular intermediates [10]. Another protein whose oligomeric status is dramatically affected by point mutations is hemoglobin, which extensively self-associates in its sickle-cell variant (HbS).

Formation of protein oligomers can be induced in some cases by point mutations, as is for hp-RNase [91], or as a consequence of changes in environmental conditions (pH, temperature, protein concentration). This is true for the various domain-swapped dimers and oligomers of RNase A and BS-RNase. Indeed, several studies have been performed on these two pancreatic-type enzymes, who display 82% identity [53]. Thus, the 23 out of 124 different residues have been considered as key stabilizing or destabilizing determinants. In particular, in both variants several hinge loops residues and/or others belonging to the swappable N-terminal or C-terminal domains or interacting with them have been mutated [64, 65, 134, 135]. Many of them have been found to be key residues in stabilizing the domain-swapped oligomers, while other were found to promote oligomers dissociation towards the native proteins, i.e., monomeric RNase A or dimeric BS-RNase [134].

Other factors which affect the stability of the oligomers are pH, temperature, ionic strength of the medium. It is easily understandable that acidic or basic pHs destabilize oligomeric assemblies, as well as native monomeric proteins. The role of temperature is very important in stabilizing or destabilizing non-covalent protein oligomers. In fact the subunit motions are proportional to temperature increase, thus, a metastable adduct can be forced to quickly dissociate by increasing the temperature. Furthermore, heat allows a protein to access its denatured state [136] which is a destabilizing event 'per sè'. Anyway, a very recent study reveals that low temperatures, beyond affecting the folding of native monomers [137] can be useful to study the denaturation and dissociation of dimers through NMR procedures [138].

All these effects, as well as the role played by ionic strength, are qualitatively and quantitatively different from one case to another, because each oligomeric complex can display different shapes, dynamics, and intersubunits surfaces.

In addition, dynamic motions can differently affect the stability of oligomeric conformers belonging to the same protein. This is the case for RNase A whose dimers display a different flexibility, higher for C_D than for N_D [139, 140]. The dynamics of multimeric assemblies dissociation can lead towards different pathways, then producing different smaller products. This is true, for example, for porphobilinogen synthase (PBGS), an octamer which can dissociate to tetramers and dimers either symmetrically, through a consecutive loss of dimeric adducts, or through an asymmetric detach of one subunit per time [141].

The concentration of a protein is crucial either to oligomerize or dissociate, or even evolve toward larger multimers and fibrils. These phenomena are ruled by the K_D values associated to each oligomerization process (see equations 1 and 2 and Figure 4), thus dilution can be a mean by which oligomers and multimers can be destabilized and dissociated [93]. Consequently, macromolecular crowding in general [142] can deeply affect the propensity of a protein to oligomerize by influencing the oligomerization yields [143] and/or the dissociation pathways and kinetics [74].

Also proteolysis, occurring naturally or induced, limited or massive, is a way through which proteins' activities are naturally or artificially switched on/off, or anyhow tuned. Thus, oligomerization can also affect proteolysis: in fact, a domain that in a native monomer is exposed to the action of a proteolytic agent can be partially or totally hidden by dimerization or oligomerization, and proteolysis can be slowed down or even blocked; otherwise, a region which in the native monomer is well structured and therefore protected, or even hidden, can be destabilized, destructured and exposed after oligomerization, becoming susceptible to proteolysis. This is what happens with RNase A, whose N_D is definitely more susceptible to proteolysis than the C_D conformer [144].

Another interesting case is represented by the pore forming toxin (PFT) families, which are most often produced as soluble monomers, proteolitically cleaved by host proteases leading to their oligomerization and pore formation. This occurs, for example, with aerolysin [145], with protective antigen (PA) from anthrax [146], and with thiol-activated cytolysin (TACY) pore forming family [147].

Finally, it has to be underlined that all factors able to stabilize/destabilize protein oligomers, i.e., pH, temperature, ionic strength, protein concentration, molecular crowding, mutations, are not independent of each other. This makes the scenario more complicated than expected and clearly indicates that an 'absolute stability' of a protein can not be easily defined. This is true, for example, for RNase A domain-swapped oligomers (see Figure 9). Their high number of conformers and the possible interconversions between oligomers make the picture quite complicated [74], and only some data concerning dimers' stabilities have been reported to date [89, 148, 149]. With an apparent contradiction, RNase A-N_D was sometimes reported to be more stable than C_D while in other environmental conditions the situation is the opposite. These data clearly indicate that different combinations of all the environmental conditions reported above are crucial to stabilize or destabilize different dimeric or oligomeric structures.

5. Oligomeric proteins: functional vs aberrant interactions

5.1. Gain or loss of function(s) after protein oligomerization

Natural oligomerization of proteins has been settled by evolution in order to obtain a control of their biological features, for instance of enzyme activity. In some cases, proteins are inactive unless they dimerize or oligomerize because of the high hydrophobic surface the monomer exposes to the solvent, as reported above for PLP-enzymes. Oligomerization can be constitutive, such as for PLP-enzymes, or induced by signal molecules, like for membrane proteins. Self-or hetero-association of proteins can also be artificial, but can in any case lead to new activities or block activities that in certain situations can become unwanted, for example in the feedback control of some enzymes.

Dimerization and oligomerization can also increase or lower pre-existing activities. In this case the phenomenon can be considered as a sharp controller and tuner of important activities. These gain- or loss-of-function events occur by exposing or hiding active surfaces or, for example, by inducing positive or negative allostery impossible to be warranted by the native monomer. These events can be ruled not only by changing the oligomerization/ polymerization status of the proteins, but also by conformational changes induced by ligands, such as it occurs with Hb (see Figure 5). Gain or loss of native functions induced by oligomerization can be benign or malignant and also nature can drive either towards physiologic or pathological events, as it happens, respectively, with actin [150] or with sickle-cell anaemia associated to a point mutation in hemoglobin (HbS). Other important pathological implications, i.e., oligomers evolving towards fibrils, will be discussed later. A benign case that deserves to be mentioned is rhodopsin, the pigment involved in the phototransduction events that are crucial for vision, which physiologically organizes itself in a supra-molecular ensemble to be active [151].

An interesting case in which protein oligomerization induces a gain of function concerns RNase A. The native monomeric enzyme only degrades single-stranded (ss) RNA and is not cytotoxic, while its artificial dimers and oligomers, either forming through non-covalent 3D-DS or covalent bonds become also active against double-stranded (ds) RNA [29, 57, 73, 152]. Moreover, they can become selectively cytotoxic towards cancer cells both *in vitro* and *in vivo* [153-156], although more recent results indicate that cytotoxicity can be dependent also on the type of cell line studied [32]. This acquired cytotoxic power can be considered benign, given the selectivity towards malignant cells, and can be mainly ascribable to the possibility of oligomeric RNases to evade the ribonuclease inhibitor (RI), which is designed for tightly trap monomeric RNases [157]. This is also the main reason why only MxM BS-RNase (see Figure 8) is cytotoxic: in fact, the cytosolic reducing environment allows only the MxM domain-swapped isoform to survive as a non-covalent dimer (NCD) and to evade RI, while M=M becomes a monomeric, RI-susceptible, derivative [158]. Another example of cytotoxic protein oligomers is represented by the ones formed by p13suc1 [159]. The native cell-cycle regulatory protein is a monomer/DS-dimer mixture [160] that can form large native-like cytotoxic aggregates through 3D-DS, and whose structural determinants governing the swapping mechanism have been extensively studied as reported above [62, 68]. Thus, the artificial induction, quench, or control of the oligomerization event(s) can be useful to avoid, induce, or tune several biological properties of proteins, such as for example enzymatic activity, or the incoming of cytotoxicity.

5.2. Protein oligomerization towards fibrillization and/or amyloidosis

Protein self- or cross-association can be naturally or artificially controlled to the degree of oligomers or multimers, but can sometimes undergo uncontrolled massive aggregation, often resulting in fibers, as it has already been reported in the domain swapping section. These supra-molecular structures can be benign, such as in muscle tissues (actin [150]), or very often harmful. The latter case is true for sickle cell HbS insoluble fibrous malignant polymers, or for amyloid or amyloid-like fibrils, which are often, although not always [2, 15], associated with deleterious neurodegenerative diseases [37]. Proteins can become prone to fibrillization because of ageing, or after changes in the environmental conditions, such as crowding, pH or temperature shocks. In addition, also point mutations can induce and often speed up the fibrillization phenomena, as for TTR V30M or L55P [161, 162], or for the human prion protein [163], or also for the homo-tetrameric p53 tumor suppressor protein [164, 165]. Several lethal amyloidoses, for example linked to TTR or β_2-m, or to other toxic fibrillogenic proteins, like prions, display a premature incoming when associated to familial pathogenic mutations or overpass the species transmission barriers [166]. Then, also cancer has been recently considered a possible prio-like disease, due to the fibrillogenic behavior of some p53 mutants [167].

In this complex scenario, it is known [163] and is more and more accepted that the first oligomeric/multimeric species preceding the formation of protofibrils and fibrils are the toxic agents responsible for the incoming of the associated pathology(ies) [2]. Thus, a lot of work is presently devoted to discover the structural and functional properties of these 'toxic oligomers', which is a very difficult endeavor indeed, because these species show a great tendency to fibrillize quickly.

In the last five years, the evidence became clear that oligomers produced by the same protein can be toxic or non toxic depending on the way they are produced [2, 7, 15]. This was found, for example, with the *Escherichia Coli* hyperforin-N (HypF-N) spherical oligomers [15] and also with other amyloidogenic proteins. Several different supramolecular large structures can be detected with new techniques, such as solid-state NMR, Cryo-TEM, High-Resolution Atomic-Force Spectroscopy, Molecular Modeling. For example, TTR has been discovered to fibrillize through the formation of annular oligomers deriving from the monomer which had been in turn detached from the native tetramer [10].

Several other amyloidogenic proteins have been extensively studied. Among them, the number of those that follow the 3D-DS mechanism is continuously increasing [80], such as for hPrP, hCC, β_2-m, the properties of which have already been discussed. In the last years several proteins initially not considered to follow 3D-DS were discovered instead to undergo this mechanism. Nevertheless, the mechanism by which an oligomer/fiber may form essentially depends on the symmetry of the interfacial association and 3D-DS can be not mandatory to reach this requisite. Thus, the formation of the cross-β-spine fibrils intima could follow an end-to-end stacking mechanism, the same followed by non-amyloidogenic proteins, such as hemoglobin-S [164], or by non-harmful proteins, such as actin [170] or tubulin [171]. Anyway, considering that amyloid fibers are not oligomers, several oligomeric precursors of amyloidogenic proteins are continuously studied, despite their transient nature. Among them, human lysozime, *Sulfolobus Solfataricus* acylphosphatase (AcP), human superoxide dismutase-1 (SOD1), the latter associated with the devastating disease amyotrophic lateral sclerosis, have been deeply investigated in their propensity to undergo amyloid fibers through amyloidogenic oligomers [167]. Further detailed studies, which are beyond the topic of this chapter and focused on the nature of amyloidogenic oligomers and their differences from the non-amyloidogenic ones have been recently reviewed [2]. Several interesting news concerning the oligomers' polymorphism and the structural rearrange-

http://dx.doi.org/10.5772/57489

ments they can undergo (Figure 12) to follow a harmful fibrillogenic pathway or a non fibrillogenic harmless destiny [7] can be found in that review [2].

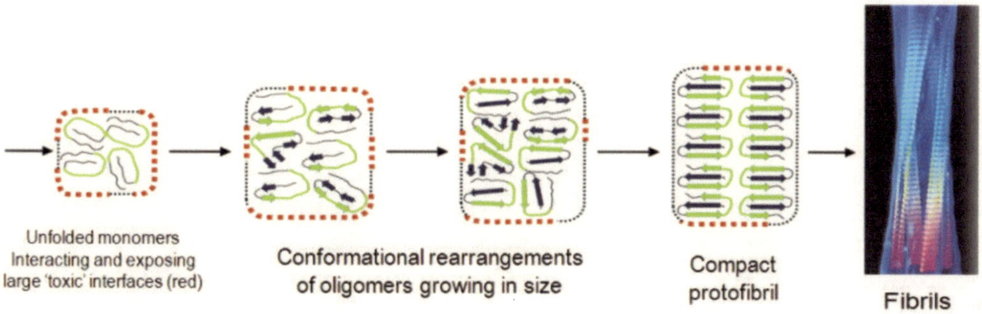

Figure 12. Evolution of partially or totally unfolded protein monomers towards ordered fibrils through oligomeric intermediates displaying different sizes and conformations. (Modified from [2] and [88]).

6. Oligomeric proteins and industry

Besides artificial chemistry products, industry has often used natural or artificially modified bio-products to produce bio-materials or bio-fibers. In this context, nano-particles and nano-materials certainly represent a new fascinating frontier to be developed.

Peptides and proteins can be driven towards controlled oligomerization, polymerization and/or fibrillization to form products displaying useful physico-chemical and mechanical features to be adopted for new or renewed industrial applications. Incidentally, several efforts have been performed in the engineering of short peptides and small biomolecules [168-170], while definitely less is known on the industrial application of protein oligomers, multimers and/or fibrils. The most common materials formed from peptides and proteins are hydrogels, which can be applied in tissue engineering and in drug delivery [169]. These materials are typically formed by hydrated cross-linked fibers that somehow resemble the agarose or polyacrylamide gels extensively used in biochemical laboratories.

The 'intima' of these hydrogel structures and their morphology define size, biocompatibility, and mechanical, elastic or rigid properties. Anyway, only some hints concerning the interesting 'industrial' applications of protein oligomers will be given here.

For example, the 3D-DS mechanism has been reported as to be applicable in material design [94]. Indeed, domain-swapped oligomeric peptides have been produced to obtain hydrogels [171], while also proteins can be artificially designed to undergo 3D-DS to form oligomers which could be useful to produce biomaterials [172]. Beyond small oligomers, the ability of some proteins to become fibrillogenic through propagative or runaway domain swapping [80, 94] could, or should, be exploited in order to engineer variants that could form harmless 3D-

DS fibers with special morphological and mechanical properties. These supra-molecular structures could also become materials devoid of direct biological applications, but with industrial and ecologic relevance. Indeed, proteins that for example combine 3D-DS with a covalent stabilization of their fibrillar products through the formation of novel disulfides (like β2-m and recombinant PrP) [110, 112], should be useful to obtain reversible products that could be easily unstructured and recycled under reducing conditions, and, thus, without negative ecologic consequences.

Many examples of protein 'benign' (non-amyloidogenic) fibers could become precious for industry applications. An example can be represented by the trimeric hexon protein. displaying a novel triple β-spiral fibrous fold with implications for the design of a new class of artificial, silk-like fibrous materials [173].

Anyway, all the potential industrial applications of protein oligomers, multimers, and fibrils are certainly far to be completely explored, and what we have here reported about 3D-DS applications confirms that further deep investigations deserve to be performed.

7. Conclusions

All the notions reported in this chapter indicate that the complexity of the protein oligomerization topic augments every day. In particular, the increasing number of studies focused on natural protein oligomerization and the improvement of the quality of the investigation techniques have greatly enlarged the complexity of the analysis of the structural and functional features of protein oligomers. Furthermore, the sharpening of the strategies used by chemistry to obtain cross-linked artificial oligomers allowed industry and laboratories to obtain less heterogeneous products, which were, in addition, scarcely modified with respect to the native monomeric protein(s).

Again, the discovery that some proteins can naturally oligomerize by combining covalent linkings with weak interactions made the scenario even more complicated. This is the case, for example, of proteins that form domain-swapped fibrils additionally stabilized by newly forming disulfides. The same is true also for membrane proteins that associate and sometimes covalently stabilize their interaction as a response to effectors which can play the role of activators or quench cell signals.

Thus, the aim of this chapter is to focus the attention of the reader on the principal features of protein oligomerization. We kept separated, when possible, the covalent linking oligomers from the non-covalent protein self-association products, as well as the natural, constitutive or induced, events from the artificial ones. When we were not able to separate these aspects well, we tried to give a picture as clear as possible. On the other hand, this underlines how much increasing interest has and how many further studies deserves the topic of protein oligomerization.

http://dx.doi.org/10.5772/57489

Author details

Giovanni Gotte* and Massimo Libonati

*Address all correspondence to: giovanni.gotte@univr.it

Department of Life and Reproduction Sciences, Biological Chemistry Section, University of Verona, Verona, Italy

References

[1] Levy ED, Teichmann S. Structural, evolutionary, and assembly principles of protein oligomerization. Prog Mol Biol Transl Sci. 2013;117:25-51.

[2] Bemporad F, Chiti F. Protein misfolded oligomers: experimental approaches, mechanism of formation, and structure-toxicity relationships. Chem Biol. 2012;19:315-27.

[3] Haass C, Selkoe DJ. Soluble protein oligomers in neurodegeneration: lessons from the Alzheimer's amyloid beta-peptide. Nat Rev Mol Cell Biol. 2007;8:101-12.

[4] Sakono M, Zako T. Amyloid oligomers: formation and toxicity of Abeta oligomers. FEBS J. 2010;277:1348-58.

[5] Libonati M, Gotte G. Oligomerization of bovine ribonuclease A: structural and functional features of its multimers. The Biochemical Journal. 2004;380:311-27.

[6] Cozza G, Moro S, Gotte G. Elucidation of the ribonuclease a aggregation process mediated by 3D domain swapping: A computational approach reveals possible new multimeric structures. Biopolymers. 2008;89:26-39.

[7] Glabe CG. Structural classification of toxic amyloid oligomers. The Journal of Biological Chemistry. 2008;283:29639-43.

[8] Fandrich M. Oligomeric intermediates in amyloid formation: structure determination and mechanisms of toxicity. Journal of Molecular Biology. 2012;421:427-40.

[9] Barghorn S, Nimmrich V, Striebinger A, Krantz C, Keller P, Janson B, et al. Globular amyloid beta-peptide oligomer - a homogenous and stable neuropathological protein in Alzheimer's disease. J Neurochem. 2005;95:834-47.

[10] Pires RH, Karsai A, Saraiva MJ, Damas AM, Kellermayer MS. Distinct annular oligomers captured along the assembly and disassembly pathways of transthyretin amyloid protofibrils. PLoS One. 2012;7:e44992.

[11] Lashuel HA, Hartley D, Petre BM, Walz T, Lansbury PT, Jr. Neurodegenerative disease: amyloid pores from pathogenic mutations. Nature. 2002;418:291.

[12] Fandrich M. On the structural definition of amyloid fibrils and other polypeptide aggregates. Cell Mol Life Sci. 2007;64:2066-78.

[13] Nelson R, Sawaya MR, Balbirnie M, Madsen AO, Riekel C, Grothe R, et al. Structure of the cross-beta spine of amyloid-like fibrils. Nature. 2005;435:773-8.

[14] Sipe JD, Cohen AS. Review: history of the amyloid fibril. Journal of Structural Biology. 2000;130:88-98.

[15] Campioni S, Mannini B, Zampagni M, Pensalfini A, Parrini C, Evangelisti E, et al. A causative link between the structure of aberrant protein oligomers and their toxicity. Nature Chemical Biology. 2010;6:140-7.

[16] D'Alessio G, Parente A, Guida C, Leone E. Dimeric structure of seminal ribonuclease. FEBS Letters. 1972;27:285-8.

[17] Urry DW. What is elastin; what is not. Ultrastruct Pathol. 1983;4:227-51.

[18] Hartman FC, Wold F. Cross-linking of bovine pancreatic ribonuclease A with dimethyl adipimidate. Biochemistry. 1967;6:2439-48.

[19] Wang D, Wilson G, Moore S. Preparation of cross-linked dimers of pancreatic ribonuclease. Biochemistry. 1976;15:660-5.

[20] Green NS, Reisler E, Houk KN. Quantitative evaluation of the lengths of homobifunctional protein cross-linking reagents used as molecular rulers. Protein Science: a publication of the Protein Society. 2001;10:1293-304.

[21] Fracasso G, Bellisola G, Castelletti D, Tridente G, Colombatti M. Immunotoxins and other conjugates: preparation and general characteristics. Mini Rev Med Chem. 2004;4:545-62.

[22] Ayers NA, Nadeau OW, Read MW, Ray P, Carlson GM. Effector-sensitive cross-linking of phosphorylase b kinase by the novel cross-linker 4-phenyl-1,2,4-triazoline-3,5-dione. The Biochemical Journal. 1998;331 (Pt 1):137-41.

[23] Sheehan JCH, J.J. The cross-linking of gelatin using a water-soluble carbodiimide. Journal of the American Chemical Society. 1957;79.

[24] Lopez-Alonso JP, Diez-Garcia F, Font J, Ribo M, Vilanova M, Scholtz JM, et al. Carbodiimide EDC Induces Cross-Links That Stabilize RNase A C-Dimer against Dissociation: EDC Adducts Can Affect Protein Net Charge, Conformation, and Activity. Bioconjug Chem. 2009.

[25] Ciglic MI, Jackson PJ, Raillard SA, Haugg M, Jermann TM, Opitz JG, et al. Origin of dimeric structure in the ribonuclease superfamily. Biochemistry. 1998;37:4008-22.

[26] Lin SH, Konishi Y, Denton ME, Scheraga HA. Influence of an extrinsic cross-link on the folding pathway of ribonuclease A. Conformational and thermodynamic analysis of cross-linked (lysine7-lysine41)-Ribonuclease A. Biochemistry. 1984;23:5504-12.

[27] Nenci A, Gotte G, Bertoldi M, Libonati M. Structural properties of trimers and tetramers of ribonuclease A. Protein Science: a publication of the Protein Society. 2001;10:2017-27.

[28] Gotte G, Libonati M. Oligomerization of ribonuclease A: two novel three-dimensional domain-swapped tetramers. The Journal of Biological Chemistry. 2004;279:36670-9.

[29] Gotte G, Laurents DV, Libonati M. Three-dimensional domain-swapped oligomers of ribonuclease A: identification of a fifth tetramer, pentamers and hexamers, and detection of trace heptameric, octameric and nonameric species. Biochimica et Biophysica Acta. 2006;1764:44-54.

[30] Simons BL, King MC, Cyr T, Hefford MA, Kaplan H. Covalent cross-linking of proteins without chemical reagents. Protein Science: a publication of the Protein Society. 2002;11:1558-64.

[31] Simons BL, Kaplan H, Fournier SM, Cyr T, Hefford MA. A novel cross-linked RNase A dimer with enhanced enzymatic properties. Proteins. 2007;66:183-95.

[32] Vottariello F, Costanzo C, Gotte G, Libonati M. "Zero-length" dimers of ribonuclease A: further characterization and no evidence of cytotoxicity. Bioconjug Chem. 2010;21:635-45.

[33] Fancy DA, Kodadek T. Chemistry for the analysis of protein-protein interactions: rapid and efficient cross-linking triggered by long wavelength light. Proceedings of the National Academy of Sciences of the United States of America. 1999;96:6020-4.

[34] Lumb MJ, Danpure CJ. Functional synergism between the most common polymorphism in human alanine:glyoxylate aminotransferase and four of the most common disease-causing mutations. The Journal of Biological Chemistry. 2000;275:36415-22.

[35] Cellini B, Montioli R, Paiardini A, Lorenzetto A, Maset F, Bellini T, et al. Molecular defects of the glycine 41 variants of alanine glyoxylate aminotransferase associated with primary hyperoxaluria type I. Proceedings of the National Academy of Sciences of the United States of America. 2010;107:2896-901.

[36] Levy ED, Boeri Erba E, Robinson CV, Teichmann SA. Assembly reflects evolution of protein complexes. Nature. 2008;453:1262-5.

[37] Eisenberg D, Jucker M. The amyloid state of proteins in human diseases. Cell. 2012;148:1188-203.

[38] Monaco HL, Rizzi M, Coda A. Structure of a complex of two plasma proteins: transthyretin and retinol-binding protein. Science. 1995;268:1039-41.

[39] Blake C, Serpell L. Synchrotron X-ray studies suggest that the core of the transthyretin amyloid fibril is a continuous beta-sheet helix. Structure. 1996;4:989-98.

[40] Montioli R, Cellini B, Bertoldi M, Paiardini A, Voltattorni CB. An engineered folded PLP-bound monomer of Treponema denticola cystalysin reveals the effect of the dimeric structure on the catalytic properties of the enzyme. Proteins. 2009;74:304-17.

[41] Zhang X, Roe SM, Hou Y, Bartlam M, Rao Z, Pearl LH, et al. Crystal structure of alanine:glyoxylate aminotransferase and the relationship between genotype and enzymatic phenotype in primary hyperoxaluria type 1. Journal of Molecular Biology. 2003;331:643-52.

[42] Burkhard P, Dominici P, Borri-Voltattorni C, Jansonius JN, Malashkevich VN. Structural insight into Parkinson's disease treatment from drug-inhibited DOPA decarboxylase. Nature Structural Biology. 2001;8:963-7.

[43] Flydal MI, Martinez A. Phenylalanine hydroxylase: function, structure, and regulation. IUBMB Life. 2013;65:341-9.

[44] Grueninger D, Treiber N, Ziegler MO, Koetter JW, Schulze MS, Schulz GE. Designed protein-protein association. Science. 2008;319:206-9.

[45] Marianayagam NJ, Sunde M, Matthews JM. The power of two: protein dimerization in biology. Trends Biochem Sci. 2004;29:618-25.

[46] Hebert TE, Bouvier M. Structural and functional aspects of G protein-coupled receptor oligomerization. Biochem Cell Biol. 1998;76:1-11.

[47] Terrillon S, Bouvier M. Roles of G-protein-coupled receptor dimerization. EMBO Rep. 2004;5:30-4.

[48] Renatus M, Stennicke HR, Scott FL, Liddington RC, Salvesen GS. Dimer formation drives the activation of the cell death protease caspase 9. Proceedings of the National Academy of Sciences of the United States of America. 2001;98:14250-5.

[49] Agre P, Kozono D. Aquaporin water channels: molecular mechanisms for human diseases. FEBS Letters. 2003;555:72-8.

[50] Bennett MJ, Choe S, Eisenberg D. Domain swapping: entangling alliances between proteins. Proceedings of the National Academy of Sciences of the United States of America. 1994;91:3127-31.

[51] Steere B, Eisenberg D. Characterization of high-order diphtheria toxin oligomers. Biochemistry. 2000;39:15901-9.

[52] Bennett MJ, Schlunegger MP, Eisenberg D. 3D domain swapping: a mechanism for oligomer assembly. Protein science : a publication of the Protein Society. 1995;4:2455-68.

[53] Benito A, Laurents DV, Ribo M, Vilanova M. The structural determinants that lead to the formation of particular oligomeric structures in the pancreatic-type ribonuclease family. Curr Protein Pept Sci. 2008;9:370-93.

[54] Crestfield AM, Stein WH, Moore S. On the aggregation of bovine pancreatic ribonuclease. Archives of Biochemistry and Biophysics. 1962;Suppl 1:217-22.

[55] Mazzarella L, Capasso S, Demasi D, Di Lorenzo G, Mattia CA, Zagari A. Bovine seminal ribonuclease: structure at 1.9 A resolution. Acta Crystallogr D Biol Crystallogr. 1993;49:389-402.

[56] Gronenborn AM. Protein acrobatics in pairs--dimerization via domain swapping. Current Opinion in Structural Biology. 2009;19:39-49.

[57] Libonati M. Molecular aggregates of ribonucleases. Some enzymatic properties. Ital J Biochem. 1969;18:407-17.

[58] Libonati M, Bertoldi M, Sorrentino S. The activity on double-stranded RNA of aggregates of ribonuclease A higher than dimers increases as a function of the size of the aggregates. The Biochemical Journal. 1996;318 (Pt 1):287-90.

[59] Piccoli R, D'Alessio G. Relationships between nonhyperbolic kinetics and dimeric structure in ribonucleases. The Journal of biological chemistry. 1984;259:693-5.

[60] Piccoli R, Tamburrini M, Piccialli G, Di Donato A, Parente A, D'Alessio G. The dual-mode quaternary structure of seminal RNase. Proceedings of the National Academy of Sciences of the United States of America. 1992;89:1870-4.

[61] Rousseau F, Schymkowitz JW, Itzhaki LS. The unfolding story of three-dimensional domain swapping. Structure. 2003;11:243-51.

[62] Rousseau F, Schymkowitz JW, Wilkinson HR, Itzhaki LS. Intermediates control domain swapping during folding of p13suc1. The Journal of Biological Chemistry. 2004;279:8368-77.

[63] Ercole C, Spadaccini R, Alfano C, Tancredi T, Picone D. A new mutant of bovine seminal ribonuclease with a reversed swapping propensity. Biochemistry. 2007;46:2227-32.

[64] Picone D, Di Fiore A, Ercole C, Franzese M, Sica F, Tomaselli S, et al. The role of the hinge loop in domain swapping. The special case of bovine seminal ribonuclease. The Journal of Biological Chemistry. 2005;280:13771-8.

[65] Ercole C, Avitabile F, Del Vecchio P, Crescenzi O, Tancredi T, Picone D. Role of the hinge peptide and the intersubunit interface in the swapping of N-termini in dimeric bovine seminal RNase. Eur J Biochem. 2003;270:4729-35.

[66] Yang S, Levine H, Onuchic JN. Protein oligomerization through domain swapping: role of inter-molecular interactions and protein concentration. Journal of Molecular Biology. 2005;352:202-11.

[67] Rousseau F, Schymkowitz JW, Wilkinson HR, Itzhaki LS. The structure of the transition state for folding of domain-swapped dimeric p13suc1. Structure. 2002;10:649-57.

[68] Rousseau F, Schymkowitz JW, Wilkinson HR, Itzhaki LS. Three-dimensional domain swapping in p13suc1 occurs in the unfolded state and is controlled by conserved proline residues. Proceedings of the National Academy of Sciences of the United States of America. 2001;98:5596-601.

[69] Di Donato A, D'Alessio G. Interchain disulfide bridges in ribonuclease BS-1. Biochem Biophys Res Commun. 1973;55:919-28.

[70] Berisio R, Sica F, De Lorenzo C, Di Fiore A, Piccoli R, Zagari A, et al. Crystal structure of the dimeric unswapped form of bovine seminal ribonuclease. FEBS Letters. 2003;554:105-10.

[71] Liu Y, Hart PJ, Schlunegger MP, Eisenberg D. The crystal structure of a 3D domain-swapped dimer of RNase A at a 2.1-A resolution. Proceedings of the National Academy of Sciences of the United States of America. 1998;95:3437-42.

[72] Liu Y, Gotte G, Libonati M, Eisenberg D. A domain-swapped RNase A dimer with implications for amyloid formation. Nature Structural Biology. 2001;8:211-4.

[73] Gotte G, Bertoldi M, Libonati M. Structural versatility of bovine ribonuclease A. Distinct conformers of trimeric and tetrameric aggregates of the enzyme. Eur J Biochem. 1999;265:680-7.

[74] Lopez-Alonso JP, Gotte G, Laurents DV. Kinetic analysis provides insight into the mechanism of ribonuclease A oligomer formation. Archives of Biochemistry and Biophysics. 2009;489:41-7.

[75] Liu Y, Gotte G, Libonati M, Eisenberg D. Structures of the two 3D domain-swapped RNase A trimers. Protein Science: a publication of the Protein Society. 2002;11:371-80.

[76] Liu Y, Eisenberg D. 3D domain swapping: as domains continue to swap. Protein Science: a publication of the Protein Society. 2002;11:1285-99.

[77] Park C, Raines RT. Dimer formation by a "monomeric" protein. Protein Science: a publication of the Protein Society. 2000;9:2026-33.

[78] Gotte G, Vottariello F, Libonati M. Thermal aggregation of ribonuclease A. A contribution to the understanding of the role of 3D domain swapping in protein aggregation. The Journal of Biological Chemistry. 2003;278:10763-9.

[79] Geiger R, Gautschi M, Thor F, Hayer A, Helenius A. Folding, quality control, and secretion of pancreatic ribonuclease in live cells. The Journal of Biological Chemistry. 2011;286:5813-22.

[80] Bennett MJ, Sawaya MR, Eisenberg D. Deposition diseases and 3D domain swapping. Structure. 2006;14:811-24.

[81] Gotte G, Mahmoud Helmy A, Ercole C, Spadaccini R, Laurents DV, Donadelli M, et al. Double domain swapping in bovine seminal RNase: formation of distinct N- and

C-swapped tetramers and multimers with increasing biological activities. PLoS One. 2012;7:e46804.

[82] Koharudin LM, Liu L, Gronenborn AM. Different 3D domain-swapped oligomeric cyanovirin-N structures suggest trapped folding intermediates. Proceedings of the National Academy of Sciences of the United States of America. 2013;110:7702-7.

[83] Chen KE, Richards AA, Ariffin JK, Ross IL, Sweet MJ, Kellie S, et al. The mammalian DUF59 protein Fam96a forms two distinct types of domain-swapped dimer. Acta Crystallogr D Biol Crystallogr. 2012;68:637-48.

[84] Adinolfi S, Piccoli R, Sica F, Mazzarella L. BS-RNase tetramers: an example of domain-swapped oligomers. FEBS Letters. 1996;398:326-32.

[85] Spadaccini R, Ercole C, Gentile MA, Sanfelice D, Boelens R, Wechselberger R, et al. NMR studies on structure and dynamics of the monomeric derivative of BS-RNase: new insights for 3D domain swapping. PLoS One. 2012;7:e29076.

[86] Sica F, Pica A, Merlino A, Russo Krauss I, Ercole C, Picone D. The multiple forms of bovine seminal ribonuclease: Structure and stability of a C-terminal swapped dimer. FEBS Letters. 2013;587:3755-62.

[87] Lopez-Alonso JP, Bruix M, Font J, Ribo M, Vilanova M, Jimenez MA, et al. NMR spectroscopy reveals that RNase A is chiefly denatured in 40% acetic acid: implications for oligomer formation by 3D domain swapping. Journal of the American Chemical Society. 2010;132:1621-30.

[88] Dobson CM. Protein misfolding, evolution and disease. Trends Biochem Sci. 1999;24:329-32.

[89] Gotte G, Donadelli M, Laurents DV, Vottariello F, Morbio M, Libonati M. Increase of RNase a N-terminus polarity or C-terminus apolarity changes the two domains' propensity to swap and form the two dimeric conformers of the protein. Biochemistry. 2006;45:10795-806.

[90] Russo N, Antignani A, D'Alessio G. In vitro evolution of a dimeric variant of human pancreatic ribonuclease. Biochemistry. 2000;39:3585-91.

[91] Canals A, Pous J, Guasch A, Benito A, Ribo M, Vilanova M, et al. The structure of an engineered domain-swapped ribonuclease dimer and its implications for the evolution of proteins toward oligomerization. Structure. 2001;9:967-76.

[92] Merlino A, Avella G, Di Gaetano S, Arciello A, Piccoli R, Mazzarella L, et al. Structural features for the mechanism of antitumor action of a dimeric human pancreatic ribonuclease variant. Protein Science: a publication of the Protein Society. 2009;18:50-7.

[93] Schlunegger MP, Bennett MJ, Eisenberg D. Oligomer formation by 3D domain swapping: a model for protein assembly and misassembly. Adv Protein Chem. 1997;50:61-122.

[94] Nagarkar RP, Hule RA, Pochan DJ, Schneider JP. Domain swapping in materials design. Biopolymers. 2010;94:141-55.

[95] Barrientos LG, Gronenborn AM. The highly specific carbohydrate-binding protein cyanovirin-N: structure, anti-HIV/Ebola activity and possibilities for therapy. Mini Rev Med Chem. 2005;5:21-31.

[96] Barrientos LG, Louis JM, Botos I, Mori T, Han Z, O'Keefe BR, et al. The domain-swapped dimer of cyanovirin-N is in a metastable folded state: reconciliation of X-ray and NMR structures. Structure. 2002;10:673-86.

[97] Margoliash E, Lustgarten J. Interconversion of horse heart cytochrome C monomer and polymers. The Journal of Biological Chemistry. 1962;237:3397-405.

[98] Hirota S, Hattori Y, Nagao S, Taketa M, Komori H, Kamikubo H, et al. Cytochrome c polymerization by successive domain swapping at the C-terminal helix. Proceedings of the National Academy of Sciences of the United States of America. 2010;107:12854-9.

[99] O'Neill JW, Manion MK, Maguire B, Hockenbery DM. BCL-XL dimerization by three-dimensional domain swapping. Journal of Molecular Biology. 2006;356:367-81.

[100] Denisov AY, Sprules T, Fraser J, Kozlov G, Gehring K. Heat-induced dimerization of BCL-xL through alpha-helix swapping. Biochemistry. 2007;46:734-40.

[101] Vendome J, Posy S, Jin X, Bahna F, Ahlsen G, Shapiro L, et al. Molecular design principles underlying beta-strand swapping in the adhesive dimerization of cadherins. Nature Structural & Molecular Biology. 2011;18:693-700.

[102] Hadjithomas M, Moudrianakis EN. Experimental evidence for the role of domain swapping in the evolution of the histone fold. Proceedings of the National Academy of Sciences of the United States of America. 2011;108:13462-7.

[103] Zegers I, Deswarte J, Wyns L. Trimeric domain-swapped barnase. Proceedings of the National Academy of Sciences of the United States of America. 1999;96:818-22.

[104] Schiering N, Casale E, Caccia P, Giordano P, Battistini C. Dimer formation through domain swapping in the crystal structure of the Grb2-SH2-Ac-pYVNV complex. Biochemistry. 2000;39:13376-82.

[105] Perutz MF, Johnson T, Suzuki M, Finch JT. Glutamine repeats as polar zippers: their possible role in inherited neurodegenerative diseases. Proceedings of the National Academy of Sciences of the United States of America. 1994;91:5355-8.

[106] Knaus KJ, Morillas M, Swietnicki W, Malone M, Surewicz WK, Yee VC. Crystal structure of the human prion protein reveals a mechanism for oligomerization. Nature Structural Biology. 2001;8:770-4.

[107] Janowski R, Kozak M, Jankowska E, Grzonka Z, Grubb A, Abrahamson M, et al. Human cystatin C, an amyloidogenic protein, dimerizes through three-dimensional domain swapping. Nature Structural Biology. 2001;8:316-20.

[108] Luhrs T, Zahn R, Wuthrich K. Amyloid formation by recombinant full-length prion proteins in phospholipid bicelle solutions. Journal of Molecular Biology. 2006;357:833-41.

[109] Hafner-Bratkovic I, Bester R, Pristovsek P, Gaedtke L, Veranic P, Gaspersic J, et al. Globular domain of the prion protein needs to be unlocked by domain swapping to support prion protein conversion. The Journal of Biological Chemistry. 2011;286:12149-56.

[110] Lee S, Eisenberg D. Seeded conversion of recombinant prion protein to a disulfide-bonded oligomer by a reduction-oxidation process. Nature Structural Biology. 2003;10:725-30.

[111] Wahlbom M, Wang X, Lindstrom V, Carlemalm E, Jaskolski M, Grubb A. Fibrillogenic oligomers of human cystatin C are formed by propagated domain swapping. The Journal of Biological Chemistry. 2007;282:18318-26.

[112] Liu C, Sawaya MR, Eisenberg D. beta(2)-microglobulin forms three-dimensional domain-swapped amyloid fibrils with disulfide linkages. Nature Structural & Molecular Biology. 2011;18:49-55.

[113] Staniforth RA, Giannini S, Higgins LD, Conroy MJ, Hounslow AM, Jerala R, et al. Three-dimensional domain swapping in the folded and molten-globule states of cystatins, an amyloid-forming structural superfamily. The EMBO Journal. 2001;20:4774-81.

[114] Sanders A, Jeremy Craven C, Higgins LD, Giannini S, Conroy MJ, Hounslow AM, et al. Cystatin forms a tetramer through structural rearrangement of domain-swapped dimers prior to amyloidogenesis. Journal of Molecular Biology. 2004;336:165-78.

[115] Nilsson M, Wang X, Rodziewicz-Motowidlo S, Janowski R, Lindstrom V, Onnerfjord P, et al. Prevention of domain swapping inhibits dimerization and amyloid fibril formation of cystatin C: use of engineered disulfide bridges, antibodies, and carboxymethylpapain to stabilize the monomeric form of cystatin C. The Journal of Biological Chemistry. 2004;279:24236-45.

[116] Orlikowska M, Jankowska E, Kolodziejczyk R, Jaskolski M, Szymanska A. Hinge-loop mutation can be used to control 3D domain swapping and amyloidogenesis of human cystatin C. Journal of Structural Biology. 2011;173:406-13.

[117] Olafsson I, Grubb A. Hereditary cystatin C amyloid angiopathy. Amyloid. 2000;7:70-9.

[118] Eakin CM, Attenello FJ, Morgan CJ, Miranker AD. Oligomeric assembly of native-like precursors precedes amyloid formation by beta-2 microglobulin. Biochemistry. 2004;43:7808-15.

[119] Byeon IJ, Louis JM, Gronenborn AM. A protein contortionist: core mutations of GB1 that induce dimerization and domain swapping. Journal of Molecular Biology. 2003;333:141-52.

[120] Kirsten Frank M, Dyda F, Dobrodumov A, Gronenborn AM. Core mutations switch monomeric protein GB1 into an intertwined tetramer. Nature Structural Biology. 2002;9:877-85.

[121] Li J, Hoop CL, Kodali R, Sivanandam VN, van der Wel PC. Amyloid-like fibrils from a domain-swapping protein feature a parallel, in-register conformation without native-like interactions. The Journal of Biological Chemistry. 2011;286:28988-95.

[122] Guo Z, Eisenberg D. Runaway domain swapping in amyloid-like fibrils of T7 endonuclease I. Proceedings of the National Academy of Sciences of the United States of America. 2006;103:8042-7.

[123] Bader R, Seeliger MA, Kelly SE, Ilag LL, Meersman F, Limones A, et al. Folding and fibril formation of the cell cycle protein Cks1. The Journal of Biological Chemistry. 2006;281:18816-24.

[124] Laidman J, Forse GJ, Yeates TO. Conformational change and assembly through edge beta strands in transthyretin and other amyloid proteins. Accounts of Chemical Research. 2006;39:576-83.

[125] Serag AA, Altenbach C, Gingery M, Hubbell WL, Yeates TO. Arrangement of subunits and ordering of beta-strands in an amyloid sheet. Nature Structural Biology. 2002;9:734-9.

[126] Sawaya MR, Sambashivan S, Nelson R, Ivanova MI, Sievers SA, Apostol MI, et al. Atomic structures of amyloid cross-beta spines reveal varied steric zippers. Nature. 2007;447:453-7.

[127] Goldschmidt L, Teng PK, Riek R, Eisenberg D. Identifying the amylome, proteins capable of forming amyloid-like fibrils. Proceedings of the National Academy of Sciences of the United States of America. 2010;107:3487-92.

[128] Teng PK, Anderson NJ, Goldschmidt L, Sawaya MR, Sambashivan S, Eisenberg D. Ribonuclease A suggests how proteins self-chaperone against amyloid fiber formation. Protein Science: a publication of the Protein Society. 2012;21:26-37.

[129] Torrent M, Odorizzi F, Nogues MV, Boix E. Eosinophil cationic protein aggregation: identification of an N-terminus amyloid prone region. Biomacromolecules. 2010;11:1983-90.

[130] Sambashivan S, Liu Y, Sawaya MR, Gingery M, Eisenberg D. Amyloid-like fibrils of ribonuclease A with three-dimensional domain-swapped and native-like structure. Nature. 2005;437:266-9.

[131] Mizuno H, Fujimoto Z, Koizumi M, Kano H, Atoda H, Morita T. Structure of coagulation factors IX/X-binding protein, a heterodimer of C-type lectin domains. Nature Structural Biology. 1997;4:438-41.

[132] Vatzaki EH, Allen SC, Leonidas DD, Trautwein-Fritz K, Stackhouse J, Benner SA, et al. Crystal structure of a hybrid between ribonuclease A and bovine seminal ribonuclease--the basic surface, at 2.0 A resolution. Eur J Biochem. 1999;260:176-82.

[133] Sica F, Di Fiore A, Merlino A, Mazzarella L. Structure and stability of the non-covalent swapped dimer of bovine seminal ribonuclease: an enzyme tailored to evade ribonuclease protein inhibitor. The Journal of Biological Chemistry. 2004;279:36753-60.

[134] Ercole C, Colamarino RA, Pizzo E, Fogolari F, Spadaccini R, Picone D. Comparison of the structural and functional properties of RNase A and BS-RNase: a stepwise mutagenesis approach. Biopolymers. 2009;91:1009-17.

[135] Merlino A, Picone D, Ercole C, Balsamo A, Sica F. Chain termini cross-talk in the swapping process of bovine pancreatic ribonuclease. Biochimie. 2012;94:1108-18.

[136] Meersman F, Heremans K. Temperature-induced dissociation of protein aggregates: accessing the denatured state. Biochemistry. 2003;42:14234-41.

[137] Babu CR, Hilser VJ, Wand AJ. Direct access to the cooperative substructure of proteins and the protein ensemble via cold denaturation. Nature Structural & Molecular Biology. 2004;11:352-7.

[138] Jaremko M, Jaremko L, Kim HY, Cho MK, Schwieters CD, Giller K, et al. Cold denaturation of a protein dimer monitored at atomic resolution. Nature Chemical Biology. 2013;9:264-70.

[139] Merlino A, Vitagliano L, Ceruso MA, Mazzarella L. Dynamic properties of the N-terminal swapped dimer of ribonuclease A. Biophys J. 2004;86:2383-91.

[140] Merlino A, Ceruso MA, Vitagliano L, Mazzarella L. Open interface and large quaternary structure movements in 3D domain swapped proteins: insights from molecular dynamics simulations of the C-terminal swapped dimer of ribonuclease A. Biophys J. 2005;88:2003-12.

[141] Selwood T, Jaffe EK. Dynamic dissociating homo-oligomers and the control of protein function. Archives of Biochemistry and Biophysics. 2012;519:131-43.

[142] Zhou HX, Rivas G, Minton AP. Macromolecular crowding and confinement: biochemical, biophysical, and potential physiological consequences. Annual Review of Biophysics. 2008;37:375-97.

[143] Ercole C, Lopez-Alonso JP, Font J, Ribo M, Vilanova M, Picone D, et al. Crowding agents and osmolytes provide insight into the formation and dissociation of RNase A oligomers. Archives of Biochemistry and Biophysics. 2011;506:123-9.

[144] Nenci A, Gotte G, Maras B, Libonati M. Different susceptibility of the two dimers of ribonuclease A to subtilisin. Implications for their structure. Biochimica et Biophysica Acta. 2001;1545:255-62.

[145] Abrami L, Fivaz M, van der Goot FG. Adventures of a pore-forming toxin at the target cell surface. Trends in Microbiology. 2000;8:168-72.

[146] Milne JC, Furlong D, Hanna PC, Wall JS, Collier RJ. Anthrax protective antigen forms oligomers during intoxication of mammalian cells. The Journal of Biological Chemistry. 1994;269:20607-12.

[147] Billington SJ, Jost BH, Songer JG. Thiol-activated cytolysins: structure, function and role in pathogenesis. FEMS, Microbiology Letters. 2000;182:197-205.

[148] Bucci E, Vitagliano L, Barone R, Sorrentino S, D'Alessio G, Graziano G. On the thermal stability of the two dimeric forms of ribonuclease A. Biophysical Chemistry. 2005;116:89-95.

[149] Vottariello F, Giacomelli E, Frasson R, Pozzi N, De Filippis V, Gotte G. RNase A oligomerization through 3D domain swapping is favoured by a residue located far from the swapping domains. Biochimie. 2011;93:1846-57.

[150] Rafelski SM, Theriot JA. Crawling toward a unified model of cell mobility: spatial and temporal regulation of actin dynamics. Annual Review of Biochemistry. 2004;73:209-39.

[151] Dell'Orco D. A physiological role for the supramolecular organization of rhodopsin and transducin in rod photoreceptors. FEBS letters. 2013;587:2060-6.

[152] Gotte G, Testolin L, Costanzo C, Sorrentino S, Armato U, Libonati M. Cross-linked trimers of bovine ribonuclease A: activity on double-stranded RNA and antitumor action. FEBS Letters. 1997;415:308-12.

[153] Bartholeyns J, Baudhuin P. Inhibition of tumor cell proliferation by dimerized ribonuclease. Proceedings of the National Academy of Sciences of the United States of America. 1976;73:573-6.

[154] Matousek J, Gotte G, Pouckova P, Soucek J, Slavik T, Vottariello F, et al. Antitumor activity and other biological actions of oligomers of ribonuclease A. The Journal of Biological Chemistry. 2003;278:23817-22.

[155] Merlino A, Russo Krauss I, Perillo M, Mattia CA, Ercole C, Picone D, et al. Toward an antitumor form of bovine pancreatic ribonuclease: the crystal structure of three non-covalent dimeric mutants. Biopolymers. 2009;91:1029-37.

[156] Cafaro V, Bracale A, Di Maro A, Sorrentino S, D'Alessio G, Di Donato A. New muteins of RNase A with enhanced antitumor action. FEBS Letters. 1998;437:149-52.

[157] Kobe B, Deisenhofer J. Mechanism of ribonuclease inhibition by ribonuclease inhibitor protein based on the crystal structure of its complex with ribonuclease A. Journal of Molecular Biology. 1996;264:1028-43.

[158] Kim JS, Soucek J, Matousek J, Raines RT. Mechanism of ribonuclease cytotoxicity. The Journal of Biological Chemistry. 1995;270:31097-102.

[159] Rousseau F, Wilkinson H, Villanueva J, Serrano L, Schymkowitz JW, Itzhaki LS. Domain swapping in p13suc1 results in formation of native-like, cytotoxic aggregates. Journal of Molecular Biology. 2006;363:496-505.

[160] Schymkowitz JW, Rousseau F, Irvine LR, Itzhaki LS. The folding pathway of the cell-cycle regulatory protein p13suc1: clues for the mechanism of domain swapping. Structure. 2000;8:89-100.

[161] Thylen C, Wahlqvist J, Haettner E, Sandgren O, Holmgren G, Lundgren E. Modifications of transthyretin in amyloid fibrils: analysis of amyloid from homozygous and heterozygous individuals with the Met30 mutation. The EMBO Journal. 1993;12:743-8.

[162] McCutchen SL, Colon W, Kelly JW. Transthyretin mutation Leu-55-Pro significantly alters tetramer stability and increases amyloidogenicity. Biochemistry. 1993;32:12119-27.

[163] Apostol MI, Sawaya MR, Cascio D, Eisenberg D. Crystallographic studies of prion protein (PrP) segments suggest how structural changes encoded by polymorphism at residue 129 modulate susceptibility to human prion disease. The Journal of Biological Chemistry. 2010;285:29671-5.

[164] Levy CB, Stumbo AC, Ano Bom AP, Portari EA, Cordeiro Y, Silva JL, et al. Co-localization of mutant p53 and amyloid-like protein aggregates in breast tumors. The International Journal of Biochemistry & Cell Biology. 2011;43:60-4.

[165] Ano Bom AP, Rangel LP, Costa DC, de Oliveira GA, Sanches D, Braga CA, et al. Mutant p53 aggregates into prion-like amyloid oligomers and fibrils: implications for cancer. The Journal of Biological Chemistry. 2012;287:28152-62.

[166] Apostol MI, Wiltzius JJ, Sawaya MR, Cascio D, Eisenberg D. Atomic structures suggest determinants of transmission barriers in mammalian prion disease. Biochemistry. 2011;50:2456-63.

[167] Silva JL, Rangel LP, Costa DC, Cordeiro Y, De Moura Gallo CV. Expanding the prion concept to cancer biology: dominant-negative effect of aggregates of mutant p53 tumour suppressor. Bioscience Reports. 2013;33.

[168] Bucciantini M, Giannoni E, Chiti F, Baroni F, Formigli L, Zurdo J, et al. Inherent toxicity of aggregates implies a common mechanism for protein misfolding diseases. Nature. 2002;416:507-11.

[169] Dykes G, Crepeau RH, Edelstein SJ. Three-dimensional reconstruction of the fibres of sickle cell haemoglobin. Nature. 1978;272:506-10.

[170] Holmes KC, Popp D, Gebhard W, Kabsch W. Atomic model of the actin filament. Nature. 1990;347:44-9.

[171] Downing KH, Nogales E. Crystallographic structure of tubulin: implications for dynamics and drug binding. Cell Structure and Function. 1999;24:269-75.

[172] Chiti F, Dobson CM. Amyloid formation by globular proteins under native conditions. Nature Chemical Biology. 2009;5:15-22.

[173] Rajagopal K, Schneider JP. Self-assembling peptides and proteins for nanotechnological applications. Current Opinion in Structural Biology. 2004;14:480-6.

[174] Bonzani IC, George JH, Stevens MM. Novel materials for bone and cartilage regeneration. Current Opinion in Chemical Biology. 2006;10:568-75.

[175] Terzaki K, Kalloudi E, Mossou E, Mitchell EP, Forsyth VT, Rosseeva E, et al. Mineralized self-assembled peptides on 3D laser-made scaffolds: a new route toward 'scaffold on scaffold' hard tissue engineering. Biofabrication. 2013;5:045002.

[176] Nagarkar RP, Hule RA, Pochan DJ, Schneider JP. De novo design of strand-swapped beta-hairpin hydrogels. Journal of the American Chemical Society. 2008;130:4466-74.

[177] Ogihara NL, Ghirlanda G, Bryson JW, Gingery M, DeGrado WF, Eisenberg D. Design of three-dimensional domain-swapped dimers and fibrous oligomers. Proceedings of the National Academy of Sciences of the United States of America. 2001;98:1404-9.

[178] van Raaij MJ, Mitraki A, Lavigne G, Cusack S. A triple beta-spiral in the adenovirus fibre shaft reveals a new structural motif for a fibrous protein. Nature. 1999;401:935-8.

Chapter 9

Oligomerization of Proteins and Neurodegenerative Diseases

Dai Mizuno and Masahiro Kawahara

Additional information is available at the end of the chapter

http://dx.doi.org/10.5772/57482

1. Introduction

Oligomerization of amino acids by binding their peptide bonds (-CONH-) forms proteins (or peptides), which are the major components of our bodies. Although the primary sequence (the linear sequence of amino acids) of the protein mainly determines its characteristics, its secondary structures (the conformations) are also critical determinants of their shapes and functions. The conformation (random coil, α-helix, and ß-sheet) is restricted by the circumstances nearby proteins. The hydrogen bond between the amino acids in the peptide chain forms the α-helix structure. Meanwhile, the ß-sheets (ß-plated sheets) consist of ß-strands which are laterally connected peptide bonds with hydrogen bonds.

Recent neurochemical evidence indicates that the oligomerization of proteins and the formation of ß-sheet structures are linked with several neurodegenerative diseases such as Alzheimer's disease (AD), prion diseases, triplet repeat diseases, dementia with Lewy bodies (DLB). The disease-related proteins, such as ß-amyloid protein (AßP) in AD, prion protein in prion diseases, polyglutamine in triplet repeat disease, α-synuclein in DLB, are identical in each disease (Table 1). However, all of these amyloidogenic proteins share common characteristics in the formation of amyloid with ß-sheet structures, and in the exhibition of cytotoxicity. Therefore, a new concept termed "conformational disease" was proposed, suggesting that protein conformation is an important determinant of its toxicity, and consequently, the development of the related disease [1].

These conformational diseases are included in amyloidosis. At 1853, Virchow found the abnormal accumulates in tissues and named "amyloid", since they exhibited similar characteristics with *amylum*. At 1968, amyloid was determined to be the oligomers of proteins with ß-sheet structures. The accumulation of amyloid causes various diseases (amyloidosis) including familial amyloid polyneuropathy (FAP), amyloid-light chain amyloidosis, dialysis

Disease	The primary sequence of amyloidogenic protein or its fragment peptide	Metal	ß-sheet formation	Cyto-toxicity	Pore-formation
Alzheimer's disease	*AßP(1-42) and AßP(1-42)truncated C-terminal* *DAEFRHDSGYEVHHQKLVFFAEDVGSNKGAIIGLMVGG VVIA*	Al, Zn, Cu, Fe	+	+	+
Prion disease	Prion protein: *PrP106-126* MANLGCWMLVLFVATWSDLGLCKKRPKPGGWNTGG SRYPGQGSPGGNRYPPQGGGGWGQPHGGGWGQPH GGGWGQPHGGGWGQPHGGGWGQGGGTHSQWN KPSKP*KTNMKHMAGAAAAGAVVGG*LGGYVLGSAMS RPIIHFGSDYEDRYYRENMHRYPNQVYYRPMDEYSNQ NNFVHDCVNITIKQHTVTTTTKGENFTETDVKMMERVV EQMCITQYERESQAYYQRGSSMVLFSSPPVILLISFLIFL IVG	Zn, Cu, M, Fe	+	+	+
Dementia with Lewy bodies (DLB)	α-synuclein; *NAC* (a *fragment of α-synuclein*) MDVFMKGLSKAKEGVVAAAEKTKQGVAEAAGKTKEG VLYVGSKTKEGVVHGVTTVAEKTKEQVSNVGGAVVTG VTAVAHKTVEGAGNFAAATGLVKKDQKNESGFGPEG TMENSENMPVNPNNETYEMPPEEEYQDYDPEA	Cu, Fe, Al	+	+	+
Triplet- repeat disease	*Polyglutamine* MATLEKLMKAFESLKSF*QQQQQQQQQQQQQQQQQ QQQQQQ*PPPPPPPPPPPQLPQPPPQAQPLLPQPQPPPP PPPPPPGP	Fe	+	+	+
Diabetes mellitus	*Human amylin* KCNTATCATQRLANFLVHSSNNFGAILSSTNVGSNTY	Cu, Al	+	+	+

* The sequence of fragment peptide of each amyloidogenic protein (PrP106-126, NAC, polyglutamine) is indicated in italic form.

Table 1. Characteristics of amyloidogenic proteins and the related peptides

amyloidosis, *etc* [2]. All of theses diseases share common properties about the deposition of amyloids in various tissues or organs with protease-resistant, insoluble fibril-like structures (amyloid fibrils), and stained by congo-red, ß-sheet specific dye. However, the component of amyloid is different in each disease. For example, the major component of amyloid in FAP patients is transthyretin, and ß2-microglobulin deposits in patients with dialysis amyloidosis. There are no effective treatments for amyloidosis.

In this chapter, we review the implication of protein oligomerization in the pathogenesis of these neurodegenerative diseases. Considering that the amyloidogenic proteins are commonly present in our brain, factors which influence oligomerization play crucial roles in their pathogenesis. As such factors, we focus on trace elements such as Al, Zn, Cu, and Fe. Metals have a property of firmly binding to metal-binding residues of proteins, such as tyrosine (Tyr) or histidine (His) or phosphorylated amino acids, and cause cross-linking of the proteins (Fig. 1). Furthermore, all of these amyloidogenic proteins were reported to have the ability to bind metals as shown in Table 1. Our and other numerous studies reported that oligomers cause neurodegeneration by induction of Ca^{2+} dyshomeostasis through the formation of amyloid channels on neuronal membranes [3,4]. The beneficial characteristics of carnosine (ß alanyl histidine) as a drug for the treatment for these neurodegenerative diseases are also discussed.

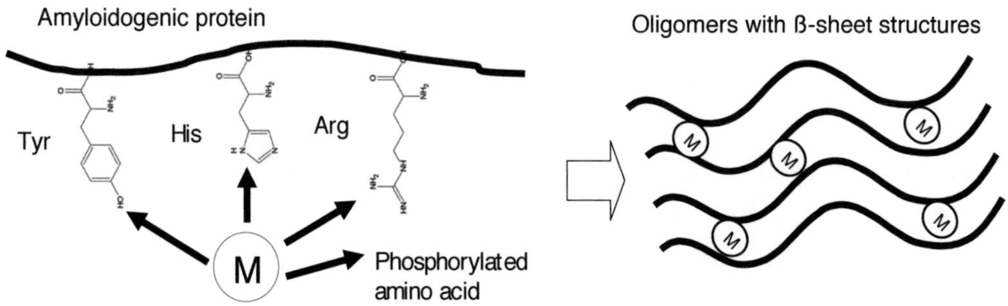

M stands for metal.

Figure 1. Trace elements acts cross-linkers of amyloidogenic proteins.

2. Alzheimer's disease and oligomerization of AßP

2.1. Amyloid cascade hypothesis

Alzheimer's disease (AD) is a severe type of senile dementia, affecting a large portion of elderly people worldwide. It is characterized by profound memory loss and inability to form new memories. The pathological hallmarks of AD are the presence of numerous extracellular deposits (senile plaques) and intraneuronal neurofibrillary tangles (NFTs). The degeneration of synapses and neurons in the hippocampus or cerebral cortex is also observed. The major components of NFTs are phosphorylated tau proteins, and that of senile plaques are ß-amyloid proteins (AßPs) [5]. Although the precise cause of AD remains elusive, numerous biochemical, cell biological, and genetic studies have supported the idea termed "amyloid cascade hypothesis" that the AßP accumulation and the consequent neurodegeneration play a central role in AD [6]. Moreover, recent studies on the identified AßP species have indicated that the oligomerization of AßP and the conformational changes are critical in the neurodegeneration process [7].

AßP is a small peptide of 39–43 amino acid long. It is derived from the proteolytic cleavage of a large precursor protein (amyloid precursor protein; APP). AßP is secreted by the cleavage of its N-terminal by ß-secretase (BACE), followed by the intra-membrane cleavage of its C-terminal by γ-secretase. Genetic studies of early-onset cases of familial AD indicated that APP mutations and AßP metabolism are associated with AD. It was also revealed that mutations in the presenilin genes account for the majority of cases of early-onset familial AD. Presenilins have been revealed to be one of γ -secretases, and their mutations also influence the production of AßP and its neurotoxicity.

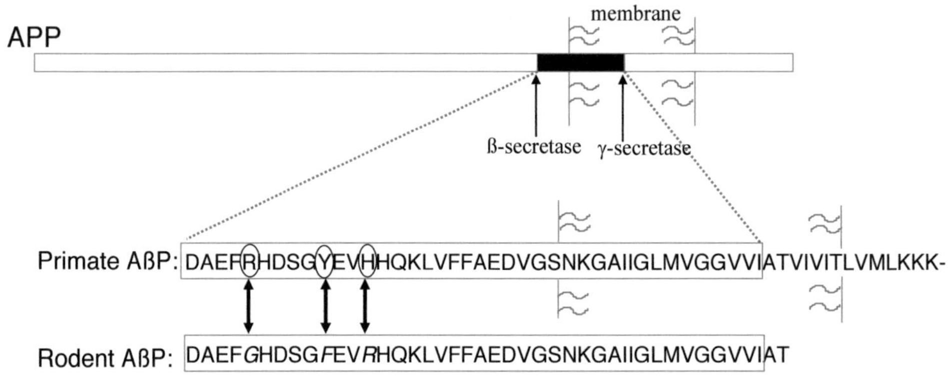

AßP is secreted from its precursor protein, APP by transmembrane cleavage. Sequences of primate AßP(1-42) and rodent AßP(1-42) are shown. The comparison between the sequence of primate (human or monkey) Aß(1-42) and rodent (rat or mouse) Aß(1-42) is depicted.

Figure 2. Structure of AßP

Yankner *et al.* reported that the first 40 amino acid residues of AßP (AßP(1–40)) caused the death of cultured rat hippocampal neurons or the neurodegeneration in the brains of experimental animals. Thereafter, it was agreed upon that the aggregation and the subsequent conformational change of AßP contribute to its neurotoxicity. AßP is a hydrophobic peptide with an intrinsic tendency to self-assemble to form insoluble oligomers with ß-pleated sheet structures. Pike *et al.* revealed that aged AßP(1–40) (aggregated under incubation at 37°C for several days) were considerably more toxic to cultured neurons as compared to freshly prepared AßP(1–40). Simmons *et al.* revealed ß-sheet contents of AßP observed by circular dichroism (CD) spectroscopy correlates with its neurotoxicity.

Furthermore, the longer peptide variant, AßP(1–42), has the characteristics of immediate polymerization compared to AßP(1–40). AßP(1–42) enhances the aggregation of AßP(1–40) and becomes a seed of the amyloid fibrils. AßP (1–42) is more abundant in the brains of AD patients as compared to those of age matched controls. The mutations of APP and those of presenilin genes induce the increased production of AßP (1–42) in the transfected cell lines.

Recent approaches using size-exclusion chromatography, gel electrophoresis, and atomic force microscopy have demonstrated that there are several stable types of soluble oligomers: naturally occurring soluble oligomers (dimers or trimers), ADDLs (AßP-derived diffusible ligands), AßP globulomers, or protofibrils. Hartley separated aggregated AßP(1–40) into low–molecular-weight (mainly monomer), protofibrillar, and fibril fractions by size-exclusion chromatography, and found that the protofibrillar fraction caused marked changes in the electrical activity of cultured neurons and neurotoxicity. Walsh *et al.* found that the intracere-

http://dx.doi.org/10.5772/57482

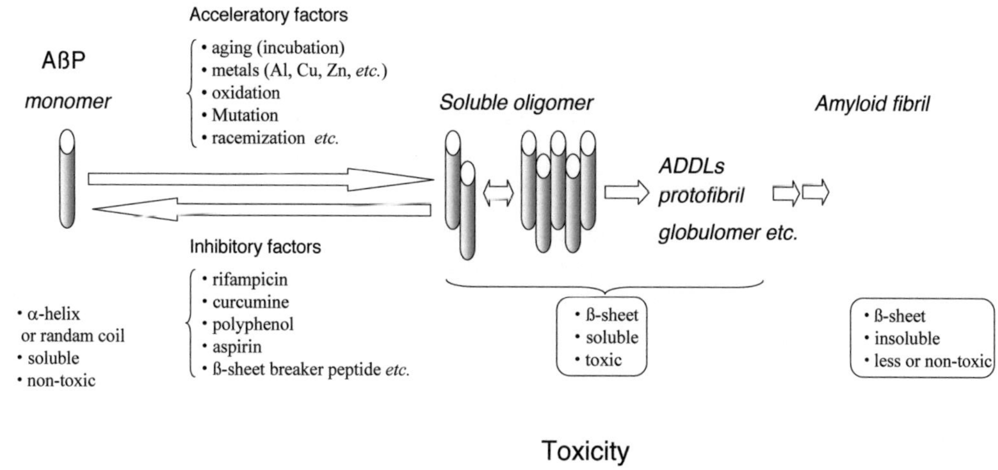

AßP monomers exhibit random or a-helix structures. However, under aging conditions or in the presence of some acceleratory factors, Aß self-aggregates and forms several types of oligomers (SDS-soluble oligomers, ADDLS, globulomers, or protofibrils *etc.*) and finally forms insoluble aggregates termed amyloid fibrils. Oligomeric soluble Aßs are toxic, although the monomeric and fibril ones are rather nontoxic.

Figure 3. Oligomerization of AßP

bral administration of the conditioned medium with cultured cells transfected with the human APP gene inhibited long-term potentiation (LTP), which is a form of synaptic information storage well known as a paradigm of memory mechanisms. They also demonstrated that LTP was blocked by SDS-stable low-molecular-weight oligomers (dimers, trimers, or tetramers) but not AßP monomers or larger aggregates. The natural AßP oligomers (derived from the cerebrospinal fluid of AD patients) cause the loss of dendritic spines, synapses, and LTP blockage. Klein and the colleagues reported that AßP-derived diffusible ligands (ADDLs) obtained from sedimentation by clustering are highly toxic to cultured neurons. They also reported that ADDLs inhibited LTP and exhibited adverse effects on synaptic plasticity such as abnormal spine morphology, decreased spine density, and decreased synaptic proteins. Based on these and other numerous findings, it is widely accepted that AßP oligomers are synaptotoxic and neurotoxic, but not monomer or fibrils. These studies further strengthened and modified the amyloid cascade hypothesis, which suggest that AßP oligomers are neurotoxic and crucial for the pathogeneis of AD [8,9].

2.2. Metal-induced oligomerization of AßP

Considering that AßP is secreted from APP into the brain of young people or of normal subjects, factors which influence (accelerate or delay) the oligomerization may become important determinants of the pathogenesis of AD. Various factors, such as the concentration of peptides, the oxidations, mutations, and racemization of AßP, pH, composition of solvents, temperature, and trace elements, can influence the oligomerization processes. A considerable amount of asparagines (Asp) or serine (Ser) residues of AßP accumulated in senile plaques are racemized. Tomiyama *et al.* reported that racemized D-Asp[23]-AßP easily aggregates compared to the L-type. Meanwhile, several substances such as rifampicin, curcumin, and aspirin have been reported to inhibit AßP oligomerization *in vitro*. Rifampicin, a drug used to treat Hansen's disease, may be an interesting inhibitor of oligomerization since patients with Hansen's disease have a low susceptibility to AD. Aspirin and other NSAIDs (non-steroidal anti-inflammatory drugs) inhibit the AßP oligomerization and simultaneously attenuate inflammation.

Among these factors, trace elements such as aluminum (Al), zinc (Zn), copper (Cu), iron (Fe) are of particular interest. The accumulation of AßP is rarely observed in the brains of rodents (rats or mice) as compared to humans or monkeys. As shown in Fig.2, the amino acid sequence of human and rodent AßP are similar, yet they differ by three amino acids. However, rodent AßP exhibits less tendency to oligomerization compared to human AßP [10]. Considering that these three amino acids (Arg^5, Tyr^{10}, and His^{13}) have the ability to bind metals and that trace metals have cross-linking ability, trace elements might play important roles in the accumulation of AßP in the human brain.

Exley *et al.* first demonstrated that Al induces a conformational change in AßP(1-40) by CD spectroscopy. Furthermore, exposure to Al causes the accumulation of AßP in cultured neurons or in brains of experimental animals or human. Pratico *et al.* found that Al-fed mice transfected with the human APP gene (Tg 2576) exhibited pathological changes similar to those of the AD brain, including a marked increase in the amount of AßP both in the secreted form and the accumulated form: an increased deposition of senile plaques was also observed [11]. The neuropathological case study of the accidental Al-exposure that occurred in 1988 at Camelford (Cornwall, U.K.) indicated that the exposure to Al, even if it is short-term, could cause the accumulation of AßP and exhibit severe amyloid angiopathy [12]. Since there have been studies indicating the link between Al in drinking water and the pathogenesis of AD, Al-induced oligomerization may directly implicated in AD pathogenesis [13].

Bush *et al.* demonstrated that Zn^{2+} and Cu^{2+}caused the oligomerization of AβP [14]. However, the metal-induced oligomerization of AßP and other amyloidogenic proteins are complex and controversial. The morphology of AßP oligomers treated with Al, Cu, Fe, and Zn were reported to be quite different. Zatta and his colleagues demonstrated that metals including Al, Cu, Fe, Zn differentially alter the oligomerization of AßP and its toxicity. We have shown that Al enhances the polymerization of AßP(1-40) and forms SDS-stable oligomers *in vitro* by immunoblotting and precipitation [15,16]. The oligomerized AßP(1-40) is heat- or SDS-stable but redissolves on adding deferoxamine, a chelator of Al. The oligomerization induced by Al is more marked than that induced by other metals, including Zn, Fe, Cu, and Cd. Furthermore, while Zn-aggregated AßPs are rarely observed on the surface of cultured neurons several days after its exposure, Al-aggregated AßPs bind tightly to the surface of cultured neurons and form fibrillar deposits. These results suggest that Al-induced AßP oligomers have a strong affinity to membrane surfaces and undergo minimal degradation by proteases compared to Zn-induced oligomers. Furthermore, AßP coupled with Al was reported to be highly toxic compared to normal AßP.

Considering the implications of metals in AD pathogenesis, chelation therapy for AD treatment is of great interest. Clioquinol (quinoform), a chelator of Cu^{2+} or Zn^{2+}, inhibits oligomerization of AßP and attenuates the accumulation of amyloid in the brains of experimental animals. Clinical trials using its analogue PBT2 are under investigation. DFO, a chelator of Al and Fe, attenuates the decline of daily living skills in AD patients. Silicates, which couple with Al and reduce its toxicity, are also candidates for chelation therapy in AD [17].

2.2. Oligomerization-induced neurotoxicity of AßP

There is a considerable interest regarding the mechanisms by which AßP oligomers cause neurotoxicity. Exposure to AßP causes various adverse effects on neuronal survivals such as the production of reactive oxygen species, the induction of cytokines, the induction of endoplasmic reticulum (ER) stresses, and the abnormal increase of intracellular calcium levels ([Ca^{2+}]i), *etc* [18]. Although these effects may be interwoven, the disruption of Ca^{2+} homeostasis is regarded to be an important determinant considering it occurs upstream of the other effects [19,20]. Ca^{2+} ions are essential for the normal brain functions. They are involved with key enzymes such as kinases, phosphatases, and proteases. Therefore, its influx is severely controlled, and the intracellular Ca^{2+} levels ($[Ca^{2+}]_i$) are strictly conserved by Ca^{2+} channels, *etc*. Ca^{2+} is also implicated in the phosphorylation of the tau protein or in APP sequestration. Increasing evidence indicates that presenilins are involved in capacitative Ca^{2+} entry or endoplasmic reticulum (ER) Ca^{2+} signaling, and that their mutations affect Ca^{2+}-regulated functions including AßP production [21].

There is considerable interest regarding the mechanism by which AßPs interact with neurons and disrupt Ca^{2+} homeostasis. In 1993, Arispe *et al.* first demonstrated that AßP(1–40) directly incorporates into artificial lipid bilayer membranes and forms cation-selective ion channels [22]. The channels termed "amyloid channels" were revealed to be giant multi-level pores and can allow a large amount of Ca^{2+} to pass through. Their activity was blocked by Zn^{2+} ions, which are abundantly present in the brain. Furthermore, soluble AßP oligomers but not amyloid fibrils were reported to increase the membrane permeability. Durell *et al.* proposed a 3-D structural model of the amyloid channels obtained from a computer simulation of the secondary structure of AßP(1–40) in membranes that showed 5- to 8-mers aggregating to form pore-like structures on the membranes. The multimeric (tetramer to hexamer) pore-like structures of AßPs on reconstituted membranes were observed using atomic force microscopy. Jang *et al.* established a model of amyloid channels on the membranes and observed that pentamer AßPs form pores, and their dimensions, shapes, and subunit organizations are in good agreement with AFM studies [23]. These results strongly support the hypothetical idea termed "amyloid channel hypothesis", which suggests that the direct incorporation of AßPs and the subsequent imbalances of Ca^{2+} and other ions through amyloid channels might be the primary event in AßP neurotoxicity. In this respect, AßP might share the mechanism of toxicity with a similar mechanism underlying the toxicity of various antimicrobial or antifungal peptides that also exhibit channel-forming activity and cell toxicity.

To determine whether AßPs form channels on neuronal cell membranes as well as artificial lipid bilayers, we employed membrane patches from a neuroblastoma cell line (GT1-7 cells), which exhibit several neuronal characteristics such as the extension of neuritis and the expression of neuron-specific proteins or receptors [24]. After exposing the excised membrane patches of GT1-7 cells in the bath solution to AßP(1–40), the current derived from the amyloid channels appeared. The amyloid channels formed on the GT1-7 cell membranes were cation-selective, multilevel, voltage-independent, long-lasting ones; the channel activity was inhibited by the addition of Zn^{2+}, and recovered by a zinc chelator, *o*-phenanthroline [25]. These features were considerably similar to those observed on artificial lipid bilayers. Meanwhile, AßP(40–1), a peptide bearing the reversed sequence of AßP(1–40), did not form any channels. Thus, we can conclude that AßPs are directly incorporated into neuronal membranes to form calcium-permeable pores. In order to test the amyloid channel hypothesis, we examined whether AßP altered the $[Ca^{2+}]_i$ levels in neurons by a high-resolution multi-site video imaging system with fura-2 as the cytosolic free calcium reporter fluorescent probe. This multisite fluorometry system enables the simultaneous long-term observation of temporal changes in $[Ca^{2+}]_i$ of more than 50 neurons. We could observe AßP-induced abnormal increase in $[Ca^{2+}]_i$ in GT1-7 cells [26-28] as well as in primary cultured rat hippocampal neurons [29]. Shortly after exposure to AßP (1–40), a marked increase in $[Ca^{2+}]_i$ occurred among many, but not all neurons. We also observed apoptotic death of cultured neurons after the exposure to AßPs and the consequent rise in the $[Ca^{2+}]_i$ levels.

Considering the results of our study together with those of the other studies, we propose the following hypothetical scheme of neurodegeneration induced by oligomerization of AßP (Fig. 4).

AßPs are normally secreted from APP into the cerebrospinal fluid and are usually degraded proteolytically by neprilysin within a short period. However, upregulation of the AßP secretion from APP, or an increased ratio of AßP(1–42) to AßP(1–40) may render AßPs liable to be retained in the brain. It has been demonstrated that APP or presenilin gene mutations promote this process. AßP possesses positive charges at neutral pH. Therefore, the net charge of the outer membrane surface may be a determinant when secreted AßPs bind to cellular membranes (Fig.4 (A)). The distribution of phospholipids on cellular membranes is usually asymmetrical and negatively charged phospholipids such as PS exist on the inner membrane surfaces. Disruption of the assymetrical distribution is the first hallmark of apoptotic cell death [30]. Therefore, the binding of AßP to neuronal membranes seldom occur in normal and young brains. This idea may explain why AD occurs in aged subjects meanwhile AßPs are secreted in the brains of young subjects. After incorporation into the membrane, the conformation of AßPs change and the accumulated AßPs aggregate on the membranes (Fig. 4(B)). The ratio of cholesterol to phospholipids in the membrane may alter membrane fluidity, thereby affecting the process from step (A) to (B). AßP oligomerization *in vitro* will also enhance the channel formation velocity. Considering that natural oligomers (dimers or trimers) are more toxic as compared to monomers or fibrils, it is provable that these oligomers might form tetrameric or hexameric pores and exhibit neurotoxicity. Micro-circumstances on the membranes, such as rafts, are suitable locations that facilitate this process. Finally, aggregated AßP oligomers form ion channels (Fig.4 (C)) leading to the various neurodegenerative processes. The processes required for channel formation (from steps (A) to (C)) may require a long life span and determine the rate of the entire process. Unlike endogenous Ca^{2+} channels, these AßP channels are not regulated by usual blockers. Thus, once formed on membranes, a continuous flow of

$[Ca^{2+}]_i$ is initiated. However, zinc ions (Zn^{2+}), which are secreted into synaptic clefts in a neuronal activity-dependent manner, inhibit AßP-induced Ca^{2+} entry, and thus have a protective function in AD.

Once AßP channels are formed on neuronal membranes, homeostasis of Ca^{2+} and other-ion will be disrupted. Disruption of Ca^{2+} homeostasis triggers several apoptotic pathways such as the activation of calpain, the induction of caspase, and promote numerous degenerative processes, including the production of reactive oxygen species (ROS) and the phosphorylation of tau, thereby accelerating neuronal death. Mutations of presenilins cause disturbances in the capacitive Ca^{2+} entry and may influence these pathways. Free radicals also induce membrane disruption, by which unregulated Ca^{2+} influx is further amplified. The disruption of Ca^{2+} homeostasis also influences the production and processing of APP. Thus, a vicious cycle of neurodegeneration is initiated. This hypothesis explains the long delay in AD development; AD occurs only in senile subjects despite the fact that Aßs are normally secreted also in younger or in normal subjects. Various environmental factors, such as foods or trace metals, as well as genetic factors will influence these processes and contribute to AD pathogenesis [31].

3. Prion diseases and other amyloidosis

The disease-related amyloidogenic proteins exhibit similarities in the formation of ß-pleated sheet structures, abnormal deposition as amyloid fibrils in the tissues, and introduction of apoptotic degeneration. Prion diseases, including human kuru, Creutzfeldt-Jakob disease, and bovine spongiform encephalopathy (BSE), are associated with the conversion of a normal prion protein (PrP^C) to an abnormal scrapie isoform (PrP^{SC}) [32]. The ß-sheet region of PrP^{SC} is suggested to play a crucial role in its transmissible degenerative processes. A peptide fragment of PrP corresponding to residues 106–126 (PrP106–126) has been reported to cause death in cultured hippocampal neurons. We investigated the oligomerization of PrP106-126 and its neurotoxicity on primary cultured rat hippocampal neurons [33]. As AßP, PrP106-126 formed amyloid-like fibrils with ß–sheet structures by observation with atomic force microspope and by thioflavin T staining during the aging process. The oligomerization and formation ß-sheet structure enhanced the neurotoxicity of PrP106-126. The co-existence of Zn or Cu inhibited ß-sheet formation of PrP106-126 and attenuated its neurotoxicity. Furthermore, the thickness of PrP106-126 fibrils was decreased in the presence of Zn or Cu.

Electrophysiological and morphological studies have revealed that PrP106-126 exhibits similarities in the formation of amyloid channels as well as AßP [34]. Lin *et al.* reported that PrP106–126 forms cation-permeable pores in artificial lipid bilayers. The activity of PrP channels was also blocked by Zn^{2+}. Kourie *et al.* investigated the detailed characteristics of channels formed by PrP106–126, concluding that it was directly incorporated into lipid bilayers and formed cation-selective, copper-sensitive ion channels. They also revealed that quinacrine, a potent therapeutic drug, possibly blocks amyloid channels induced by PrP106-126.

The oligomerization and fibrillation of α-synuclein has been implicated in the formation of abnormal inclusions, termed Lewy bodies, and the etiology of dementia with Lewy bodies

(DLB). Non-amyloid component (NAC), a fragment peptide of α-synuclein, accumulates in Alzheimer's senile plaques and causes apoptotic neuronal death. Lashuel *et al.* demonstrated by electron microscope observation that α-synuclein forms annular pore-like structures [35].

The elongation of a polyglutamine-coding CAG triplet repeat in the responsible genes is based on the pathogenesis of triplet-repeat disease such as Huntington's disease or Machado-Joseph disease. Hirakura *et al.* reported that polyglutamine formed ion channels in lipid bilayers.

Lal *et al.* investigated the oligomerization and conformational changes of AßP, synuclein, amylin, and other amyloidogenic proteins using gel electrophoresis and AFM imaging, and demonstrated that these amyloidogenic proteins form annular channel-like structures on bilayer membranes [36]. We have demonstrated that these amyloidogenic peptide also cause the elevations in $[Ca^{2+}]_i$ as well as AßP [3,31]. Considering these results together as shown in Table 1, it is suggested that the oligomerization of disease-related amyloidogenic proteins and the introduction of apoptotic degeneration by disruption of calcium homeostasis *via* unregulated amyloid channels may be the molecular basis of neurotoxicity of these diseases.

4. Conclusion

This hypothesis about the pathogenesis of conformational diseases may help in the development of drugs for these diseases. We focus carnosine (ß-alanyl histidine) as such a protective drug. Carnosine is a naturally occurring dipeptide and is commonly present in vertebrate tissues, particularly within the skeletal muscles and nervous tissues [37]. It is found at high concentrations in the muscles of animals or fish which exhibit high levels of exercise, such as horses, chickens, and whales. Thus, it is believed that carnosine plays important roles in the buffering capacities of muscle tissue and the administration of carnosine has been reported to induce hyperactivity in animals.

Secretion from synapses of AßP, and its direct incorporation into membranes and formation of oligomeric amyloid channels are depicted. Details are discussed in the text.

In the brain, a considerable amount of carnosine is localized in the neurons of the olfactory bulb. It is secreted into synaptic clefts along with the excitatory neurotransmitter glutamate during neuronal excitation. Carnosine reportedly has several beneficial effects including the antioxidant activity, the chelating ability to metal ions, the inhibition of the Maillard reaction. Furthermore, carnosine is reported to have anti-crosslinking properties. Attanasio *et al.* reported that carnosine inhibited the fibrillation of alpha-crystallin. It was also demonstrated that carnosine inhibited the oligomerization and subsequent neurotoxicity of AßP. Corona *et al.* showed that dietary supplementation of carnosine attenuated mitochondrial dysfunction and the accumulation of AßP in Alzheimer's model mice [38]. We also showed that carnosine attenuated the neuronal death induced by prion protein fragment peptide (PrP106-126) by changing its conformation [33]. Carnosine level is significantly reduced in the serum of AD patients. These results suggest possible beneficial effects of carnosine as a treatment for AD and prion diseases. We also demonstrated that carnosine attenuates Zn-induced neuronal

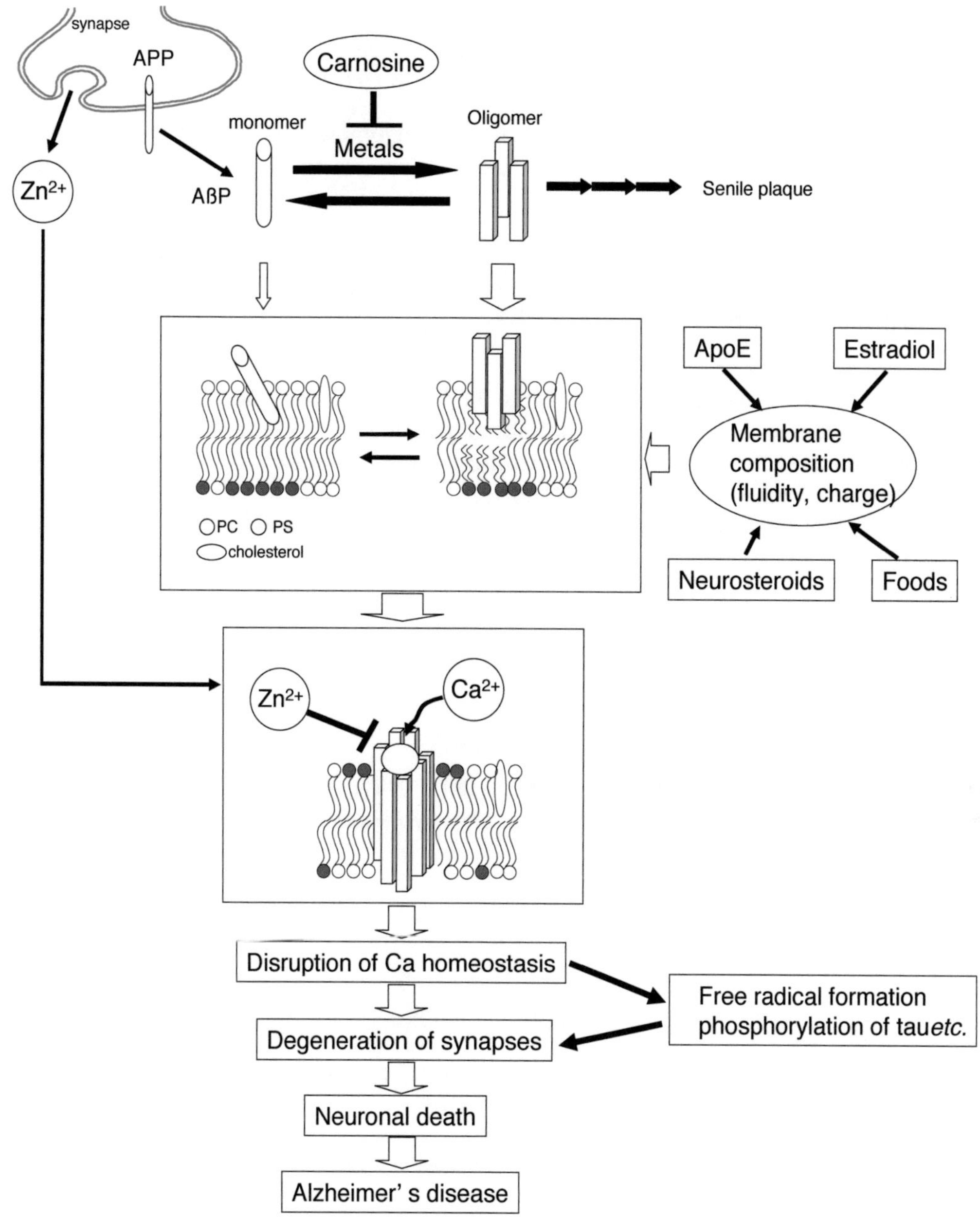

Figure 4. Amyloid channel hypothesis

death and becomes a candidate for drugs of vascular dementia [39,40]. All of these functions of carnosine (e.g., antioxidant, anti-glycating, anti-crosslinking, and scavenging toxic aldehydes) are related to the aging processes. The level of carnosine varies during development and is low in the aged animals. Therefore, it is highly possible that carnosine protects against external toxins and acts as an endogenous protective substance against neuronal injury, senescence, and aging. We have applied patents for carnosine and related compounds as drugs for vascular type of senile dementia (Patent No. 5382633, Patent No. JP5294194).

In conclusion, further research into the role of protein oligomerization and Ca homeostasis via amyloid channels might lead to the development of new treatments for neurodegenerative diseases.

Acknowledgements

The authors would like to thank Mr. M. Yanagita, Ms. A. Komuro, and Ms. N. Kato for their technical assistance. This work was partially supported by a Grant-in Aid for Scientific Research from the Ministry of Education, Culture, Sports, Science and Technology of Japan and by a Grant from Cooperation for Innovative Technology and Advanced Research in Evolutional Area (CITY AREA) from the Miyazaki Prefectural Industrial Support Foundation.

Author details

Dai Mizuno* and Masahiro Kawahara

Department of Bio-Analytical Chemistry, Musashino University, Research Institute of Pharmaceutical Sciences, Musashino University, Nishitokyo-shi, Tokyo, Japan

References

[1] Carrell, R.W. & Lomas, D.A. Conformational disease. *The Lancet*, 1997; 350: 134-138.

[2] Loo D *et al.* Proteomics in molecular diagnosis: typing of amyloidosis. *J Biomed Biotechnol.* 2011; 754109. doi: 10.1155/2011/754109.

[3] Kawahara M. Role of calcium dyshomeostasis via amyloid channels in the pathogenesis of Alzheimer's disease. *Current Pharmaceutical Design* 2010; 16: 2779-2789.

[4] Demuro A *et al.* Calcium dysregulation and membrane disruption as a ubiquitous neurotoxic mechanism of soluble amyloid oligomers. *J Biol Chem* 2005; 280: 17294-300.

[5] Selkoe, D.J. The molecular pathology of Alzheimer's disease. *Neuron* 1991; 6: 487-498.

[6] Hardy J and Selkoe DJ. The amyloid hypothesis of Alzheimer's disease: progress and problems on the road to therapeutics. *Science* 2002; 297: 353-6.

[7] Kawahara M, Negishi-Kato M, Sadakane Y. Calcium dyshomeostasis and neurotoxicity of Alzheimer's ß-amyloid protein. *Expert Rev Neurother*. 2009; 9: 681-93.

[8] Walsh DM and Selkoe DJ. Aß oligomers - a decade of discovery. *J. Neurochem*. 2007; 101: 1172-1184.

[9] Klyubin I *et al.* Alzheimer's disease Aβ assemblies mediating rapid disruption of synaptic plasticity and memory. *Mol Brain*. 2012; 5: doi: 10.1186/1756-6606-5-25.

[10] Dyrks T *et al.* Amyloidogenicity of rodent and human ß A4 sequences. *FEBS letters* 1993; 324: 231-236.

[11] Pratico D *et al.* Aluminum modulates brain amyloidosis through oxidative stress in APP transgenic mice. *FASEB J* 2002; 16: 1138-40.

[12] Exley C and Esiri MM. Severe cerebral congophilic angiopathy coincident with increased brain aluminium in a resident of Camelford, Cornwall, UK. *J Neurol Neurosurg Psychiatry* 2006; 77: 877-9.

[13] Kawahara M and Kato-Negishi M. Link between aluminum and the pathogenesis of Alzheimer's disease: the integration of the aluminum and amyloid cascade hypotheses, *Int. J. Alzheimer Dis* 2011; doi: 10.4061/2011/276393.

[14] Bush AI. The metal theory of Alzheimer's disease. *J Alzheimers Dis.* 2013; 33 Suppl 1: S277-81.

[15] Kawahara M *et al.* Aluminum promotes the aggregation of Alzheimer's β-amyloid protein *in vitro*. *Biochem Biophys Res Commun* 1994; 198: 531-535.

[16] Kawahara M *et al.* Effects of aluminum on the neurotoxicity of primary cultured neurons and on the aggregation of β-amyloid protein. *Brain Res Bull* 2001; 55: 211-217.

[17] Faux NG *et al.* PBT2 rapidly improves cognition in Alzheimer's Disease: additional phase II analyses. *J Alzheimers Dis*. 2010;20(2):509-16.

[18] Small DH, Mok SS and Bornstein JC. Alzheimer's disease and Aß toxicity: from top to bottom. *Nat Rev Neurosci.* 2001; 2: 595-8.

[19] Brorson JR *et al.* The Ca^{2+} influx induced by beta-amyloid peptide 25-35 in cultured hippocampal neurons results from network excitation. *J Neurobiol*. 1995; 26:325-38.

[20] Camandola S and Mattson MP. Aberrant subcellular neuronal calcium regulation in aging and Alzheimer's disease. *Biochim Biophys Acta*. 2011;1813:965-73.

[21] Green KN and LaFerla FM. Linking calcium to Aß and Alzheimer's disease. *Neuron* 2008; 59: 190-4.

[22] Arispe N, Pollard HB, and Rojas E. Alzheimer disease amyloid ß protein forms calcium channels in bilayer membranes: Blockade by tromethamine and aluminum. *Proc Natl Acad Sci USA*, 1993; 90: 567-571.

[23] Jang H *et al.* β-Barrel topology of Alzheimer's β-amyloid ion channels. *J. Mol. Biol.* 2010; 404: 917-34.

[24] Mellon PL *et al.* Immortalization of hypothalamic GnRH neurons by genetically targeted tumorigenesis. *Neuron,* 1990; 5: 1-10.

[25] Kawahara M, Arispe N, Kuroda Y, Rojas E. Alzheimer's disease amyloid β-protein forms Zn^{2+}-sensitive, cation-selective channels across excised membrane patches from hypothalamic neurons. *Biophys J,* 1997; 73: 67-75.

[26] Kawahara M, Arispe N, Kuroda Y, Rojas E. Alzheimer's β-amyloid, human islet amylin and prion protein fragment evoke intracellular free-calcium elevations by a common mechanism in a hypothalamic GnRH neuronal cell-line. *J Biol Chem,* 2000; 275: 14077-14083.

[27] Kawahara M and Kuroda Y Molecular mechanism of neurodegeneration induced by Alzheimer's β-amyloid protein: channel formation and disruption of calcium homeostasis. *Brain Res. Bull.,* 2000; 53: 389-397.

[28] Kawahara M and Kuroda Y. Intracellular calcium changes in neuronal cells induced by Alzheimer's β-amyloid protein are blocked by estradiol and cholesterol. *Cell Mol Neurobio,* 2001; 21: 1-13.

[29] Kato-Negishi M and Kawahara M. Neurosteroids block the increase in intracellular calcium level induced by Alzheimer's β-amyloid protein in long-term cultured rat hippocampal neurons. *Neuropsychiatr Dis Treat,* 2008; 4: 209-218.

[30] Mountz JD et al. Molecular imaging: new applications for biochemistry. J Cell Biochem Suppl. 2002;39:162-71.

[31] Kawahara M *et al.* Membrane incorporation, channel formation, and disruption of calcium homeostasis by Alzheimer's ß-amyloid protein, *Int J Alzheimer Dis,* 2011; 304583.

[32] Prusiner SB. Prions. *Proc. Natl. Acad. Sci. USA.,* 1998; 95: 13363-83.

[33] Kawahara M, Koyama H, Nagata T, Sadakane Y. Zinc, copper, and carnosine attenuate neurotoxicity of prion fragment PrP106-126, *Metallomics* 2011; 3: 726-734.

[34] Kourie JI and Culverson A. Prion peptide fragment PrP[106-126]forms distinct cation channel types. *J. Neurosci. Res.,* 2000 ; 62 : 120-33.

[35] Lashuel HA and Lansbury PT Jr. Are amyloid diseases caused by protein aggregates that mimic bacterial pore-forming toxins? *Q Rev. Biophys.* 2002 ; 39 : 167-201.

[36] Lal R *et al.* Amyloid β ion channel: 3D structure and relevance to amyloid channel paradigm. *Biochim Biophys Acta,* 2007; 1768: 1966-75.

[37] Hipkiss AR. Carnosine and its possible roles in nutrition and health, *Adv Food Nutr Res,* 2009 ; 57 : 87-154.

[38] Corona C *et al*. Effects of dietary supplementation of carnosine on mitochondrial dysfunction, amyloid pathology, and cognitive deficits in 3xTg-AD mice. *PLoS One* 2011; 6(3):e17971.

[39] Koyama H, Konoha K, Sadakane Y, Ohkawara S, Kawahara M. Zinc neurotoxicity and the pathogenesis of vascular-type dementia: Involvement of calcium dyshomeostasis and carnosine, *J. Clin Toxicol.* 2012; S3-002, doi: 10.4172/2161-0495.

[40] Mizuno D and Kawahara M: The molecular mechanism of zinc neurotoxicity and the pathogenesis of vascular type dementia, *Intern J Mol Sci. 2013;* 14: 22067-81.

Chapter 10

Structure and Function of Stefin B Oligomers – Important Role in Amyloidogenesis

Ajda Taler-Verčič, Mira Polajnar and Eva Žerovnik

Additional information is available at the end of the chapter

http://dx.doi.org/10.5772/57570

1. Introduction

Many functional proteins act as oligomers. Oligomerization is a well-controlled and regulated process. Those proteins cannot adopt their physiological functions without oligomerization, while the protein misfolding and subsequent unphysiological oligomerization influence the primary protein functions and produce a "gain in toxic function" as the prefibrillar oligomers are toxic for the cells. Misfolding, oligomerization and aggregation are the reasons for the so called conformational diseases. Several of them are neurodegenerative, but some also affect other vital organs. The accumulation of intracellular protein aggregates (various inclusions) and of extracellular protein deposits cause severe cellular degeneration, such as neurodegeneration of affected neurons. Different proteins form rather similar but not identical fibrillar structures, all showing cross-β-structure, where continuous β-sheets run perpendicular to the long axis of the fibrils.

Neurodegenerative diseases like Alzheimer's disease, Parkinson's disease, Huntington's disease, prion diseases and many others have in common protein aggregation to amyloid fibrils. Type II diabetes is also an amyloid disease although it is not neurodegenerative. Nowadays it is believed that ordered prefibrillar oligomers or protofibrils may be responsible for cell death and that mature fibrils may even be neuroprotective. Most amyloid prone proteins form different oligomers in the lag phase of the amyloid fibril formation. Conformational diseases (Table 1) are difficult to diagnose in the early stage, because they are usually a-symptomatic during their development. Even when they can be diagnosed in the early stage there is an ethical reason for avoiding diagnosis (i.e. to hesitate letting know the patient or the relatives), because there are still no therapies which would slow down or even stop the progression of this type of diseases. Many different proteins are being studied in order to understand the common molecular mechanism of the conformational disease and to develop appropriate treatments. Studies use *in vitro* (different spectroscopic methods), *ex vivo* (different cell cultures) and *in vivo* (mouse models) systems to clarify the changes of the protein conformation and the down-stream effects during the whole process.

Conformational diseases (organ/systemic))	**Protein**
Alzheimer's disease, Down's syndrome (Trisomy 21), Hereditary cerebral angiopathy (Dutch)	Amyloid precursor protein (Aβ 1-42)
Kuru, Gerstmann-Straussler-Scheinker Syndrome (GSS), Creutzfeld-Jacob disease, Scrapie (sheep), Bovine spongiform encephalopathy ("mad cow")	Prion protein
Type II diabetes mellitus (adult onset)	Islet amyloid polypeptide (amylin)
Dialysis-associated amyloidosis	β_2-microglobulin
Senile cardiac amyloidosis	Atrial natriuretic factor
Familial amyloid polyneuropathy	Transthyretin
Reactive amyloidosis familial Mediterranean fever	Serum amyloid A
Familial amyloid polyneuropathy (Finnish)	Gelsolin
Macroglobulinemia	Gamma-1 heavy chain
Primary systemic amyloidosis	Ig-lambda, Ig-kappa
Familial polyneuropathy – Iowa (Irish)	Apolipoprotein A1
Hereditary cerebral myopathy – Iceland	Cystatin C
Nonneuropathic hereditary amyloid with renal disease	Fibrinogen alpha, Lysozyme
Familial British dementia	FBDP
Familial Danish dementia	FDDP
Diffuse lewy body disease, Parkinson's disease	α-synuclein
Fronto-temporal dementia	Tau
Amyotropic lateral sclerosis	Superoxide dismutase-1
Triplet repeat diseases: (Huntinghton's, Spinocerebellar ataxias	Polyglutamine tracts (Huntingtin)
Spinal and bulbar muscural atropy	Androgen receptor
Spinocerebellar ataxias	Ataxins
Spinocerebellar ataxia 17	TATA box-binding protein

Table 1. Diseases of protein misfolding: amyloidoses and non-amyloidoses (reviewed in [1])

Both the amyloid forming proteins involved in certain conformational disease and model proteins are used (Table 2). Stefin B is a model protein for studying amyloid fibril formation.

SH3 domain p 85 phosphatidylinositol 3-kinase	Fibronectin type III phosphoglycenite linase acylphosphatase
HypF N-terminal domain (*E. coli*)	Amphoterin (human)
Apomyoglobin (equine)	Apocytochrome c
Endostatin (human)	Met aminopeptidase
Stefin B (human)	ADA2H
Fibroblast growth factor (*N. viridescens*)	Apolipoprotein CII
VI domain (murine)	B1 domain of IgG binding protein
Curlin CgsA subunit	Monellin
Serpins	

Table 2. Nondisease related amyloid forming proteins/peptides – model proteins (reviewed in [1]) serpins leads to diseases when there are mutations inducing aggregation or change of conformation (e.g. see serpinopathies and the Conformational dementias by *David A. Lomas and Robin W. Carrell* [2].

2. Human stefin B

Stefins are endogenous cysteine protease inhibitors [3], which are ubiquitously expressed in human tissues [4]. They are specific for the papain-family of cysteine proteases and classified as the IH clan in the MEROPS scheme [5]. They are mainly intracellular inhibitors, although have also been found outside the cell in body fluids [6]. They do not have a signal peptide, and they bear a cystatin motif QXVXG, which is the main site involved in binding to target enzymes.

Human stefin B (also termed cystatin B) possesses 98 amino acid residues (Mr= 11 kDa) and no carbohydrate groups or disulphide bridges [7], although it contains one free cysteine [8]. It is an intracellular protein, which tightly and reversibly binds to papain-like cysteine proteases. We will simply call it stefin B from now on.

Stefin B main function is protection against inappropriate proteolysis of lysosomal cysteine proteases [9]. It is an inhibitor of cathepsins B, L, H and S [7, 10, 11]. However, it exerts some additional functions. It was found to interact with five known non-protease proteins (neuro-filament light chain (NFL), brain β-spectrin, RACK-1, human myotubularin related protein 8 (Mtrp) and human T-cell activation protein (Tcrp)) [12]. NFL and β-spectrin are specific to the nervous system. Stefin B (cystatin B) knock-out mice show neurological disorder (loss of the cerebellar granule cells, because of apoptotic bodies, chromatin condensation and some other changes). This suggests that stefin B has an essential anti apoptotic role in the cerebellum [12].

Stefin B is overexpressed in patients with hepatocellular carcinoma and is therefore in combination with some other proteins used as a marker for this disease (it is elevated in the serum already at the early stage of the disease development and therefore easy to detect) [13]. In different types of cancer (human colorectal cancer [14], gastric cancer [15], esophageal carcinoma [16], prostatic adenocarcinoma [17], bladder cancer [18]...) both lower expression level and higher activity due to higher expression level have been detected.

Stefin B is localized both in the cytosol and in the nucleus [19, 20]. It is expressed in neurons and glial cells in the brain, but with slightly different localizations; in neurons it localizes to the nucleus while in astrocytes it is localized in the nucleus and in the cytoplasm [21]. It inhibits cathepsin L in the nucleus, whose substrates are transcription factors. Interaction with nucleosome – histones H2A.Z, H2B and H3, and cathepsin L in the nucleus has also been reported [19]. Therefore, it likely regulates transcription. Furthermore, stefin B was reported to regulate cell cycle progression into the S phase – entry into the S phase is delayed [19]. Increased expression of stefin B in the nucleus of T98G astrocytoma cells delays caspase-3 and caspase-7 activation and this delay is independent of cathepsin inhibition [22].

Stefin B plays an important role in the immune system. It upregulates the release of nitric oxide from interferon-γ-activated macrophages [23]. The protein is involved in innate immune response to bacterial challenge of the leech *Theromyzon tessulatum* [24]. It has an essential role in protection of the central nervous system from apoptosis [25] and from oxidative stress [26]. Increased level of stefin B was found in the senile plaques of Alzheimer's and Parkinson's diseases and in samples of patients suffering from senile dementia [27].

Stefin B functions are summarized in Table 3.

Loss of functions because of alterations in the cystatin B gene (dodecameric repeat expansions in the promoter region or point mutations in the coding gene) is the cause of progressive myoclonus epilepsy of type 1 also known as Unverricht-Lundborg disease (EPM1) [25, 28]. EPM1 mutans are polymeric and aggregate prone *in vivo* [29]. *In vitro* and *ex vivo* G4R, R68X, [30, 31], G50E and Q71P [32, 33] have been studied. Only G4R mutant folds like the wild type protein [31], all others lack tertiary structure and are partially unfolded [31, 33]. G4R and R68X also form amyloid fibrils [31]. Both G50E and Q71P mutants are more susceptible to cleavages by proteases, because of the partially unfolded structure [33], which likely contributes to loss of function in cells. Stefin B deficiency triggers neurodegeneration by impaired redox homeostasis [26]. G4R and R68X do not have any inhibitory activity, while G50E and Q71P are much less active than the wild type protein [32]. All three missense mutants, except for G4R form rather large but diffuse aggregates in cells when over-expressed, while G4R forms small aggregates similar to those of the wild type protein [32].

Cystatin C, a secreted and extracellular protein, is another cysteine protease inhibitor and mutations in its gene are the cause of hereditary cerebral amyloid angiopathy [34]. Human cystatin C is a risk factor for late onset Alzheimer's disease. The protein co-deposits with amyloid β peptide (Aβ peptide) amyloid plaques in patients with Alzheimer's disease [35]. Moreover, cystatin C binds to both the whole amyloid precursor protein (APP) and to Aβ peptide and it inhibits Aβ peptide amyloid fibril formation *in vitro* [36].

Stefin B functions
inhibition of cathepsins B, L, H and S [7, 10, 11];
interaction with five non protease proteins: neurofilament light chain (NFL), brain β-spectrin, RACK-1, human myotubularin related protein 8 (Mtrp) and human T-cell activation protein (Tcrp) (NFL and β-spectrin are specific to the nervous system) – essential anti apoptotic role in the cerebellum [12];
involved in immune-response to bacterial challenge of the leech *Theromyzon tessulatum* [24];
increased level in the senile plaques of Alzheimer's and Parkinson's and of patients suffering from senile dementia [27];
essential role in some of the neurons in the central nervous system, protecting the cells against apoptosis [25];
protects thymocytes against cell death [37];
cell specific expression [21];
inhibits Aβ peptide amyloid fibril formation in oligomer specific manner [38];
positive/negative progression of cancer [13-18];
upregulation of nitric oxide release from interferon-γ-activated macrophages [23];
found mainly in the nucleus of proliferating cells and both in the nucleus and cytoplasm of differentiated cells;
regulates cell cycle [19];
increased expression in the nucleus delays caspase activation [22];
protecting cells from oxidative stress [26].

Table 3. Stefin B has many different functions.

3. Stefin B oligomers *in vitro* and in cells

Stefin B can adopt different oligomeric states *in vitro* and in cells. On the size-exclusion chromatography (SEC) the wild type protein elutes as a set of well-defined oligomers apart from monomers, dimers, tetramers and even higher oligomers (Figure 1) [39]. Y31 isoform is predominantly dimeric [31], while Y31 P79S mutant is tetrameric [40]. All oligomers can be isolated as separate peaks by SEC and stay stable for weeks at pH 7.0 and 4 °C [38].

SEC results have been confirmed up to decamers by electrospray-ionization mass spectrometry (ESI-MS) [41]. In cells stefin B is present both in monomeric and oligomeric forms. Oligomers size ranges between 10 and 250 kDa. The higher oligomeric species are resistant to 1% SDS and 8 M urea and partially resistant to reducing agents (DTT treatment). The low molecular species comprise monomers, dimers, trimers and pentamers. Stefin B polymers *in vivo* seem to grow by monomer addition and not by domain-swapped dimer addition. The protein binds to many different proteins of various sizes [29]. Already the endogenous stefin B forms small punctate aggregates in cells and after overexpression the aggregates amount increases [30].

Oligomers have been observed for many other proteins, especially the amyloidogenic ones. Oligomeric species appear in the lag phase of the reaction and are usually more stable under

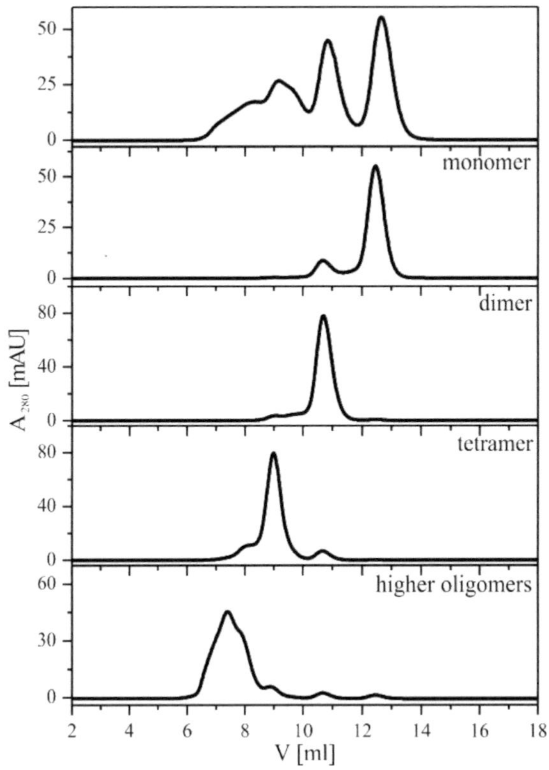

Figure 1. Elution profile of stefin B wild type protein (S3E31 form) at pH 7.0 from SEC. The whole sample (top trace) and separated oligomers as indicated, are shown.

non-amyloidogenic conditions. Size and morphology of those oligomers can be determined by SEC, mass spectrometry, dynamic light scattering and analytical ultracentrifugation. Some examples are given below.

Serpins are serine protease inhibitors also used as a model for studies of amyloid fibril formation. They have key regulatory functions in the inflammatory, complement, coagulation and fibrinolytic cascades [2]. The mechanism of their fibril formation is known in crystallographic details [42, 43]. They form amyloid fibrils through domain-swapping [44]. Serpins preferably oligomerize than interact with other proteins [45]. They form both smaller oligomers and condensed longer polymers [46]. Monomers of some serpins are meta-stable with the rate limiting step representing the transition to the dimer. Once dimers are formed, they can connect to each other to form tetramers or recruit monomers to form trimers and also much longer oligomers [47].

Aβ peptide forms a set of oligomers from monomer to hexadecamer; all even and odd numbered oligomers have been detected by ESI MS [48]. Also for many other amylod forming proteins the whole set of oligomeric species has been observed (insulin [49], β_2-microglobulin [50, 51], SH3 domain [52]...).

http://dx.doi.org/10.5772/57570

4. Inhibitory activity of stefin B oligomers

For a long time it has been thought that monomer is the only active form of stefin B. All other oligomeric forms would serve as a reservoir of monomers or possess additional functions. Now we have shown that all types of oligomers are active against cysteine protease papain (Figure 2). Dimers which are presumably domain-swapped (discussed in the structural characterization part) show the lowest activity in the enzyme to inhibitor 1:2 ratio. They, as well as other oligomers – tetramers and higher, are fully active in the enzyme to inhibitor 1:4 ratio (Fig. 2). SEC-isolated oligomers are all >95% pure and the amount of monomer present in other oligomers samples has been tested to show that the inhibitory effect is not due to monomer contamination. Higher oligomers are a mixture of various oligomers, starting from hexamers. The highest peak represents octamers, which are the prevalent species and therefore we propose that the average size of the higher oligomer is an octamer (this was used for determining the molar ratio). The sample of the higher oligomers contains a small amount of all other oligomers and we suspect that some inhibitory effect could be due to this contamination, but possibly not the whole effect. The results are in agreement with the detection of different oligomers in cells [29] and explain that all detected species could retain inhibitory function.

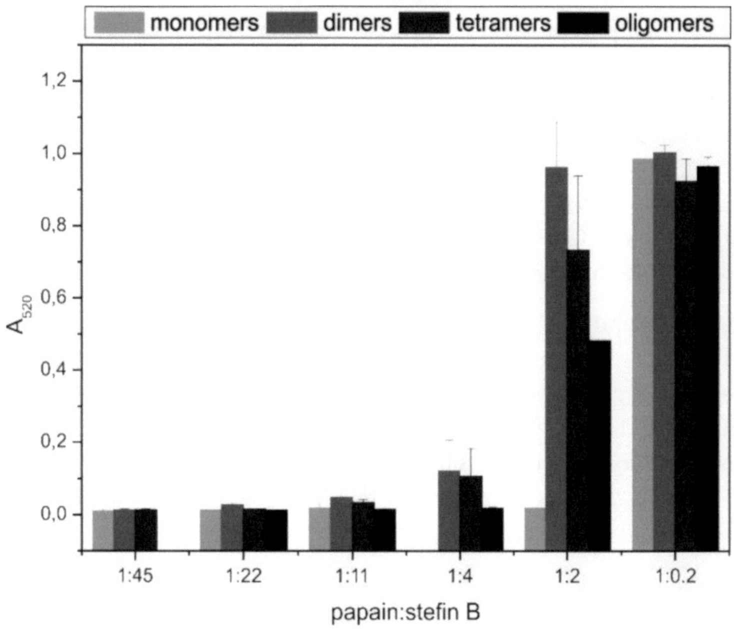

Figure 2. Different stefin B oligomers all exert inhibitory activity against the cysteine protease papain. To evaluate the inhibitory activity of stefin B, BANA test was performed [53]. Stefin B monomers, dimers, tetramers and oligomers were prepared in eight different molar ratios [E]:[I]– 1:45, 1:22, 1:11, 1:4, 1:2, 1:0.2. Higher absorbance means lower inhibitory activity. The average oligomeric state (the highest peak from SEC) in the sample of higher oligomers are the octamers.

5. Structural characterization of different oligomeric states of stefin B

Crystal or solution structure of stefin B monomer and oligomers have been so far characterized from monomer in complex with the target enzyme [54], dimer under amyloid forming conditions [55], tetramer [40] to model of amyloid fibrils [56]. Domain-swapped dimer of stefin A has also been determined by heteronuclear NMR [57].

Stefin B was first crystallized in complex with the cysteine protease papain (Figure 3A) [54]. The structure of the monomeric stefin B consists of a five stranded β-sheet wrapped around a five turn α-helix and with an additional carboxy terminal strand running along the convex side of the β-sheet. This type of the tertiary structure is conserved throughout the cystatin family (Fig. 3A)

High resolution structure of stefin B dimer was determined by heteronuclear NMR [55]. Furthermore, changes within the dimer under amyloid forming conditions were observed and flexible residues were defined. Even under amyloid forming conditions the structure remains largely folded, the main differences are in the flexible loops regions. Prolines on positions 74 and 79 play an important role in the orientations of the loop between strands 4 and 5 (the same loop is involved in "hand-shaking" in the tetramer formation). Four different dimer states have been observed in solution.

When the proline at position 79 was mutated into serine, the tetrameric form of the protein became favourable [40]. This tetramer has been crystallized and its structure determined (Figure 3B) [40]. The tetramer consists of two domain-swapped dimers connected with the so-called "hand-shaking" mechanism to each other. Hand-shaking occurs along with *trans* to *cis* isomerization of the proline at position 74. This proline is conserved throughout the cystatin family. Both domains preserve the fold of the monomer (the same as observed in the complex with papain); each domain is composed of the N-terminal part from one chain (strand 1, the α-helix and strand 2) and C-terminal part from the other chain (strands 3, 4 and 5), between both domains there is a linker region (two peptides, belonging each to one chain).

The structured core of human stefin B amyloid fibrils has been determined (Figure 3C) [56]. Based on the H/D exchange rates the structure could be divided onto protected region (inside the fibril core) and unprotected region (striking out of the fibril). Strands 2, 3, 4 and 5 are protected, while strand 1, α-helix and loops between strands 3 and 4 and 4 and 5 are unprotected. The loop between strands 2 and 3 is the same as the one involved in the domain-swapping mechanism. The fibril core would therefore be made of domain-swapped dimers but with the loop between strands 4 and 5 in position of a tetramer.

6. Initial oligomers on the way to fibril formation

That stefin B forms amyloid fibrils *in vitro* under relatively mild conditions was first shown in 2002 [58] and various solution conditions were probed, fibrils and protofibrils were imaged by AFM and TEM [59]. The fibril formation of stefin B was compared to that of stefin A, a much

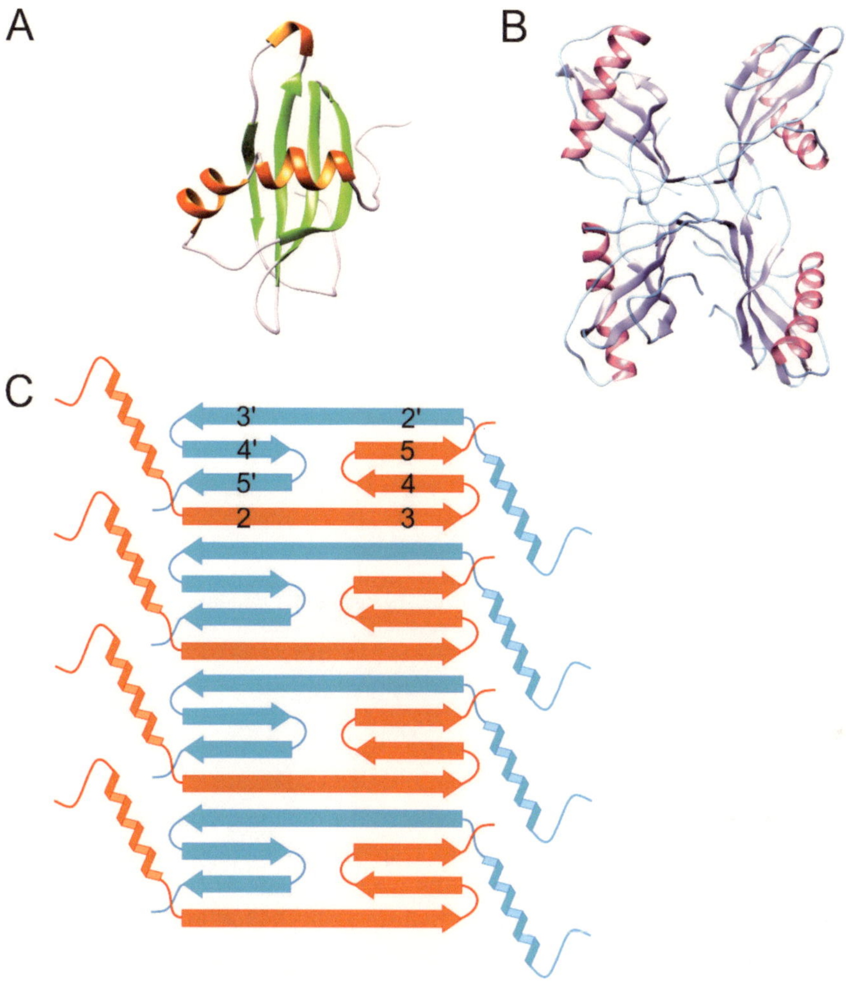

Figure 3. Stefin B structure. (A) stefin B monomer (PDB 1stf) [54], (B) stefin B tetramer (PDB 2oct) [40], (C) secondary structure representation of stefin B fibril [56]

more stable homologue [60], to proline mutants [61], EPM1 mutants [62] and to chimeras between stefins A and B [63].

A model for the mechanism of stefin B amyloid fibril formation has been proposed based on temperature and concentration dependence of the kinetics [64]. Following changes during the lag phase of the reaction by SEC and ESI MS we were able to improve the model of amyloid fibril formation (Figure 4), further explaining the role of various off-pathway oligomers [41]. Dimers seem to be the building block from which fibril formation starts and the fibrils grow [56]. When the process was started from any kind of oligomers, these transformed into dimers [41]. NMR studies have shown that there are actually four types of dimers observed in the lag phase under amyloid forming conditions [55].

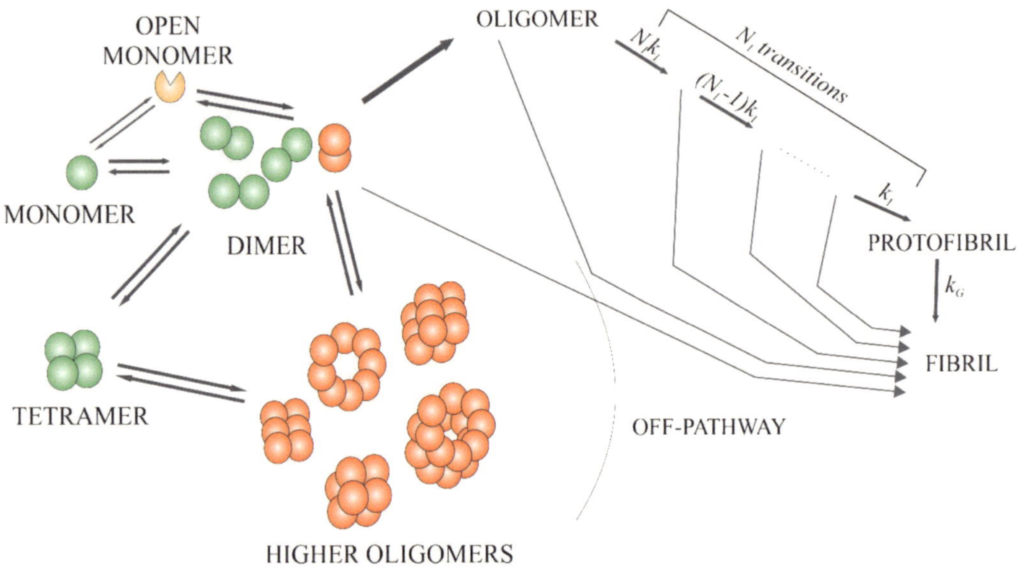

Figure 4. The proposed mechanism for amyloid fibril formation by stefin B [41]. Non-toxic species are coloured green, toxic are coloured red and potentially toxic are coloured orange.

7. Oligomeric state related toxicity

Stefin B higher oligomers are equally toxic to cells as the molten globule state (mutant P74S Y31) [39]. Higher oligomers still have native like CD spectra and increased ANS binding, which would mean that they are properly folded. The toxicity seems to correlate with the ANS binding (ANS binding is indicator of exposed hydrophobic patches in the protein structure) [39]. MTS test has shown that isolated dimers and tetramers are not toxic (Figure 5A), while higher oligomers show some toxicity to neuroblastoma cells (SH-SY5Y). In accordance, low ordered oligomers, monomers, dimers and tetramers do not increase caspase-3-like activity (Figure 5B) [39]. All those results correlate with insertion into lipid membranes, where toxic species insert more effectively than not toxic. Monomers do not get internalized into the cytoplasm when added to the cell medium, while higher oligomers do [39].

Confirming the above results, in another study [65] it was demonstrated that a mutated P74S Y31 form of stefin B, which is molten globule at neutral pH and prefibrillar aggregates at pH 3, which are also in the molten globule conformation, are most toxic.

A growing body of evidence shows that the soluble oligomers formed during amyloid fibril formation exert toxicity and likely cause neurodegeneration [66]. Small soluble oligomers are the cause of synaptic disfunction, whereas large and insoluble aggregates are likely the reservoir of those toxic species [66]. In the case of Aβ peptide a significant correlation has been found between the levels of soluble oligomers and the degree of synaptic alteration, neurodegeneration and cognitive decline, while the correlation between insoluble deposits and those symptoms has not been demonstrated [67]. Non fibrilar dimers or oligomers of α-synuclein could play a major role in Parkinson's disease progression [68]. In accordance with a generic process of amyloid type of protein aggregation, not only oligomers of amyloidogenic proteins related to amyloid diseases exert toxicity, also HypF-N from *E. coli*, SH3 domain of PI3 kinase and some other proteins form oligomers that are toxic to fibroblasts and neurons, whereas amyloid fibrils of the same proteins show only low toxicity [69, 70].

Conformation dependent antibodies distinguish between soluble oligomers [71-73] and amyloid fibrils [74, 75]. Many of those antibodies recognize the generic epitopes regardless of the amino acid sequence and this suggests that different proteins form structurally similar oligomers and fibrils. It was shown that the antibody recognizing Aβ peptide fibrils, recognize also fibrils of transthyretin, islet amyloid polypeptide, β_2-microglobulin and polyglutamine and that the same antibody does not recognize soluble oligomers of those proteins nor the native forms [75]. Antibodies recognizing soluble oligomers also inhibit their toxicity which suggests that soluble oligomers do not share only structural properties but also the mechanism of toxicity [72].

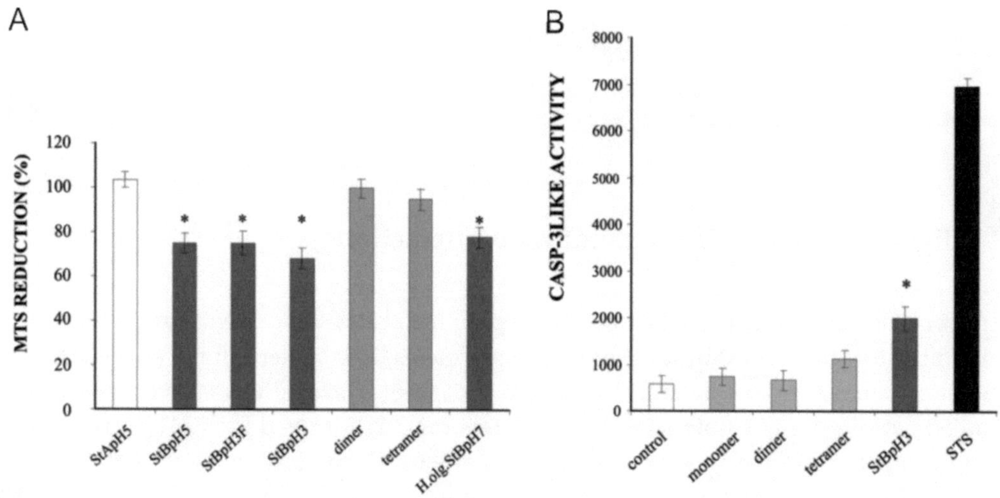

Figure 5. Viability of cells exposed to prefibrillar aggregates and oligomers of stefin B. (A) MTS test-viability of the SHSY5Y cells was measured after exposure for 16 hours to serum free medium, which contained prefibrillar oligomers of stefin B – StBpH3, StBpH5, StBpH3F (filtered), and separated dimers, tetramers and higher-order oligomers of stefin B at pH 7 – StBpH7; each at a final concentration of 44 µM. As a control for the effect of a non-toxic protein, cells were exposed for the corresponding length of time to 44 µM of soluble stefin A prepared at pH 5 (StApH5). All the results are relative to the cells alone, with no added substance or protein in the medium. (B) DEVD-ase activity after exposing SHSY5Y cells to prefibrillar aggregates. 200 µl of concentrated StBpH3 aggregates were added to 800 µL of cell medium to a final concentration of 44 µM (final pH was checked to be neutral). The cells were treated for four hours with 0.5 µM staurosporine (STS). Statistics-as described [39].

Both oligomers and amyloid fibrils are harmful for the cells; oligomers present mostly the toxic effect, while amyloid fibrils present steric obstacles.

• the disruption of the tissue architecture and functions promoted by the invasion of the extracellular space of organ by amyloid [76, 77]
• the destabilization of intracellular and extracellular membranes by oligomers [78, 79]
• the apoptotic cell death and receptor-mediated toxicity triggered by the oligomer interaction with various neuronal receptors [80]
• the oligomer-mediated impairment of the P/Q-type calcium currents [81]
• the impaired maturation of autophagosomes to lysosome mediated by the oligomer accumulation [82]
• the dysfunction of autophagy, a lysosomal pathway for degrading organelles and proteins [83]
• the oxidative damage-induced disruption of the cell viability promoted by the incorporation of redox metals into amyloid fibrils and subsequent generation of reactive oxygen species [84-88]
• the general disorganization of cellular protein homeostasis associated with the exhaustion of the cell defence mechanisms, such as chaperone system [89, 90]
• proteasome inhibition [91]
• the loss of crucial protein functions and/or gain of toxic functions

Table 4. Cellular degeneration after protein oligomerization/deposition (summarized in [92])

8. Oligomeric state related membrane interactions

Amyloid forming proteins interact with biological membranes and even form pores, which are similar to those formed by the pore-forming proteins [1, 93]. Interactions between stefin B and membranes have been extensively studied [65, 94, 95]. Three different proteins have been studied: wt protein, Y31 isoform and G4R mutant (mutant involved in EPM1 disease). Also some comparisons have been made between native and prefibrillar states of the protein (prepared with incubation at pH 3.3 or 4.8) (Figure 6) [94]. Prefibrillar oligomers may be organized in such a way that they are more amphipatic than the native protein and therefore acquire a higher surface-seeking potential. All forms of the protein insert into acidic lipid membranes, cause permeabilization of unilamelar vesicles and destabilize the membranes [94, 95]. Nevertheless there are some significant differences between them concerning pore formation. The mutant G4R does not form pores, but breaks the membrane very fast (in a few minutes), while both the wt protein and Y31 isoform form pores; wt pores are cation selective (and may not be deleterious for the cell) and Y31 pores are anion selective [95]. Wt sample was separated into different oligomers apart from monomers, into dimers, tetramers and higher oligomers. Monomers, dimers and tetramers insert into lipid membranes with approximately the same critical pressure, while the higher oligomers (average size is around octamers) insert more effectively [65].

The exact mechanism of amyloid induced toxicity is yet controversial; nevertheless the pore formation into plasma or mitochondrial membrane is now the leading theory of pathogenesis. Aβ peptide needs anionic phospholipids for binding and insertion into membranes [96-101]. Monomeric form does not interact with membranes. Amylin (islet amyloid polypeptide) readily forms pores into planar lipid bilayer, preferring negatively charged lipids and furthermore the channel activity is inversely proportional to the amount of negative surface charge in the membrane [102]. Native form of prion protein does not form pores, while the mutant prion protein forms irreversible pores in the lipid bilayer. They claim that the pore formation is the main cause of the prion disease [103]. α-synuclein, a natively unfolded protein, also forms pores into lipid membranes but in contrast to others described it is not known how does the oligomeric form increase membrane permeability without forming discrete channels [104]. In the literature we can find contradictory results what effects amyloid proteins exert on membranes. For the same proteins we can find reports that they affect membranes by a non-channel mechanism and other reports that they affect membranes by the formation of ion channels. Two mechanisms of pore formation into membranes by the amyloid forming proteins are summarized in Table 5.

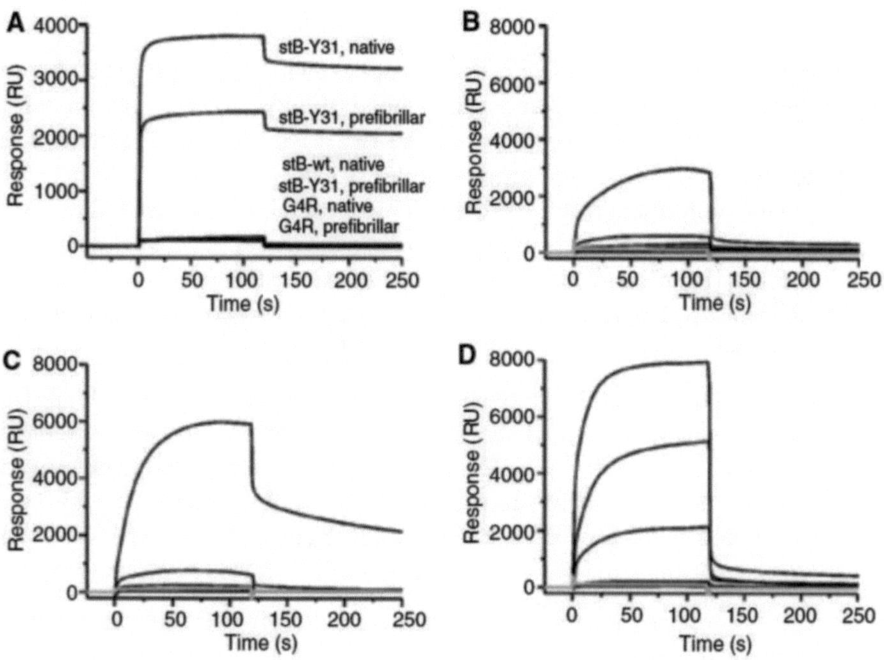

Figure 6. Binding of stefin B variants to liposomes as measured by SPR. (A) Binding to PC LUV. A comparison of binding of 70 μM stefin variants to PC LUV immobilized on the surface of a L1 sensor chip. (B–D) A comparison of the binding of prefibrillar form of stefin variants to PG LUV. The concentration of the protein was 10, 20, 40, 50, 60 and 70 μM (curves from the bottom to the top) in each case. The thick gray line represents binding of native stefin variants at 70μM. (B) StB-wt; (C) StBY31; (D) G4R [95].

Membrane poration followed by nonspecific membrane leakage – increased conductivity of the membranes by a non-channel mechanism	Specific ion transport through ion channel followed by destabilized ionic homeostasis
Aβ40 peptide [105]	Aβ40 peptide [106, 107]
Aβ42 peptide [105]	Aβ42 peptide [108]
α-synuclein [105]	Aβ22-35 peptide [109]
IAPP or amylin [105, 110]	α-synuclein [111]
Polyglutamine [105]	IAPP or amylin [112]
Prion (106-126) H1 [105]	stefin B [95]
SOD1 [113]	

Table 5. The mechanisms underlying globular peptides induced cell dysfunction

9. Copper binding to stefin B and its inhibition of amyloid fibril formation

Divalent metal ions (Cu (II), Zn (II), Fe (II)) are often observed to colocalize with amyloid plaques *in vivo* in much higher concentrations than usually present in the normal environment. This has led to the hypothesis that this metal ions bind to mature fibrils and influence the fibril formation reaction [114].

Indeed it was proven that human stefin B is a copper binding protein [114]. It shows a picomolar affinity to copper at pH 7 and a nanomolar affinity at pH 5. Both the wild type protein and the Y31 isoform bind copper, while the Y31 P79S, which is tetrameric, does not. Monomers and dimers are able to bind copper, while other oligomers are not. Copper binding does not change the conformation of the protein; however it inhibits amyloid fibril formation (Figure 8). Of note it does not prevent aggregation to prefibrillar oligomers, which can be even more toxic to cells. The protein binds two Cu^{2+} ions. Copper binding promotes protein dimerization at neutral pH, while at acidic pH protein undergoes dimerization, already without added copper [114].

Later-on, NMR was used to detect the three possible copper binding sites of stefin B. The first is in the α-helix facing away from the β-sheets together with the loop between strands 4 and 5. The second potential binding site is the C-terminal together with the loop between strands 3 and 4 (this binding site is more likely in the dimeric form of the protein). The third binding site is in the dimer only and is the loop between strands 2 and 3, which is the stretched loop in the domain-swapped dimer together with the few residues at the N-terminal [55].

It was confirmed that copper affects the stefin B fibril formation. In more details, it slows down the elongation phase. It was also shown that the presence of copper destabilizes the protein structure and therefore it was concluded that it preferably binds to the slightly unfolded state of the protein. The final fibril morphology does not differ if copper ions are present or not [55].

Metal ions have different effect on amyloid forming proteins. They can speed up the fibril formation, reduce the lag phase, retard the fibrillization or even stop the process completely [115-117].

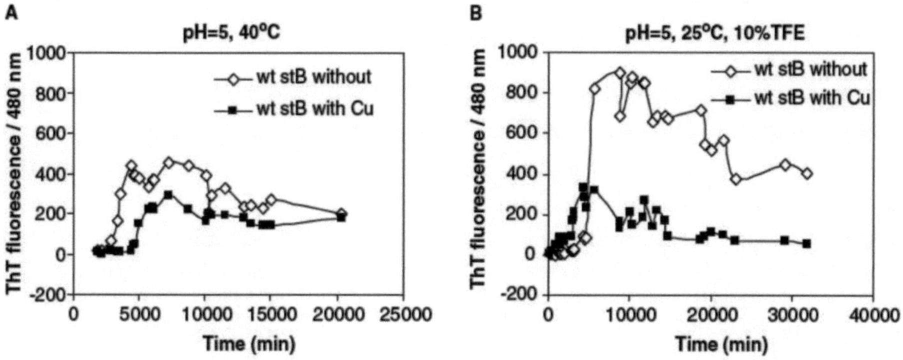

Figure 7. Inhibition of fibrillation of stefin B by Cu^{2+} as probed by ThT fluorescence. Final protein concentration was in all cases 45 µM and final concentration of Cu^{2+} 46 µM, leading to 1 : 1 of protein to Cu^{2+} ratio. (A) Stefin B wildtype (E31 isoform) at pH 5, 40 °C, 0 and 50 µM Cu^{2+} in the buffer. (B) Stefin B wildtype (E31 isoform) at pH 5, 10% TFE, 25 °C, 0 and 50 µM Cu^{2+} in the buffer [114].

Binding of metal ions to prion protein increases proteolysis resistance and structural changes which might play an important role in the conversion process [118].

SOD1 aggregation is enhanced and modulated by Ca^{2+} ions; at physiological pH Ca^{2+}induces conformational changes that increase β-sheet content, they can also divert the aggregation from amyloid fibrils to amorphous aggregates [119]. Cu^{2+} and Zn^{2+} accelerate deposition of Aβ40 peptide and Aβ42 peptide, which results in the amorphous aggregates [120, 121]. Fe^{3+} induces the deposition of fibrillar amyloid plaques of Aβ40 peptide and Aβ42 peptide [120]. Prion protein can directly influence neuronal zinc concentrations [122]. In prion plaques there was found a significant disregulation of copper and manganese, copper was depleted and manganese was enriched [123]. Fe^{3+} and Al^{3+} enhance both formation of mixed oligomers and recruitment of α-synuclein in pre-formed tau oligomers [124]. Copper and selenium inhibit amylin fibril formation, while aluminium and manganese promote it [125]. Copper amylin complex has an anti-aggregating and anti-apoptotic properties, quenching the metal catalysed ROS [126]. Zinc also inhibits amylin fibril formation, furthermore, it favours the formation of amylin hexamers and inhibits the formation of dimers [127].

10. Interaction between oligomers of stefin B and Aβ peptide *in vitro* and in cells-"amateur chaperones"

"Professional chaperones" (heat shock protein family) prevent protein misfolding and subsequent protein aggregation. "Amateur chaperones" bind amyloidogenic proteins and may affect their aggregation process. Both types of chaperones colocalize with pathological lesions of Alzheimer's disease, may be involved in Aβ peptide conformational changes, clearance of Aβ peptide from the brain. Both types of chaperones may be involved in the aggregation, accumulation, persistence and clearance of Aβ peptide and may therefore serve as a potential targets for medical treatment of those patients [128].

Aβ peptides are produced by proteolytic cleavage of the amyloid precursor protein (APP) by α-secretase, β-secretase and γ-secretase. Cathepsin B, which can be inhibited by stefin B, is likely contributor to β-secretase activity [129]. Cysteine protease inhibitors reduce both Aβ peptide level in the brain and β-secretase activity *in vivo* [129]. Stefin B has an effect on production of Aβ peptides and furthermore it also inhibits Aβ peptide fibril formation [38].

Two isoforms of stefin B (Y31 and wt E31) have been studied. Y31 isoform is predominantly a dimer [31] while the wild type protein exist as mixture of monomers, dimer, tetramers and even higher oligomers [39]. ThT fluorescence and transmission electron microscopy (TEM) have shown that Y31 isoform completely inhibits Aβ peptide fibril formation (Figure 9I). The direct interaction between those two proteins has also been shown by SPR measurements, where concentration dependent interaction has been reported and by ESI MS where the complex between dimer stB and monomer Aβ peptide has been detected (Figure 9II) [38]. Furthermore, isolated oligomers of the wild type protein have been studied and it was shown that only the tetramer inhibits Aβ peptide fibril formation and that the higher oligomers show only a weak inhibition. Stefin B also colocalizes with the Aβ peptide aggregates in cells (shown by confocal microscopy) and with the C-terminal fragment of APP (comprising Aβ peptide sequence) (shown by immunoprecipitation).

Molecule/protein	Type of interaction
stefin B	Inhibits Aβ peptide fibril formation, interaction *in vitro* and in cells [38].
laminin	Inhibits Aβ peptide fibril formation, induce depolymerisation of preformed fibrils [130].
polyphenols	Inhibits Aβ peptide fibril formation [131].
heme	Prevents Aβ peptide aggregation [132].
GroEL	Prevents Aβ peptide aggregation [133].
apolipoprotein E	Slows down the oligomerization of Aβ peptide [134].
myelin basic protein	Inhibitor of Aβ peptide fibrilar assembly [135].
cystatin C	Concentration dependent inhibition of Aβ peptide fibril formation [36].
ferulic acid	Inhibits Aβ peptide oligomer formation from monomers and at the same time accelerates fibril formation from already formed oligomers [136].
S14G humanin	Inhibits aggregation into fibrils and disaggregates preformed fibrils, reduces cytotoxicity effect of Aβ peptide [137].
albumin	Increases the lag phase and decreases the total amount of fibrils [138]
crocetin	Inhibits Aβ peptide fibril formation, destabilizes preformed fibrils, stabilize Aβ peptide oligomers and prevents their conversion into fibril [139].

Table 6. Molecules interacting with Aβ peptide

http://dx.doi.org/10.5772/57570

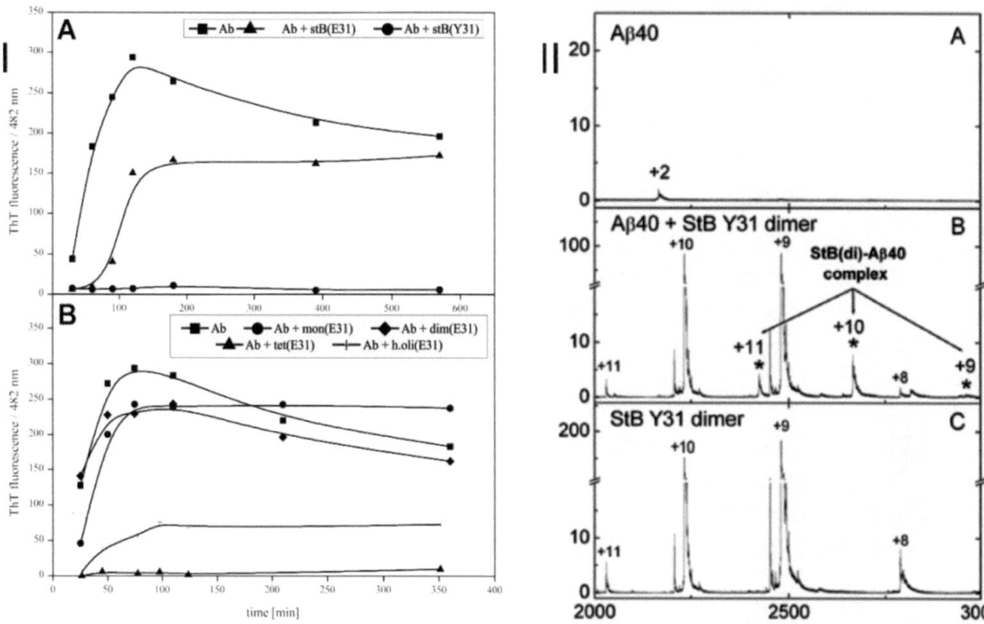

Figure 8. (I) Inhibition of Aβ peptide fibril formation by stefin B measured by ThT fluorescence and (II) the detection of the complex by ESI MS. (I) The Aβ peptide concentration was 17 μM throughout, pH 7.3, 40 °C. (I, A), Aβ peptide alone, 1:1 molar ratio of Aβ to Y31 stefin B (complete inhibition) and 1:1 molar ratio of Aβ to E31 stefin B. (I, B), Aβ peptide alone, and 1:1 molar ratios to E31 stefin B monomers, dimers, tetramers, and higher oligomers. The protein concentrations of stefin B were 17 μM. (II) Complex detected by ESI-MS. (II, A) ESI-MS spectra of Aβ(1–40) peptide, (II, C) stefin B dimer (Y31 variant) and (II, B) their mixtures were recorded: (II, A), 2 μM Aβ(1–40) peptide; (II, B), a mixture of 2 μM Aβ peptide and 2 μM stefin B; and (II, C), 2 μM stefin B. Peaks corresponding to the Aβ peptide-stefin B complex are denoted with an asterisk and numbers above the peaks denote charge state of the ions [38].

In order to show if binding between domain-swapped dimers and Aβ peptide is a generic property, other cystatins have been used. It was shown that only cystatin C dimers inhibit Aβ peptide fibril formation (30% inhibition) while stefin A dimers exhibited no such effect [140].

Several molecules interacting with Aβ peptide have been reported by now (Table 6).

11. Conclusions and perspectives

Stefin B has been so far found as a good model system for studying amyloid fibril formation. It exhibits nearly all features shared with other amyloid forming proteins: it forms mature fibrils under mildly acidic conditions or even at neutral pH at somewhat higher temperature, forms membrane pores and therefore promotes membrane leaking, binds copper ions and its' oligomers are toxic. It is not a model protein only, but also could be termed an "amateur chaperone" affecting Aβ peptide fibril formation both *in vitro* and *in vivo* [38].

We are trying to extend our *in vitro* knowledge to cell cultures to contribute even more to the understanding of conformational disease. It is hoped that new knowledge of protein oligomerization and aggregation on the molecular and cellular levels will contribute to the development of new therapeutic strategies for patients with various conformational diseases, including the neurodegenerative ones.

Acknowledgements

All this work has been done by several contributors over past years. We are thankful to Manca Kenig, Sabina Rabzelj, Slavko Čeru, Katja Škerget, Aida Smajlović and Saša Jenko Kokalj, similarly to some collegues: Vito Turk, Nataša Kopitar Jerala (both JSI, Slovenia), Selma Berbić (University of Tuzla, BiH). Many collaborators from different fields helped us to achieve different perspectives on the problem; Peep Palumaa (TTU, Estonia) – mass spectrometry, Dušan Turk (JSI, Slovenia) – structural biology, Magda Tušek Žnidarič (NIB, Slovenia) – transmission electron microscopy, Andrej Vilfan (JSI, Slovenia) – calculations and mathematical models, Rosemary A. Staniforth (University of Sheffield, UK) – NMR, Miha Škarabot (JSI, Slovenia) – atomic force microscopy and many others.

This work was supported by the program P1-0140 and the projects J7-4050 (led by E. Z.) via the Slovenian Research Agency (ARRS) and ARRS young investigators grants (A. T.-V.).

Author details

Ajda Taler-Verčič[1,2], Mira Polajnar[1,2] and Eva Žerovnik[1,2*]

*Address all correspondence to: eva.zerovnik@ijs.si

1 Department of Biochemistry, Molecular and Structural Biology, Jožef Stefan Institute, Ljubljana, Slovenia

2 Jožef Stefan International Postgraduate School, Ljubljana, Slovenia

References

[1] Kagan B.L.,J. Thundimadathil, Amyloid peptide pores and the beta sheet conformation. Adv Exp Med Biol 2010; 677: p. 150-167.

[2] Lomas D.A.,R.W. Carrell, Serpinopathies and the conformational dementias. Nat Rev Genet 2002; 3(10): p. 759-768.

[3] Turk V.,W. Bode, The cystatins: protein inhibitors of cysteine proteinases. FEBS Lett 1991; 285(2): p. 213-219.

[4] Jerala R., M. Trstenjak, B. Lenarcic, V. Turk, Cloning a synthetic gene for human stefin B and its expression in E. coli. FEBS Lett 1988; 239(1): p. 41-44.

[5] Rawlings N.D., D.P. Tolle, A.J. Barrett, MEROPS: the peptidase database. Nucleic Acids Res 2004; 32(Database issue): p. D160-164.

[6] Abrahamson M., A.J. Barrett, G. Salvesen, A. Grubb, Isolation of six cysteine proteinase inhibitors from human urine. Their physicochemical and enzyme kinetic properties and concentrations in biological fluids. J Biol Chem 1986; 261(24): p. 11282-11289.

[7] Brzin J., M. Kopitar, V. Turk, W. Machleidt, Protein inhibitors of cysteine proteinases. I. Isolation and characterization of stefin, a cytosolic protein inhibitor of cysteine proteinases from human polymorphonuclear granulocytes. Hoppe Seylers Z Physiol Chem 1983; 364(11): p. 1475-1480.

[8] Pol E.,I. Bjork, Role of the single cysteine residue, Cys 3, of human and bovine cystatin B (stefin B) in the inhibition of cysteine proteinases. Protein Sci 2001; 10(9): p. 1729-1738.

[9] Turk B., V. Turk, D. Turk, Structural and functional aspects of papain-like cysteine proteinases and their protein inhibitors. Biol Chem 1997; 378(3-4): p. 141-150.

[10] Bromme D., R. Rinne, H. Kirschke, Tight-binding inhibition of cathepsin S by cystatins. Biomedica biochimica acta 1991; 50(4-6): p. 631-635.

[11] Turk V., J. Brzin, M. Kotnik, B. Lenarcic, T. Popovic, A. Ritonja, M. Trstenjak, L. Begic-Odobasic, W. Machleidt, Human cysteine proteinases and their protein inhibitors stefins, cystatins and kininogens. Biomedica biochimica acta 1986; 45(11-12): p. 1375-1384.

[12] Di Giaimo R., M. Riccio, S. Santi, C. Galeotti, D.C. Ambrosetti, M. Melli, New insights into the molecular basis of progressive myoclonus epilepsy: a multiprotein complex with cystatin B. Human molecular genetics 2002; 11(23): p. 2941-2950.

[13] Lee M.J., G.R. Yu, S.H. Park, B.H. Cho, J.S. Ahn, H.J. Park, E.Y. Song, D.G. Kim, Identification of cystatin B as a potential serum marker in hepatocellular carcinoma. Clin Cancer Res 2008; 14(4): p. 1080-1089.

[14] Sheahan K., S. Shuja, M.J. Murnane, Cysteine protease activities and tumor development in human colorectal carcinoma. Cancer research 1989; 49(14): p. 3809-3814.

[15] Plebani M., L. Herszenyi, R. Cardin, G. Roveroni, P. Carraro, M.D. Paoli, M. Rugge, W.F. Grigioni, D. Nitti, R. Naccarato, et al., Cysteine and serine proteases in gastric cancer. Cancer 1995; 76(3): p. 367-375.

[16] Shiraishi T., M. Mori, S. Tanaka, K. Sugimachi, T. Akiyoshi, Identification of cystatin B in human esophageal carcinoma, using differential displays in which the gene ex-

pression is related to lymph-node metastasis. International journal of cancer. Journal international du cancer 1998; 79(2): p. 175-178.

[17] Mirtti T., K. Alanen, M. Kallajoki, A. Rinne, K.O. Soderstrom, Expression of cystatins, high molecular weight cytokeratin, and proliferation markers in prostatic adenocarcinoma and hyperplasia. The Prostate 2003; 54(4): p. 290-298.

[18] Feldman A.S., J. Banyard, C.L. Wu, W.S. McDougal, B.R. Zetter, Cystatin B as a tissue and urinary biomarker of bladder cancer recurrence and disease progression. Clinical cancer research : an official journal of the American Association for Cancer Research 2009; 15(3): p. 1024-1031.

[19] Ceru S., S. Konjar, K. Maher, U. Repnik, I. Krizaj, M. Bencina, M. Renko, A. Nepveu, E. Zerovnik, B. Turk, N. Kopitar-Jerala, Stefin B interacts with histones and cathepsin L in the nucleus. J Biol Chem 2010; 285(13): p. 10078-10086.

[20] Riccio M., R. Di Giaimo, S. Pianetti, P.P. Palmieri, M. Melli, S. Santi, Nuclear localization of cystatin B, the cathepsin inhibitor implicated in myoclonus epilepsy (EPM1). Experimental cell research 2001; 262(2): p. 84-94.

[21] Brannvall K., H. Hjelm, L. Korhonen, U. Lahtinen, A.E. Lehesjoki, D. Lindholm, Cystatin-B is expressed by neural stem cells and by differentiated neurons and astrocytes. Biochem Biophys Res Commun 2003; 308(2): p. 369-374.

[22] Sun T., V. Turk, B. Turk, N. Kopitar-Jerala, Increased expression of stefin B in the nucleus of T98G astrocytoma cells delays caspase activation. Front Mol Neurosci 2012; 5: p. 93.

[23] Verdot L., G. Lalmanach, V. Vercruysse, S. Hartmann, R. Lucius, J. Hoebeke, F. Gauthier, B. Vray, Cystatins up-regulate nitric oxide release from interferon-gamma-activated mouse peritoneal macrophages. J Biol Chem 1996; 271(45): p. 28077-28081.

[24] Lefebvre C., C. Cocquerelle, F. Vandenbulcke, D. Hot, L. Huot, Y. Lemoine, M. Salzet, Transcriptomic analysis in the leech Theromyzon tessulatum: involvement of cystatin B in innate immunity. Biochem J 2004; 380(Pt 3): p. 617-625.

[25] Pennacchio L.A., D.M. Bouley, K.M. Higgins, M.P. Scott, J.L. Noebels, R.M. Myers, Progressive ataxia, myoclonic epilepsy and cerebellar apoptosis in cystatin B-deficient mice. Nat Genet 1998; 20(3): p. 251-258.

[26] Lehtinen M.K., S. Tegelberg, H. Schipper, H. Su, H. Zukor, O. Manninen, O. Kopra, T. Joensuu, P. Hakala, A. Bonni, A.E. Lehesjoki, Cystatin B deficiency sensitizes neurons to oxidative stress in progressive myoclonus epilepsy, EPM1. J Neurosci 2009; 29(18): p. 5910-5915.

[27] Ii K., H. Ito, E. Kominami, A. Hirano, Abnormal distribution of cathepsin proteinases and endogenous inhibitors (cystatins) in the hippocampus of patients with Alzheimer's disease, parkinsonism-dementia complex on Guam, and senile dementia and in

the aged. Virchows Archiv. A, Pathological anatomy and histopathology 1993; 423(3): p. 185-194.

[28] Pennacchio L.A., A.E. Lehesjoki, N.E. Stone, V.L. Willour, K. Virtaneva, J. Miao, E. D'Amato, L. Ramirez, M. Faham, M. Koskiniemi, J.A. Warrington, R. Norio, A. de la Chapelle, D.R. Cox, R.M. Myers, Mutations in the gene encoding cystatin B in progressive myoclonus epilepsy (EPM1). Science 1996; 271(5256): p. 1731-1734.

[29] Cipollini E., M. Riccio, R. Di Giaimo, F. Dal Piaz, G. Pulice, S. Catania, I. Caldarelli, M. Dembic, S. Santi, M. Melli, Cystatin B and its EPM1 mutants are polymeric and aggregate prone in vivo. Biochim Biophys Acta 2008; 1783(2): p. 312-322.

[30] Ceru S., R. Layfield, T. Zavasnik-Bergant, U. Repnik, N. Kopitar-Jerala, V. Turk, E. Zerovnik, Intracellular aggregation of human stefin B: confocal and electron microscopy study. Biol Cell 2010; 102(6): p. 319-334.

[31] Rabzelj S., V. Turk, E. Zerovnik, In vitro study of stability and amyloid-fibril formation of two mutants of human stefin B (cystatin B) occurring in patients with EPM1. Protein science : a publication of the Protein Society 2005; 14(10): p. 2713-2722.

[32] Polajnar M., S. Ceru, N. Kopitar-Jerala, E. Zerovnik, Human stefin B normal and patho-physiological role: molecular and cellular aspects of amyloid-type aggregation of certain EPM1 mutants. Front Mol Neurosci 2012; 5: p. 88.

[33] Polajnar M., R. Vidmar, M. Vizovisek, M. Fonovic, N. Kopitar-Jerala, E. Zerovnik, Influence of partial unfolding and aggregation of human stefin B (cystatin B) EPM1 mutants G50E and Q71P on selective cleavages by cathepsins B and S. Biol Chem 2013; 394(6): p. 783-790.

[34] Olafsson I., L. Thorsteinsson, O. Jensson, The molecular pathology of hereditary cystatin C amyloid angiopathy causing brain hemorrhage. Brain pathology 1996; 6(2): p. 121-126.

[35] Levy E., M. Sastre, A. Kumar, G. Gallo, P. Piccardo, B. Ghetti, F. Tagliavini, Codeposition of cystatin C with amyloid-beta protein in the brain of Alzheimer disease patients. Journal of neuropathology and experimental neurology 2001; 60(1): p. 94-104.

[36] Sastre M., M. Calero, M. Pawlik, P.M. Mathews, A. Kumar, V. Danilov, S.D. Schmidt, R.A. Nixon, B. Frangione, E. Levy, Binding of cystatin C to Alzheimer's amyloid beta inhibits in vitro amyloid fibril formation. Neurobiol Aging 2004; 25(8): p. 1033-1043.

[37] Kopitar-Jerala N., A. Schweiger, R.M. Myers, V. Turk, B. Turk, Sensitization of stefin B-deficient thymocytes towards staurosporin-induced apoptosis is independent of cysteine cathepsins. FEBS Lett 2005; 579(10): p. 2149-2155.

[38] Skerget K., A. Taler-Vercic, A. Bavdek, V. Hodnik, S. Ceru, M. Tusek-Znidaric, T. Kumm, D. Pitsi, M. Pompe-Novak, P. Palumaa, S. Soriano, N. Kopitar-Jerala, V. Turk, G. Anderluh, E. Zerovnik, Interaction between oligomers of stefin B and amyloid-beta in vitro and in cells. J Biol Chem 2010; 285(5): p. 3201-3210.

[39] Ceru S., S.J. Kokalj, S. Rabzelj, M. Skarabot, I. Gutierrez-Aguirre, N. Kopitar-Jerala, G. Anderluh, D. Turk, V. Turk, E. Zerovnik, Size and morphology of toxic oligomers of amyloidogenic proteins: a case study of human stefin B. Amyloid : the international journal of experimental and clinical investigation : the official journal of the International Society of Amyloidosis 2008; 15(3): p. 147-159.

[40] Jenko Kokalj S., G. Guncar, I. Stern, G. Morgan, S. Rabzelj, M. Kenig, R.A. Staniforth, J.P. Waltho, E. Zerovnik, D. Turk, Essential role of proline isomerization in stefin B tetramer formation. Journal of molecular biology 2007; 366(5): p. 1569-1579.

[41] Taler-Vercic A., T. Kirsipuu, M. Friedemann, A. Noormagi, M. Polajnar, J. Smirnova, M.T. Znidaric, M. Zganec, M. Skarabot, A. Vilfan, R.A. Staniforth, P. Palumaa, E. Zerovnik, The role of initial oligomers in amyloid fibril formation by human stefin B. Int J Mol Sci 2013; 14(9): p. 18362-18384.

[42] Dunstone M.A., W. Dai, J.C. Whisstock, J. Rossjohn, R.N. Pike, S.C. Feil, B.F. Le Bonniec, M.W. Parker, S.P. Bottomley, Cleaved antitrypsin polymers at atomic resolution. Protein Sci 2000; 9(2): p. 417-420.

[43] Huntington J.A., N.S. Pannu, B. Hazes, R.J. Read, D.A. Lomas, R.W. Carrell, A 2.6 A structure of a serpin polymer and implications for conformational disease. Journal of molecular biology 1999; 293(3): p. 449-455.

[44] Guo Z.,D. Eisenberg, Runaway domain swapping in amyloid-like fibrils of T7 endonuclease I. Proc Natl Acad Sci U S A 2006; 103(21): p. 8042-8047.

[45] Carrell R.W., A. Mushunje, A. Zhou, Serpins show structural basis for oligomer toxicity and amyloid ubiquity. FEBS Lett 2008; 582(17): p. 2537-2541.

[46] Mushero N.,A. Gershenson, Determining serpin conformational distributions with single molecule fluorescence. Methods Enzymol 2011; 501: p. 351-377.

[47] Zhou A.,R.W. Carrell, Dimers initiate and propagate serine protease inhibitor polymerisation. Journal of molecular biology 2008; 375(1): p. 36-42.

[48] Kloniecki M., A. Jablonowska, J. Poznanski, J. Langridge, C. Hughes, I. Campuzano, K. Giles, M. Dadlez, Ion mobility separation coupled with MS detects two structural states of Alzheimer's disease Abeta1-40 peptide oligomers. Journal of molecular biology 2011; 407(1): p. 110-124.

[49] Nettleton E.J., P. Tito, M. Sunde, M. Bouchard, C.M. Dobson, C.V. Robinson, Characterization of the oligomeric states of insulin in self-assembly and amyloid fibril formation by mass spectrometry. Biophys J 2000; 79(2): p. 1053-1065.

[50] Smith A.M., T.R. Jahn, A.E. Ashcroft, S.E. Radford, Direct observation of oligomeric species formed in the early stages of amyloid fibril formation using electrospray ionisation mass spectrometry. Journal of molecular biology 2006; 364(1): p. 9-19.

[51] Smith D.P., L.A. Woods, S.E. Radford, A.E. Ashcroft, Structure and dynamics of oligomeric intermediates in beta2-microglobulin self-assembly. Biophys J 2011; 101(5): p. 1238-1247.

[52] Ruzafa D., B. Morel, L. Varela, A.I. Azuaga, F. Conejero-Lara, Characterization of oligomers of heterogeneous size as precursors of amyloid fibril nucleation of an SH3 domain: an experimental kinetics study. PLoS One 2012; 7(11): p. e49690.

[53] Lenarcic B., I. Krizaj, P. Zunec, V. Turk, Differences in specificity for the interactions of stefins A, B and D with cysteine proteinases. FEBS Lett 1996; 395(2-3): p. 113-118.

[54] Stubbs M.T., B. Laber, W. Bode, R. Huber, R. Jerala, B. Lenarcic, V. Turk, The refined 2.4 A X-ray crystal structure of recombinant human stefin B in complex with the cysteine proteinase papain: a novel type of proteinase inhibitor interaction. The EMBO journal 1990; 9(6): p. 1939-1947.

[55] Paramore R., G.J. Morgan, P.J. Davis, C.A. Sharma, A. Hounslow, A. Taler-Vercic, E. Zerovnik, J.P. Waltho, M.J. Cliff, R.A. Staniforth, Mapping local structural perturbations in the native state of stefin B (cystatin B) under amyloid forming conditions. Frontiers in molecular neuroscience 2012; 5: p. 94.

[56] Morgan G.J., S. Giannini, A.M. Hounslow, C.J. Craven, E. Zerovnik, V. Turk, J.P. Waltho, R.A. Staniforth, Exclusion of the native alpha-helix from the amyloid fibrils of a mixed alpha/beta protein. J Mol Biol 2008; 375(2): p. 487-498.

[57] Staniforth R.A., S. Giannini, L.D. Higgins, M.J. Conroy, A.M. Hounslow, R. Jerala, C.J. Craven, J.P. Waltho, Three-dimensional domain swapping in the folded and molten-globule states of cystatins, an amyloid-forming structural superfamily. The EMBO journal 2001; 20(17): p. 4774-4781.

[58] Zerovnik E., M. Pompe-Novak, M. Skarabot, M. Ravnikar, I. Musevic, V. Turk, Human stefin B readily forms amyloid fibrils in vitro. Biochim Biophys Acta 2002; 1594(1): p. 1-5.

[59] Zerovnik E., M. Skarabot, K. Skerget, S. Giannini, V. Stoka, S. Jenko-Kokalj, R.A. Staniforth, Amyloid fibril formation by human stefin B: influence of pH and TFE on fibril growth and morphology. Amyloid : the international journal of experimental and clinical investigation : the official journal of the International Society of Amyloidosis 2007; 14(3): p. 237-247.

[60] Jenko S., M. Skarabot, M. Kenig, G. Guncar, I. Musevic, D. Turk, E. Zerovnik, Different propensity to form amyloid fibrils by two homologous proteins-Human stefins A and B: searching for an explanation. Proteins 2004; 55(2): p. 417-425.

[61] Smajlovic A., S. Berbic, C. Schiene-Fischer, M. Tusek-Znidaric, A. Taler, S. Jenko-Kokalj, D. Turk, E. Zerovnik, Essential role of Pro 74 in stefin B amyloid-fibril formation: dual action of cyclophilin A on the process. FEBS Lett 2009; 583(7): p. 1114-1120.

[62] Kenig M., S. Berbic, A. Krijestorac, L. Kroon-Zitko, M. Tusek, M. Pompe-Novak, E. Zerovnik, Differences in aggregation properties of three site-specific mutants of recombinant human stefin B. Protein Sci 2004; 13(1): p. 63-70.

[63] Kenig M., S. Jenko-Kokalj, M. Tusek-Znidaric, M. Pompe-Novak, G. Guncar, D. Turk, J.P. Waltho, R.A. Staniforth, F. Avbelj, E. Zerovnik, Folding and amyloid-fibril formation for a series of human stefins' chimeras: any correlation? Proteins 2006; 62(4): p. 918-927.

[64] Skerget K., A. Vilfan, M. Pompe-Novak, V. Turk, J.P. Waltho, D. Turk, E. Zerovnik, The mechanism of amyloid-fibril formation by stefin B: temperature and protein concentration dependence of the rates. Proteins 2009; 74(2): p. 425-436.

[65] Ceru S.,E. Zerovnik, Similar toxicity of the oligomeric molten globule state and the prefibrillar oligomers. FEBS Lett 2008; 582(2): p. 203-209.

[66] Haass C.,D.J. Selkoe, Soluble protein oligomers in neurodegeneration: lessons from the Alzheimer's amyloid beta-peptide. Nature reviews. Molecular cell biology 2007; 8(2): p. 101-112.

[67] McLean C.A., R.A. Cherny, F.W. Fraser, S.J. Fuller, M.J. Smith, K. Beyreuther, A.I. Bush, C.L. Masters, Soluble pool of Abeta amyloid as a determinant of severity of neurodegeneration in Alzheimer's disease. Annals of neurology 1999; 46(6): p. 860-866.

[68] El-Agnaf O.M., S.A. Salem, K.E. Paleologou, L.J. Cooper, N.J. Fullwood, M.J. Gibson, M.D. Curran, J.A. Court, D.M. Mann, S. Ikeda, M.R. Cookson, J. Hardy, D. Allsop, Alpha-synuclein implicated in Parkinson's disease is present in extracellular biological fluids, including human plasma. FASEB journal : official publication of the Federation of American Societies for Experimental Biology 2003; 17(13): p. 1945-1947.

[69] Bucciantini M., G. Calloni, F. Chiti, L. Formigli, D. Nosi, C.M. Dobson, M. Stefani, Prefibrillar amyloid protein aggregates share common features of cytotoxicity. The Journal of biological chemistry 2004; 279(30): p. 31374-31382.

[70] Malisauskas M., J. Ostman, A. Darinskas, V. Zamotin, E. Liutkevicius, E. Lundgren, L.A. Morozova-Roche, Does the cytotoxic effect of transient amyloid oligomers from common equine lysozyme in vitro imply innate amyloid toxicity? The Journal of biological chemistry 2005; 280(8): p. 6269-6275.

[71] Hrncic R., J. Wall, D.A. Wolfenbarger, C.L. Murphy, M. Schell, D.T. Weiss, A. Solomon, Antibody-mediated resolution of light chain-associated amyloid deposits. Am J Pathol 2000; 157(4): p. 1239-1246.

[72] Kayed R., E. Head, J.L. Thompson, T.M. McIntire, S.C. Milton, C.W. Cotman, C.G. Glabe, Common structure of soluble amyloid oligomers implies common mechanism of pathogenesis. Science 2003; 300(5618): p. 486-489.

[73] Moretto N., A. Bolchi, C. Rivetti, B.P. Imbimbo, G. Villetti, V. Pietrini, L. Polonelli, S. Del Signore, K.M. Smith, R.J. Ferrante, S. Ottonello, Conformation-sensitive antibod-

ies against alzheimer amyloid-beta by immunization with a thioredoxin-constrained B-cell epitope peptide. J Biol Chem 2007; 282(15): p. 11436-11445.

[74] Kayed R., E. Head, F. Sarsoza, T. Saing, C.W. Cotman, M. Necula, L. Margol, J. Wu, L. Breydo, J.L. Thompson, S. Rasool, T. Gurlo, P. Butler, C.G. Glabe, Fibril specific, conformation dependent antibodies recognize a generic epitope common to amyloid fibrils and fibrillar oligomers that is absent in prefibrillar oligomers. Mol Neurodegener 2007; 2: p. 18.

[75] O'Nuallain B.,R. Wetzel, Conformational Abs recognizing a generic amyloid fibril epitope. Proc Natl Acad Sci U S A 2002; 99(3): p. 1485-1490.

[76] Merlini G.,V. Bellotti, Molecular mechanisms of amyloidosis. N Engl J Med 2003; 349(6): p. 583-596.

[77] Tan S.Y.,M.B. Pepys, Amyloidosis. Histopathology 1994; 25(5): p. 403-414.

[78] Caughey B.,P.T. Lansbury, Protofibrils, pores, fibrils, and neurodegeneration: separating the responsible protein aggregates from the innocent bystanders. Annu Rev Neurosci 2003; 26: p. 267-298.

[79] Lashuel H.A., D. Hartley, B.M. Petre, T. Walz, P.T. Lansbury, Jr., Neurodegenerative disease: amyloid pores from pathogenic mutations. Nature 2002; 418(6895): p. 291.

[80] Ferreira S.T., M.N. Vieira, F.G. De Felice, Soluble protein oligomers as emerging toxins in Alzheimer's and other amyloid diseases. IUBMB Life 2007; 59(4-5): p. 332-345.

[81] Nimmrich V., C. Grimm, A. Draguhn, S. Barghorn, A. Lehmann, H. Schoemaker, H. Hillen, G. Gross, U. Ebert, C. Bruehl, Amyloid beta oligomers (A beta(1-42) globulomer) suppress spontaneous synaptic activity by inhibition of P/Q-type calcium currents. J Neurosci 2008; 28(4): p. 788-797.

[82] Nixon R.A., Autophagy in neurodegenerative disease: friend, foe or turncoat? Trends Neurosci 2006; 29(9): p. 528-535.

[83] Powers E.T., R.I. Morimoto, A. Dillin, J.W. Kelly, W.E. Balch, Biological and chemical approaches to diseases of proteostasis deficiency. Annu Rev Biochem 2009; 78: p. 959-991.

[84] Barnham K.J., G.D. Ciccotosto, A.K. Tickler, F.E. Ali, D.G. Smith, N.A. Williamson, Y.H. Lam, D. Carrington, D. Tew, G. Kocak, I. Volitakis, F. Separovic, C.J. Barrow, J.D. Wade, C.L. Masters, R.A. Cherny, C.C. Curtain, A.I. Bush, R. Cappai, Neurotoxic, redox-competent Alzheimer's beta-amyloid is released from lipid membrane by methionine oxidation. J Biol Chem 2003; 278(44): p. 42959-42965.

[85] Cuajungco M.P., L.E. Goldstein, A. Nunomura, M.A. Smith, J.T. Lim, C.S. Atwood, X. Huang, Y.W. Farrag, G. Perry, A.I. Bush, Evidence that the beta-amyloid plaques of Alzheimer's disease represent the redox-silencing and entombment of abeta by zinc. J Biol Chem 2000; 275(26): p. 19439-19442.

[86] Huang X., C.S. Atwood, M.A. Hartshorn, G. Multhaup, L.E. Goldstein, R.C. Scarpa, M.P. Cuajungco, D.N. Gray, J. Lim, R.D. Moir, R.E. Tanzi, A.I. Bush, The A beta peptide of Alzheimer's disease directly produces hydrogen peroxide through metal ion reduction. Biochemistry 1999; 38(24): p. 7609-7616.

[87] Huang X., M.P. Cuajungco, C.S. Atwood, M.A. Hartshorn, J.D. Tyndall, G.R. Hanson, K.C. Stokes, M. Leopold, G. Multhaup, L.E. Goldstein, R.C. Scarpa, A.J. Saunders, J. Lim, R.D. Moir, C. Glabe, E.F. Bowden, C.L. Masters, D.P. Fairlie, R.E. Tanzi, A.I. Bush, Cu(II) potentiation of alzheimer abeta neurotoxicity. Correlation with cell-free hydrogen peroxide production and metal reduction. J Biol Chem 1999; 274(52): p. 37111-37116.

[88] Opazo C., X. Huang, R.A. Cherny, R.D. Moir, A.E. Roher, A.R. White, R. Cappai, C.L. Masters, R.E. Tanzi, N.C. Inestrosa, A.I. Bush, Metalloenzyme-like activity of Alzheimer's disease beta-amyloid. Cu-dependent catalytic conversion of dopamine, cholesterol, and biological reducing agents to neurotoxic H(2)O(2). J Biol Chem 2002; 277(43): p. 40302-40308.

[89] Macario A.J.,E. Conway de Macario, Sick chaperones, cellular stress, and disease. N Engl J Med 2005; 353(14): p. 1489-1501.

[90] Muchowski P.J.,J.L. Wacker, Modulation of neurodegeneration by molecular chaperones. Nat Rev Neurosci 2005; 6(1): p. 11-22.

[91] Almeida C.G., R.H. Takahashi, G.K. Gouras, Beta-amyloid accumulation impairs multivesicular body sorting by inhibiting the ubiquitin-proteasome system. J Neurosci 2006; 26(16): p. 4277-4288.

[92] Uversky V.N., Mysterious oligomerization of the amyloidogenic proteins. Febs J 2010; 277(14): p. 2940-2953.

[93] Anderluh G.,E. Zerovnik, Pore formation by human stefin B in its native and oligomeric states and the consequent amyloid induced toxicity. Front Mol Neurosci 2012; 5: p. 85.

[94] Anderluh G., I. Gutierrez-Aguirre, S. Rabzelj, S. Ceru, N. Kopitar-Jerala, P. Macek, V. Turk, E. Zerovnik, Interaction of human stefin B in the prefibrillar oligomeric form with membranes. Correlation with cellular toxicity. Febs J 2005; 272(12): p. 3042-3051.

[95] Rabzelj S., G. Viero, I. Gutierrez-Aguirre, V. Turk, M. Dalla Serra, G. Anderluh, E. Zerovnik, Interaction with model membranes and pore formation by human stefin B: studying the native and prefibrillar states. Febs J 2008; 275(10): p. 2455-2466.

[96] Alarcon J.M., J.A. Brito, T. Hermosilla, I. Atwater, D. Mears, E. Rojas, Ion channel formation by Alzheimer's disease amyloid beta-peptide (Abeta40) in unilamellar liposomes is determined by anionic phospholipids. Peptides 2006; 27(1): p. 95-104.

[97] Bokvist M., F. Lindstrom, A. Watts, G. Grobner, Two types of Alzheimer's beta-amyloid (1-40) peptide membrane interactions: aggregation preventing transmembrane

anchoring versus accelerated surface fibril formation. Journal of molecular biology 2004; 335(4): p. 1039-1049.

[98] Hertel C., E. Terzi, N. Hauser, R. Jakob-Rotne, J. Seelig, J.A. Kemp, Inhibition of the electrostatic interaction between beta-amyloid peptide and membranes prevents beta-amyloid-induced toxicity. Proc Natl Acad Sci U S A 1997; 94(17): p. 9412-9416.

[99] Hirakura Y., M.C. Lin, B.L. Kagan, Alzheimer amyloid abeta1-42 channels: effects of solvent, pH, and Congo Red. J Neurosci Res 1999; 57(4): p. 458-466.

[100] McLaurin J.,A. Chakrabartty, Characterization of the interactions of Alzheimer beta-amyloid peptides with phospholipid membranes. Eur J Biochem 1997; 245(2): p. 355-363.

[101] Terzi E., G. Holzemann, J. Seelig, Interaction of Alzheimer beta-amyloid peptide(1-40) with lipid membranes. Biochemistry 1997; 36(48): p. 14845-14852.

[102] Janson J., R.H. Ashley, D. Harrison, S. McIntyre, P.C. Butler, The mechanism of islet amyloid polypeptide toxicity is membrane disruption by intermediate-sized toxic amyloid particles. Diabetes 1999; 48(3): p. 491-498.

[103] Lin M.C., T. Mirzabekov, B.L. Kagan, Channel formation by a neurotoxic prion protein fragment. J Biol Chem 1997; 272(1): p. 44-47.

[104] Zakharov S.D., J.D. Hulleman, E.A. Dutseva, Y.N. Antonenko, J.C. Rochet, W.A. Cramer, Helical alpha-synuclein forms highly conductive ion channels. Biochemistry 2007; 46(50): p. 14369-14379.

[105] Kayed R., Y. Sokolov, B. Edmonds, T.M. McIntire, S.C. Milton, J.E. Hall, C.G. Glabe, Permeabilization of lipid bilayers is a common conformation-dependent activity of soluble amyloid oligomers in protein misfolding diseases. J Biol Chem 2004; 279(45): p. 46363-46366.

[106] Kawahara M., N. Arispe, Y. Kuroda, E. Rojas, Alzheimer's disease amyloid beta-protein forms Zn(2+)-sensitive, cation-selective channels across excised membrane patches from hypothalamic neurons. Biophys J 1997; 73(1): p. 67-75.

[107] Lin H., Y.J. Zhu, R. Lal, Amyloid beta protein (1-40) forms calcium-permeable, Zn2+-sensitive channel in reconstituted lipid vesicles. Biochemistry 1999; 38(34): p. 11189-11196.

[108] Rhee S.K., A.P. Quist, R. Lal, Amyloid beta protein-(1-42) forms calcium-permeable, Zn2+-sensitive channel. J Biol Chem 1998; 273(22): p. 13379-13382.

[109] Di Scala C., J.D. Troadec, C. Lelievre, N. Garmy, J. Fantini, H. Chahinian, Mechanism of cholesterol-assisted oligomeric channel formation by a short Alzheimer beta-amyloid peptide. J Neurochem 2013.

[110] Green J.D., L. Kreplak, C. Goldsbury, X. Li Blatter, M. Stolz, G.S. Cooper, A. Seelig, J. Kistler, U. Aebi, Atomic force microscopy reveals defects within mica supported lip-

id bilayers induced by the amyloidogenic human amylin peptide. Journal of molecular biology 2004; 342(3): p. 877-887.

[111] Fantini J.,N. Yahi, The driving force of alpha-synuclein insertion and amyloid channel formation in the plasma membrane of neural cells: key role of ganglioside-and cholesterol-binding domains. Adv Exp Med Biol 2013; 991: p. 15-26.

[112] Zhao J., Y. Luo, H. Jang, X. Yu, G. Wei, R. Nussinov, J. Zheng, Probing ion channel activity of human islet amyloid polypeptide (amylin). Biochim Biophys Acta 2012; 1818(12): p. 3121-3130.

[113] Oladzad Abbasabadi A., A. Javanian, M. Nikkhah, A.A. Meratan, P. Ghiasi, M. Nemat-Gorgani, Disruption of mitochondrial membrane integrity induced by amyloid aggregates arising from variants of SOD1. International journal of biological macromolecules 2013; 61C: p. 212-217.

[114] Zerovnik E., K. Skerget, M. Tusek-Znidaric, C. Loeschner, M.W. Brazier, D.R. Brown, High affinity copper binding by stefin B (cystatin B) and its role in the inhibition of amyloid fibrillation. Febs J 2006; 273(18): p. 4250-4263.

[115] Atwood C.S., R.D. Moir, X. Huang, R.C. Scarpa, N.M. Bacarra, D.M. Romano, M.A. Hartshorn, R.E. Tanzi, A.I. Bush, Dramatic aggregation of Alzheimer abeta by Cu(II) is induced by conditions representing physiological acidosis. The Journal of biological chemistry 1998; 273(21): p. 12817-12826.

[116] Raman B., T. Ban, K. Yamaguchi, M. Sakai, T. Kawai, H. Naiki, Y. Goto, Metal ion-dependent effects of clioquinol on the fibril growth of an amyloid {beta} peptide. The Journal of biological chemistry 2005; 280(16): p. 16157-16162.

[117] Zou J., K. Kajita, N. Sugimoto, Cu(2+) Inhibits the Aggregation of Amyloid beta-Peptide(1-42) in vitro We thank JEOL for the AFM measurement. This work was supported in part by Grants-in-Aid from the Japanese Ministry of Education, Science, Sports, and Culture, and a Grant from "Research for the Future" Program of the Japan Society for the Promotion of Science to N.S. Angewandte Chemie 2001; 40(12): p. 2274-2277.

[118] Kuczius T.,R. Kelsch, The effect of copper and zinc binding on the solubility and resistance to proteolysis of physiological prion protein PrP depends on the tissue source and the PrP glycotypes. Journal of cellular biochemistry 2013.

[119] Leal S.S., I. Cardoso, J.S. Valentine, C.M. Gomes, Calcium ions promote superoxide dismutase 1 (SOD1) aggregation into non fibrillar amyloid: a link to toxic effects of calcium overload in amyotrophic lateral sclerosis (ALS)? The Journal of biological chemistry 2013.

[120] Ha C., J. Ryu, C.B. Park, Metal ions differentially influence the aggregation and deposition of Alzheimer's beta-amyloid on a solid template. Biochemistry 2007; 46(20): p. 6118-6125.

[121] Tougu V., A. Karafin, K. Zovo, R.S. Chung, C. Howells, A.K. West, P. Palumaa, Zn(II)-and Cu(II)-induced non-fibrillar aggregates of amyloid-beta (1-42) peptide are transformed to amyloid fibrils, both spontaneously and under the influence of metal chelators. J Neurochem 2009; 110(6): p. 1784-1795.

[122] Watt N.T., H.H. Griffiths, N.M. Hooper, Neuronal zinc regulation and the prion protein. Prion 2013; 7(3): p. 203-208.

[123] Johnson C.J., P.U. Gilbert, M. Abrecht, K.L. Baldwin, R.E. Russell, J.A. Pedersen, J.M. Aiken, D. McKenzie, Low copper and high manganese levels in prion protein plaques. Viruses 2013; 5(2): p. 654-662.

[124] Nubling G., B. Bader, J. Levin, J. Hildebrandt, H. Kretzschmar, A. Giese, Synergistic influence of phosphorylation and metal ions on tau oligomer formation and coaggregation with alpha-synuclein at the single molecule level. Mol Neurodegener 2012; 7: p. 35.

[125] Mirhashemi S.M.,M.E. Shahabaddin, Evaluation of aluminium, manganese, copper and selenium effects on human islets amyloid polypeptide hormone aggregation. Pakistan journal of biological sciences: PJBS 2011; 14(4): p. 288-292.

[126] Lee E.C., E. Ha, S. Singh, L. Legesse, S. Ahmad, E. Karnaukhova, R.P. Donaldson, A.M. Jeremic, Copper(ii)-human amylin complex protects pancreatic cells from amylin toxicity. Physical chemistry chemical physics : PCCP 2013; 15(30): p. 12558-12571.

[127] Salamekh S., J.R. Brender, S.J. Hyung, R.P. Nanga, S. Vivekanandan, B.T. Ruotolo, A. Ramamoorthy, A two-site mechanism for the inhibition of IAPP amyloidogenesis by zinc. Journal of molecular biology 2011; 410(2): p. 294-306.

[128] Wilhelmus M.M., R.M. de Waal, M.M. Verbeek, Heat shock proteins and amateur chaperones in amyloid-Beta accumulation and clearance in Alzheimer's disease. Mol Neurobiol 2007; 35(3): p. 203-216.

[129] Hook G., V.Y. Hook, M. Kindy, Cysteine protease inhibitors reduce brain beta-amyloid and beta-secretase activity in vivo and are potential Alzheimer's disease therapeutics. Biol Chem 2007; 388(9): p. 979-983.

[130] Morgan C.,N.C. Inestrosa, Interactions of laminin with the amyloid beta peptide. Implications for Alzheimer's disease. Braz J Med Biol Res 2001; 34(5): p. 597-601.

[131] Porat Y., A. Abramowitz, E. Gazit, Inhibition of amyloid fibril formation by polyphenols: structural similarity and aromatic interactions as a common inhibition mechanism. Chem Biol Drug Des 2006; 67(1): p. 27-37.

[132] Zhao L.N., Y. Mu, L.Y. Chew, Heme prevents amyloid beta peptide aggregation through hydrophobic interaction based on molecular dynamics simulation. Phys Chem Chem Phys 2013; 15(33): p. 14098-14106.

[133] Yagi-Utsumi M., T. Kunihara, T. Nakamura, Y. Uekusa, K. Makabe, K. Kuwajima, K. Kato, NMR characterization of the interaction of GroEL with amyloid beta as a model ligand. FEBS Lett 2013; 587(11): p. 1605-1609.

[134] Ly S., R. Altman, J. Petrlova, Y. Lin, S. Hilt, T. Huser, T.A. Laurence, J.C. Voss, Binding of apolipoprotein E inhibits the oligomer growth of amyloid-beta peptide in solution as determined by fluorescence cross-correlation spectroscopy. J Biol Chem 2013; 288(17): p. 11628-11635.

[135] Kotarba A.E., D. Aucoin, M.D. Hoos, S.O. Smith, W.E. Van Nostrand, Fine mapping of the amyloid beta-protein binding site on myelin basic protein. Biochemistry 2013; 52(15): p. 2565-2573.

[136] Cui L., Y. Zhang, H. Cao, Y. Wang, T. Teng, G. Ma, Y. Li, K. Li, Ferulic Acid Inhibits the Transition of Amyloid-beta42 Monomers to Oligomers but Accelerates the Transition from Oligomers to Fibrils. Journal of Alzheimer's disease : JAD 2013.

[137] Zhang W., Y. Du, M. Bai, Y. Xi, Z. Li, J. Miao, S14G-humanin inhibits Abeta1-42 fibril formation, disaggregates preformed fibrils, and protects against Abeta-induced cytotoxicity in vitro. Journal of peptide science : an official publication of the European Peptide Society 2013; 19(3): p. 159-165.

[138] Stanyon H.F.,J.H. Viles, Human serum albumin can regulate amyloid-beta peptide fiber growth in the brain interstitium: implications for Alzheimer disease. The Journal of biological chemistry 2012; 287(33): p. 28163-28168.

[139] Ahn J.H., Y. Hu, M. Hernandez, J.R. Kim, Crocetin inhibits beta-amyloid fibrillization and stabilizes beta-amyloid oligomers. Biochem Biophys Res Commun 2011; 414(1): p. 79-83.

[140] Taler-Vercic A.,E. Zerovnik, Binding of amyloid peptides to domain-swapped dimers of other amyloid-forming proteins may prevent their neurotoxicity. BioEssays : news and reviews in molecular, cellular and developmental biology 2010; 32(12): p. 1020-1024.

Section 3

Computational Approaches

Chapter 11

The Assembly of Protein Oligomers — Old Stories and New Perspectives with Graph Theory

Claire Lesieur

Additional information is available at the end of the chapter

http://dx.doi.org/10.5772/58576

1. Introduction

Proteins are biological entities made of a chain of amino acids bound to one another in a specific order, called the primary structure or the amino acid sequence of the protein. Based on the sequence and the environment, the protein acquires a tridimensional shape called tertiary structure (3D-structure), conformation or fold, suitable for its biological function. The functional shape is the native structure of the protein. The set of reactions leading to the native structure is the folding of the protein. The vast majority of proteins are oligomers which function only after the association of several copies of their chains. Homo-oligomers have chains with identical sequences and hetero-oligomers have chains with different sequences. The number of associated chains defines the quaternary structure of the oligomer, or its stoichiometry [1]. According to the Protein Database (PDB) where all known 3D structures of proteins are stored, the most observed quaternary structure in all taxa is the dimer (Fig. 1A). A taxon is a set of living organisms grouped because of some shared characteristics. Besides dimers, there exists a large variety of assemblies in terms of quaternary structures and point group symmetries (Fig. 1). In forming a protein oligomer, subunit association has to be considered in addition to folding.

Folding involves the formation of interactions/bonds between atoms of the amino acids of a single chain. These are intramolecular (within a single molecule) amino acid interactions (Fig. 2A). Chain association involves the formation of interactions/bonds between atoms of the amino acids provided by at least two individual chains. These are intermolecular (between two molecules) amino acid interactions (Fig. 2B). Here the protein chain is considered as the molecule.

The twenty natural amino acids share four atoms called the backbone atoms and are distinguished by a set of atoms called the side chain atoms. These atoms can make different types

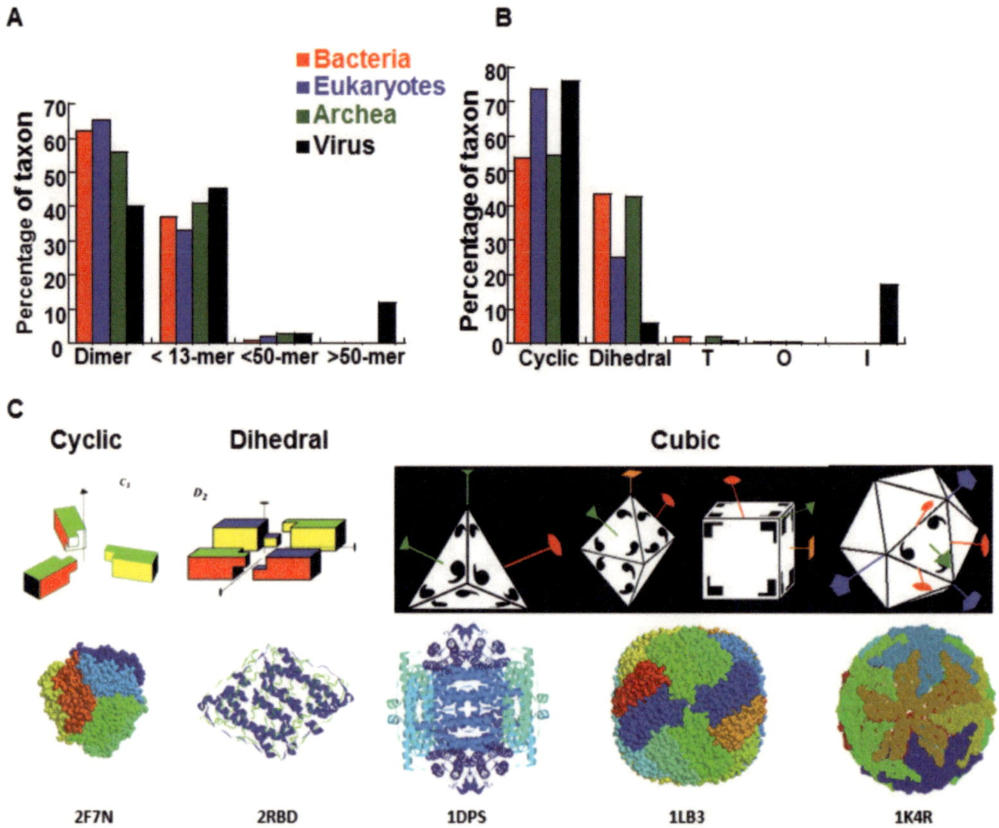

Figure 1. Protein oligomer features. The data in A and B are obtained by screening the PDB. **A.** Distribution of taxa according to quaternary structures. **B.** Distribution of taxa according to point group symmetries. **C.** Protein oligomers belong to three different point group symmetries. Cn, cyclic with n rotational axes. One ferritin is given as an example for a C3 symmetry (PDB 2F7N). Dn, dihedral with n rotational axes plus 2-fold perpendicular axes. One ferritin is given as an example for a D2 symmetry (PDB 2RBD). Cubic. T, tetrahedral with four 3-fold axes and six 2-fold axes. Ferritin is given as an example (PDB 1DPS) [2]. O, octoahedral, octahedron or hexahedron with six 4-fold axes, eight 3-fold axes and twelve 2-fold axes. One ferritin is given as an example (PDB 1LB3). I, icosahedral with twelve 5-fold axes, twenty 3-fold axes and thirty 2-fold axes. Ferritin is given as an example (PDB 1K4R). Icosahedral is not a cubic point group symmetry but has been conflated to the cubic point group symmetry in chemistry [3].

of chemical bonds. First, the amino acids are linked to one another by a covalent bond involving two backbone atoms and called the peptide bond. The covalent bonds are thus used to make a chain of amino acids arranged in a specific order, the primary sequence of the protein. Cysteine and methionine amino acids are the only amino acids that can make a supplementary covalent bond, called a disulfide bond, using the sulfur atom of their side chains. There can be intramolecular disulfide bonds (between two cysteines of one chain) or intermolecular disulfide bonds (between two cysteines, each one produced by a distinct chain), the latter making a covalent oligomer. Some collagen trimers are stabilized by inter-chain disulfide bonds [4]. The collagen oligomers have been reviewed recently [5, 6]. Some other examples of covalent oligomers can be found in the chapter on protein oligomerization by Giovanni Gotte

http://dx.doi.org/10.5772/58576

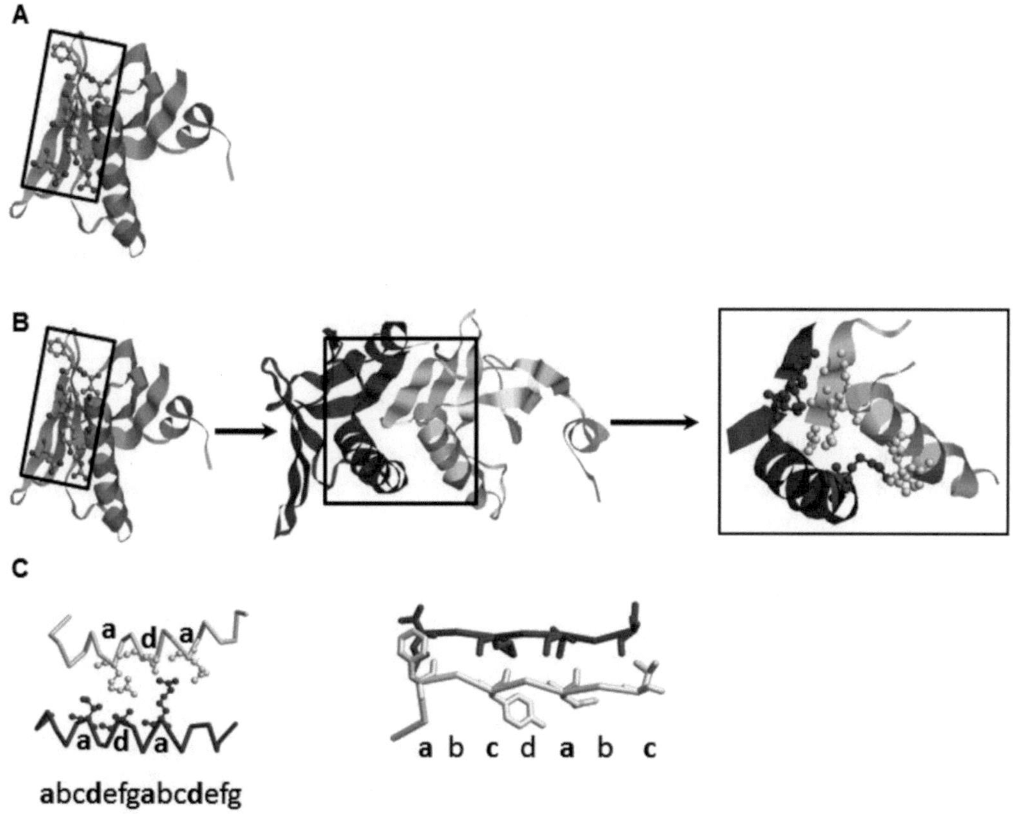

Figure 2. Interactions in proteins.A.Protein monomer. Monomeric proteins perform their biological function with a single chain. The formation and the stability of their native 3D structure involve only intramolecular atomic interactions (interaction within a chain). **B. Protein oligomer.** Oligomeric proteins need to assemble several copies of their chain to perform their biological function. The formation and the stability of their native structure involve intramolecular atomic interactions (interaction within one chain) to acquire their fold and intermolecular atomic interactions (interaction between chains) to acquire their quaternary structure. The pictures are generated with Rasmol. The protein chains are shown in ribbons of different colors. For a few amino acids, all atoms are indicated in balls and sticks to highlight intra and inter atomic interactions. **C. Recognition modes.** A protein interface is made of two set of atoms, one per chain, spatially organized to yield a chemical and geometrical complementarity. Two simple cases are presented. On the left are two interacting α-helices and on the right are two interacting β-strands. Because of the geometry, the interacting amino acids produced particular sequence motifs/pattern **a**bc**d**efg with *a* and *d* residues interacting in the α-helical interface and **a**bc**d**a**b**d**c**d, with a and c residues interacting in the β-strands.

and Massimo Libonati. Disulfide bonds can significantly increase the stability of a chain or of an oligomer, but it is not necessarily true and that needs to be measured case by case [7]. Covalent bonds are strong interactions as it takes a large amount of energy to break them (110-50 kcal/mol). In living organisms, an enzyme (protease) is necessary to cut a covalent bond.

Second, the tertiary and quaternary structures of protein as well as the folding and the chain association involve mostly non-covalent bonds between the atoms of the amino acids, called weak bonds because it takes a small amount of energy to break them (1-7 kcal/mol). These are hydrogen bonds, hydrophobic bonds, electrostatic bonds (between charges), polar bonds (between dipoles) and van der Waals interactions. Under physiological conditions, the weak bonds continuously form and break. The secondary structures of proteins, α-helices and β-sheets (intramolecular β-strand interactions) are stabilized by hydrogen bonds between atoms of the backbone of the amino acids. Likewise for intermolecular β-sheets but the hydrogen bonds are between atoms of the backbone of amino acids located on different chains. At last, but not least, worth noted amino acids in terms of folding and association is the proline. Its side chain geometry is particular and can adopt two positions named *cis* and *trans*, affecting the relative position of its neighboring amino acids accordingly. The consequence is the existence of two different local tridimensional states. The transition between the *cis and trans* conformation is called a *cis-trans* isomerization and is known to slow down the folding of a protein, and to also affect the association of chains indirectly [8-11].

The zone of contact between two associated chains is called the protein interface. The protein interface is made of intermolecular amino acid interactions. Every chain provides a domain that recognizes another domain, or the same domain, on another chain and associates with it. The association is based on the chemical and geometrical complementarities of the two domains. These complementarities are constructed on the spatial layout of the intermolecular amino acid interactions (Fig. 2C). These layouts are referred to as recognition modes and have been extensively studied [12-20]. Yet the rules that would enable us to predict recognition modes from sequences still remain elusive.

Understanding and predicting the modes of recognition of protein interfaces is essential for several reasons. First, because oligomers are involved in many cellular activities and when default interactions occur there are numerous consequences among which certain diseases. Second, because it is important to distinguish biologically significant interfaces from non-specific interfaces observed in protein crystals in order to properly assess biological assemblies in x-ray structures [21-24]. Along the same line, it is still not trivial to determine experimentally whether a protein is an oligomer and if so its quaternary structure, so any predictive quaternary structure tool is helpful. Third, because the knowledge on protein interfaces is used in synthetic biology to engineer artificial oligomers for several purposes from drug delivery devices to the development of new material [25-29]. As an example, one can read the chapter by Keqin Zhang on silkworm and spider protein fibers and their potential use in the fabric industry.

2. Overview of protein assembly

2.1. Intermolecular amino acid interactions

Protein interfaces have been extensively investigated [14, 30-35]. But because protein interfaces are large and rather flat in nature, they lack the spatial constraints achieved by a limited number of sequences. Thus the sequences of protein interfaces rarely share trivial profiles or patterns, in contrast to protein-small molecule interfaces or residues involved in enzymatic active sites [36, 37].

Thus, it has been clear very early on that looking at the 3D structures of protein interfaces was a necessary alternative to sequence analysis [34, 38-43]. The increasing number of 3D-structures available for oligomeric proteins has also favored such investigations and the development of computational methods to identify and study the amino acids at protein interfaces on large scale datasets. In the chapter, we essentially review these computational advances but discuss little experimental progress. One can read the chapter on protein oligomerization by Giovanni Gotte and Massimo Libonati for information on experimental approaches or read information provided in the following publication [44].

The first benefit of computational approaches is the facility in discriminating intermolecular amino acid interactions from intramolecular amino acid interactions systematically and efficiently by relatively simple algorithms.

2.1.1. Identification of the intermolecular contacts at protein interfaces

Many algorithms are available to identify the amino acids involved in intermolecular interactions from the x-ray coordinates of the 3D-structure of a protein oligomer (reviewed in [13, 21, 45]). The coordinates are accessible at the protein database (PDB, http://www.rcsb.org/pdb/home/home.do)[46]. There are databases were interfaces have been classified according to their 3D organization, residue conservation and residue types [47]. Some databases are used to complement cellular networks (interactomes) with structural information on the binding modes between cellular partners [12, 18-20, 48].

The classical algorithms are based on three different measures: (i) accessibility surface area, (ii) voronoi cells and (iii) arithmetic distances. More novel algorithms use graph theory measures such as centrality and coefficient clustering.

Accessible surface area (ASA). The first method calculates the solvent accessible surface area by rolling a probe of a given radius around the Van der Waal's surface of the protein atoms whose centre is the accessible surface [49]. Typically, the probe has the same radius as water (1.4 Å) and hence the surface described is referred to as the solvent accessible surface. The ASA are calculated for the monomer and for the oligomer and the interface residues are obtained by the difference in their ASA. ASA is currently used to discriminate biological contacts (large ASA, 1600 ± 400 Å^2)) from crystal ones (small ASA <400 Å^2) [50, 51]. The PDB entries are now processed accordingly and provide both biological assembly and asymmetric unit coordinates. The biological assembly entry includes a remark to explain whether the oligomeric state is "author provided" (experimentally shown to be an oligomer) or "software determined" or both. Alternatively, the biological assembly can be downloaded directly from the Structure Server PQS (Protein Quaternary structure) at EBI (http://pqs.ebi.ac.uk) [50]. ASA can be calculated from different servers and programs such as PISA (Protein, Interfaces, Structures and Assemblies, http://www.ebi.ac.uk/msd-srv/prot_int/) or Naccess (http://www.bioinf.manchester.ac.uk/naccess/), both essentially implemented from the Lee & Richards method [52].

Voronoi cells. The second method selects interfacial residues based on the Voronoi diagram or its closely related power diagram [53-55]. The Voronoi diagram associates to each atom its Voronoi cell, namely the convex polyhedron that contains all points of space closer to that atom than to any other atom. Instead of the Euclidean distance |ax| between a point x and an atom centered in a, this diagram use the power distance p(x) of x with respect to the ball of radius r that represents the atom,$p(x)=|ax|^2-r^2$. The Voronoi cell of an atom then comprises all points of space that have a power distance to that atom less than to any other atom. Its facets belong to the radical plane, which contains the intersection of the spheres if they do intersect. The Voronoi (or power) diagram offers a natural definition of contacts: two atoms are in contact if and only if their Voronoi cells share a facet. The use of Voronoi diagrams has been extended for assessing the reconstruction of protein assembly with the impressive example of the Nuclear Pore Complex [56].

Arithmetic distances. The third method also requires the 3D-structure (available in the PDB) and calculates Euclidian distances between atoms of the amino acids of different chains to detect only intermolecular atomic interactions [23–25]. The selection is the pairs of atoms which are within a cut-off distance from each other classically around 5.0 Å such that any type of chemical bonds between the atoms are considered (H-bonds, electrostatic interactions, van der Waals forces, salt bridges and hydrophobic attractions). The pairs of atoms selected as part of the interface, depend on the choice of the cut off distance. This is a serious issue because a distance cannot fully describe a spatial arrangement and there is a chance that the geometry of the interface is not faithfully represented by the set of selected pairs [57]. The need to use a cut off distance for the selection prevents from having a natural read of the geometry of the interface. Better alternatives select pairs of atoms in interactions as the nearest neighbor atoms instead of using a cut-off [40, 58-60]. This measure is more capable of reading the whole geometry of the interface and therefore supplies a more accurate set of pairs of the intermolecular contacts. Differences in the set of atoms selected according to distances are illustrated in figure 3.

In addition, residue conservation or spatial chemical conservation can be implemented to yield a set of intermolecular amino acid interactions based on structural and sequence information [47]. The method requires the PDBs and a multiple sequence alignment as input data. Individual residues are represented in terms of regional alignments that reflect both their structural environment and their evolutionary variation, as defined by the alignment of homologous sequences. Multiple alignments use either the Shannon or the Von Neumann entropy [61]. Conservation scores are also efficient in discriminating genuine biological assemblies from crystal contacts [22, 24]. There exist several algorithms, the most efficient are mapping conservation score to the 3D-structures such as Evolutionary Trace [62-65].

Hotspots-In the mid-nineties, the specific energetic contribution of the side chain atoms to protein interfaces was investigated using Alanine Scanning Mutagenesis because the mutation by alanine reduced the interactions to backbone atoms [66]. It was found that only a small subset of interfacial residues were sensitive to an alanine mutation indicating that the energetic contribution of the interfacial residues was not distributed uniformly. Some key 'hot spot' residues contributed dominantly to the binding free energy. Thorn and Bogan deposited hot spots from alanine scanning mutagenesis experiments in the ASEdb database (http://nic.ucsf.edu/asedb/) [67]. BID (The Binding Interface Database) is another database of experimental hot spots, which collects all available experimental data related to hot spots in protein

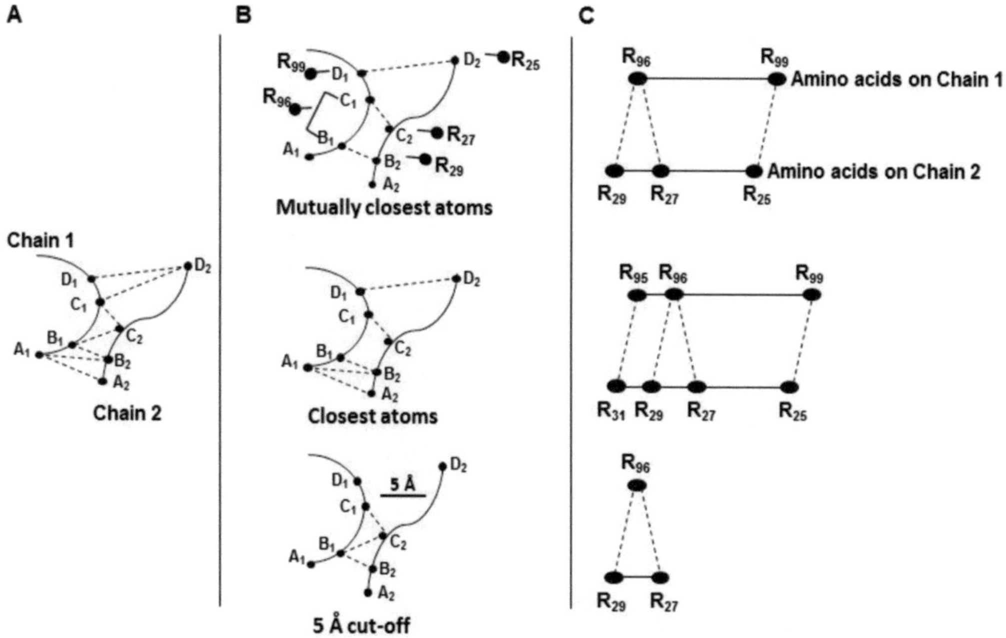

Figure 3. Selection of the amino acids in interactions at interfaces. A. Schematic of an interface between chain 1 and chain 2. Each chain is symbolized by a line and the chain respective atoms are indicated by black dot and letters with indices corresponding to the chain. Distances and so potential interactions between atoms are indicated by dotted lines. Only few interactions are indicated for the sake of clarity. **B. Selection of atoms in interactions at the interface.** The same schematic is reproduced after selection of the atoms in interaction at the interface. The top schematic is a selection based on mutually closest atoms, the middle one is a selection of all closest atoms and the bottom one is a selection of atoms at distances shorter than a cut-off of 5 Å. **C. Coarse-grained graphs of the interface.** Based on the selected atoms, a graph of the interactions between amino acid is drawn. The top, middle and bottom graphs correspond to the top, middle and bottom selections illustrated in B, respectively.

interfaces(http://tsailab.chem.pacific.edu/wikiBID/index.php/Main_Page)[68]. In parallel, there have been several experimental evidences not based on ala-scanning mutagenesis but on kinetics of assembly that showed the role of only some amino acids of the interfaces in regulating the chain association [69, 70]. Now hotspot (or hot spot) is a colloquial term that distinguishes a residue relevant for interface formation from others.

There exist several predictors of hotspots combined or not with evolution conservation based on ASA, voronoi and distances, some reviewed in [21, 47].

Algorithms based on graph theory-More recently, methods based on graph theory have also been proposed to identify hotspots. A graph or a network is a mathematical representation of pairwise relation between objects. A graph is made of vertices (or nodes) and lines, called edges (or links) that connect them. Proteins have been described as networks, with the amino acids of a protein chain considered as the nodes and the interactions between amino acids as the links. These networks, referred to as protein structure networks or amino acid networks, describe the entire protein and are used to infer global characteristics of the protein. Likewise, protein interfaces have been described as networks of hot spots in interactions with the hot spots as nodes and the interactions between hot spots as links. Now, it is important to realize that protein interfaces are not networks but sub-networks (sub graph) as they describe local properties of the protein, namely the interfaces. A good overview of network measures can be found in [71].

The sub-graphs are built of pairs of amino acids in intermolecular interactions (Fig. 3C). As for the previous methods, the 3D-structures of the protein oligomer are used to build the graph and different measures are used to infer the amino acid in interactions. Basically, the atoms are considered as points in space, each chain of a protein oligomer constituting a distinct set of points. All distances between atoms of the different set are calculated and any two atoms within a given cut off distance or closest atoms are considered linked. A coarse-grain graph is built by replacing the interacting atoms by their respective interacting amino acids.

The measure of path length and centrality is used (Central nodes). In graphs, the notion of links between amino acids goes beyond chemical/physical bonds which are based on arithmetic distances. In a graph, any two amino acids are connected by a geodesic distance which is the shortest distance between them, measured as the minimum number of links that need to be crossed to connect them by the shortest path. This distance is called the path length and is symbolized by the letter *l*. The mean path length, <*l*>, represents the average over the shortest paths between all pairs of nodes and offers a measure of the network's overall navigability. This introduces the notion of contacts through communication routes in addition to the more classical notion of geometrical/chemical contacts. This novel notion will have vast applications in the field of protein structure and protein dynamics. It has been already used for investigating protein allosteric mechanisms as discussed later [72-77].

Several measures of the centrality of a graph (closeness, betweenness) are associated with geodesic distances. Basically the numbers of shortcut paths going through every node are calculated and the most central nodes are those with the highest numbers of shortcut paths going through them. In other words, the centrality is finding nodes at the crossroad of communication routes. Centrality measures have been used to identify residues important for the function and for the fold of proteins [78-80]. It has also been used to identify hot spot residues in protein interfaces [81, 82]. One has to be careful to keep in mind that a central node needs not to be at the center of the protein or at the center of the interface, but it is necessarily at a crossroad of many paths.

The measure of degree and clustering coefficient is used. The number of links a node has, in the context, the number of interactions an amino acid has with other amino acids (number of contact amino acids), is called the degree and is symbolized by the letter *k*. The mean degree <*k*> represents the average degree over all the nodes of the network. The degree distribution of networks is informative on the characteristics of the network [71]. Networks with a power law degree distribution are called scale-free, a name that is rooted in statistical physics literature. It indicates the absence of a typical degree for the nodes in the network (one that could be used to characterize the rest of the nodes). This is in strong contrast to random networks, which have Poisson degree distributions, and for which the degree of all nodes is in the vicinity of the average degree <*k*>, which can be considered typical. There are also exponential degree distributions which are single-scale networks. Scale free networks are made of many nodes with few links and few nodes with many links, called hubs. Hubs are absent in random and single scale networks. Hubs are communication devices that allow most nodes of the networks to be connected with others. This has noticeable consequences in terms of the vulnerability of the networks to changes on the hubs or elsewhere [83]. This is discussed later in the chapter. Proteins are essentially random networks or single-scale networks [84].

The clustering coefficient, called *C*, is based on the degree of the nodes and it measures the probability that a node A which is connected to B, itself connected to C, has to be connected to C as well. Calculated over all nodes of the networks, it identifies clusters of nodes highly connected to one another and hence it discriminates different clusters as distinct communities. The calculation of clustering coefficient is detailed in [71]. In protein oligomers, the protein interface is made of many bonds between two adjacent chains and few bonds between non-adjacent chains. Hence the interface makes a cluster in terms of graph and hotspots have been successfully identified by clustering coefficient [85].

Protein interfaces are either analyzed based on all interfacial residues or on hotspots only.

2.2. Features of the intermolecular contacts at protein interfaces

To infer the features of protein interfaces, the method is simple: a dataset of protein oligomers/ protein interfaces is built, an algorithm is applied to each one of the interfaces to identify intermolecular contacts and the features of the intermolecular contacts are analyzed using statistics. Classically the parameters to describe a protein interface are: (i) interface size, expressed either in number of amino acids or in ASA, (ii) the number of regions of interfaces over the full-length chain, (iii) the chemical properties of the amino acids (amino acid frequency, the interface propensity, namely the frequency of a residue a_i in the protein interface divided by its frequency in a reference set, generally the full-length chain). The size of protein interfaces is an important parameter because it may vary depending on the strength of the association [86].

Evolutionary conservation, protein folds, secondary structures, quaternary structures or crystallographic B-factors can also be considered depending on the question and the criteria used to build the dataset. The idea is to find enough specific features to distinguish the residues of protein interfaces from the rest of the residues.

2.2.1. *Dataset based on features of the full-length protein*

Many dataset are built on proteins sharing properties at the level of their full-length chains (function, organism, superfamily, folds, and quaternary structures) but without necessarily sharing features at the level of their interfaces [39, 87-89]. In particular, the geometries of the interfaces are not necessarily looked at and therefore interfaces with different geometries are often compared. But it is generally assumed that proteins related in terms of folds or functions associate in similar ways. However, a screen over a large dataset of dimers, performed by Keskin *et al.* has shown that a non-negligible amount of protein oligomers have interfaces sharing features although they have different folds and functions [90]. This set is referred in the paper as "type II", following the term "type I" used for protein interfaces sharing features and derived from protein oligomers having similar fold and/or functions. To establish the determinants of the construction of an interface it is simpler to look at type II interfaces because the pressure of evolution over the fold and the function of the protein chain is alleviated compared to type I interfaces.

Globally, the results of studies on protein interface dataset (mainly type I) revealed the importance of hydrophobic interactions in the formation of protein interfaces, greater residue conservation and chemical property similar to surface residues but packing like core residues [34, 38]. The two latter properties are coherent with the fate of a protein interface. The topology of a soluble protein is defined by surface residues which are accessible to the solvent and core residues which are, on the contrary, buried and inaccessible to the solvent. The amino acids of a protein interface have the solubility requirement of surface residues because the domains of the interface are initially accessible to the solvent to allow binding. To have stable binding, the domains need to minimize void and maximize packing as for the core residues.

If the role of hydrophobic residues is consistent over any dataset, the importance of polar and charged residues in interfaces varies very much between datasets. Altogether this indicates that hydrophobic residues are involved in promiscuous interactions while polar and charged residues yield alternative recognition modes and hence provide each type of interfaces its specificity.

Up to date, there is no single property sufficiently unambiguous to identify the protein interface from the rest of the protein, and considerable disagreement exists on which properties are actually useful. Conservation is an excellent example of a property both widely used and widely debated. De Vries and Bonvin as well as Neuvirth raise the matter of having so many algorithms and the absence of consensus on the parameters truly relevant to the formation of a protein interface [45, 91]. This may well explain the contradictory results on protein interface properties.

Most studies are performed on the features of individual hot spots. Yet protein interfaces result from intermolecular pairwise interactions and are likely encoded at the pair's level. Supporting this view, the few studies investigating the features of pairs of hotspots show sufficient specificity of the residue pair preferences for accurate prediction [40, 58, 92, 93].

2.2.2. Dataset based on features of the interfaces

To investigate properties responsible for interface formation, an alternative is to build a dataset based on the features of the interfaces and not on the features of the full-length chain. For instance, one can build a dataset of proteins sharing the same geometry of interface.

2.2.2.1. β-strand geometry

Interfaces made of two interacting β-strands (intermolecular β-strands) have been largely studied because it is present in many conformational diseases such as Alzheimer's disease, Parkinson's disease or serpinopathies [94-98]. Supporting the view of comparing interfaces with identical geometries, the proteins involve in conformational diseases share no functions,

http://dx.doi.org/10.5772/58576

fold, and quaternary structures yet they have a common local fold involved in the intermolecular contacts that lead to fiber formation. Their pathological form, whether a fiber or an oligomer, involves interactions between two β-strands, each provided by a different chain (intermolecular β-strands). These intermolecular β-strands share several structural properties. They are recognized by the same antibody A11 [99]. Their formation depends on interactions between atoms of the backbone, result which has led to the proposal that aggregation is a generic property of the polypeptide chain [100, 101]. They adopt a cross β structure which can be predicted from sequences by the PIRA (Parallel 'In Register' Arrangement) model, a network made of single pairs of residues [102-107]. Different predictors of the aggregation-prone sequences involved in the fiber formation are now available [96, 98, 108-111].

We have studied a dataset of 1056 interfaces present in 755 protein oligomers not known to be involved in conformational diseases [59]. As others, we found no specificity at the level of individual hot spots. The chemical properties of the individual hot spots and their distribution on the sequence characterize only the secondary structure and the solubility of the β-strands. In contrast the interaction pairs provide the interface some specificity. Interestingly, the interfaces are best described by two sets of interaction pairs, pairs involving backbone atoms made essentially of hydrophobic and/or small residues and pairs involving at least one atom of the side chain, preferentially made of charged, polar, long and medium residues. The backbone pairs have properties common to intramolecular β-strand interactions and intermolecular β-strands involved in fiber formation in terms of amino acid preferences. Thus hydrophobic amino acids whether in pairs or as individual are not giving any specificity to interfaces. That explains that they always appear in any dataset. On the other hand the side chain pairs have particular geometrical characteristic in terms of number of atoms, branching and length. They also show preferred chemical pairing different from those measured for β-fibers [98]. However this result is only based on comparison with the literature and it could be due to differences in the datasets and/or the algorithms.

The geometry of the side chains has been so far neglected when it appears in our study as a key parameter. Using Steiner Minimal Tree approach (SMT), MacGregor *Smith et al* proposed an elegant geometrical representation of the amino acids that was successfully applied to the problem of protein folding [112]. It will be interesting to extend this approach onto protein interfaces to see if the specificity of protein interfaces may be provided by the geometry of the amino acids rather than their sole chemistry. Similar double layer of interactions, has been observed at the interfaces between colicins and their cognate immunity proteins [113]. One set of the intermolecular residues was common to all colicin-immunity members and produced a low binding affinity between the colicin and its cognate immunity protein while the other set was made of variable residues providing high affinity and specificity to the colicin for a particular cognate. Double layer of interactions has also been reported in monomeric proteins (intramolecular networks) [84].

As mentioned earlier, proteins and protein interfaces are now described as networks of amino acids in interaction or as sub-networks of hot spots in interactions, respectively. This relatively new concept offers the possibility of looking at the layout of interactions in addition to the amino acid properties. It is clear now that the network of interactions is as much important as the components of the network in providing the protein its properties in terms of folding, function, evolution and interface formation [59, 78, 84, 114-117].

We have observed in our 1056 β-strand interface dataset that the side chain pairs also have specific network features. The side chain hot spot sub-networks have nodes with more contacts than the backbone or the PIRA (Parallel In Register Alignment) networks, used to predict fiber sequences [98]. Yet the β-strand interfaces have no hubs and maintain a low interconnectedness (little communication between residues of the interface), probably a mechanism to resist the effect of mutation by secluding the nodes of the networks (Fig. 7). Simultaneously, the side chain residues make less than three contacts avoiding stringency on the choice of amino acids capable of making an interface and providing the β-strand interface high sequence plasticity. Robustness and plasticity of networks are well explored by graph theory and there are several very inspiring papers on that topic [83, 118-120]. This point is discussed in more details later in the chapter.

2.2.2.2. α-coiled interfaces

To date the only interfaces accurately predicted from sequences are α-coiled interfaces [121-124]. Intermolecular residues follow a so-called knobs-into-holes regular packing producing the α-coiled coil helix-helix assembly [125]. In the simplest case (dimer), the α-coiled coil sequence displays a repeat pattern of seven amino acids so-called heptad repeat, labeled *abcdefg*, with hydrophobic residues at the *a* and *d* positions (Fig. 2C). These hydrophobic intermolecular contacts constitute the seam of the core of the knobs-into-holes interface. The repeats can be shorter than 20 residues or span many hundreds of amino acids.

There are obvious reasons to why it has been possible to understand α-coiled coil interfaces when other geometries still elude us. First, α-helices are geometrically more constraint than β-strands and second backbone interactions do not participate in α-helix interfaces because the hydrogen bond networks are made intra-molecularly. Hence, there is no "backbone noise" information that interferes with the side chain information.

2.2.3. Interfaces and quaternary structures

The quaternary structures of the protein oligomers and the features of their interfaces are related and different methods are currently developed aimed at understanding such relations [126, 127].

In some cases such relation is more or less understood. For example, in higher-order α-coiled coil oligomers (above dimer) additional (peripheral) knobs-into-holes take place and broaden the helical contacts [128]. Such multiple repeats lead to multi-faceted helices, which combine repeats of different amino acid compositions to accommodate quaternary structures accordingly [129, 130]. Thus, it is possible by analyzing amino acid sequences to predict the quaternary structures of α-coiled coil oligomers [129-132].

The relation between the interfaces features and the quaternary structure is less understood in β-strand interfaces with few exception as the legume lectin family (81-82). Combining clustering algorithms with sequence alignments, motifs of sequentially and structurally conserved residues are detected at the β-strand interfaces of lectins. The different motifs are built on a subset of residues at the interface that provide a specific 3D-orientation of the β-strands. Consensus patterns at the interfaces have been found for the different quaternary structures of the lectins. Briefly, there are nine different kinds of quaternary structures in legume lectins including Canonical, ECorL-type, GS4-type, DBL-type, ConAtype, PNA-type, GS1-type, DB58-type, and Arcelin-5-type (monomeric). Seven different consensus is observed including types II (canonical), X1 (DB58-type), X2 (noncanonical interface of ConA), X3 (ECorL-type, handshake), X4 GS4-type, back to back), and the unusual interfaces of PNA and GS1.

For a long time, there are experimental evidences of sequences that are responsible for the quaternary structure of protein. These sequences, called registration sequence, are located upstream the interface region and promote oligomerization of monomeric protein when genetically added [133]. Proline and histidine residues located upstream of interfaces have also been shown to regulate the association between chains [8, 11, 134-136]. Collagen α-fibers and silkworm/spider β-fibers contain several repeats composed of proline residues which also participate in the quaternary structures. Whether these residues belong to the interfaces or are systematically located outside the interface regions has not yet been established.

In summary, residues located within the interfaces and outside are participating in the quaternary structures and the chain assembly. This implies that protein assembly is regulated at two levels, at the level of intramolecular interactions (residues outside interfaces) and at the level of intermolecular interactions (residues in interfaces). Thus it is necessary to also investigate the residues involved in intramolecular interactions to discriminate those participating in folding reactions from those participating in both folding and interface formation. The latter residues are probably coordinating the whole assembly process by regulating communication between folding and association steps.

3. Intermolecular and intramolecular amino acid interactions: The mechanism of protein assembly

As mentioned at the beginning, protein assembly or protein oligomerization entails folding and association reactions. Thus, to have a full picture of the mechanism of assembly, besides studying intermolecular amino acid interactions, it is necessary to investigate intramolecular amino acid interactions and to apprehend how both types of interactions are coordinated. Different models of assembly have been recently reviewed [137, 138].

First, let's consider the simple case of the formation of a dimer. There are two routes to a dimer. One is through the three states model where unfolded monomers *U* (state 1) fold into monomers *M* (state 2) which subsequently associate into dimers *D (state 3)*. The alternative route is through the two states model in which unfolded monomers *U* (state 1) associate into folded dimers *D* (state 2). Intramolecular and intermolecular interactions occur sequentially in the three states model but concomitantly in the two states model. One can anticipate that folding and association are going to be related but independent in the three states model but concerted in the two states model. In terms of networks, one can speculate that the three states model suggests a protein organized in two sub-graphs remotely connected, one governing the intramolecular interactions and the other the intermolecular reactions. On the contrary the two states model suggests two connected sub-graphs.

Discriminating the route of assembly is crucial in term of drug design strategy. In the three states model, it is likely that the interface in the folded monomer and in the folded dimer is similarly organized. Thus, it is consistent to use the x-ray structure of the native protein oligomer as a template to design assembly inhibitors. In contrast, in the two states model the interface in the unfolded monomer is different from the interface in the folded dimer so assembly inhibitors designed on the native structure of the protein oligomer are unlikely to recognize the unfolded monomer and block the assembly at the monomeric stage (or at early stage). This is one illustration on why it is important to anticipate the mechanism of assembly.

3.1. Can the mechanism of assembly be predicted by investigating evolutionary relationship?

D'Alesio offers good historical reviews on this question [139, 140]. The interfaces of dimers assembling by a two states model are found to share patterns with intramolecular interactions observed in monomers. It is proposed that such dimers have evolved from mutations in an existing monomer that led to its unfolding, followed by further mutations that yielded a viable dimer with intermolecular interactions similar to the intramolecular interactions present in the initial monomer. This mechanism is reminiscent of the domain swapping mechanism which is well presented in the chapter by Giovanni Gotte and Massimo Libonati or reviewed in [141]. In such a situation the evolution to the dimer depends initially on the evolution of a viable monomer towards unfolding induced by random mutation. There wouldn't be any folded monomer in the assembly route because it wouldn't bear the mutations in a folded state. The evolutionary route between the dimer and the monomer suggests the presence of epistatic mutations (mutations that have different effects in combination and individually). Here the fold and association steps are related and dependent on one another, in term of evolution and mechanism of assembly. In contrast, dimers assembling through a three states model were not found to share motifs with monomers. The folding of the monomer might then be a natural route towards association, and folding and association would appear evolutionary and mechanistically related but independent.

Next, let's look beyond the simple case of dimers. Possible relations between evolution and assembly mechanism have been further exploited by looking at protein oligomers sharing the same functions (superfamily) but adopting different quaternary structures. In one study the authors exploit the symmetry of oligomers to establish a relationship between evolution and assembly mechanisms [142]. From an initial screen of 5375 PDB structures, they found tetramers with D_2 symmetry having homologous dimers with C_2 point group symmetry and hexamers with D_3 point group symmetry homologous dimers with C_2 symmetry or homologue trimers with C_3 point group symmetry. In total 49 protein oligomers with a symmetry relation from C_n to D_n are reported. They found evolutionary links between the C_n and D_n counterparts and isolated experimentally *the in vitro* disassembly C_n intermediates for ten of the D_n oligomers. In addition, for five cases, the C_n intermediates were also shown to be formed during *in vitro* reassemblies. They concluded that the evolutionary and assembly pathways were related and that assembly intermediates could be predicted solely from the atomic structure.

But this conclusion might be over optimistic because based on very little cases (49/5375) and because protein structure evolution and protein assembly are plastic in terms of mechanisms such that to date it remains difficult to establish either route.

For example a single mutation has been found responsible for a transition from Cn to Dn point group symmetry, it is not obvious how such global change could have been anticipated by simply considering the full-length 3D-structure of the wild-type protein [143]. On the other hand it tells that the protein assembly is regulated by local properties since a single mutation is enough to alter the global assembly. This strongly suggests that the solution lies in understanding the local properties and how they propagate information to regulate the global shape.

Hemoglobin is another complex example of a protein sharing a function but distinct quaternary structures and for which evolutionary and assembly routes are not easily drawn even if the structures are available (Fig. 4).

The *Synechocystis cyanoglobin* produces a monomeric hemoglobin (PDB 1S69) with a C_1 point group symmetry, the human hemoglobin is tetrameric (PDB 2HHB) with a C_2 point group symmetry, the *Oligobrachia mashikoi* produces a 12-mer hemoglobin (PDB 2ZS1) with a D_3 point group symmetry while the giant earthworm hemoglobin contains 144 chains (PDB 2GTL) with a D_6 point group symmetry. In such case, the different hemoglobin point group symmetries and quaternary structures may result from coding constraints of their respective organisms rather than from a relation in terms of evolution or assembly intermediates. As described by Crick in the early 60s, symmetric assemblies require fewer distinct kinds of specific interaction interfaces compared to asymmetric assemblies [144]. Likewise, higher symmetries require fewer distinct interfaces compared to lower symmetries and thus, the smaller a genome the more often its protein structural complexity may rely on high symmetry. This is consistent with the large occurrences of proteins with icosahedral symmetry in viruses while most eukaryotic molecular machines have C1 symmetry (Fig. 1B). Now, some high point group symmetry oligomers have been also discovered in eukaryotes, bacteria and archea with the vault proteins assembling 78 copies (2zuo, 2zv4, 2zv5), the encapsulin (3dkt) and the vault from *Pyrococcus furiosus* (2E0Z). Clearly one has to be cautious in interpreting data and statistics derived from the PDB.

In addition, hemoglobin is also an interesting example of how a unique function is provided by a combinatory of assemblies using the same protein fold. There are many other examples (e.g. ferritin, rubisco) but probably α-coiled coil oligomers offer the largest combinatory of assemblies. It was recently shown that they formed before LUCA (last universal common ancestor), by independent routes and most likely as the result of all possible geometric solutions to packing helices in a stable way [145]. Again, that illustrates how diverse evolution routes are.

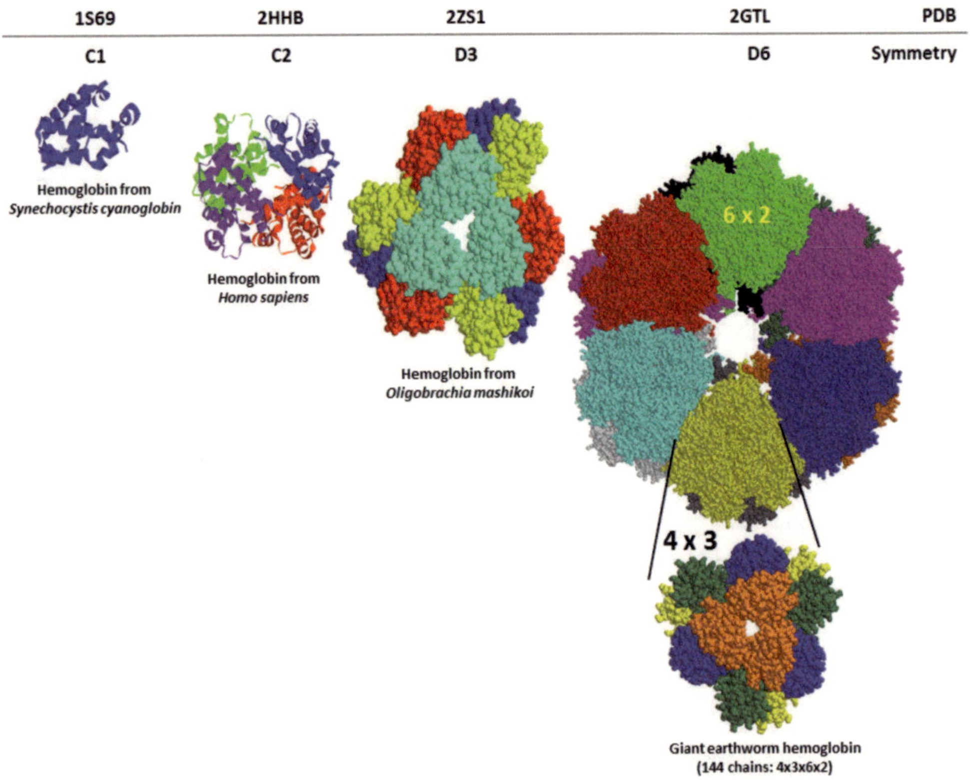

Figure 4. Plasticity of quaternary structures fulfilling a single biological function: the hemoglobin example. The hemoglobin chain exists as a single fold which is copied and assembled with different stoichiometries (number of chains) and different symmetries across species to maintain the same biological function. Few cases are represented from a hemoglobin monomer to a 144-mer assembly. The structures are shown in ribbons except when spacefill is better to illustrate the symmetry of the assembly. The pictures are generated with Rasmol. The PDB codes and the symmetries of the hemoglobins are indicated above their respective structures.

The relation between evolution and assembly routes assumes that an oligomer evolves/ assembles from a monomeric entity. But reverse situation exists as for the native tachylectin-2 monomer which has been proposed to have evolved from a pentameric ancestor through short, functional gene segments that, at later stages, duplicated, fused, and rearranged [146]. The authors propose that new folds evolved through the structural plasticity of assembly intermediates.

This last example illustrates quite ironically that protein folds and quaternary structures still hold surprises and a direct relation between the evolution of protein oligomers and the mechanism of their assembly is not readily systematic. Both evolution and assembly certainly involve multiple parameters making their prediction rather challenging.

3.2. Can the mechanism of assembly be predicted by experimental approaches?

The two and the three state models are depicted in the figure 5A.

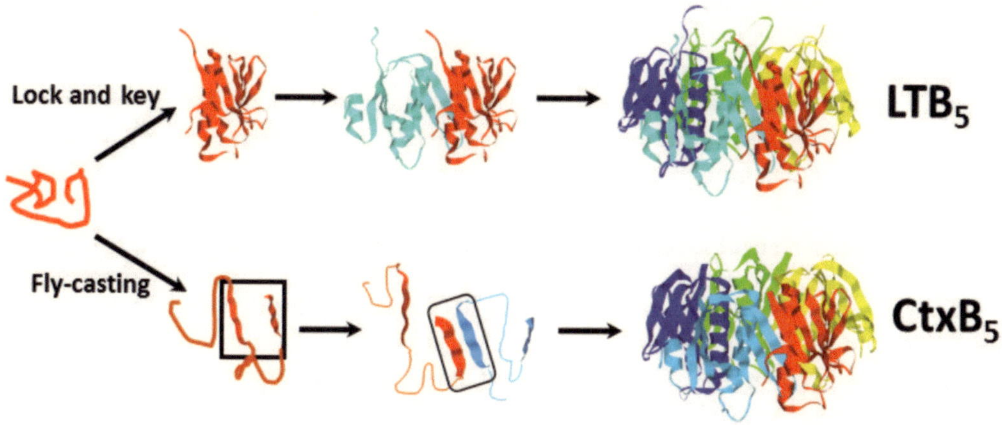

Figure 5. Mechanisms of assembly. Two mechanisms of assembly have been described and experimentally observed. The protein chain folds before association in the lock and key mechanism, also called the three states model because the protein can be observed in three states, unfolded monomer, folded monomer and native oligomer (top route). The protein chains associate in a more or less partially folded state, and only subsequently acquire native folded conformation, in the fly-casting mechanism also referred to as the two states model because because the protein exist only in two states in a dimer, unfolded monomer or native dimer. These two models are illustrated with the assemblies of the two related AB_5 toxins, the heat labile enterotoxin B pentamer (LTB_5) and the cholera toxin B pentamer ($CtxB_5$). The two toxins share 94 % sequence identity and almost superimposable atomic structures but nevertheless assemble through two different mechanisms.

The three states model is the oldest and most classical mechanism observed, it is generally referred to as the lock and key mechanism. There are plenty of experimental evidences of both the two and three states mechanisms. Non-native oligomers, namely oligomers with native quaternary structures but not native folds have been isolated experimentally for a long time and are common intermediates of assemblies [69, 70, 133, 134, 147-153]. Such intermediates are typical border line cases as they might be produced by a lock and key mechanism or by a fly-casting mechanism. There are clear examples of protein associating by a fly-casting assembly with unfolded monomers able to associate [8, 134, 151, 154, 155]. The RING domain protein family of scaffolding oligomers presents an interesting case of the formation of a stable partially folded assembly tetramer along the oligomerization route to a native 24-mer [156-158]. The C_4 symmetry tetramer populates because of its fast formation from monomers and its slow disappearance into a D_4 24-mer (6 x 4). The transition to the D4 symmetry 24-mer is rate-limiting, because of the slow folding Proline cis/trans isomerization that regulates the association of two monomers via the ligation of Zn sites. Likewise dimer, trimer and tetramer assembly intermediates are isolated along the route to the native cholera toxin B pentamer ($CtxB_5$) because the formation of one of the toxin interface is regulated by a *cis-trans* Proline isomerization [8]. The CtxB assembly intermediates acquire some of their native secondary structure along with association because their main interface involves the formation of an intermolecular β-sheet, this folding/association step is regulated by histidine residues [134].

Proline and histidine residues are rare at interfaces but are often found upstream the region of interfaces and are indirectly acting on their formation, as mentioned at the beginning of the chapter (see introduction). Registration sequences that control the quaternary structure of protein oligomers are also located outside interfaces. In fact, several cases of residues located outside interfaces have been shown to be involved indirectly in the chain association and a variety of small amino acid modules have been proposed to act upon assembly by different processes. Basically they introduce the flexibility required to modulate the 3D position of interface domains so to increase the chance of successful encounters [138, 159].

3.3. Can the mechanism of assembly be predicted by computational approaches?

The two states model was revisited by Wolynes' laboratory showing that an unfolded protein has a greater capture radius for a specific binding site than the folded state with its restricted conformational freedom [160]. In this scenario of binding, the unfolded state binds weakly at a relatively large distance followed by folding as the protein approaches the binding site: the "fly-casting mechanism" (Fig. 5A). In 2004, Wolynes introduces the notion that certain characteristics of the atomic structure like the interface size and hydrophobicity, the ratio of the number of interfacial contacts to the number of intramonomeric contacts enabled to determine whether a homodimer assembled into a fly casting or lock and key mechanism [161]. A large ratio of interfacial to monomeric contacts is typical of a two-state model and of the fly-casting mechanism.

Computational approaches also provide evidences supporting the lock and key mechanism, the fly-casting mechanism and a series of in-between mechanisms attesting of back and forth between folding and association reactions and whose idea lies on an "induced-fit" principle during which intermolecular contacts "catalyze" folding (allostery, conformational gating, induced fit) [162-165].

Recently, molecular dynamic (MD) simulations have been combined to network analysis to provide detail understanding of the route of assembly. For example, coarse-grained transition networks (CGTNs) can be derived from MD simulation to show the transition between oligomers of different sizes [166, 167]. In a recent report, the role of the sequences in the aggregation kinetics and assembly mechanisms was described in great details [168]. Briefly, MD is performed and the state of each conformation/state observed in the MD is defined by a set of digit. Based on the MD, a transition matrix $N \, x \, N$ is built with N states and with the matrix elements defined by the occurrences of any transition between two states. The matrix transition is converted into a graph called KTN (Kinetic Transition Network) with the nodes corresponding to the states and the edges to the transitions. Such graphs provide measures of the population of different states and the probability of transition between them. Energy barriers are associated to the transitions and disconnectivity graphs are constructed to evaluate the energy barrier to go from one conformation to another with min-cut algorithms. The dynamics of aggregation was also evaluated using FPTD (First Passage Time Distribution) which informs on the most populated states and kinetics. Although such approach has not yet been applied to a protein assembly on a full-length protein, there is no doubt that such combination of molecular dynamics with graph theory would provide new directions in predicting protein assembly mechanisms.

http://dx.doi.org/10.5772/58576

In fact, graph theory is the ideal tool to investigate the residues involved in intramolecular interactions, the residues involved in intermolecular interactions and their cross-talk communications. Basically sub-graphs or clusters are produced and allosteric communication between the different clusters is investigated. This has been used in enzyme/ligand intermolecular interactions and in interfaces [74]. It appears that the intramolecular networks maintained the robustness of the structure while the interface residues are more plastic to accommodate the flexible motion required for association.

Obviously folding and association reactions intertwine to orchestrate the protein assembly. This means that the key factors for protein assembly is the balance between intra and inter molecular interactions.

4. Local key contacts regulate global conformations

There are cases of proteins sharing functions, high sequence identities, folds and quaternary structures but following distinct assembly mechanisms. For example the two related AB_5 toxins, heat labile enterotoxin (LTB_5) and cholera toxin ($CtxB_5$) have 94 % sequence identity, almost superimposable 3D structures, and identical quaternary structures but nevertheless assemble through different mechanisms under identical experimental conditions (Fig. 5A). LTB_5 follows a lock and key mechanism whereas $CtxB_5$ assembles through a fly-casting mechanism [8, 134, 135]. Out of 103 amino acids, 11 are different among which only two in the interface. The cpn10 heptamers are another of such example [70].

The role of only few residues in controlling an assembly or a disassembly mechanism is also evidenced in diseases called conformational diseases where a single amino acid mutation is enough to redirect the protein native conformation to an aberrant conformation such as a fiber, through unfolding/refolding steps [169-175]. Consequently the protein loses its function leading to the disease.

This tends to show that the assembly of a protein is in fact regulated by only few amino acids, indicating that little differences are enough to go from a fly-casting to an induced-fit mechanism. This is in good agreement with allosteric mechanisms and the MWC (Monod, Wyman, Changeux) theory that unifies fly-casting and an induced-fit routes into a single mechanism [176]. Accordingly, protein assembly can be expressed as a single scheme with transitions between the fly-casting and the induced-fit mechanisms depending on thermodynamic equilibrium and kinetic rates (Fig. 6).

There exist several evidences of such transitions in biology, some of which are illustrated on figure 7. For example, in the course of evolution proteins may change their folding and/or assembly routes upon random mutations. Or proteins very similar in sequences and structures may favor different routes because of small amino acid differences in their sequence and/ or environmental factors. This illustrates the plasticity of proteins in terms of mechanism of formation and in terms of quaternary structures but also supports the fact that it is the local characteristics (few amino acids) that impact on the global structure of a protein.

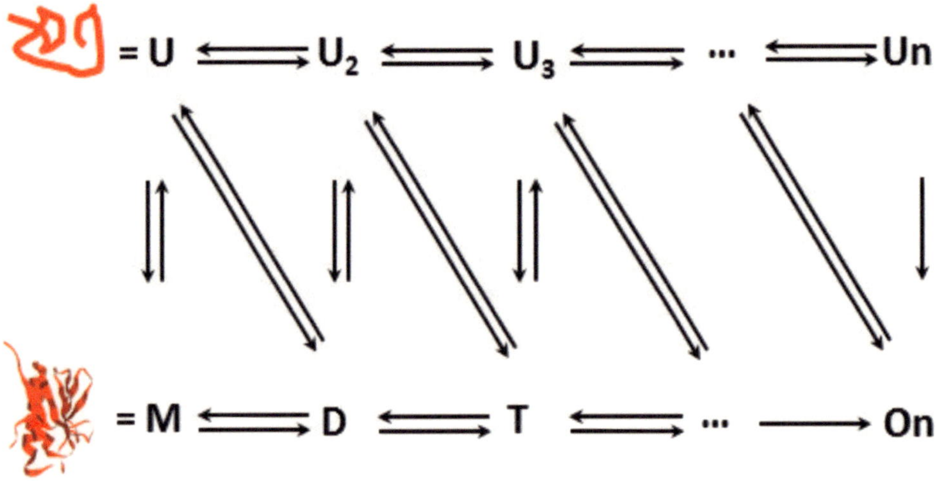

Figure 6. Kinetic scheme of protein assembly. A protein monomer may exist in an unfolded state ***U*** and folds into a folded state ***M***. A protein oligomer may assemble from (top reactions) unfolded monomers ***U*** which associate in "partially"folded states ***Ui*** with *i* going from 2 to *n*, *n* being the number chains; until they assemble into a non-native oligomeric state *Un* which finally folds into a native folded oligomer *On*. Alternatively, a protein oligomer may assemble from (bottom reactions) unfolded monomers ***U*** which fold and associate into dimers *D*, trimers *T*, etc until they reach the native oligomeric state *On*. Each of the conformational state exists in equilibrium and may go from one state to another according to k_{on} and k_{off} rates of the reaction. Only the formation of the native state are considered irreversible. The population of every species and the transition species depend on kinetic parameters.

How to identify few amino acids as key determinant for a protein fold or an assembly and what properties they must have to affect the mechanism of assembly and/or its final output? Graph theory is at present probably one of the most suitable approach to investigate such questions. For example, the effects of the mutations involved in conformational diseases have been considered in terms of network. Recently, a novel approach using graph-based signatures has shown that the impact of a mutation correlated with the atomic-distance patterns surrounding an amino acid residue [177]. They showed that the signatures can be used to predict stability changes of a wide range of mutations occurring in the tumor suppressor protein p53.

We have also investigated the effects of mutation on graph features and the possible consequences in terms of the disease mechanism [59]. As briefly mentioned earlier, we have seen that the networks of the 1056 intermolecular β-strands present in "healthy" protein oligomers, avoid hubs (highly connected residues) to be robust to mutation. The intermolecular β-strands are essentially disconnected graphs so any mutation would not spread damages far in the network. We compared these "healthy" networks with the β-strand interface of the p53 tetramer, which has known familial mutations related to dissociation of the tetramer, fiber formation, and associated with cancer [169]. The p53 network has a higher interconnectedness because its nodes have higher degrees (ie more contacts) than those in the "healthy" networks, with the consequences that a single node modification (ie a mutation) is enough to reorganize the interactions in the entire network. Thus the higher connectivity of the p53 network leads to a greater sensitivity to rewiring (rearrangement of links upon node modification) than the

http://dx.doi.org/10.5772/58576

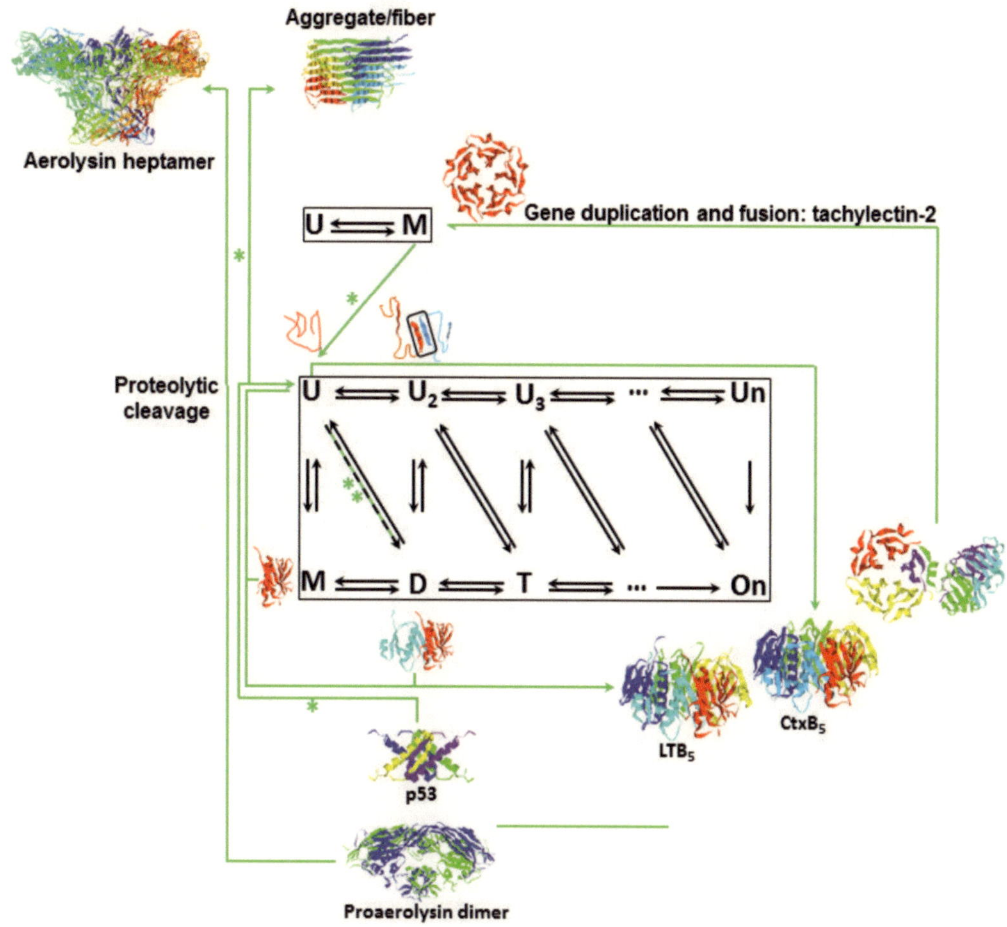

Figure 7. Evidences of transitions between different states and different reaction paths. The kinetic scheme described in figure 6 is reported here at the center of the figure. The transitions between different protein conformations or different reaction paths are indicated by green arrows. The x-ray structures of protein cases undergoing such transition are shown as illustration of these transitions. The transition can take place during evolution and because of mutation (e.g. tachylectin-2, two states model). It can take place because of mutation and lead to conformational diseases as for the p53 tumor suppressor p53. The main route may depend on a difference of few amino acids as for the two related toxins $CtxB_5$ and LTB_5. Or else, the protein as the pore-forming toxin aerolysin may adopt different quaternary structures and go from one to another because of environmental factors (e.g. pH, proteolyic cleavage, cell receptor etc…).

disconnected graph observed in healthy proteins. In some cases, such ample rewiring probably promotes chain dissociation, first step to fiber formation. We have now started to investigate why the p53 network has a higher interconnectedness than the "healthy" networks. The p53 tetramer has a D_2 point group symmetry and its interfaces adopt a local central symmetry because the two interacting domains have identical sequences and the residues are paired in an anti-parallel manner. In contrast 60 % of the protein interfaces of "healthy" proteins have domains made of different sequences and their β-interfaces have no local symmetry.

Let's consider three intermolecular networks, one with different amino acid sequences and no local symmetry, a second with identical sequence arranged in a parallel manner (horizontal axe symmetry) and a third with identical sequence arranged in an anti-parallel manner (rotational axe symmetry) (Fig. 8).

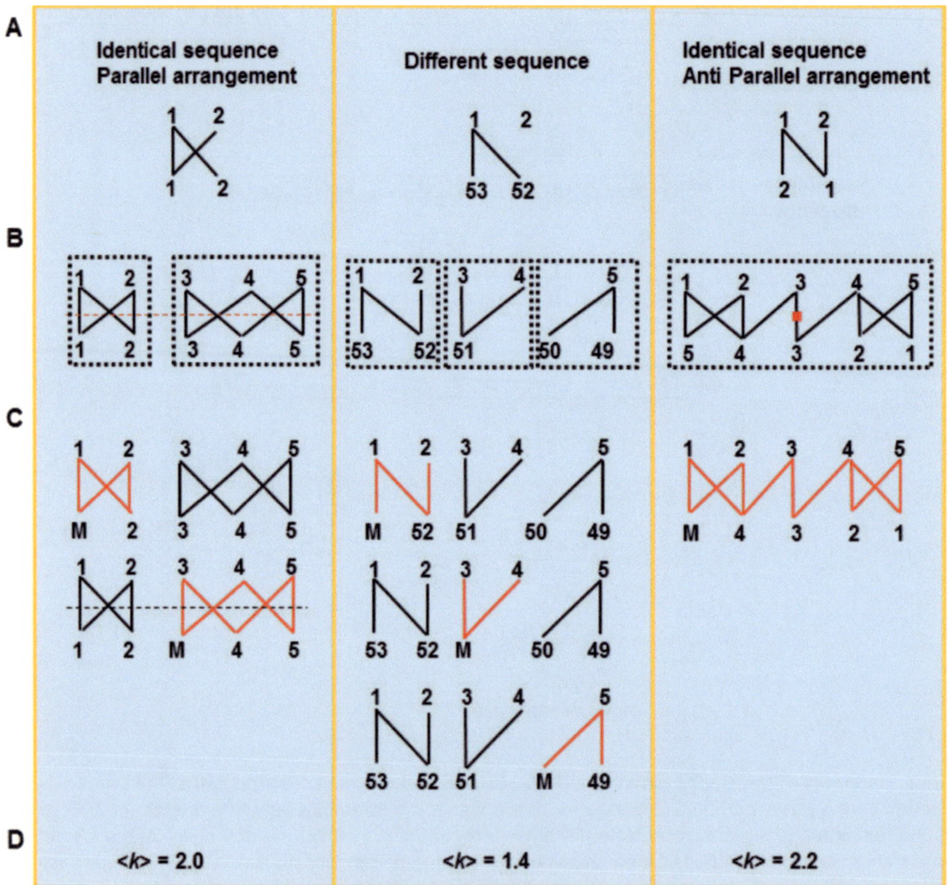

Figure 8. Effect of local symmetry on network features.A. Protein interfaces may be formed by association of domains with identical sequences (left and right pannels) or different sequences (middle pannel). In the former cases, the two domains maybe aligned in a parallel or antiparallel manner (in register or out register arrangement). The identical sequences will have an intrinsic symmetry in their amino acid pairing, namely if the residue 1 is in interaction with the residue 2 then the residue 2 is in interaction with the residue 1. Such symmetrical constraint will produce some motifs in the network which are not necessarily present in "asymmetrical" interfaces made of domains with two different sequences. This is illustrated on a simple network. **B. Connected components.** Considering a slightly more complex network one can see that the motifs results from the elements of symmetry, a horizontal axial or a rotational axe-symmetry for a parallel or antiparallel arrangement, respectively. The dotted boxes indicated the connected components, namely the residues which are connected to each other. One can see the effect of the symmetry on the total number of connected component. **C. Propagation of changes.** The effect of a single node modification on the network, indicated by a M for mutation, is considered. Assuming there is an effect as long as there is a link between two nodes, the symmetry enables the changes (red link) to propagate within the network. **D.** The average degree <*k*> of the nodes is given for each of the networks.

Let's now look at the consequences on the network features. The first consequence is a multiplicity of the number of interactions in interfaces with local symmetry and when the sequences are identical and therefore an intrinsic increase of the network interconnectedness (Fig. 8B). As observed for the p53 case, such increase would lead to network sensitivity to rewiring which in terms of protein may introduce a vulnerability to chain dissociation or chain reorganization. The second consequence is the decrease of the number of distinct connected components (Fig. 8B). This again would increase the propagation of changes within the network upon mutation because the amino acids are not secluded from one another. It is interesting that local symmetry is enough to improve the communication within the network without altering significantly the average degree $<k>$. This means that even without hubs the protein interfaces become highly connected by long paths. This preliminary analysis suggests that interfaces made of domains with different sequences might be more resistant to fold plasticity because of an absence of sequence symmetry. Protein oligomers which undergo a transition to pathological assemblies (fiber or oligomers) probably have global and local properties that make them amenable to fold plasticity. How the local properties alter the global properties remain to be explored.

5. Conclusion

The novel results obtained by graph theory are that the layout of the interactions, called the network topology, is extremely important for understanding the formation of an interface and the plasticity of fold and quaternary changes. It is also important to understand that the keys are not in any hot spot features but are in the residues whose local properties spread enough global effects to regulate/affect the full-length chain structure. In other words, the formation of interfaces and the quaternary plasticity lay on the residues that control allosteric transitions, mechanisms now revisited using propagation measures in networks. This local to global transition is also investigated by mathematical concepts in the chapter by Laurent Vuillon and Claire Lesieur.

The take home message of the chapter is to exhibit the usefulness of computational approaches to efficiently complement experimental approaches and gain insight in protein assembly. Obviously, future challenges are on understanding how intramolecular and intermolecular interactions are coordinated and the determinants of allosteric transitions. Graph theory and networks approaches open new venues to explore such problems and are certainly going to provide important breakthrough. Briefly, graph theory can help in identifying intramolecular and intermolecular key interactions as well as in investigating their communication means by analyzing the topology of the networks, isolating appropriate clusters and determining propagation route (allostery).

Now, it may be yet too early to grasp what are the network measures most relevant to the problem of protein assembly and how they can be interpreted in terms of protein's needs. For example, proteins and protein interfaces have been described as random networks with Poisson degree distributions centered to a characteristic average $<k>$ degree [178]. This means all nodes have on average the same number of links (or contacts) and there are no hubs. Proteins have also been described as single-scale network with exponential degree distribution and no hubs again [59, 84]. In some case, the random network is attributed to the backbone interactions while the single-scale network is attributed to the side chain interactions. A network made of a minimum number of contacts seems rather coherent as proteins probably minimize the number of links (bonds) per amino acid to reduce the "building" cost in terms of bonds and the sequence stringency.

Simultaneously, proteins are described as small world because they have small average path length <l>. Small <l> generally indicates that most nodes, namely amino acids, of the network are within the reach of each other. Such node accessibility would suggest that a single modification anywhere in the network (ie any mutation in a protein) would easily spread changes in the whole network, a hazardous situation for a protein and in contradiction with the fact that protein folds and functions resist most mutations. Small world networks generally have hubs, highly connected nodes that govern the network communication routes. But proteins are random or single-scale networks and as such are not expected to have hubs, at least not hubs with many more links than other nodes. The absence of hubs is good as it reduces the protein vulnerability to mutation. We have measured <l> from 10 to 19 in protein interfaces for networks made of about 300 nodes (unpublished). For comparison the world wide web has similar <l>=19, but 800 million nodes. Maybe it just happens that some worlds are smaller than other.

It is therefore not so simple to deconvoluate the topology of a network with a small average <l> depending if it is a random, single scale or scale-free (power law degree distribution) network. Theoretical developments aiming at this understanding are proposed and allow considering distribution of connected components, distribution of clustering coefficient and approximation of <l>. Such work will help analyzing the network measures obtained for amino acid networks [179].

One problem of network is the number of interactions and nodes generated to describe a protein network and how to discriminate a hierarchy within these set of interactions to understand the determinant ones. To this goal, one elegant strategy is to experimentally measure kinetics and affinity to prioritize interactions in networks [180]. Such approaches would complement MD simulations and help discriminating the good from the bad.

Acknowledgements

A special thanks to Kave Salamatian and Laurent Vuillon for numerous fruitful discussions on network and graph theories. We thank the federation of research FR2914 MSIF (Modelization, Simulations, Interaction Fundamentals) for supporting our work on interdisciplinary research.

http://dx.doi.org/10.5772/58576

Author details

Claire Lesieur*

Address all correspondence to: claire.lesieur@agim.eu

AGIM-FRE3405 UJF-CNRS, Grenoble, France

References

[1] Goodsell DS, Olson AJ. Structural symmetry and protein function. Annu Rev Biophys Biomol Struct. 2000;29:105-53. PubMed PMID: 10940245.

[2] Grant RA, Filman DJ, Finkel SE, Kolter R, Hogle JM. The crystal structure of Dps, a ferritin homolog that binds and protects DNA. Nat Struct Biol. 1998 Apr;5(4):294-303. PubMed PMID: 9546221.

[3] Lloyd DR. Cubic Icosahedra? A Problem in Assigning Symmetry. Journal of Chemical Education. 2010;87(8):823-6.

[4] Cheung DT, DiCesare P, Benya PD, Libaw E, Nimni ME. The presence of intermolecular disulfide cross-links in type III collagen. J Biol Chem. 1983 Jun 25;258(12):7774-8. PubMed PMID: 6863264.

[5] Ricard-Blum S. The collagen family. Cold Spring Harbor perspectives in biology. 2011 Jan;3(1):a004978. PubMed PMID: 21421911. Pubmed Central PMCID: 3003457.

[6] Gordon MK, Hahn RA. Collagens. Cell and tissue research. 2010 Jan;339(1):247-57. PubMed PMID: 19693541. Pubmed Central PMCID: 2997103.

[7] Lesieur C, Frutiger S, Hughes G, Kellner R, Pattus F, van der Goot FG. Increased stability upon heptamerization of the pore-forming toxin aerolysin. J Biol Chem. 1999 Dec 17;274(51):36722-8. PubMed PMID: 10593978.

[8] Lesieur C, Cliff MJ, Carter R, James RF, Clarke AR, Hirst TR. A kinetic model of intermediate formation during assembly of cholera toxin B-subunit pentamers. J Biol Chem. 2002 May 10;277(19):16697-704. PubMed PMID: 11877421.

[9] McLaughlin SH, Bulleid NJ. Molecular recognition in procollagen chain assembly. Matrix Biol. 1998 Feb;16(7):369-77. PubMed PMID: 9524357.

[10] Reimer U, Scherer G, Drewello M, Kruber S, Schutkowski M, Fischer G. Side-chain effects on peptidyl-prolyl cis/trans isomerisation. J Mol Biol. 1998 Jun 5;279(2):449-60. PubMed PMID: 9642049.

[11] Tacnet P, Cheong EC, Goeltz P, Ghebrehiwet B, Arlaud GJ, Liu XY, et al. Trimeric reassembly of the globular domain of human C1q. Biochim Biophys Acta. 2008 Mar; 1784(3):518-29. PubMed PMID: 18179779.

[12] Tuncbag N, Gursoy A, Nussinov R, Keskin O. Predicting protein-protein interactions on a proteome scale by matching evolutionary and structural similarities at interfaces using PRISM. Nature Protocols. 2011;6(9):1341-54.

[13] Tuncbag N, Kar G, Keskin O, Gursoy A, Nussinov R. A survey of available tools and web servers for analysis of protein–protein interactions and interfaces. Briefings in Bioinformatics. 2009;10(3):217.

[14] Guney E, Tuncbag N, Keskin O, Gursoy A. HotSprint: database of computational hot spots in protein interfaces. Nucleic acids research. 2008;36(suppl 1):D662-D6.

[15] Shoemaker BA, Panchenko AR. Deciphering protein-protein interactions. Part II. Computational methods to predict protein and domain interaction partners. PLoS Comput Biol. 2007 Apr 27;3(4):e43. PubMed PMID: 17465672. Pubmed Central PMCID: 1857810.

[16] Wass MN, David A, Sternberg MJ. Challenges for the prediction of macromolecular interactions. Curr Opin Struct Biol. 2011 Jun;21(3):382-90. PubMed PMID: 21497504.

[17] Juan D, Pazos F, Valencia A. High-confidence prediction of global interactomes based on genome-wide coevolutionary networks. Proc Natl Acad Sci U S A. 2008 Jan 22;105(3):934-9. PubMed PMID: 18199838. Pubmed Central PMCID: 2242690.

[18] Mosca R, Ceol A, Stein A, Olivella R, Aloy P. 3did: a catalog of domain-based interactions of known three-dimensional structure. Nucleic Acids Res. 2014 Jan 1;42(1):D374-9. PubMed PMID: 24081580.

[19] Mosca R, Pons T, Ceol A, Valencia A, Aloy P. Towards a detailed atlas of protein-protein interactions. Curr Opin Struct Biol. 2013 Dec;23(6):929-40. PubMed PMID: 23896349.

[20] Mosca R, Ceol A, Aloy P. Interactome3D: adding structural details to protein networks. Nature methods. 2013 Jan;10(1):47-53. PubMed PMID: 23399932.

[21] Janin J, Bahadur RP, Chakrabarti P. Protein-protein interaction and quaternary structure. Q Rev Biophys. 2008 May;41(2):133-80. PubMed PMID: 18812015.

[22] Valdar WS, Thornton JM. Conservation helps to identify biologically relevant crystal contacts. J Mol Biol. 2001 Oct 19;313(2):399-416. PubMed PMID: 11800565.

[23] Ponstingl H, Henrick K, Thornton JM. Discriminating between homodimeric and monomeric proteins in the crystalline state. Proteins. 2000 Oct 1;41(1):47-57. PubMed PMID: 10944393.

[24] Elcock AH, McCammon JA. Identification of protein oligomerization states by analysis of interface conservation. Proc Natl Acad Sci U S A. 2001 Mar 13;98(6):2990-4. PubMed PMID: 11248019. Pubmed Central PMCID: 30594.

[25] Lai YT, King NP, Yeates TO. Principles for designing ordered protein assemblies. Trends Cell Biol. 2012 Dec;22(12):653-61. PubMed PMID: 22975357.

[26] Woolfson DN, Bartlett GJ, Bruning M, Thomson AR. New currency for old rope: from coiled-coil assemblies to alpha-helical barrels. Curr Opin Struct Biol. 2012 Aug; 22(4):432-41. PubMed PMID: 22445228.

[27] Channon K, Bromley EH, Woolfson DN. Synthetic biology through biomolecular design and engineering. Curr Opin Struct Biol. 2008 Aug;18(4):491-8. PubMed PMID: 18644449.

[28] Ringler P, Schulz GE. Self-assembly of proteins into designed networks. Science. 2003 Oct 3;302(5642):106-9. PubMed PMID: 14526081.

[29] King NP, Lai Y-T. Practical approaches to designing novel protein assemblies. Current opinion in structural biology. 2013.

[30] Gursoy A, Keskin O, Nussinov R. Topological properties of protein interaction networks from a structural perspective. Biochemical Society Transactions. 2008;36:1398-403.

[31] Tuncbag N, Gursoy A, Guney E, Nussinov R, Keskin O. Architectures and functional coverage of protein-protein interfaces. J Mol Biol. 2008 Sep 5;381(3):785-802. PubMed PMID: 18620705. Pubmed Central PMCID: 2605427.

[32] Shulman-Peleg A, Shatsky M, Nussinov R, Wolfson HJ. Spatial chemical conservation of hot spot interactions in protein-protein complexes. BMC Biol. 2007;5:43. PubMed PMID: 17925020.

[33] Ma B, Nussinov R. Trp/Met/Phe hot spots in protein-protein interactions: potential targets in drug design. Current topics in medicinal chemistry. 2007;7(10):999-1005.

[34] Jones S, Thornton JM. Principles of protein-protein interactions. Proc Natl Acad Sci U S A. 1996 Jan 9;93(1):13-20. PubMed PMID: 8552589.

[35] Laskowski RA, Thornton JM. Understanding the molecular machinery of genetics through 3D structures. Nature Reviews Genetics. 2008;9(2):141-51.

[36] Toogood PL. Inhibition of protein-protein association by small molecules: approaches and progress. Journal of medicinal chemistry. 2002 Apr 11;45(8):1543-58. PubMed PMID: 11931608.

[37] Gadek TR, Nicholas JB. Small molecule antagonists of proteins. Biochemical pharmacology. 2003 Jan 1;65(1):1-8. PubMed PMID: 12473372.

[38] Lo Conte L, Chothia C, Janin J. The atomic structure of protein-protein recognition sites. J Mol Biol. 1999 Feb 5;285(5):2177-98. PubMed PMID: 9925793.

[39] Chakrabarti P, Janin J. Dissecting protein-protein recognition sites. Proteins. 2002 May 15;47(3):334-43. PubMed PMID: 11948787.

[40] Ofran Y, Rost B. Analysing six types of proteinâ€"protein interfaces. Journal of molecular biology. 2003;325(2):377-87.

[41] Bashton M, Chothia C. The geometry of domain combination in proteins. J Mol Biol. 2002 Jan 25;315(4):927-39. PubMed PMID: 11812158.

[42] Xu D, Tsai CJ, Nussinov R. Hydrogen bonds and salt bridges across protein-protein interfaces. Protein Eng. 1997 Sep;10(9):999-1012. PubMed PMID: 9464564.

[43] Selbig J, Argos P. Relationships between protein sequence and structure patterns based on residue contacts. Proteins. 1998 May 1;31(2):172-85. PubMed PMID: 9593191.

[44] Shoemaker BA, Panchenko AR. Deciphering protein-protein interactions. Part I. Experimental techniques and databases. PLoS Comput Biol. 2007 Mar 30;3(3):e42. PubMed PMID: 17397251. Pubmed Central PMCID: 1847991.

[45] de Vries SJ, Bonvin AM. How proteins get in touch: interface prediction in the study of biomolecular complexes. Curr Protein Pept Sci. 2008 Aug;9(4):394-406. PubMed PMID: 18691126.

[46] Guex N, Peitsch MC. SWISS-MODEL and the Swiss-PdbViewer: an environment for comparative protein modeling. Electrophoresis. 1997 Dec;18(15):2714-23. PubMed PMID: 9504803.

[47] Shulman-Peleg A, Shatsky M, Nussinov R, Wolfson HJ. MultiBind and MAPPIS: webservers for multiple alignment of protein 3D-binding sites and their interactions. Nucleic Acids Res. 2008 Jul 1;36(Web Server issue):W260-4. PubMed PMID: 18467424.

[48] Orchard S, Ammari M, Aranda B, Breuza L, Briganti L, Broackes-Carter F, et al. The MIntAct project--IntAct as a common curation platform for 11 molecular interaction databases. Nucleic Acids Res. 2014 Jan 1;42(1):D358-63. PubMed PMID: 24234451.

[49] Chothia C, Janin J. Principles of protein-protein recognition. Nature. 1975 Aug 28;256(5520):705-8. PubMed PMID: 1153006.

[50] Henrick K, Thornton JM. PQS: a protein quaternary structure file server. Trends Biochem Sci. 1998 Sep;23(9):358-61. PubMed PMID: 9787643.

[51] Janin J, Rodier F. Protein-protein interaction at crystal contacts. Proteins. 1995 Dec; 23(4):580-7. PubMed PMID: 8749854.

[52] Lee B, Richards FM. The interpretation of protein structures: estimation of static accessibility. J Mol Biol. 1971 Feb 14;55(3):379-400. PubMed PMID: 5551392.

[53] Poupon A. Voronoi and Voronoi-related tessellations in studies of protein structure and interaction. Curr Opin Struct Biol. 2004 Apr;14(2):233-41. PubMed PMID: 15093839.

[54] Cazals F, Proust F, Bahadur RP, Janin J. Revisiting the Voronoi description of protein-protein interfaces. Protein Sci. 2006 Sep;15(9):2082-92. PubMed PMID: 16943442.

[55] Bouvier B, Grunberg R, Nilges M, Cazals F. Shelling the Voronoi interface of protein-protein complexes reveals patterns of residue conservation, dynamics, and composition. Proteins. 2009 Aug 15;76(3):677-92. PubMed PMID: 19280599.

[56] Dreyfus T, Doye V, Cazals F. Probing a continuum of macro-molecular assembly models with graph templates of complexes. Proteins. 2013 Nov;81(11):2034-44. PubMed PMID: 23609891.

[57] Faure G, Bornot A, de Brevern AG. Protein contacts, inter-residue interactions and side-chain modelling. Biochimie. 2008;90:626-39.

[58] Ofran Y, Rost B. Protein–protein interaction hotspots carved into sequences. PLoS computational biology. 2007;3(7):e119.

[59] Feverati G, Achoch M, Vuillon L, Lesieur C. Intermolecular β-Strand Networks Avoid Hub Residues and Favor Low Interconnectedness: A Potential Protection Mechanism against Chain Dissociation upon Mutation. PloS one. 2014;9(4):e94745.

[60] Feverati G, Lesieur C. Oligomeric interfaces under the lens: gemini. PloS one. 2010;5(3):e9897.

[61] Caffrey DR, Somaroo S, Hughes JD, Mintseris J, Huang ES. Are protein-protein interfaces more conserved in sequence than the rest of the protein surface? Protein Sci. 2004 Jan;13(1):190-202. PubMed PMID: 14691234.

[62] Lichtarge O, Bourne HR, Cohen FE. An evolutionary trace method defines binding surfaces common to protein families. J Mol Biol. 1996 Mar 29;257(2):342-58. PubMed PMID: 8609628.

[63] Armon A, Graur D, Ben-Tal N. ConSurf: an algorithmic tool for the identification of functional regions in proteins by surface mapping of phylogenetic information. J Mol Biol. 2001 Mar 16;307(1):447-63. PubMed PMID: 11243830.

[64] Pupko T, Bell RE, Mayrose I, Glaser F, Ben-Tal N. Rate4Site: an algorithmic tool for the identification of functional regions in proteins by surface mapping of evolutionary determinants within their homologues. Bioinformatics. 2002;18 Suppl 1:S71-7. PubMed PMID: 12169533.

[65] Landgraf R, Xenarios I, Eisenberg D. Three-dimensional cluster analysis identifies interfaces and functional residue clusters in proteins. J Mol Biol. 2001 Apr 13;307(5): 1487-502. PubMed PMID: 11292355.

[66] Clackson T, Wells JA. A hot spot of binding energy in a hormone-receptor interface. Science. 1995;267(5196):383-6.

[67] Bogan AA, Thorn KS. Anatomy of hot spots in protein interfaces. J Mol Biol. 1998 Jul 3;280(1):1-9. PubMed PMID: 9653027.

[68] Fischer TB, Arunachalam KV, Bailey D, Mangual V, Bakhru S, Russo R, et al. The binding interface database (BID): a compilation of amino acid hot spots in protein interfaces. Bioinformatics. 2003 Jul 22;19(11):1453-4. PubMed PMID: 12874065.

[69] Guidry JJ, Shewmaker F, Maskos K, Landry S, Wittung-Stafshede P. Probing the interface in a human co-chaperonin heptamer: residues disrupting oligomeric unfolded state identified. BMC Biochem. 2003 Oct 2;4:14. PubMed PMID: 14525625.

[70] Luke K, Perham M, Wittung-Stafshede P. Kinetic folding and assembly mechanisms differ for two homologous heptamers. J Mol Biol. 2006 Oct 27;363(3):729-42. PubMed PMID: 16979655.

[71] Barabasi AL, Oltvai ZN. Network biology: understanding the cell's functional organization. Nature reviews Genetics. 2004 Feb;5(2):101-13. PubMed PMID: 14735121.

[72] Bhattacharyya M, Vishveshwara S. Probing the allosteric mechanism in pyrrolysyl-tRNA synthetase using energy-weighted network formalism. Biochemistry. 2011 Jul 19;50(28):6225-36. PubMed PMID: 21650159.

[73] De Ruvo M, Giuliani A, Paci P, Santoni D, Di Paola L. Shedding light on protein-ligand binding by graph theory: the topological nature of allostery. Biophys Chem. 2012 May;165-166:21-9. PubMed PMID: 22464849.

[74] Daily MD, Gray JJ. Allosteric communication occurs via networks of tertiary and quaternary motions in proteins. PLoS Comput Biol. 2009 Feb;5(2):e1000293. PubMed PMID: 19229311. Pubmed Central PMCID: 2634971.

[75] Tsai CJ, Del Sol A, Nussinov R. Protein allostery, signal transmission and dynamics: a classification scheme of allosteric mechanisms. Molecular bioSystems. 2009 Mar; 5(3):207-16. PubMed PMID: 19225609. Pubmed Central PMCID: 2898650.

[76] Gunasekaran K, Ma B, Nussinov R. Is allostery an intrinsic property of all dynamic proteins? Proteins: Structure, Function, and Bioinformatics. 2004;57(3):433-43.

[77] del Sol A, Tsai CJ, Ma B, Nussinov R. The origin of allosteric functional modulation: multiple pre-existing pathways. Structure. 2009 Aug 12;17(8):1042-50. PubMed PMID: 19679084. Pubmed Central PMCID: 2749652.

[78] Amitai G, Shemesh A, Sitbon E, Shklar M, Netanely D, Venger I, et al. Network analysis of protein structures identifies functional residues. J Mol Biol. 2004 Dec 3;344(4):1135-46. PubMed PMID: 15544817.

[79] Vendruscolo M, Dokholyan NV, Paci E, Karplus M. Small-world view of the amino acids that play a key role in protein folding. Phys Rev E Stat Nonlin Soft Matter Phys. 2002 Jun;65(6 Pt 1):061910. PubMed PMID: 12188762.

[80] Dokholyan NV, Li L, Ding F, Shakhnovich EI. Topological determinants of protein folding. Proc Natl Acad Sci U S A. 2002 Jun 25;99(13):8637-41. PubMed PMID: 12084924. Pubmed Central PMCID: 124342.

[81] del Sol A, O'Meara P. Small-world network approach to identify key residues in protein-protein interaction. Proteins. 2005 Feb 15;58(3):672-82. PubMed PMID: 15617065.

[82] Brinda KV, Kannan N, Vishveshwara S. Analysis of homodimeric protein interfaces by graph-spectral methods. Protein Eng. 2002 Apr;15(4):265-77. PubMed PMID: 11983927.

[83] Albert R, Jeong H, Barabasi AL. Error and attack tolerance of complex networks. Nature. 2000 Jul 27;406(6794):378-82. PubMed PMID: 10935628.

[84] Greene LH, Higman VA. Uncovering network systems within protein structures. J Mol Biol. 2003 Dec 5;334(4):781-91. PubMed PMID: 14636602.

[85] Brinda KV, Vishveshwara S. Oligomeric protein structure networks: insights into protein-protein interactions. BMC Bioinformatics. 2005;6:296. PubMed PMID: 16336694. Pubmed Central PMCID: 1326230.

[86] Dey S, Pal A, Chakrabarti P, Janin J. The subunit interfaces of weakly associated homodimeric proteins. J Mol Biol. 2010 Apr 23;398(1):146-60. PubMed PMID: 20156457.

[87] Talavera D, Robertson DL, Lovell SC. Characterization of protein-protein interaction interfaces from a single species. PloS one. 2011;6(6):e21053.

[88] Kim WK, Henschel A, Winter C, Schroeder M. The many faces of protein-protein interactions: A compendium of interface geometry. PLoS Comput Biol. 2006 Sep 29;2(9):e124. PubMed PMID: 17009862.

[89] Winter C, Henschel A, Kim WK, Schroeder M. SCOPPI: a structural classification of protein-protein interfaces. Nucleic Acids Res. 2006 Jan 1;34(Database issue):D310-4. PubMed PMID: 16381874.

[90] Keskin O, Tsai CJ, Wolfson H, Nussinov R. A new, structurally nonredundant, diverse data set of proteinâ€"protein interfaces and its implications. Protein Science. 2004;13(4):1043-55.

[91] Neuvirth H, Heinemann U, Birnbaum D, Tishby N, Schreiber G. ProMateus--an open research approach to protein-binding sites analysis. Nucleic Acids Res. 2007 Jul;

35(Web Server issue):W543-8. PubMed PMID: 17488838. Pubmed Central PMCID: 1933218.

[92] Yan C, Wu F, Jernigan RL, Dobbs D, Honavar V. Characterization of protein-protein interfaces. Protein J. 2008 Jan;27(1):59-70. PubMed PMID: 17851740.

[93] Yan C, Dobbs D, Honavar V. A two-stage classifier for identification of protein-protein interface residues. Bioinformatics. 2004 Aug 4;20 Suppl 1:i371-8. PubMed PMID: 15262822.

[94] Cheng P-N, Pham JD, Nowick JS. The Supramolecular Chemistry of β-Sheets. Journal of the American Chemical Society. 2013.

[95] Khakshoor O, Nowick JS. Artificial beta-sheets: chemical models of beta-sheets. Curr Opin Chem Biol. 2008 Dec;12(6):722-9. PubMed PMID: 18775794.

[96] López De La Paz M, Serrano L. Sequence determinants of amyloid fibril formation. Proceedings of the National Academy of Sciences of the United States of America. 2004;101(1):87.

[97] Lopez De La Paz M, Goldie K, Zurdo J, Lacroix E, Dobson CM, Hoenger A, et al. De novo designed peptide-based amyloid fibrils. Proc Natl Acad Sci U S A. 2002 Dec 10;99(25):16052-7. PubMed PMID: 12456886.

[98] Trovato A, Chiti F, Maritan A, Seno F. Insight into the structure of amyloid fibrils from the analysis of globular proteins. PLoS computational biology. 2006;2(12):e170.

[99] Kayed R, Head E, Thompson JL, McIntire TM, Milton SC, Cotman CW, et al. Common structure of soluble amyloid oligomers implies common mechanism of pathogenesis. Science. 2003;300(5618):486-9.

[100] Guijarro JI, Sunde M, Jones JA, Campbell ID, Dobson CM. Amyloid fibril formation by an SH3 domain. Proc Natl Acad Sci U S A. 1998 Apr 14;95(8):4224-8. PubMed PMID: 9539718. Pubmed Central PMCID: 22470.

[101] Dobson CM. Protein misfolding, evolution and disease. Trends Biochem Sci. 1999;24:329-32.

[102] Petkova AT, Ishii Y, Balbach JJ, Antzutkin ON, Leapman RD, Delaglio F, et al. A structural model for Alzheimer's beta-amyloid fibrils based on experimental constraints from solid state NMR. Proc Natl Acad Sci U S A. 2002 Dec 24;99(26):16742-7. PubMed PMID: 12481027. Pubmed Central PMCID: 139214.

[103] Der-Sarkissian A, Jao CC, Chen J, Langen R. Structural organization of alpha-synuclein fibrils studied by site-directed spin labeling. J Biol Chem. 2003 Sep 26;278(39): 37530-5. PubMed PMID: 12815044.

[104] Kajava AV, Aebi U, Steven AC. The parallel superpleated beta-structure as a model for amyloid fibrils of human amylin. J Mol Biol. 2005 Apr 29;348(2):247-52. PubMed PMID: 15811365.

[105] Krishnan R, Lindquist SL. Structural insights into a yeast prion illuminate nucleation and strain diversity. Nature. 2005 Jun 9;435(7043):765-72. PubMed PMID: 15944694. Pubmed Central PMCID: 1405905.

[106] Margittai M, Langen R. Template-assisted filament growth by parallel stacking of tau. Proc Natl Acad Sci U S A. 2004 Jul 13;101(28):10278-83. PubMed PMID: 15240881. Pubmed Central PMCID: 478563.

[107] Lv G, Kumar A, Giller K, Orcellet ML, Riedel D, Fernandez CO, et al. Structural comparison of mouse and human alpha-synuclein amyloid fibrils by solid-state NMR. J Mol Biol. 2012 Jun 29;420(1-2):99-111. PubMed PMID: 22516611.

[108] Fernandez-Escamilla A-M, Rousseau F, Schymkowitz J, Serrano L. Prediction of sequence-dependent and mutational effects on the aggregation of peptides and proteins. Nature Biotechnology. 2004;22(10):1302-6.

[109] Galzitskaya OV, Garbuzynskiy SO, Lobanov MY. Prediction of amyloidogenic and disordered regions in protein chains. PLoS computational biology. 2006;2(12):e177.

[110] Thompson MJ, Sievers SA, Karanicolas J, Ivanova MI, Baker D, Eisenberg D. The 3D profile method for identifying fibril-forming segments of proteins. Proceedings of the National Academy of Sciences of the United States of America. 2006;103(11):4074-8.

[111] Belli M, Ramazzotti M, Chiti F. Prediction of amyloid aggregation in vivo. EMBO Rep. 2011 Jul;12(7):657-63. PubMed PMID: 21681200. Pubmed Central PMCID: 3128957.

[112] Smith JM, Jang Y, Kim MK. Steiner minimal trees, twist angles, and the protein folding problem. Proteins: Structure, Function, and Bioinformatics. 2007;66(4):889-902.

[113] Levin KB, Dym O, Albeck S, Magdassi S, Keeble AH, Kleanthous C, et al. Following evolutionary paths to protein-protein interactions with high affinity and selectivity. Nature structural & molecular biology. 2009;16(10):1049-55.

[114] Caflisch A. Network and graph analyses of folding free energy surfaces. Curr Opin Struct Biol. 2006 Feb;16(1):71-8. PubMed PMID: 16413772.

[115] Bahar I, Chennubhotla C, Tobi D. Intrinsic dynamics of enzymes in the unbound state and relation to allosteric regulation. Curr Opin Struct Biol. 2007 Dec;17(6): 633-40. PubMed PMID: 18024008. Pubmed Central PMCID: 2197162.

[116] Chennubhotla C, Bahar I. Signal propagation in proteins and relation to equilibrium fluctuations. PLoS Comput Biol. 2007 Sep;3(9):1716-26. PubMed PMID: 17892319. Pubmed Central PMCID: 1988854.

[117] Jin Y, Turaev D, Weinmaier T, Rattei T, Makse HA. The evolutionary dynamics of protein-protein interaction networks inferred from the reconstruction of ancient networks. PLoS One. 2013;8(3):e58134. PubMed PMID: 23526967. Pubmed Central PMCID: 3603955.

[118] Liu YY, Slotine JJ, Barabasi AL. Controllability of complex networks. Nature. 2011;473:167-73.

[119] Albert R, Barabasi AL. Topology of evolving networks: local events and universality. Physical review letters. 2000 Dec 11;85(24):5234-7. PubMed PMID: 11102229.

[120] Callaway DS, Newman ME, Strogatz SH, Watts DJ. Network robustness and fragility: percolation on random graphs. Physical review letters. 2000 Dec 18;85(25):5468-71. PubMed PMID: 11136023.

[121] Lupas A, Van Dyke M, Stock J. Predicting coiled coils from protein sequences. Science. 1991 May 24;252(5010):1162-4. PubMed PMID: 2031185.

[122] Gruber M, Soding J, Lupas AN. Comparative analysis of coiled-coil prediction methods. J Struct Biol. 2006 Aug;155(2):140-5. PubMed PMID: 16870472.

[123] Bartoli L, Fariselli P, Krogh A, Casadio R. CCHMM_PROF: a HMM-based coiled-coil predictor with evolutionary information. Bioinformatics. 2009 Nov 1;25(21):2757-63. PubMed PMID: 19744995.

[124] Wolf E, Kim PS, Berger B. MultiCoil: a program for predicting two-and three-stranded coiled coils. Protein Sci. 1997 Jun;6(6):1179-89. PubMed PMID: 9194178. Pubmed Central PMCID: 2143730.

[125] Crick FHC. The packing of alpha-helices: simple coiled-coils. Acta Crystallogr. 1953;6:689–97.

[126] Poupon A, Janin J. Analysis and prediction of protein quaternary structure. Methods Mol Biol. 2010;609:349-64. PubMed PMID: 20221929.

[127] Comeau SR, Camacho CJ. Predicting oligomeric assemblies: N-mers a primer. J Struct Biol. 2005 Jun;150(3):233-44. PubMed PMID: 15890272.

[128] Walshaw J, Woolfson DN. Extended knobs-into-holes packing in classical and complex coiled-coil assemblies. J Struct Biol. 2003 Dec;144(3):349-61. PubMed PMID: 14643203.

[129] Calladine CR, Luisi BF, Pratap JV. A "mechanistic" explanation of the multiple helical forms adopted by bacterial flagellar filaments. J Mol Biol. 2013 Mar 11;425(5):914-28. PubMed PMID: 23274110. Pubmed Central PMCID: 3605589.

[130] Calladine CR, Sharff A, Luisi B. How to untwist an alpha-helix: structural principles of an alpha-helical barrel. J Mol Biol. 2001 Jan 19;305(3):603-18. PubMed PMID: 11152616.

[131] Moutevelis E, Woolfson DN. A periodic table of coiled-coil protein structures. J Mol Biol. 2009 Jan 23;385(3):726-32. PubMed PMID: 19059267.

[132] Testa OD, Moutevelis E, Woolfson DN. CC+: a relational database of coiled-coil structures. Nucleic Acids Res. 2009 Jan;37(Database issue):D315-22. PubMed PMID: 18842638.

[133] Papanikolopoulou K, Forge V, Goeltz P, Mitraki A. Formation of highly stable chimeric trimers by fusion of an adenovirus fiber shaft fragment with the foldon domain of bacteriophage t4 fibritin. J Biol Chem. 2004b Mar 5;279(10):8991-8. PubMed PMID: 14699113.

[134] Zrimi J, Ng Ling A, Giri-Rachman Arifin E, Feverati G, Lesieur C. Cholera toxin B subunits assemble into pentamers-proposition of a fly-casting mechanism. PLoS One. 2010;5(12):e15347. PubMed PMID: 21203571.

[135] Ruddock LW, Coen JJ, Cheesman C, Freedman RB, Hirst TR. Assembly of the B subunit pentamer of Escherichia coli heat-labile enterotoxin. Kinetics and molecular basis of rate-limiting steps in vitro. J Biol Chem. 1996b Aug 9;271(32):19118-23. PubMed PMID: 8702586.

[136] Dang LT, Purvis AR, Huang RH, Westfield LA, Sadler JE. Phylogenetic and functional analysis of histidine residues essential for pH-dependent multimerization of von Willebrand factor. Journal of Biological Chemistry. 2011.

[137] Hashimoto K, Nishi H, Bryant S, Panchenko AR. Caught in self-interaction: evolutionary and functional mechanisms of protein homooligomerization. Phys Biol. 2011 Jun;8(3):035007. PubMed PMID: 21572178. Pubmed Central PMCID: 3148176.

[138] Csermely P, Palotai R, Nussinov R. Induced fit, conformational selection and independent dynamic segments: an extended view of binding events. Trends Biochem Sci. 2010 Oct;35(10):539-46. PubMed PMID: 20541943. Pubmed Central PMCID: 3018770.

[139] D'Alessio G. The evolutionary transition from monomeric to oligomeric proteins: tools, the environment, hypotheses. Prog Biophys Mol Biol. 1999;72(3):271-98. PubMed PMID: 10581971.

[140] D'Alessio G. Oligomer evolution in action? Nat Struct Biol. 1995 Jan;2(1):11-3. PubMed PMID: 7719846.

[141] Eisenberg D, Jucker M. The Amyloid State of Proteins in Human Diseases. Cell. 2012;148(6):1188-203.

[142] Levy ED, Erba EB, Robinson CV, Teichmann SA. Assembly reflects evolution of protein complexes. Nature. 2008;453(7199):1262-5.

[143] Luo M, Singh RK, Tanner JJ. Structural determinants of oligomerization of delta(1)-pyrroline-5-carboxylate dehydrogenase: identification of a hexamerization hot spot. J Mol Biol. 2013 Sep 9;425(17):3106-20. PubMed PMID: 23747974. Pubmed Central PMCID: 3743950.

[144] Crick FH, Watson JD. Structure of small viruses. Nature. 1956 Mar 10;177(4506): 473-5. PubMed PMID: 13309339.

[145] Rackham OJ, Madera M, Armstrong CT, Vincent TL, Woolfson DN, Gough J. The evolution and structure prediction of coiled coils across all genomes. J Mol Biol. 2010 Oct 29;403(3):480-93. PubMed PMID: 20813113.

[146] Yadid I, Kirshenbaum N, Sharon M, Dym O, Tawfik DS. Metamorphic proteins mediate evolutionary transitions of structure. Proc Natl Acad Sci U S A. 2010 Apr 20;107(16):7287-92. PubMed PMID: 20368465. Pubmed Central PMCID: 2867682.

[147] King J, Wood WB. Assembly of bacteriophage T4 tail fibers: the sequence of gene product interaction. J Mol Biol. 1969 Feb 14;39(3):583-601. PubMed PMID: 5390559.

[148] King J. Assembly of the tail of bacteriophage T4. J Mol Biol. 1968 Mar 14;32(2):231-62. PubMed PMID: 4868421.

[149] Rennell D, Bouvier SE, Hardy LW, Poteete AR. Systematic mutation of bacteriophage T4 lysozyme. J Mol Biol. 1991 Nov 5;222(1):67-88. PubMed PMID: 1942069.

[150] Goldenberg DP, Berget PB, King J. Maturation of the tail spike endorhamnosidase of Salmonella phage P22. J Biol Chem. 1982 Jul 10;257(13):7864-71. PubMed PMID: 7045114.

[151] Perham M, Chen M, Ma J, Wittung-Stafshede P. Unfolding of heptameric co-chaperonin protein follows "fly casting" mechanism: observation of transient nonnative heptamer. J Am Chem Soc. 2005 Nov 30;127(47):16402-3. PubMed PMID: 16305220.

[152] Bascos N, Guidry J, Wittung-Stafshede P. Monomer topology defines folding speed of heptamer. Protein Sci. 2004 May;13(5):1317-21. PubMed PMID: 15075408.

[153] Tacnet P, Thielens N, Arifin Giri Rachman E, Hirst TR, Lesieur C. Cholera toxin B assembly intermediates provide some explanation to the existence of a pentameric toxin state.

[154] Pell LG, Cumby N, Clark TE, Tuite A, Battaile KP, Edwards AM, et al. A conserved spiral structure for highly diverged phage tail assembly chaperones. J Mol Biol. 2013 Jul 24;425(14):2436-49. PubMed PMID: 23542344.

[155] Aghera N, Udgaonkar JB. Kinetic studies of the folding of heterodimeric monellin: evidence for switching between alternative parallel pathways. J Mol Biol. 2012 Jul 13;420(3):235-50. PubMed PMID: 22542529.

[156] Kentsis A, Gordon RE, Borden KL. Control of biochemical reactions through supramolecular RING domain self-assembly. Proc Natl Acad Sci U S A. 2002 Nov 26;99(24):15404-9. PubMed PMID: 12438698. Pubmed Central PMCID: 137729.

[157] Kentsis A, Gordon RE, Borden KL. Self-assembly properties of a model RING domain. Proc Natl Acad Sci U S A. 2002 Jan 22;99(2):667-72. PubMed PMID: 11792829. Pubmed Central PMCID: 117363.

[158] Kentsis A, Borden KL. Construction of macromolecular assemblages in eukaryotic processes and their role in human disease: linking RINGs together. Curr Protein Pept Sci. 2000 Jul;1(1):49-73. PubMed PMID: 12369920.

[159] Pereira-Leal JB, Levy ED, Teichmann SA. The origins and evolution of functional modules: lessons from protein complexes. Philos Trans R Soc Lond B Biol Sci. 2006 Mar 29;361(1467):507-17. PubMed PMID: 16524839. Pubmed Central PMCID: 1609335.

[160] Shoemaker BA, Portman JJ, Wolynes PG. Speeding molecular recognition by using the folding funnel: the fly-casting mechanism. Proc Natl Acad Sci U S A. 2000 Aug 1;97(16):8868-73. PubMed PMID: 10908673.

[161] Levy Y, Wolynes PG, Onuchic JN. Protein topology determines binding mechanism. Proc Natl Acad Sci U S A. 2004 Jan 13;101(2):511-6. PubMed PMID: 14694192.

[162] Spaar A, Dammer C, Gabdoulline RR, Wade RC, Helms V. Diffusional encounter of barnase and barstar. Biophys J. 2006 Mar 15;90(6):1913-24. PubMed PMID: 16361332.

[163] Ehrlich LP, Nilges M, Wade RC. The impact of protein flexibility on protein-protein docking. Proteins. 2005 Jan 1;58(1):126-33. PubMed PMID: 15515181.

[164] Gabdoulline RR, Wade RC. Protein-protein association: investigation of factors influencing association rates by brownian dynamics simulations. J Mol Biol. 2001 Mar 9;306(5):1139-55. PubMed PMID: 11237623.

[165] Gabdoulline RR, Wade RC. Simulation of the diffusional association of barnase and barstar. Biophys J. 1997 May;72(5):1917-29. PubMed PMID: 9129797.

[166] Noe F, Fischer S. Transition networks for modeling the kinetics of conformational change in macromolecules. Curr Opin Struct Biol. 2008 Apr;18(2):154-62. PubMed PMID: 18378442.

[167] Wales DJ. Energy landscapes: some new horizons. Curr Opin Struct Biol. 2010 Feb; 20(1):3-10. PubMed PMID: 20096562.

[168] Barz B, Wales DJ, Strodel B. A kinetic approach to the sequence-aggregation relationship in disease-related protein assembly. J Phys Chem B. 2014 Jan 30;118(4):1003-11. PubMed PMID: 24401100. Pubmed Central PMCID: 3908877.

[169] Higashimoto Y, Asanomi Y, Takakusagi S, Lewis MS, Uosaki K, Durell SR, et al. Unfolding, aggregation, and amyloid formation by the tetramerization domain from mutant p53 associated with lung cancer. Biochemistry. 2006;45(6):1608-19.

[170] Fujita T, Kiyama M, Tomizawa Y, Kohno T, Yokota J. Comprehensive analysis of p53 gene mutation characteristics in lung carcinoma with special reference to histological subtypes. International journal of oncology. 1999;15(5):927-34.

[171] Mateu MG, Fersht AR. Nine hydrophobic side chains are key determinants of the thermodynamic stability and oligomerization status of tumour suppressor p53 tetramerization domain. The EMBO Journal. 1998;17(10):2748-58.

[172] Reixach N, Foss TR, Santelli E, Pascual J, Kelly JW, Buxbaum JN. Human-murine transthyretin heterotetramers are kinetically stable and non-amyloidogenic. A lesson in the generation of transgenic models of diseases involving oligomeric proteins. J Biol Chem. 2008 Jan 25;283(4):2098-107. PubMed PMID: 18006495.

[173] Jiang X, Buxbaum JN, Kelly JW. The V122I cardiomyopathy variant of transthyretin increases the velocity of rate-limiting tetramer dissociation, resulting in accelerated amyloidosis. Proc Natl Acad Sci U S A. 2001 Dec 18;98(26):14943-8. PubMed PMID: 11752443. Pubmed Central PMCID: 64963.

[174] Lomas DA, Carrell RW. Serpinopathies and the conformational dementias. Nature Reviews Genetics. 2002;3(10):759-68.

[175] Chiti F, Stefani M, Taddei N, Ramponi G, Dobson CM. Rationalization of the effects of mutations on peptide and protein aggregation rates. NATURE-LONDON-. 2003:805-8.

[176] Changeux JP, Edelstein SJ. Allosteric mechanisms of signal transduction. Science. 2005 Jun 3;308(5727):1424-8. PubMed PMID: 15933191.

[177] Pires DE, Ascher DB, Blundell TL. mCSM: predicting the effects of mutations in proteins using graph-based signatures. Bioinformatics. 2014 Feb 1;30(3):335-42. PubMed PMID: 24281696. Pubmed Central PMCID: 3904523.

[178] Bode C, Kovacs IA, Szalay MS, Palotai R, Korcsmaros T, Csermely P. Network analysis of protein dynamics. FEBS Lett. 2007 Jun 19;581(15):2776-82. PubMed PMID: 17531981.

[179] Newman ME, Strogatz SH, Watts DJ. Random graphs with arbitrary degree distributions and their applications. Phys Rev E Stat Nonlin Soft Matter Phys. 2001 Aug;64(2 Pt 2):026118. PubMed PMID: 11497662.

[180] Peysselon F, Ricard-Blum S. Heparin-protein interactions: From affinity and kinetics to biological roles. Application to an interaction network regulating angiogenesis. Matrix Biol. 2013 Nov 16. PubMed PMID: 24246365.

Chapter 12

Geometry and Topology in Protein Interfaces -- Some Tools for Investigations

Giovanni Feverati

Additional information is available at the end of the chapter

http://dx.doi.org/10.5772/58420

1. Introduction

The present work is motivated by the biological problem of understanding and possibly predicting the assembly of biological molecules, in particular proteins. This is one of the most common processes in living cells thus it is essential to understand its key aspects, especially in relation to the implication in several pathologies, from bacterial infections (cholera, anthrax, ...) to protein misfolding diseases (Alzheimer, Parkinson, ...) [1–4]. The stable association of different subunits requires the formation of specific intermolecular bonds, thus constituting what is called an interface. Unfortunately, in spite of extensive analyses, the identification of the patterns, in the polypeptidic chain, responsible for the establishment of an interface remains difficult.

Geometry has developed the ability to measure and characterize complex shapes but it is not a priori obvious that it may also reveal important aspects of the interactions. To understand this point, let consider the following examples.

The main geometrical elements of the modern Golden Gate Bridge in S. Francisco and the ancient Roman aqueduct bridge Pont du Gard, in the Gard department in France, are arcs. But with a main difference: the three arcs that form the suspension system of the Golden Gate Bridge are concave upward while the many arcs that form the Pont du Gard are concave downward. Indeed, in the first case the arcs resist to longitudinal tension while in the second case resist to longitudinal compression. Stones are unsuited to resist to strong tension, while they perfectly resist to huge compression. Thus, architectural elements that have to undergo strong tensions are made of wood or steel, but not of stone. Notice that the simple observation of the geometrical form, the concave upward or downward aspect of the bridges, has lead us to understand the basic interactions and formulate constraints on the possible choices of materials. This argument can be pushed much forward: the structural analysis in architecture and engineering largely rely on euclidean geometry (diagrams of forces are diagrams of vectors). Even if the elastic properties of construction materials and the action of gravity are important, the main ingredient in studying the equilibrium of forces is geometry.

A second example is taken from Einstein's general relativity, where the notion of gravitational interaction and the space-time geometry are fully identified by the equivalence principle and the Einstein's field equations. The equivalence principle was first formulated by A. Einstein in 1907, when he recognized that the local behaviour of falling bodies is equivalent to the effect of being in an accelerated reference system (this holds for local effects only). The Einstein's field equations (1915) provide with a mathematical formulation of the principle. In summary, spatial and temporal distances fully inform on the gravitational interaction.

This connection between geometry and interactions also works at the atomic scale. The perfectly planar and hexagonal symmetric form of benzene molecules, compared to the non planar and less symmetric cyclohexane molecule, is clear indication of the different nature of the corresponding Carbon-Carbon bonds and of the sp2 or sp3 hybridization respectively.

In the early 50's, F. Crick [5] observed that the formation of the coiled-coil protein interface is due to the appropriate geometrical and chemical complementarity of the two interacting domains, as in a lock and key mechanism. The key has a particular geometrical form combined to some contact points which together provide it the capacity to associate to one lock.

Moving on from all these examples, in our group we have developed tools to investigate the geometry of the interfaces in multi-chain proteins. In essence, we measure shapes and compare measures between different proteins. Our input data are protein atomic positions, as those from the Protein Data Bank (PDB) repository of protein structural data.

In particular, we compare protein interfaces of similar geometrical form. From the previous examples, there is no surprise in claiming that the geometry will provide information on the interactions at the interfaces. In our previous publications [6–8], careful statistical analyses have been performed and have led to formulate constraints on the amino acid sequences and atom pairs that are compatible with a given geometrical form. The long term perspective of our work is to rationalize the interface, namely to establish a clear understanding of its sequence-structure relationship, in order to develop interface prediction tools and help to advance in interface design.

1.1. Basic information on interfaces

The shape and the function of proteins are normally encoded within their sequences, i.e. in their amino acid compositions but it is not yet possible, by simply reading the primary sequence of a protein, to predict its three-dimensional structure or the quaternary organization, in the case of an oligomeric protein. One of the difficulties is the non linear encoding of the information in the sequence, namely the fact that the three-dimensional structure is often generated and stabilized by bonds between residues that are not contiguous or even are very far apart, along the chain. Another difficulty is due to the degeneracy between sequences and structures, consisting in the observation that several sequences can code for the same shape, that indicates a versatile role of the amino acids. The secondary structures of proteins which are mainly composed of α helices, β structures and loops are partially understood, and several prediction programs are now available. Prediction of 3D structures is mainly based on homology, namely comparison of sequences that have similar three-dimensional elements. A rich collection of prediction tools is available on [9]. In some

http://dx.doi.org/10.5772/58420

cases, the prediction also takes into account the known geometrical constraints present in amino acids.

Oligomeric proteins associate by forming an interface. Various descriptions of the interface have been proposed.

The simplest definition of interface between two adjacent polypeptidic chains A and B is provided by selecting the set of pairs of atoms, one from each chain, whose distance is lower than a given cut-off, typically fixed near 0.5 nm (cut-off interface). This definition does not provide any measure to distinguish pairs. As such, it provides little information since firstly physical interactions decrease when distance grows, secondly two interactions of equal strength may not play the same role if they are in different parts of the molecules, inserted in different local atomic environments. In another definition, the interface is identified to the surface buried between the two components A and B, namely to those atoms that belong to the surface of A and B and that loose solvent accessibility once the complex AB is formed [10] This makes use of the Van der Waals atomic radii, and leads to distinguish a rim (exposed to the solvent) from a core (inaccessible to the solvent). The interface can be defined also by constructing the Voronoi α-complex [10], namely the set of Voronoi restricted balls. The construction follows a precise mathematical procedure, and determines the volume in which an atom interacts more than its neighbours.

Both the buried surface and Voronoi restricted balls methods focus on the volume of the atoms and the importance of the specific chemical properties of each atom. They make use of a cut off and describe the interactions by using the Van der Waals radius. Differently from these descriptions of the interface, we felt the need to develop a stronger analysis of the structural organization of a protein interface, in order to evaluate the specific role of each residue and the rules of pairings.

In [11], we shown that many aspects of the structural organization of a protein interface can be effectively described by a graph, namely the ensemble of nodes and edges, constructed following the precise geometrical analysis of the three-dimensional structure of the interface known as symmetrization or symmetric minimization of distances. The algorithm and the graph theory terms are described in Methods. The graph describes how the different atoms are connected. In fact, among its edges one recognizes the known hydrogen bonds present at the interface, that are obtained as a bonus, because the symmetric minimization does not make use of them (see Methods). An example of interaction graph is given in Figure 1.

Statistical analyses have been performed on the case of the β interfaces, that are formed by two adjacent β strands, one from each subunit [6–8]. In [8] the analysis has been extended to a dataset of 755 proteins. It is known that there are three possible orientations of the adjacent β strands: they can be anti-parallel (by far, this is the most common case), parallel or oblique. The latter actually includes all the cases that do not enter into the previous ones, for example perpendicular or oblique β strands. The most significant results of these statistical analyses are summarized here (please refer to the Methods for the precise definition of the motifs).

- Two typical interaction graphs have been observed, one for the parallel and one for the anti-parallel orientation. The anti-parallel case shows a BB graph of type ladder, were rungs are typically spaced of 2 amino acids. The parallel case shows a BB graph of type zigzag, in which one recognizes a separation of 2 amino acids in each oscillation of the

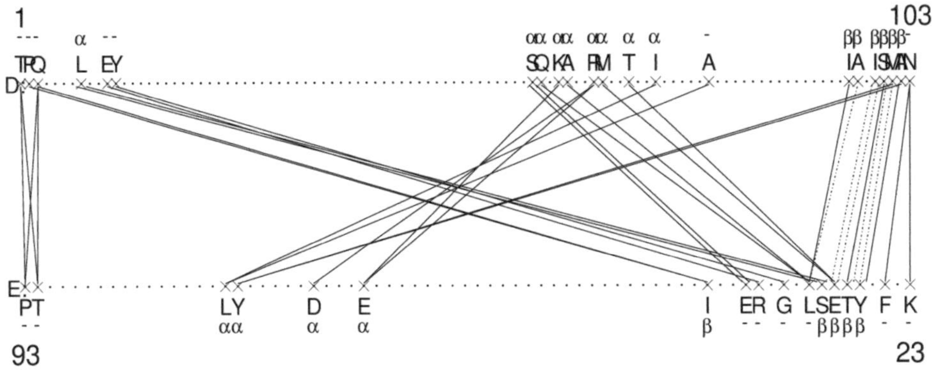

Figure 1. Full interaction graph of the level 0 of the protein 1EEI interface (see Methods). The upper horizontal line represents the sub-unit D, the lower one the sub-unit E. With the crosses we indicate the residues that participate to the interaction graph; their name and membership to specific secondary structures is indicated, when available. The dots represent the residues that do not participate to the interface. The dotted-dashed lines represent pairs of atoms both from the backbone of the residues (BB graph). The solid lines indicate that at least one atom of the pair is from the side chain (SC graph).

zigzag. In some cases, the zigzag topology reduces to a simple vertex V (defined in Methods).

Both these specific topologies identified in the BB graph correspond to the known hydrogen bond graph between backbone atoms. While it is not surprising to find them in the BB graph, the surprise comes from the fact that the position of hydrogen atoms has not been used as input by our algorithms (in most cases it is not even given in the PDB files). Thus, the symmetric minimization algorithm is able to reconstruct the backbone hydrogen bonds network by the unique input of the backbone atoms N, C_α, C, O coordinates, with an accuracy of 90%. In other words, the backbone hydrogen bonds satisfy the mathematical property of symmetrically minimized distances and the backbone non-hydrogen atoms that engage in BB hydrogen bonds are reciprocal nearest neighbours, with the indicated accuracy. This identification has been obtained using the web server RING (see Methods) to calculate the hydrogen bonds and compare with the graphs (see also [6]). Among the 120 pairs of residues in the graphs (anti-parallel and parallel cases only), 108 are recognized by RING as hydrogen bonds and 12 are not. Thus, the symmetric minimization recognizes hydrogen bonds with accuracy of 90%.

- Interfaces of the oblique family have very small or absent BB graph, thus the two BB graphs are rather specific to the respective anti-parallel and parallel cases. It is also rare to find BB graphs in non β interfaces.

  In fact, in these cases the BB graph is intra-chain, namely it develops between atoms of the same sub-unit and is almost absent in the interface.

- The amino acids are not randomly paired. Rather, the frequency with which residues are connected in the interaction graph clearly deviates from the expected frequency calculated from the average frequency of residues in the interface. This indicates that the edges in the interaction graph form in order to provide to the interface very specific features.

http://dx.doi.org/10.5772/58420

2. Methods

2.1. Symmetric minimization of distances

The description of the interface, that has been developed starting with [11] and that will be used in this paper, was introduced to help extract the structural organization of the interface. It focuses on the way atoms pair, keeping into account the local connectivity, namely the possibility that atoms interact with other atoms, according to their local arrangement. It is based on the notion of symmetric minimization of distance pairs, defined by the symmetric minimization algorithm, presented here in pseudocode. The flow chart is given in Figure 3. The needed mathematical explanations and demonstrations have been provided in [12].

Input: $A \leftarrow \{$ atoms of the first subunit $\}$
$B \leftarrow \{$ atoms of the second subunit $\}$
$d(a,b) \leftarrow$ A metric ; typically, the inter-atomic distance is used

Start: $R_0 \leftarrow \{(a,b) : a \in A, b \in B\}$
$i \leftarrow 0$
while $R_i \neq \{\ \}$ **do:**
$\min_A(a) \leftarrow \min \left\{d(a,b') : (a,b') \in R_i \text{ and } b' \in B\right\}$
$\min_B(b) \leftarrow \min \left\{d(a',b) : (a',b) \in R_i \text{ and } a' \in A\right\}$
$L_A \leftarrow \left\{(a,b) \in R_i : d(a,b) = \min_A(a)\right\}$
$L_B \leftarrow \left\{(a,b) \in R_i : d(a,b) = \min_B(b)\right\}$
$S_i \leftarrow L_A \cap L_B$
$R_{i+1} \leftarrow R_i - S_i$
$i \leftarrow i+1$
end while

Output: $S_0, \quad S_1, \quad S_2, \quad \ldots$

The sets R_0, R_i are sets of edges. The empty set is indicated with $\{\ \}$. The symbols $-$ and $\cap$ indicate set difference and set intersection respectively. The symbols $\min_A(a)$, $\min_B(b)$ indicate the functions that give the shortest distances in the neighbourhood of the indicated point. The symbols L_A, L_B indicate the shortest edges relative to the given iteration R_i. The reciprocal shortest edge sets are indicated with S_i, and called symmetrized levels i.

In summary, the symmetric minimization is a recursive method that, at the first iteration, defines the lowest level of the interface as the pairs of atoms that are reciprocal nearest neighbours. The nearest neighbour condition must be verified for both the two subunit atoms, as the name itself suggests. These pairs form the lowest level S_0, called symmetrized interface. Please see the caption of Figure 2.

The algorithm can be repeated on all the edges that have not been retained at the lowest level. This produces a new set of reciprocally shortest edges, not contained in S_0, that forms the symmetrized set S_1, named level 1. The repetition of the algorithm up to exhaustion of all the atom pairs, provides a hierarchy of levels

$$S_0, \quad S_1, \quad S_2, \quad \ldots, \quad S_M \tag{1}$$

that define the symmetrized interface (SI). Thus, inter-atomic distances between the two sub-units are ranked according to levels. Notice that, as proteins are of finite and not of infinite size, there must be a maximum level. Also, the algorithm is free of ambiguities, even in case of regular structures. In Figure 1 and in previous papers the lowest level only was analysed, while it is a purpose of this paper to start the investigation of the higher levels.

The symmetric minimization of distances is a variant of the case $k = 1$ of the known k-reciprocal nearest neighbour method (kRNN), discussed in [13] and used now in the domain of hierarchical classification and object retrieval in images. A modern presentation is in [14]. Actually, the lowest level S_0 could be equivalently obtained with both methods but at higher levels the equivalence breaks down. The purpose of kRNN is to assert the relative proximity of several images containing the same object appearing in different scenes. Each image is a point in some very high-dimensional space. Using an appropriate metric, the closest images are found and agglomerated in a cluster. At the new iteration, new images will join the cluster. The method kRNN compares high-dimensional vectors. It can be applied to atomic coordinates. Once two atoms are found to be reciprocal nearest neighbours at some iteration, they are removed from the pool (and put in a cluster) before the next iteration could start. On the other hand, in the symmetric minimization, two atoms that are reciprocal nearest neighbours are not removed from the pool: the edge they form is removed but the atoms remain. This is the role of the sets R_i in the algorithm: at the next, the same atoms may be nearest neighbours with others. In kRNN, the sets of edges R_i would simply be replaced by some A_i, B_i where A_i or B_i would be obtained by removing from the initial A and B the atoms found at each iteration. The choice of working with edges comes from the goal of describing the interactions from a geometrical point of view. Moreover, the binding energy in a protein interface accumulates between all pairs of atoms, at least in a suitable range of distances, no matter if they are nearest neighbours or not. Using edges, the information about all the neighbours of an atom is recorded and used at one or other of the levels. Using atoms, part of the edges are not evaluated and some information seems lost, at least for purposes related to protein structure.

The classification of pairs into levels reminds one of perturbative calculations, very common in physics, where the lowest order contains the strongest interaction and the higher orders introduce weaker and weaker terms. The ranking, and the set of levels in equation (1), have been computed on the basis of inter-atomic distances: the higher the level, the larger the distance between atoms. Physically speaking, moving to higher distances implies a tendency to move to weaker interactions. Here, force fields and types of atoms have not been used in the symmetric minimization, thus rising to higher rank only indicates a tendency to weaker interactions and does not hold in a strict sense.

Imagining a mechanical model of balls and sticks, and a quantity of glue, can one mount a human size model of the protein interface? Yes, if the sequence in equation (1) is followed.

http://dx.doi.org/10.5772/58420

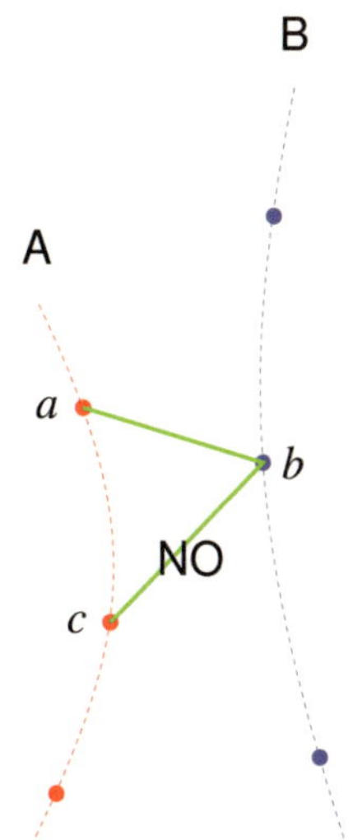

Figure 2. Example of symmetric minimization. The distance (ab) is the smallest for the atom a and, at the same time, is the smallest for b. Thus, the edge (ab) is retained in the symmetrized interface. The distance (bc) is the smallest for (c) but not for (b) thus it is not retained.

Otherwise said, the geometrical action of ranking the pairs of an interface corresponds to the sequence of actions that one has to follow to construct the interface with minimum amount of steric effects. The sequence read in the opposite sense, from the maximum level S_M to S_0, indicates the steps to access the closest atoms of the interface by removing one layer after the other.

It is important to stress that each edge of S_0 behaves as a nucleation centre, or a bud, because every edge that appears at level 1 is attached to a level 0 edge; thus, level 0 edges act as the starting points of a growth process where, at each higher level, new edges attach to the bud. The growth occurs simultaneously from the various level 0 edges. Each bud is technically a cluster. At some level, an edge appears that joins two different buds; the edge length is the resolution threshold for the two buds. Certain edges do not start a growth process; this occurs when the atoms joined by these edges are not further connected to the rest of the interface. Also, a bud must be present at level 0 and cannot appear at higher level: S_0 already contains all the buds. In other words, the various parts of the interface are already contained in S_0 and a set smaller than S_0 is possibly insufficient to reconstruct the full interface. This provides the mathematical justification for calling S_0 a framework of the interface. Indeed, as we have empirically observed in our previous publications, S_0 is the smallest set that can describe the interface. This is the most important point of the whole construction and is based on the theorems proved in [12] but is published here for the first time.

In summary, the symmetric minimization has been introduced to responds to the following needs and possesses the following features [12].

Scalable. The symmetric minimization may be applied to proteins or objects of any size, without size limits. Moreover, it can be applied also to objects at the human or interstellar scale.

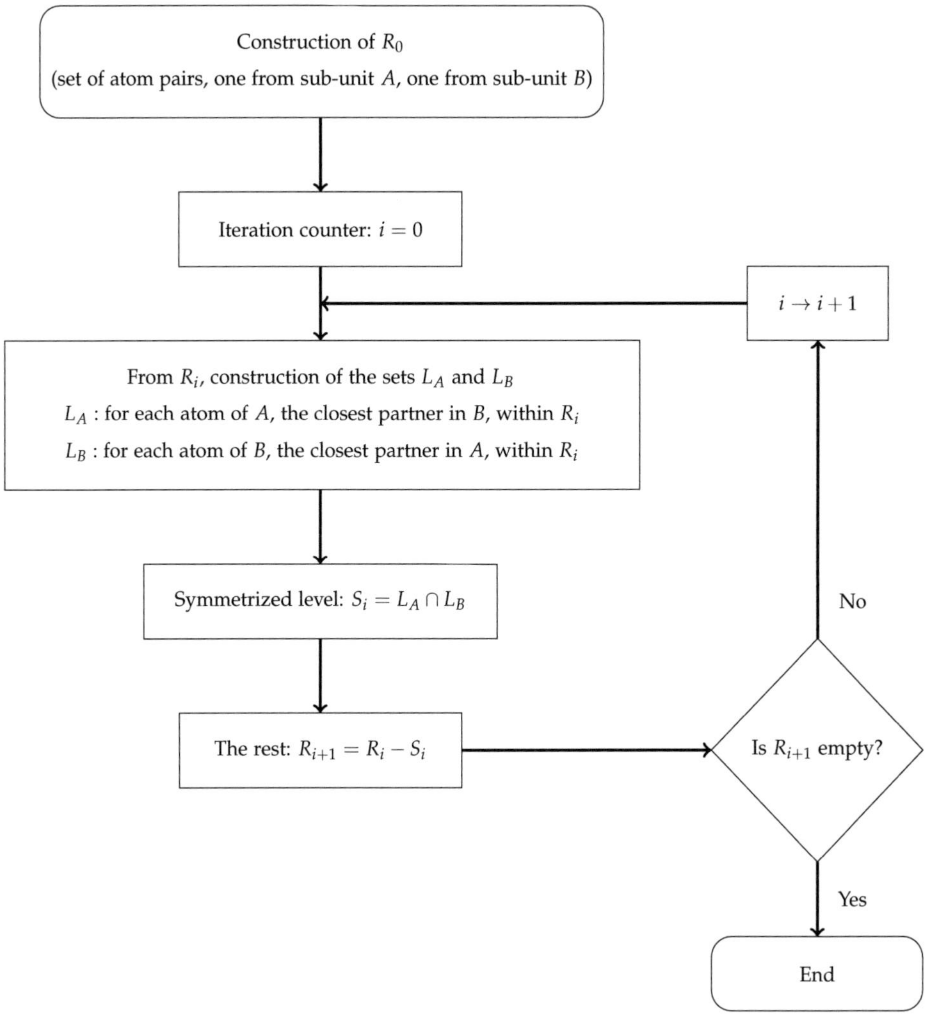

Figure 3. Flow chart of the symmetric minimization algorithm.

Local. It is based on the local arrangement of atoms (or points) and not on global features. So, it captures the differences occurring in situations like a dense atomic packing or a dilute packing.

Scale-free. No length scale has been imposed from outside.

Intrinsic scales. It defines a set of characteristic scales, intrinsic to the interface itself. Indeed, the symmetric minimization allows to divide the interface in clusters (the buds previously presented). The first edge that joins two different clusters (two buds) is a characteristic scale for the interface.

Metrics independent. It is independent on the explicit distance function adopted. Namely, the actual distance used can be different from Euclidean distance, non euclidean geometries being allowed if they use positive definite metrics (a distance).

http://dx.doi.org/10.5772/58420

2.2. Interaction graphs

The previous subsection demonstrates how the data are analysed. By construction, each set in equation (1) is a collection of edges, the extrema of which are points in some metric space. Mathematically, this corresponds to the notion of graph, namely a set of nodes (here the points in the metric space) and a set of edges joining some of the nodes. Thus, from now on each set representing a level will be naturally identified with a graph. In the present case it is important to stress that the graph has been obtained by evaluating distances, thus each edge is associated to the physical distance of the endpoints. Mathematically speaking, this is a weighted graph.

As the graph nodes were initially atoms of one or the other of two adjacent subunits, the graph is automatically bipartite: each edge has an endpoint on one subunit and the other on the second subunit. Edges between atoms of the same subunit are not considered here, as the analysis focuses on the interface.

By construction, a pair of atoms can be connected by a single edge only, as there is just one distance value between them. Thus parallel edges are absent (edges with the same end nodes) in this description that can be called full atoms.

Since the first paper on the subject [11], the levels S_i have been coarse-grained in order to facilitate the human interpretation of the data and to transfer the information to the residue scale. This is actually the scale used by biological entities to store and transfer information: DNA and RNA code for amino acids, not for atoms, and proteins form from residue chains and not from individual atoms.

In the coarse-grained representation the information appears at the amino acid resolution: all the atoms of a residue are identified and represented with the residue name itself. All edges ending on the atoms of the residue are now referred to the residue itself. In graph theory this procedure is called fusion. If $(a,b),(a',b') \in S_i$ are edges of one of the graphs, let's call F (fusion) the equivalence relation such that

$$(a,b)F(a',b') \iff a \text{ and } a' \text{ belong to the same residue and } b,b' \text{ belong to the same residue} \tag{2}$$

Then, the following quotient set defines equivalent classes

$$S_i^{aa} = \frac{S_i}{F} \tag{3}$$

whole elements are all the edges that start and end on the same residues (*aa* means amino acid resolution). By consequence, the graphs S_i^{aa} could have parallel edges; this happens when two residues are connected by more than one pair of atoms.

The subsequent analyses will always refer to this coarse-grained graphs, while the initial symmetric minimization has always been done with atoms. The graphs S_i^{aa} can be represented in a very effective way: the first subunit residues are represented as an horizontal line of equispaced dots; similarly, the second subunit residues appear as dots in an horizontal line below the previous one. Clearly, the two horizontal lines represent the chains of residues connected by the peptide bonds. Being the graph bipartite, edges can only join a node from one horizontal line to a node of the other horizontal line. In Figure 1 there is an example.

In fact we found it very convenient to distinguish two sub-graphs. The dotted-dashed lines represent pairs of atoms both from the backbone of the residues (BB graph). The solid lines indicate that at least one atom of the pair is from the side chain (SC graph). Mathematically speaking, this is called an edge-labelled graph, the two possible labels being BB and SC. In principle, there is a third edge label; indeed, the nodes of the same sub-unit form a sequence connected by the peptide bond. Albeit this type of edge is not explicitly shown in the graphs, it is implicitly present and motivates the choice of organizing the residues along straight lines.

Notice that while the full atom graphs are a fortiori different, $S_i \cap S_j = \{\}$, this is no longer true for the fused graphs: it is not a priori granted that they are different.

In summary, the sets in equation (3) characterize the interaction between subunits; each set is a bipartite and edge-labelled graph where parallel edges may occur.

The amino acid resolution graph of the interaction between subunits has shown to be extremely effective in interpreting data and designing statistical analyses. Notice that the choice of marking non interacting residues with a dot and interacting residues with a cross is purely conventional and does not add information.

2.3. Topological analysis of the graphs

The analysis will proceed with the inspection and comparison of the interaction graphs of the level 1 of the interface, with some additional information from the level 2. The focus will be on the topology, namely on the organisation of edges in the graphs, whose importance has already been shown in the previous publications [6, 11]. As already stated, the amino acid resolution will be systematically adopted.

The interface graphs will be analysed using the following motifs.

Zigzag: it is a path that alternates from one to the other of the two sub-units, namely from one to the other of the two nodes subsets of the bipartite graph, like in a zigzag seam. In principle, the smallest recognizable zigzag visits three residues; the experience with data has clearly shown that it is not useful to consider such a path in the zigzag family but it is better to classify it in the vertex family (see later). Thus, the smallest zigzag seam considered is the one with 4 residues visited, zigzag4. It is visible in Figure 5, involving the residues A, I, E, A (BB graph). Zigzag paths with 5 or more residues are comprehensively indicated as zigzag5. See Figure 27, residues T, E, L, V, A.

Vertex: it is a residue connected with two or more different amino acids. In the smallest case, the residue is connected to two other residues and indicated V, as in Figure 4, residues F, L, K. The case of more than two residues connected with the same vertex will be comprehensively indicated with V3. Three exemplars are present in Figure 28. Especially, the residue I on chain B has four connections, three of type BB and one of type SC.

Ladder: This motif occurs when 2 (or more) amino acids at separation δ on the first subunit, namely at positions $M, M + \delta$, are connected with 2 (or more) amino acids $N, N - \delta$ on the second sub-unit, respectively. Notice that the second sub-unit runs oppositely to the first. The two groups of amino acids are the ladder rails and the edges are the rungs, from which the name was adopted. In Figure 3, residues E, Y and E, W form a ladder with 2 rungs and separation $\delta = 2$. In Figure 22 the residues C, S, F (47, 49, 51) of chain A and the residues Y, E, E (110, 108, 106) of chain B form a BB ladder of separation $\delta = 2$. The two most common separations are $\delta = 1$ or 2, ladder1 and ladder2 respectively. Cases with more than 3 rungs are known. In Figure 14, the residues from 98 to 102 (of the chain D) with the residues from 29 to 25 (chain E) form a ladder1 with 5 rungs.

Multiple edges: presence of parallel edges in the graph (edges with the same endpoints). The term multiple edges seems more adequate to the present case, given the use of the word parallel to indicate the orientation of the β-strands in the interface. In Figure 8, the residues E and F are connected by 4 edges, three of type SC and one of type BB. In Figure 22, the residues S and E are connected by three BB edges.

It is important to look for these motifs starting from the largest elements, to avoid useless multiple countings. Especially, it is obvious that a zigzag of 8 nodes contains all the shortest sizes, from 4 to 7. There is no need to record all of them, the largest one being the most informative. Also, a V3 or zigzag motif automatically contains the simplest vertex V, and often more than once. Thus, the vertex V is not counted when it appears inserted in a V3 or a zigzag. In Figure 33, the zigzag5 (residues V, M, I, V, A) contains three times a V. In Figure 30, the vertex V3 (residue H) contains three V motifs.

These topological elements may appear alone or in combination and have been chosen, as in other publications, because they represent the most common motifs in interface graphs. It may happen that a graph contains more than one motif, in which case all of them will be recorded. For example, in a common situation one of the nodes of a zigzag5 is also a vertex V3. In that case, both motifs are registered. The main reason is that a systematic classification of graphs needs to be accurate and free of ambiguous search criteria, thus it is definitely preferable to accept a redundancy in the identification of motifs than introducing untested criteria. For example, often a ladder has separation of 2 in one part of the graph and separation of 1 in another part, with a edge (a rung) in common between the two parts. Thus, it is recorded as part of both a ladder of 2 and ladder of 1. Indeed, so far, no acceptable criterion has been found to discriminate if a edge must be considered part of a ladder with separation of 1 or of 2. In Figure 28, there is a BB ladder formed by M, I, K, I (chain A) and V, L, I, E (chain B). Residues M and I are separated of 2 while residues I and K, K and I are separated of 1.

2.4. The dataset

In [6] a dataset of 39 oligomeric proteins was chosen on the basis of the presence, in the three-dimensional structure, of a well recognizable β interface. Here the same set will be used. It is listed in Table 1 with the indication of the chains and the intervals participating to the β interfaces. All proteins are homomeric of stoichiometry from 3 to 8. This set is characterized by the absence of sequence homology, structural homology or functional homology. Viral and membrane proteins are absent, given their specificities. Thus, the set can be considered representative of the general behaviour of β interfaces, without reference to specific classes of proteins. In the dataset there are anti-parallel, parallel and oblique orientations of the adjacent β strands. Actually, the oblique family will be considered for comparison only and not deeply investigated. In fact its graphs are less structured and it is difficult to find similarities between proteins.

2.5. Residue interaction network generator (RING)

It is a web server [15] with software for rendering a protein structure into a network of interactions. Nodes represent single amino acids and edges represent the non-covalent bonding interactions that exist between them. In particular, this web server has been used to calculate the hydrogen bonds for the proteins of the dataset. The hydrogen bonds have not been supplied to the symmetric minimization method but just used to compare some results.

3. Results

The dataset has been described in Methods and its proteins are listed in Table 1.

The graphs of the level 0 were provided as Supporting Information to the publication [6] and are downloadable in open access, as well as the publication itself.

In this section the description focuses on the level 1 and, to a minor extent, on the level 2. The graphs of the level 1 are provided in Figures 3 to 26, and 27 to 34. Given their minor role, the graphs of the level 2 will be provided on demand.

In the following description, the results are grouped and numbered from Result 1 (R1) to Result 5 (R5).

R1. BB graph at level 1. The level 1 graphs are not identical to the level 0 ones.

In all the anti-parallel cases (Figures 3 to 26), the presence of a level 1 BB graph is observed, of size comparable with the one observed at level 0, namely with a similar number of edges. It is very small in just one case (2ojw level 1), where a single edge is present.

Similarly, in all the parallel cases (Figures 27 to 34), there is a level 1 graph, with a number of edges similar to the one found at level 0.

R2. BB graph structure at level 1. In the anti-parallel orientation, the graphs present a ladder structure in 21 out of 24 cases. Its separation is of 2 amino acids in 20 cases out of 24 and of 1 amino acid in 15 out of 24 cases[1]. The zigzag connection is absent. The vertex V is present in 13 graphs on 24. The V3 motif is present in 9 out of 24 cases. The zigzag4 is counted 12 out of 24 times. Just 4 multiple edges are observed. In all cases at least one ladder motif shows up at level 0 or at level 1.

In the parallel orientation, the graphs shows a common presence of the zigzag5 topological element, in 4 out of 8 graphs, often accompanied by one or more V3, in 5 out of 8 graphs. The zigzag5 motifs are probably not completely independent from those that appear at level 0, as they always show up together. This aspect needs a larger statistics to be confirmed or disproved. The zigzag opening is of 1 or 2 amino acids. There are also 4 ladders of separation 1. Multiple bonds appear once out of 8 cases[2].

R3. BB graph structure at level 2. A level 2 BB graph is observed in all cases, often less populated than the lower levels graphs.

In the anti-parallel orientation, there are 13 ladders out of 24. The other cases show a V or zigzag4 motif, and in one case a zigzag5 is found.

[1] At level 0 the ladder separation is of 2 residues (23 cases out of 24). There are multiple edges in 21 cases out of 24.

[2] At level 0 there are 5 zigzag5 out of 8. After, one finds 2 V and 1 zigzag4. Multiple bonds are absent.

In the parallel orientation, the zigzag topology and the vertices V3 are absent at this level. On the contrary, these are the most common topological elements at the lower levels. At this level we just find the ladder motif with separation of 1 or 2 amino acids. In some cases the BB graph reduces to one edge.

R4. Other orientation. In interfaces other that the anti-parallel or parallel β, it was systematically observed in previous publications and is confirmed here that the BB graphs are rare and poorly structured, as composed of one or two edges. This obviously holds true also at levels 1 and 2.

R5. SC graph. These graphs are more elaborated and very rich of a variety of elements. One may recognize that the motif of the ladder is present in nearly half the graphs, at all the various levels examined (0, 1, 2). The V, V3 and zigzag elements are very common in all the three orientations. A more complete analysis of the SC graphs will be realized in future publications.

4. Discussion

In Methods, four definitions of interface have been introduced.

The participating amino acids are actually very similar: in [11], paper fully dedicated to compare the level 0, S_0^{aa}, with published interface data more than 85% of similarity has been detected. At the level 0, the difference is more in terms of the description that emerges. In the buried surface or Voronoi cells based interface approach, is very natural to distinguish between rim and core: a solvent molecule that tries to penetrate the interface first has to visit the rim, and after may force into the core. The symmetrized levels instead present the growth of the interface, as in a budding process, from the set S_0. The hierarchy in (1) indicates the dynamical sequence of events that may construct the interface with minimal steric effects. Somehow, this introduces the notion of time in an otherwise static view. The distinction between rim and core is possibly given by the number of the connections of a residue: a highly connected residue must be in the core and cannot be in the rim. Vice versa, a poorly connected residue is in the rim. Indeed, the part of a residue surface that is exposed to the solvent will not be connected to other atoms while an atom that is completely buried will have connections with all its neighbours. This indicates that a minimal distinction between rim and core may appear if one compares few close levels, like S_0, S_1 and S_2. The point is that the more an atom is present at different levels, the more is connected thus the rim must correspond to those atoms that appear few times across the levels, the others being in the core.

The importance of how the information circulates in an interface has been first shown in [8]. The graphs S_i^{aa} show the presence of long range correlations that are not easily detected with approaches based on the contact surface (buried surface or Voronoi cells based interface). The motifs that allow this transfer are the zigzags, especially the long ones, and the vertices V, V3. On the contrary, a pure ladder topology, in BB or SC or both, does not have ways to correlate far apart atoms. The frequent presence of zigzag, V and V3 motifs implies the existence of constraints on the positions or the physico-chemical properties of non neighbouring atoms, and sometimes of very far apart atoms. Indeed, in Figure 27, the mere establishment of the path formed by the residues S2, E123, L5, L124, I7, V127 (where BB and SC are both present) requires to satisfy physical conditions due to the volume of the atoms and the distribution of electric charge. Each residue has its internal constraints, among which the fixed length of covalent bonds and the planarity of the surface $C_\alpha, C, 0, N$ that contains the peptide bond. The sequence of edges transfers constraints and establishes a correlation between the outer amino acids S2 and V127 even if they are not in physical contact. The exact microscopic description of the constrains is not easy to find, although. Notice that this example of information transfer along the interface has been detected thanks to a description guided by graph theory and based on edges.

In summary, the main difference that appears comparing buried surface and Voronoi cells based interface with SI is that the first two definitions focus on the spatial organization of the interface while the latter may suggest a temporal organization and allows to evaluate how the information circulate in the interface.

The result R1 clearly states that the level 1 graphs can have a BB graph of size comparable to that of level 0, thus still informative. At level 2 a smaller BB graph is observed. Preliminary results on levels higher than 2 indicate that the BB graphs are also present.

The result R2 indicates that the level 1 graphs have a structure similar to the one found at level 0, in other words that these two levels present several common elements. Instead, from R3 the level 2 graphs seem organized in a different way. The main structure of the level 2 graphs is the ladder one, in both the anti-parallel and parallel orientations, that indicates that these orientations are not distinguishable at this level, and possibly above. Preliminary results on levels higher than 2 indicate that the main distinctive motif of the parallel orientation, namely the zigzag5, is quite rare. Thus, it seems that level 0 and 1 are those that contain the most useful geometrical and topological information.

In [8] we have already used the properties in R2, R3, by implementing algorithms that, from the PDB structure, are able to characterize an interface and tell if it has a β structure, and which is its orientation. These algorithms are based on the level 0 only[3]. The analysis of level 1 graphs confirms and expands this possibility, because the information from both the levels can now be combined for a more accurate recognition.

The BB graphs at level 0 have been previously associated to structural hydrogen bonds that are present in the anti-parallel and in the parallel orientations of β strands.

It is possible that level 1 (for both the BB and SC graphs) doesn't describe proper chemical bonds but weaker dipole-dipole or Van der Waals interactions . This comparison for the level 1 has not yet been explored.

4.1. Counting the degrees of freedhom

The question that we address now is to evaluate if the description provided by the graphs is enough to reconstruct the shape of the interface or not.

To reconstruct a shape in three dimensions one needs 3 coordinates for each point: $3N$, where N is the number of points. Actually, the absolute position of the centre of mass and the spatial orientation of the object in the space are totally irrelevant thus three overall translations and

[3] Other interface arrangements are not yet recognized.

three rotations can be removed from the counting, that reduces the number of needed relative positions or distances to

$$\text{degrees of freedhom} = 3N - 6 \tag{4}$$

As an example, consider the graph of 2ojw in Figure 19; there are $N = 7$ amino acids (even the amino acids that do not participate to the graph but are included within the considered regions must be counted) thus one needs $3N - 6 = 15$ relative positions. The distances between two consecutive C_α is fixed in all polypeptides (as the distance TG in the graph). In the present case there are 5 of them. The distance between the first and the third of three consecutive C_α (distances TT and GI in the graph) is fixed by the general properties of polypeptides, that makes 3 distances. The graph has 3 edges, namely 3 other distances, thus one remains with $15 - 5 - 3 - 3 = 4$ more distances to be fixed. This indicates that this graph is insufficient to reconstruct the shape. A more general counting is possible. In [8] the average number of amino acids (18) and of edges (12) in a β interface have been evaluated. Their ratio is very close to $3/2 = 1.5$, thus it is reasonable to assume that if there are N residues in the interface, there will be nearly $2N/3$ edges (actually, multiple edges should not be counted, here; this may further reduce the number of known edges). Also, we expects $N - 2$ consecutive C_α and $N - 4$ groups of three consecutive C_α namely we have

$$\text{known distances} = (N-2) + (N-4) + \frac{2}{3}N = \frac{8}{3}N - 6 \tag{5}$$

We subtract the number of known distances to the number of degrees of freedhom and we are left with the number of distances that are needed to fix the shape

$$\text{unknown distances} = (3N-6) - \left(\frac{8}{3}N - 6\right) = \frac{1}{3}N \tag{6}$$

Thus a single level does not provide enough data but two levels provide sufficient information to fix the shape of the interface.

Of course, a full evaluation of the interface degrees of freedom needs a much more complex calculation with atoms but the present counting suggests that few lowest levels should be enough to provide an accurate description of the interface shape and the position of most of the atoms.

A more complete account of the problem of reconstructing the shape of a set of points given an incomplete set of distances is treated in [16].

4.2. Perspectives

The result R5 is clearly indicative of the major complexity of the side chain by respect to the backbone. In the paper [8] the role of the residues with multiple interactions, namely V and V3, has been studied in detail in S_0^{aa} and has been correlated to the length of the side chains. Following this observation, one could introduce a parametrization based on the length of the side chain. The discussion on the information flow in the interface points in the same direction, of dedicating a new publication to the study of the SC graphs.

stoichiometry	PDB name	chains	range on the chains	orientation of the β interface
3	1JN1	AB	120-140,1-20	oblique
3	1PM4	AB	82-95,64-80	anti-parallel
3	1SJN	AB	1-12,118-130	parallel
3	1SNR	AB	109-129,326-342	anti-parallel
3	1T0A	AB	1-15,125-141	oblique
3	1Y13	AB	9-21,161-174	anti-parallel
3	2BAZ	AB	1-8,116-130	parallel
3	2BCM	BA	43-53,17-26	anti-parallel
3	2BT9	AB	44-57,1-16	oblique
3	2GVH	AB	189-202,59-73	anti-parallel
3	2I9D	AB	148-166,17-33	anti-parallel
3	2JCA	AB	1-17,103-124	oblique
3	2P90	AB	71-88,168-180	anti-parallel
4	1J8D	AB	19-29,30-40	anti-parallel
4	1L3A	AD	118-129,88-98	parallel
4	1PVN	AD	489-496,432-438	anti-parallel
4	2A7R	AD	1-16,327-339	parallel
4	2H5X	AD	1-8,18-29	anti-parallel
4	3BF0	AB	445-468,178-197	anti-parallel
5	1B09	AB	197-206,99-112	oblique
5	2XSC	AB	62-69,8-16	oblique
5	1EEI	DE	94-103,21-33	anti-parallel
5	1EFI	DH	23-33,94-103	anti-parallel
5	1FB1	AE	125-138,218-237	anti-parallel
5	1HI9	AB	66-84,175-191	anti-parallel
5	1NQU	AE	1-6,43-54	parallel
5	1WUR	AB	186-197,93-105	anti-parallel
5	2OJW	AB	42-48,188-195	anti-parallel
5	2RCF	AB	72-83,8-21	anti-parallel
6	1U1S	AB	48-60,54-69	anti-parallel
6	2BVC	AF	211-219,33-40	oblique
6	2GJV	AB	43-56,102-112	anti-parallel
6	2Z9H	AB	5-18,77-89	anti-parallel
7	1HX5	AG	3-13,92-99	anti-parallel
7	1OEL	AG	34-43,511-524	parallel
7	1WNR	AG	1-10,87-96	anti-parallel
7	2RAQ	AB	33-46,76-91	anti-parallel
8	1Q3S	AB	46-57,515-527	parallel
8	2V9U	AB	140-148,170-177	parallel

Table 1. Table of the proteins considered in this paper, from [6]. In summary, we have 24 antiparallel, 8 parallel and 7 oblique orientations.

http://dx.doi.org/10.5772/58420

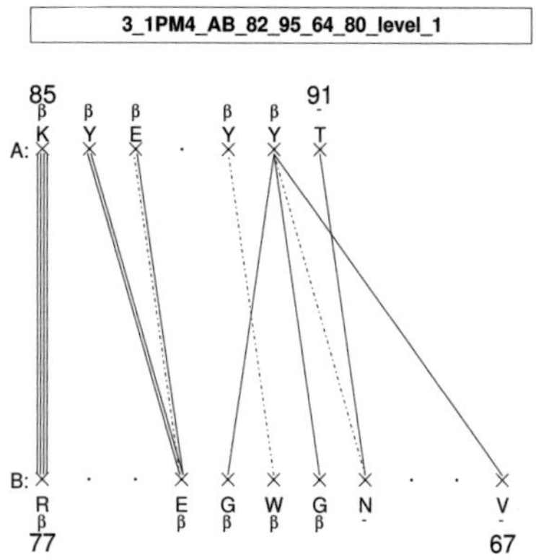

Figure 4. Anti-parallel orientation of the β-strands.

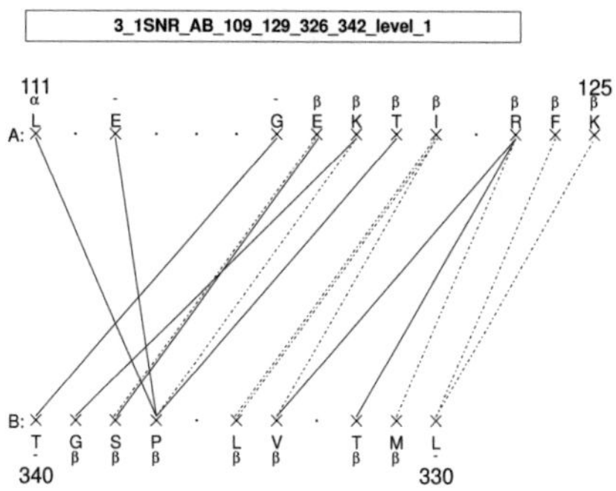

Figure 5. Anti-parallel orientation of the β-strands.

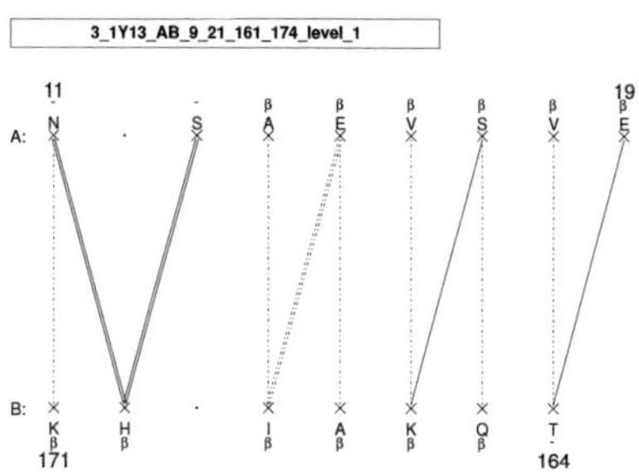

Figure 6. Anti-parallel orientation of the β-strands

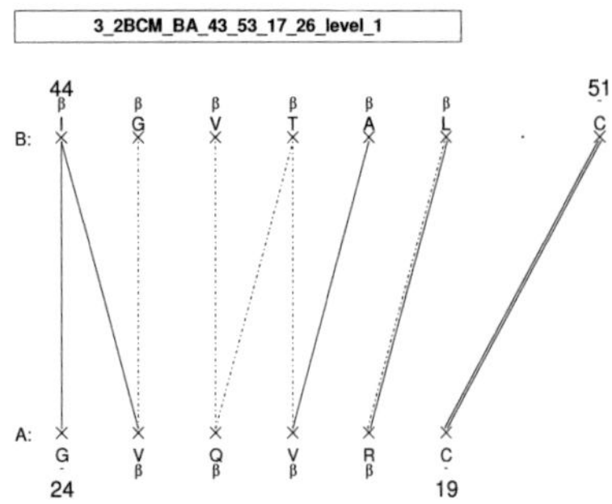

Figure 7. Anti-parallel orientation of the β-strands.

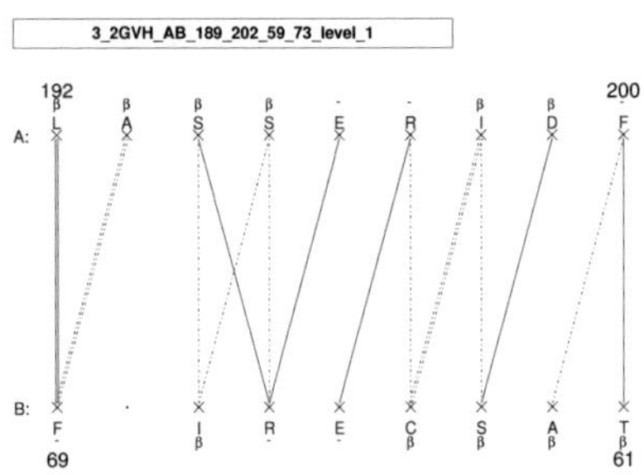

Figure 8. Anti-parallel orientation of the β-strands.

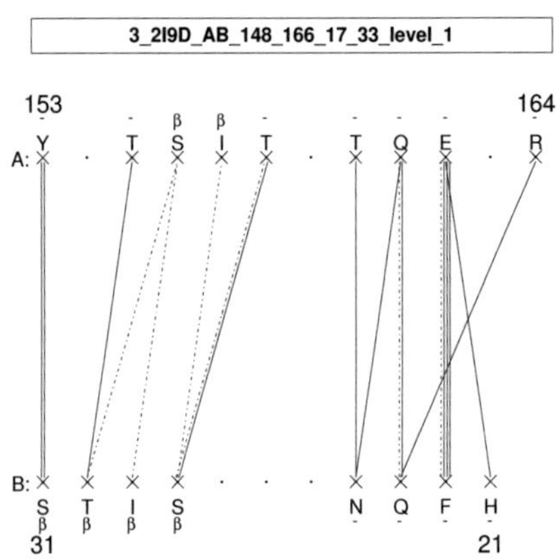

Figure 9. Anti-parallel orientation of the β-strands.

http://dx.doi.org/10.5772/58420

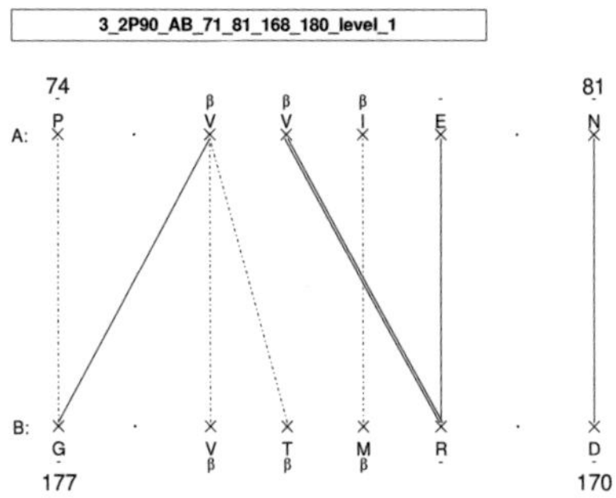

Figure 10. Anti-parallel orientation of the β-strands.

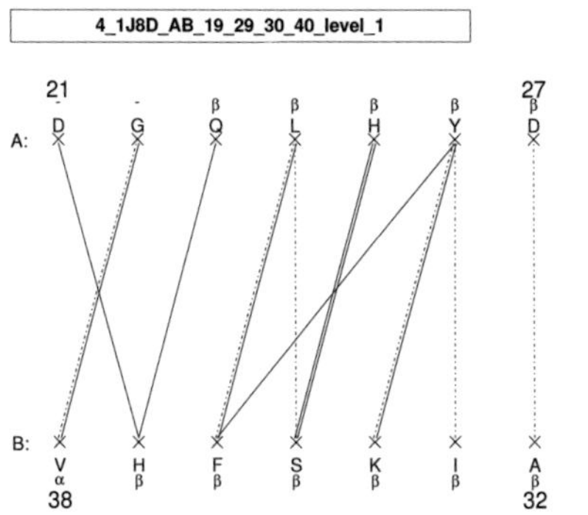

Figure 11. Anti-parallel orientation of the β-strands.

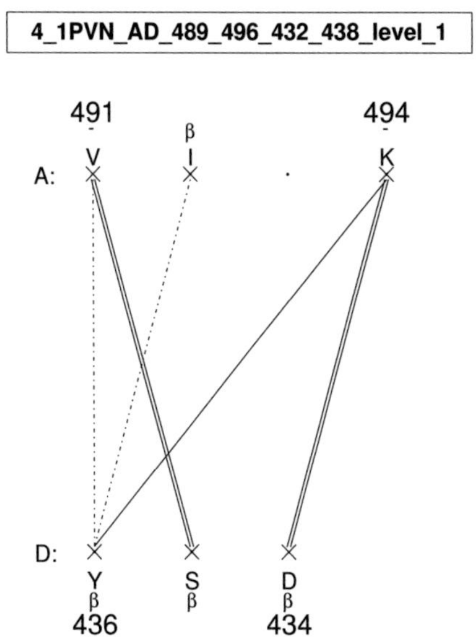

Figure 12. Anti-parallel orientation of the β-strands.

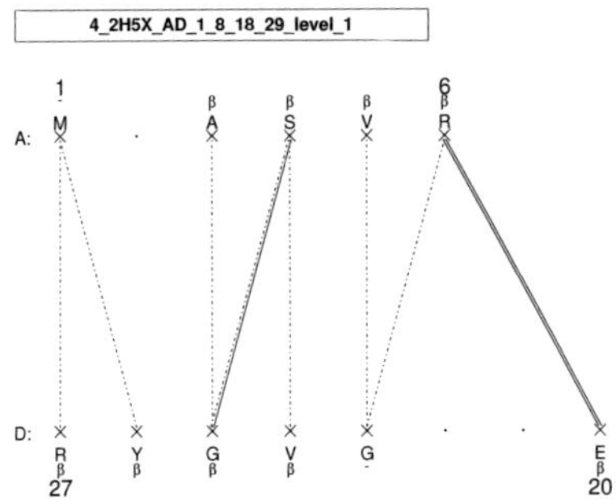

Figure 13. Anti-parallel orientation of the β-strands.

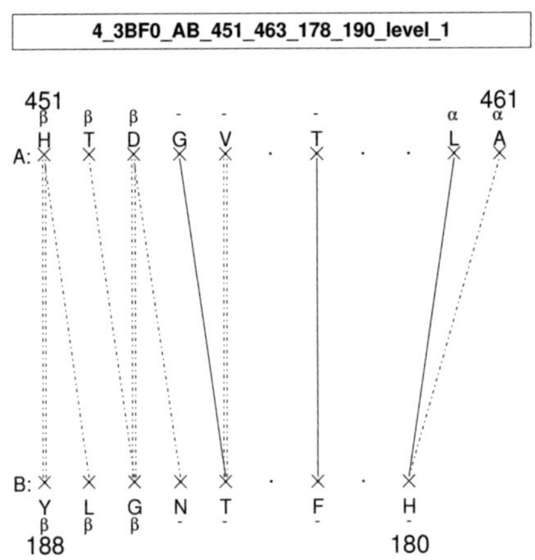

Figure 14. Anti-parallel orientation of the β-strands.

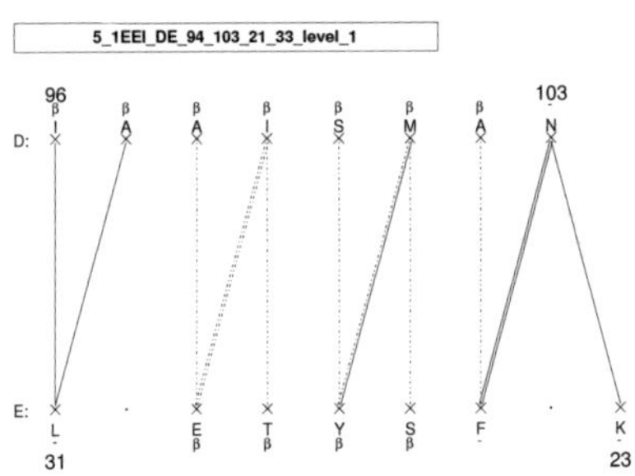

Figure 15. Anti-parallel orientation of the β-strands.

http://dx.doi.org/10.5772/58420

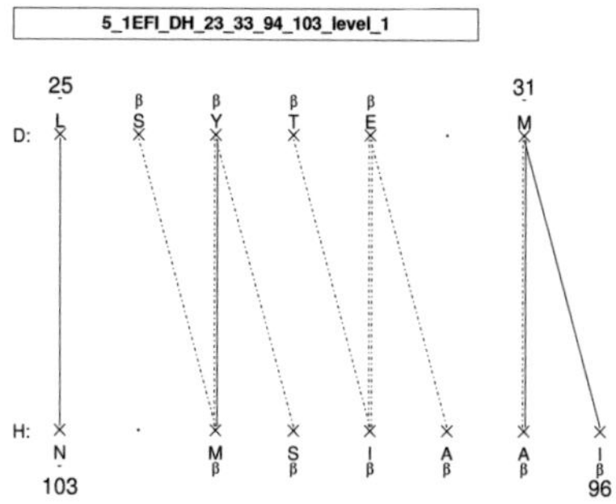

Figure 16. Anti-parallel orientation of the β-strands.

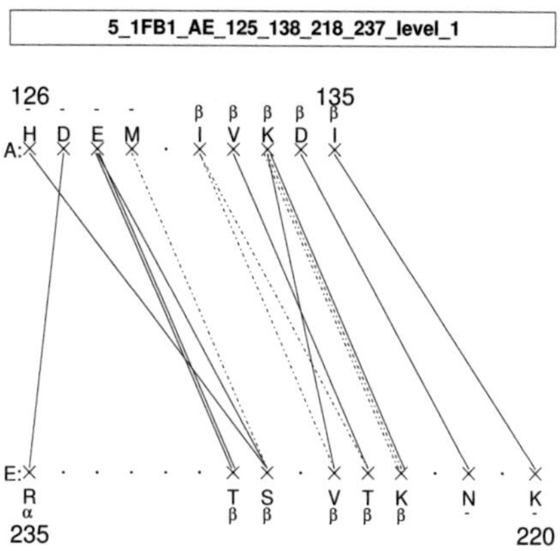

Figure 17. Anti-parallel orientation of the β-strands.

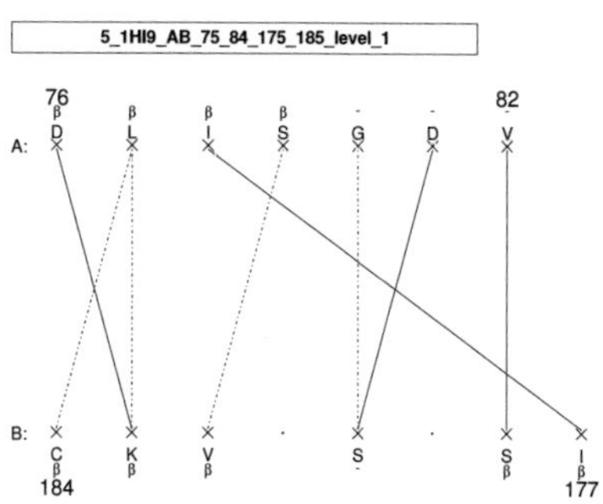

Figure 18. Anti-parallel orientation of the β-strands.

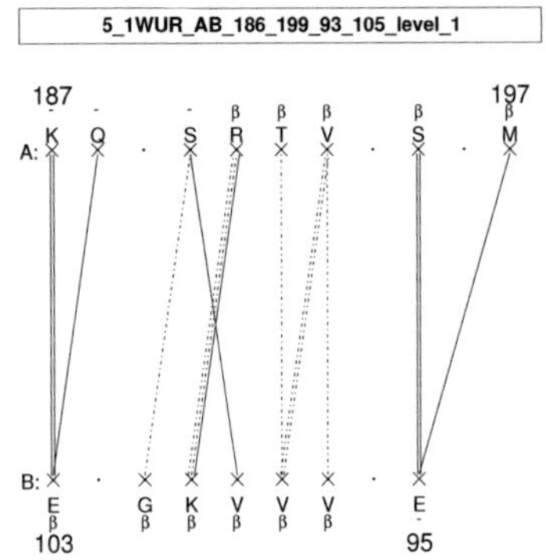

Figure 19. Anti-parallel orientation of the β-strands.

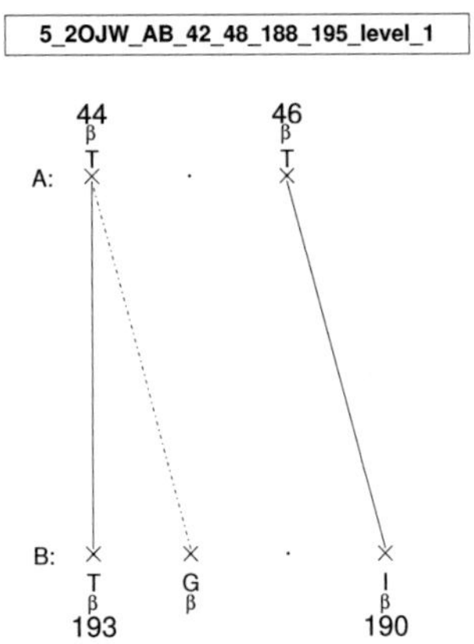

Figure 20. Anti-parallel orientation of the β-strands.

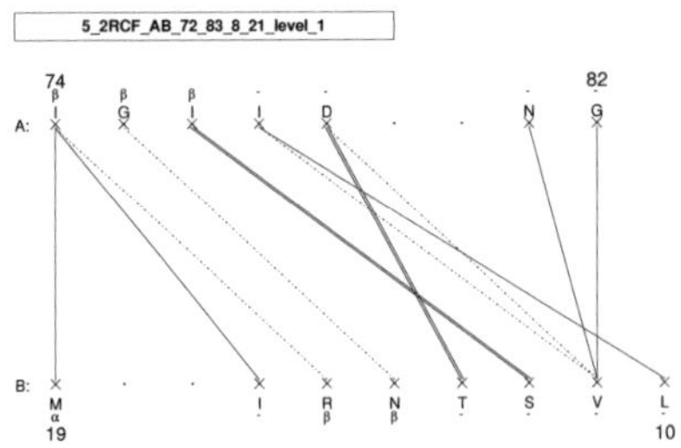

Figure 21. Anti-parallel orientation of the β-strands.

http://dx.doi.org/10.5772/58420

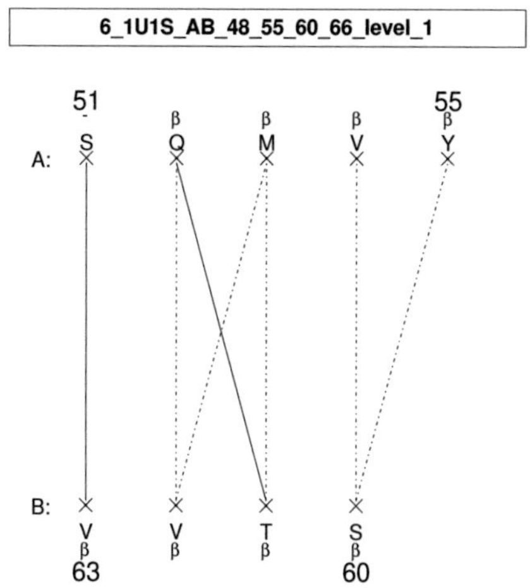

Figure 22. Anti-parallel orientation of the β-strands.

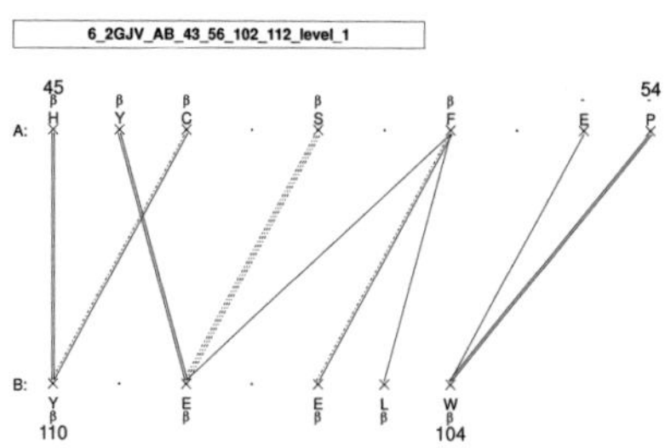

Figure 23. Anti-parallel orientation of the β-strands.

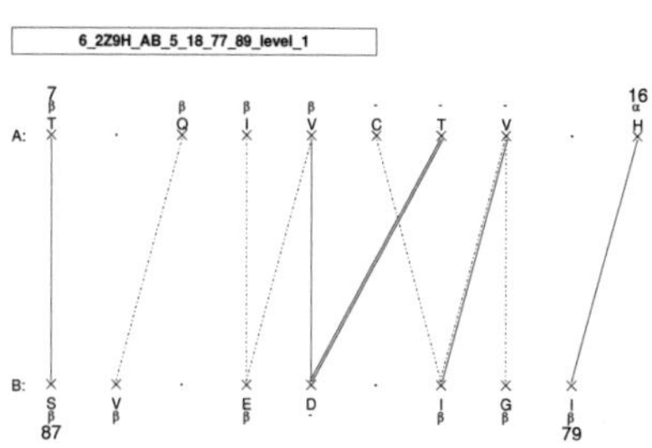

Figure 24. Anti-parallel orientation of the β-strands.

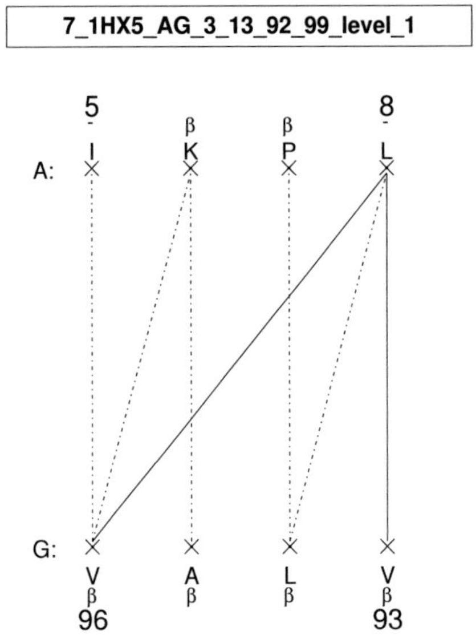

Figure 25. Anti-parallel orientation of the β-strands.

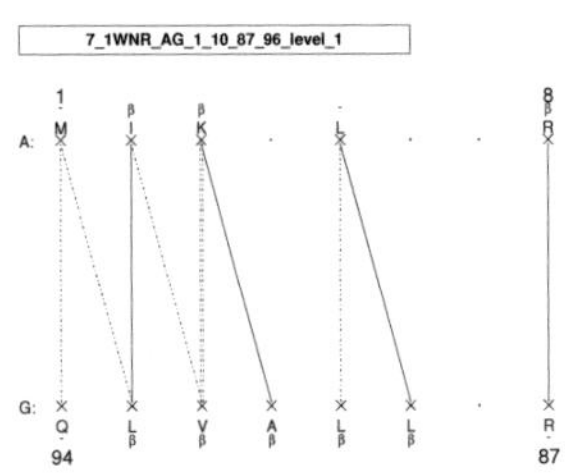

Figure 26. Anti-parallel orientation of the β-strands.

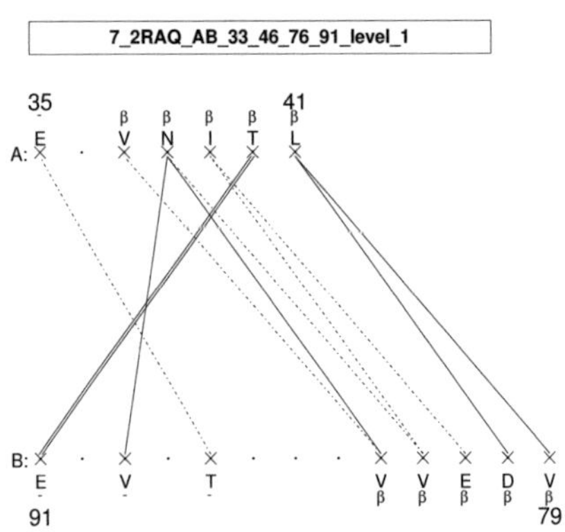

Figure 27. Anti-parallel orientation of the β-strands.

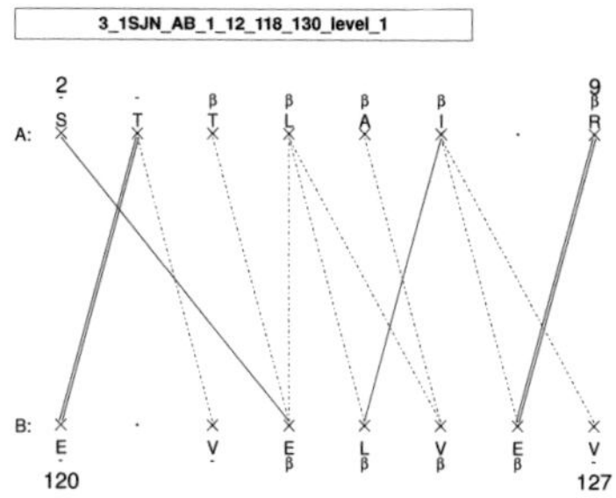

Figure 28. Parallel orientation of the β-strands.

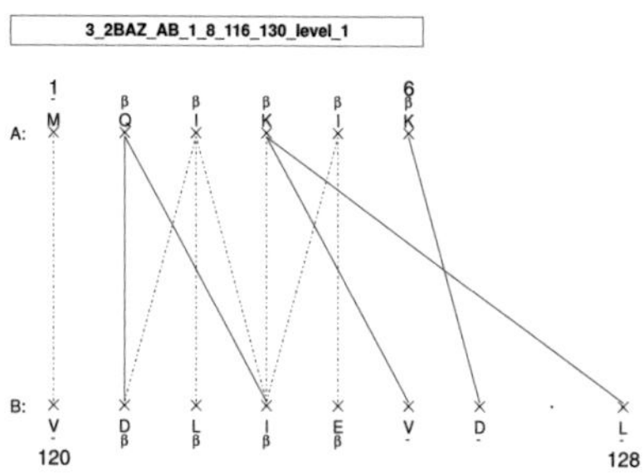

Figure 29. Parallel orientation of the β-strands.

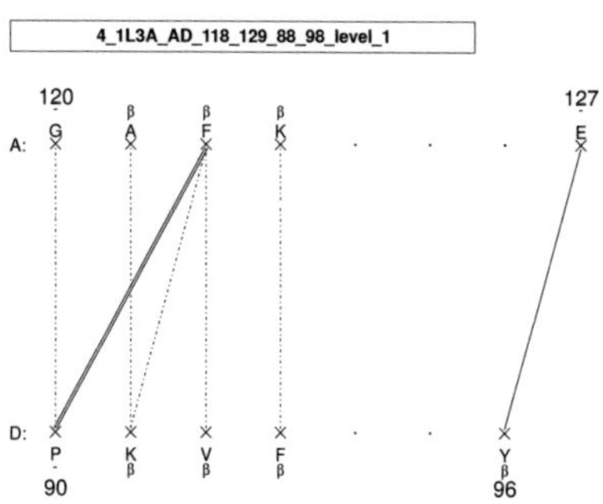

Figure 30. Parallel orientation of the β-strands.

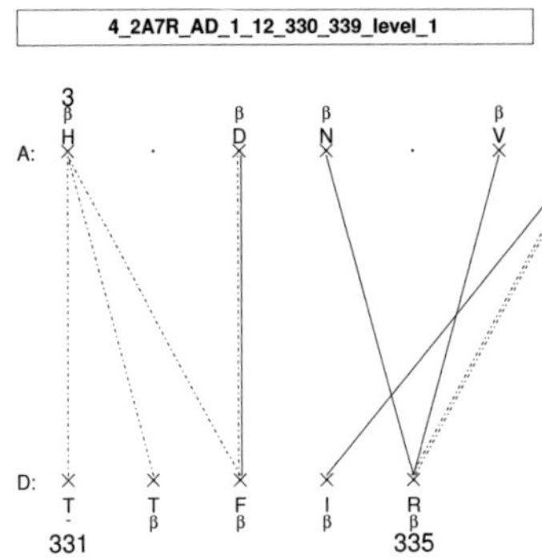

Figure 31. Parallel orientation of the β-strands.

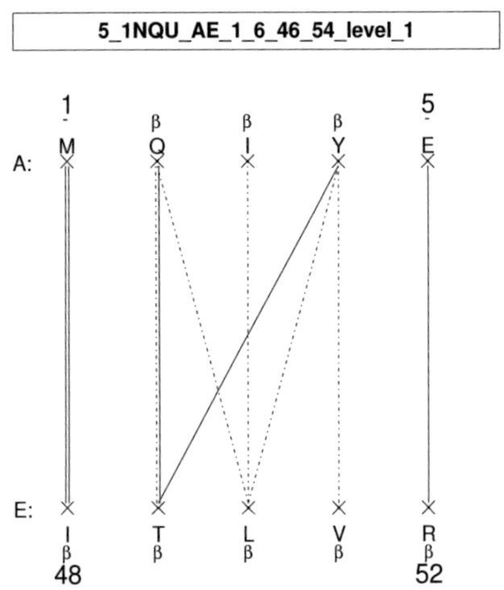

Figure 32. Parallel orientation of the β-strands.

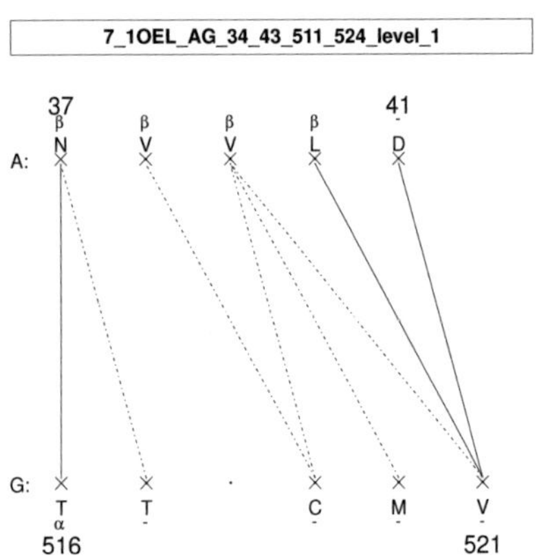

Figure 33. Parallel orientation of the β-strands.

http://dx.doi.org/10.5772/58420

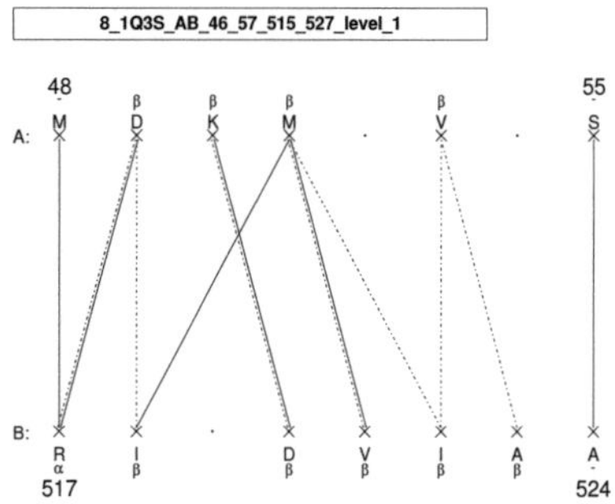

Figure 34. Parallel orientation of the β-strands.

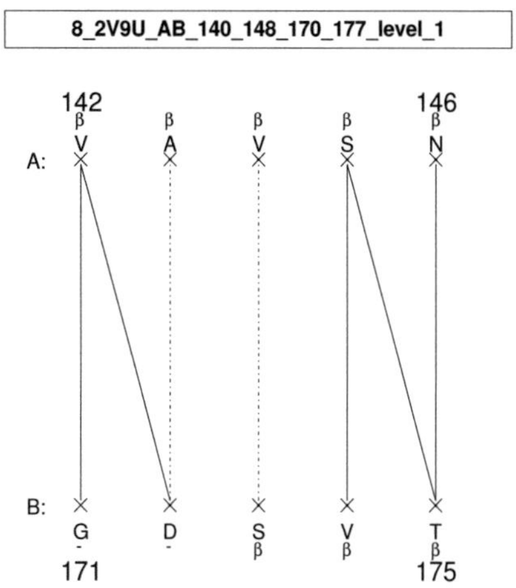

Figure 35. Parallel orientation of the β-strands.

5. Conclusion

The whole analysis presented so far has been triggered by the problem of investigating biological interfaces, namely interfaces that form during the biochemical activity in a cell, between or inside proteins, protein-DNA or protein-RNA complexes and so on.

In the introduction, motivations are given for the possibility of using geometry to understand the interactions. In the Results and Discussion sections, the topological, but intrinsically geometrical, properties of the interaction graphs have been presented with the first goal of learning how to distinguish the two main orientations and the second goal of estimating aspects that have been systematically neglected in all the previous publications.

Both these aspects have been clearly addressed in this paper. The tools to distinguish the two orientations are now more precise.

The results are the consistency of the information from level 1 and level 0 and the limited amount of information from level 2. Both the results suggest to enrich previous statistics and models for prediction with some input from level 1.

These results are confirmed by the naive counting of the degrees of freedhom: at least in a coarse-grained view, the two levels S_0^{aa} and S_1^{aa} provide enough information to reconstruct the shape of the interface (completeness).

In Methods, the notion of S_0^{aa} as the minimal description of the interface, or framework, already used in all our previous publications, has been presented here with solid mathematical arguments: as all buds are present in S_0^{aa} and no bud can appear later, it is legitimate to call S_0^{aa} a framework because a smallest set would miss some buds, namely some parts of the interface. We find that this and the completeness of the interface add important values to the validity of the methods.

Also, it is important to stress the ability of the symmetric minimization to detect the BB hydrogen bonds from the knowledge of the positions of non-hydrogen atoms only: in this case, geometry intrinsically reveals the chemical interactions, without making use of a cut off or other external scales.

Acknowledgements

This work is founded by the region Rhone-Alpes.

It's a pleasure to thanks Claire Lesieur for most valuable comments and suggestions, and Laurent Vuillon, for his help.

With sadness, the author remembers the friend and colleague Laurent Fournier, recently deceased, for his useful suggestions to improve the C++ codes used in the analyses.

Author details

Giovanni Feverati

Fédération de recherche MSIF, University of Savoie and CNRS, Annecy-le-Vieux, France

References

[1] Iacovache I, van der Goot G F, Pernot L. Pore formation: An ancient yet complex form of attack. Biochim Biophys Acta, vol. 1778, num. 7-8, p. 1611-23 (2008).

[2] Lesieur C, Vecsey-Semjen B, Abrami L, Fivaz M, van der Goot G F. Membrane insertion: The strategies of toxins (review). Mol Membr Biol 14: 45-64 (1997).

[3] Kirkitadze M D, Bitan G, Teplow D B. Paradigm shifts in Alzheimer's disease and other neurodegenerative disorders: the emerging role of oligomeric assemblies. J Neurosci Res 69: 567-577 (2002).

[4] Harrison R S, Sharpe P C, Singh Y, Fairlie D P. Amyloid peptides and proteins in review. Rev Physiol Biochem Pharmacol 159: 1-77 (2007).

[5] Crick F H C. The packing of alpha-helices: simple coiled-coils. Acta Crystallogr 6: 689-697 (1953).

[6] Feverati G, Achoch M, Zrimi J, Vuillon L, Lesieur C. β-strand interfaces of non-dimeric protein oligomers are characterized by scattered charge residues pattern. PLoS ONE 7(4): e32558 (2012). The article and the Supporting Information are freely downloadable from Plos One: http://www.plosone.org/article/info%3Adoi%2F10.1371%2Fjournal.pone.0032558#s5.

[7] Zrimi J, Ng Ling A, Giri-Rachman Arifint E, Feverati G, Lesieur C. Cholera toxin B subunits assemble into pentamers: proposition of a fly-casting mechanism. Plos One 5(12) e15347 (2010).

[8] Feverati G, Achoch M, Vuillon L, Lesieur C. Residue interaction networks of β-strand interfaces from healthy protein oligomers have few HUBs (multiple contact residues) and low interconnectedness: a potential protection mechanism against network rewiring and chain dissociation. Under revision by Plos One.

[9] ExPASy: Bioinformatics resource portal, http://www.expasy.org/proteomics/protein_structure

[10] Janin J, Bahadur R P, Chakrabarti P. Protein-protein interaction and quaternary structure. Q Rev Biophys 41: 133-180 (2008).

[11] Feverati G, Lesieur C. Oligomeric interfaces under the lens: Gemini. PLoS ONE 5(3): e9897 (2010).

[12] Feverati G, Lesieur C, Vuillon L. Symmetrization: ranking and clustering in protein interfaces. Referred proceeding of the conference "Mathematics of distances and applications", MDA 2012, Varna. Publisher: ITHEA, Sofia. Editors: M. Deza, M. Petitjean, K. Markov.

[13] Lance G N, Williams W T. A General Theory of Classificatory Sorting Strategies. 1. Hierarchical Systems The Computer Journal (1967) 9 (4): 373-380.

[14] Murtagh F and Contreras P. Methods of Hierarchical Clustering. Data Mining and Knowledge Discovery. Wiley-Interscience, Vol. 2, No. 1, pp. 86âĂŞ97 (2012).

[15] Martin A J M, Vidotto M, Boscariol F, Di Domenico T, Walsh I, Tosatto S C E. Residue Interaction Network Generator (RING). http://protein.bio.unipd.it/ring/

[16] Liberti L, Lavor C, Maculan N, Mucherino A. Euclidean Distance Geometry and Applications SIAM Review 56[1]:3-69. doi 10.1137/120875909

Chapter 13

From Tilings to Fibers – Bio-mathematical Aspects of Fold Plasticity

C. Lesieur and L. Vuillon

Additional information is available at the end of the chapter

http://dx.doi.org/10.5772/58577

1. Introduction

Protein oligomers are made by the association of protein chains via intermolecular amino acid interactions (interaction between subunits) forming so called protein interfaces. This chapter proposes mathematical concepts to investigate the shape constraints on the protein interfaces in order to promote oligomerization. First, we focus on tiling the plane (2 dimensions) by translation with abstract shapes. Using the fundamental Theorem of Beauquier-Nivat, we show that the shapes of the tiles must be either like a square or like a hexagon to tile the whole plane. Second, we look in more details at the tiling of a cylinder and discuss its relevancy in constructing protein fibers. The universality of such "building" properties are investigated through biological examples. This chapter is written four-hand by a mathematician and a biologist in order to present bio-mathematical aspects of fiber constructions.

Proteins are made by polymerization of 20 different amino acids via a covalent bond called the peptide bond. The ordered amino acids read along the peptide bonds constitute the protein backbone and are referred to as the primary structure or the primary sequence. The secondary structure involves hydrogen bonding to form α-helix or pleated-sheet structures. The intricate folding of the polypeptide chain in a globular protein via long range atomic interactions is referred to as the tertiary structure. Some proteins (hemoglobin, for example) have quaternary structure— several polypeptide chains are nested together. These proteins are called protein oligomers. The different chains associate by the formation of contact zones called protein interfaces, made of contacts between atoms of the amino acids (Fig. 1A). The majority of proteins are organized as oligomers and not as single monomers (for a review see chapter in this book by Gotte). Protein oligomers adopt different symmetries and different stoichiometries (number of repeated monomers). According to the PDB (Protein Data Bank [30]) where all available atomic structures of proteins are stored, protein oligomers exist in cyclic, dihedral and cubic point group symmetries. Most proteins have C_n or D_n symmetry (see [17, 22, 24]) and almost only viral proteins adopt cubic symmetry [35]. High stoichiometry complexes in eukaryotes have most often a C_1 symmetry (identity). Some proteins adopt helical symmetry and construct fibers (see [24]).

In this chapter, we investigate the construction of biological fibers by using a mathematical model for fibers. Essentially the formation of the fiber is considered as the tiling of an infinite height cylinder derived from an initial two dimensional tiling of the plane followed by the translation of a single tile. This single tile could be replaced by a n-mer in order to construct more complex fibers. Of course proteins have tridimensional shapes (3D-structure) and the biological fiber could have cross section and more complex internal organization not considered here because the idea is to construct a complete mathematical model for fibers applicable for any 3D-structure of the protein. This "simplification" is supported biologically because there exists many severe human diseases associated with the formation of fibers by proteins structurally and functionally unrelated. Notorious examples, are Alzheimer ($A\beta$-amyloid), Parkinson (synuclein), cerebral amyloid angiopathy (cystatin C-amyloidosis) and type II diabetes (IAPP, amylin). It is important to realize that fiber formation is also observed in cancer (p53), cardiovascular (transthyretin, serpin) and inflammatory diseases (serpin) (reviewed in [2, 6, 9, 26, 27]). These proteins (indicated in bracket next to the disease) have the fold plasticity to undergo a transition from an oligomeric state to a fiber state, change that leads to the loss of the protein function and the pathologies, called conformational diseases (Fig. 1B). Because this transition is shared by unrelated proteins, it is reasonable to assume that the change is based on a generic binding properties of the interface. The idea of the work is to define the shape constraints of the interfaces of a fiber (infinite height cylinder) compared to the interfaces of an oligomer (finite height cylinder) to ultimately address the problem of conformational diseases. The 2D tiling is relevant as the answer lies in the properties of surfaces. As mentioned, we are particularly interested in determining invariant properties, i.e. property of the surface that are true for any 3D-structure of the protein. In other words, the goal is to trace the local properties (the interface) necessary to carry out a global change. The relation between local properties and quaternary structures is discussed in details in the work by Claverie, Hofunung and Monod and the Monod, Changeux, Wyman (MCW) model ([8, 28]).

Clearly the transition to a fiber involves other determinants besides those of the protein interface, as for example the spatial positioning of the interface relative to the main body of the protein, a typical folding problem for which 3D-tiling would be more appropriate. Nevertheless this problem is not addressed in the present chapter which focus essentially on the transition from an oligomer to a fiber seen from the protein interface view point. For more information on the mechanisms of assembly and 3D determinant, one can read the chapter by Claire Lesieur titled "The Assembly of Protein Oligomers: Old Stories and New Perspectives with Graph Theory". For fiber formation, readers can referred to the work by Gebauer [11, 12].

This chapter proposes a mathematical approach of the problem using the principles of tiling and considering the protein interfaces as boundaries of a tile. The chapter is divided into 9 parts: (1) After this introduction. (2) We focus on usual C_n and D_n symmetries that build finite height cylinders and we explain the construction of interfaces between chains. In paragraph (3) we give a formal description of the boundary of a tile by using abstract geometrical objects called polyominoes, we state the fundamental Theorem of Beauquier-Nivat which explains the properties of the boundary of a tile associated with its capacity to tile a plan. (4) We present the regular tilings of the plane which are tilings constructed by translation of all integral combinations of 2 vectors. (5.1) These two directions of periodicity on the tilings allow us to construct fibers (that are infinite cylinders) and to

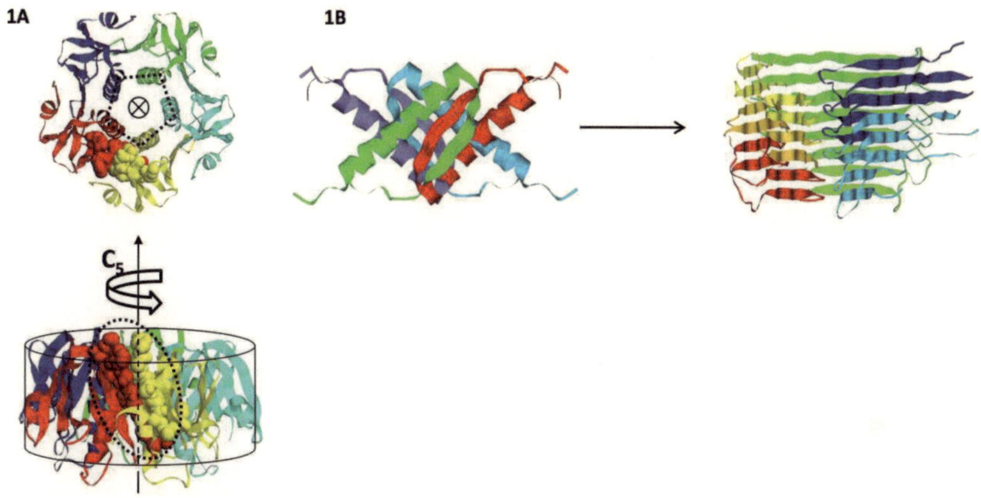

Figure 1. Protein oligomers and fibers. A. Protein oligomer. As an example, the x-ray structure of the cholera toxin B pentamer ($CtxB_5$) is shown (PDB code 3CHB). Each chain is indicated in a different ribbon color. The top and bottom images represent a top view and a side view of the pentamer, respectively.The atomic interactions involved in a protein interface are highlighted in space fill representation showing all atoms. Each chain participates for one domain of the interface, also called segment or side. ($CtxB_5$) has a C_5 symmetry that is well described by a cylinder (bottom image). **B. Oligomer to fiber transition.** Each monomer is indicated by a different color.

impose the use of either 2 interfaces or 3 interfaces. (5.2) We extend our construction to tile the fiber by n-mers instead of a single chain. (5.3) We explain how the fold plasticity is able to make a transition from finite cylinder to fiber in the p53 case. (6.1) We investigate the case of non regular tilings in order to design fibers with only 1 direction of periodicity and this implies strong constraints on the shape of the tile. (6.2) We also investigate transition from fiber to tiling of the whole three dimensional space. (7) We give a conclusion by comparing the fiber case to the whole tiling of the 3D space. All along this chapter biological illustrations are presented in parallel to the mathematical descriptions.

2. C_n and D_n symmetries on oligomerization

Oligomers with C_n symmetry can be constructed using a single chain replicated with a rotational axe of order n that is by a rotation of angle $360/n$ degrees (in other terms n equals to the number of monomers in the oligomers). Such oligomer adopts a cylindrical surface with each chain associating with 2 adjacent chains via a single interface (Fig. 1A). The interface is made of 2 interacting domains also called segments or sides, each one provided by one chain (Fig. 1A). It is important to clearly distinguish the notion of having one interface from the notion of having multiple regions or multiple patches of interfaces. Interfaces can be made of amino acids that are not contiguous along the backbone making up several "regions" (Fig. 2A). Yet the different regions still constitute a single interface because they are only able to bind to a single side. As an analogy, let's consider a poster to be stuck on a wall. You may choose to put several patches of glue on the reverse side of the poster or only one : it will still bind to only one surface. The C_n oligomers have only 1 interface and they polymerize following a single direction that is why they can describe a cylindrical surface (fig. 1A). An important notion is the number of adjacent chains in the oligomerization. In the C_n symmetry, each chain has exactly 2 adjacent chains (namely the Mth chain has for adjacent chains the $(M-1)$th chain and the $(M+1)$th chain (this calculation is done modulo n)) (see fig. 3). To summarize C_n is an oligomerization with 1 interface (possibly with distinct regions) and with 2 adjacent chains at each chain (see Fig. 3 and Fig. 4).

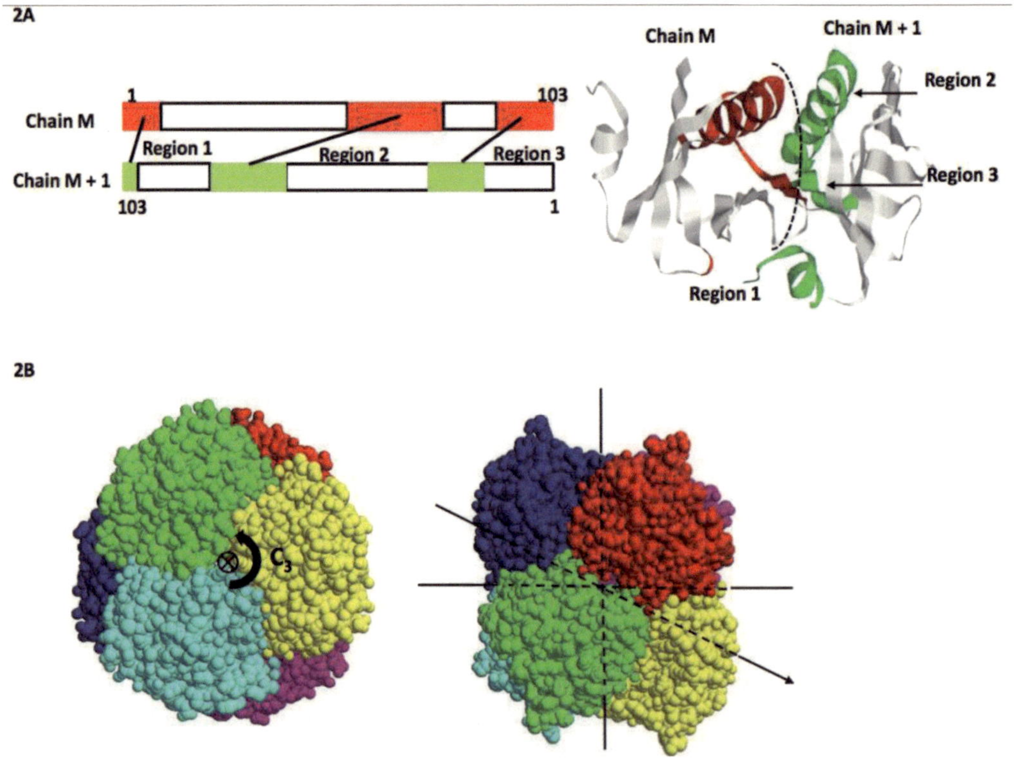

Figure 2. Protein interfaces. A. Left, rectangular schematic of the backbone of a protein oligomer. Two interacting chains are indicated, chain M and $M+1$. Chain M participates in one side of the interface indicated by the red boxes and chain $M+1$ participates in the complementary sides indicated in green. In the example the protein has 3 distinct regions of interface. Right, Illustration of the 3 regions of interface on the x-ray structure of CtxB5 (see fig. 1A). **B. D_n symmetry oligomer.** Example with the protein 3GVF (PDB code), a D_3 symmetry oligomer made of 6 chains organized as 3 dimers.

A slightly more complicated symmetry is the D_n symmetry oligomer which is reconstructed by a rotational axe plus 2 axes of symmetry perpendicular to the rotational axe. Such symmetry might look like to polymerize in 2 directions because it has one region of interface orthogonal to another one so the former can bind in a perpendicular direction. However, the "orthogonal" interface cannot grow besides forming a dimer so a D_n symmetry is in fact similar to a C_n symmetry considering the polymerization of a dimer instead of a monomer (Fig. 2B).

http://dx.doi.org/10.5772/58577

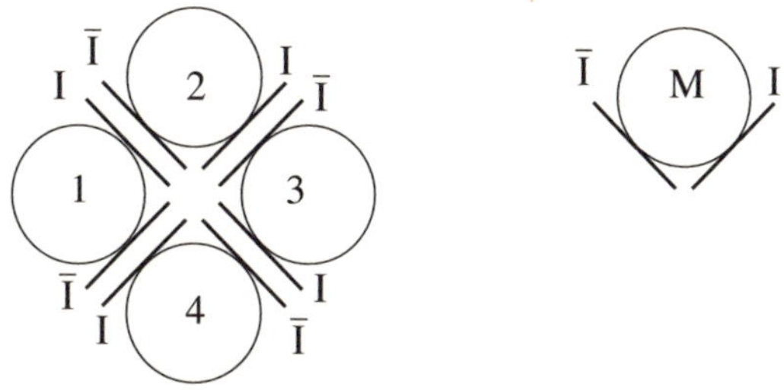

Figure 3. **Abstract view of C_4 symmetry.** A tetramer formed by a single interface between the part I and $\bar{I}$. Each chain is adjacent via the interface to exactly 2 other chains

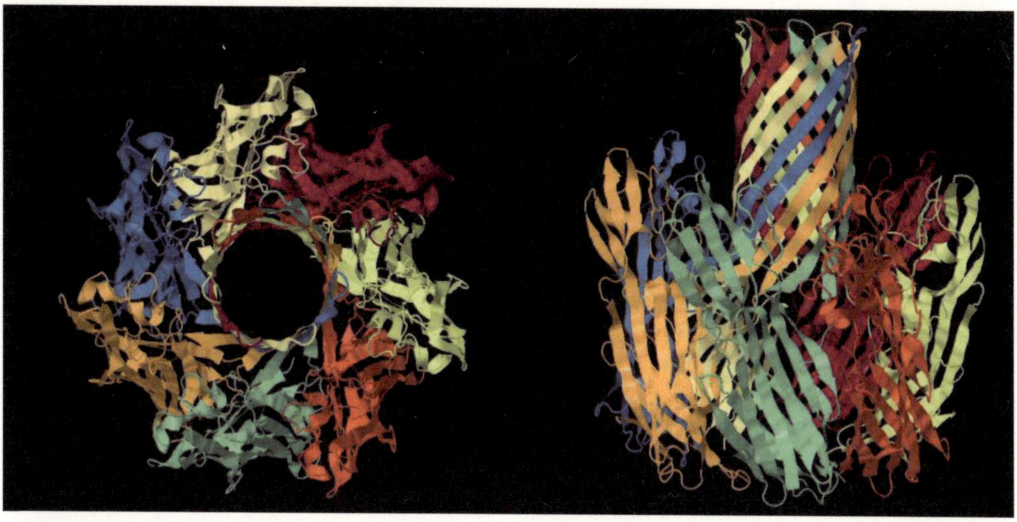

Figure 4. **Heptamer with 4H56 PDB code with each chain adjacent to 2 other chains.** Remark that in the beta barrel the adjacency between pairs of chains is the same as in the other part of the heptamer. This implies an oligomerization with exactly 1 interface.

3. Definitions and notation

A *polyomino* is a fundamental object in theory of tilings introduced by Golomb [16]. Polyominoes are building blocks for tiling and will be for us an abstract version of a single chain in an oligomerization process.

A polyomino is composed of juxtaposition of unit squares with corners on the $\mathbb{Z}^2$ grid (the corners of each unit square are integer points) such that every 2 adjacent squares in the polyomino share a unit segment; and also such that we can reach every pair of squares of the polyomino by a path inside the polyomino moving by unitary steps on adjacent unit squares (see Fig. 5).

The most simple polyomino has only one unit square and is called "mino"(Fig. 6).

There are 2 distinct polyominoes constructed by 2 adjacent unit squares and are called "domino". Remark that a domino can be horizontal or vertical this is why we find 2 dominoes (Fig. 7).

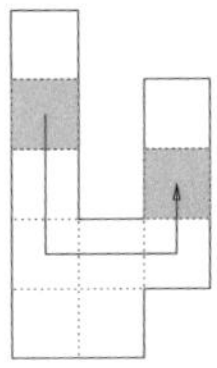

Figure 5. Path inside a polyomino

Figure 6. The most simple polyomino: the mino

Figure 7. The 2 dominoes

There are 6 polyominoes constructed by 3 2-by-2 adjacent unit squares: an horizontal bar of 3 unit squares, a vertical bar of 3 unit squares and four "triominoes *L*" with 3 adjacent unit squares that form an "*L*" shape and the 3 rotations by 90, 180 and 270 degrees (Fig. 8).

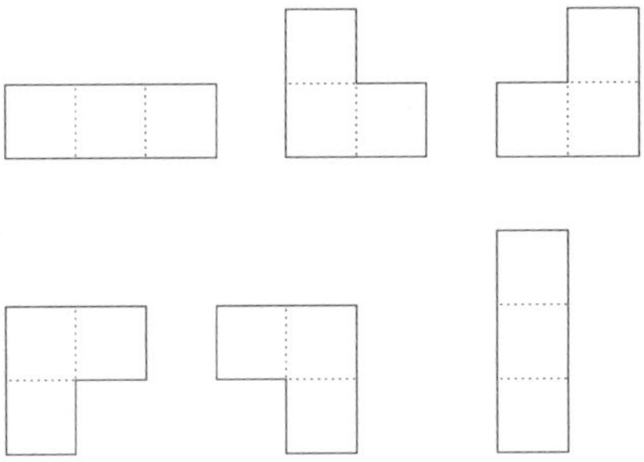

Figure 8. The 6 triominoes

In fact, the number of polyominoes with four unit squares (namely the tetraminoes) is 19.

In addition, the usual definition adds that there is no hole in the interior of a polyomino (Fig. 9).

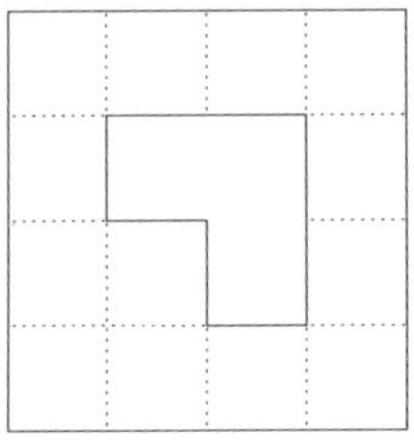

Figure 9. Union of unit squares with a hole in the interior

The number of polyominoes with n unit squares is given by the sequence number A001168 according to the Sloane Integer Sequences Encyclopedia (http://oeis.org/A001168): 1, 2, 6, 19, 63, 216, 760, 2725, 9910, 36446, 135268, 505861, 1903890, 7204874, 27394666,

http://dx.doi.org/10.5772/58577

104592937, 400795844, 1540820542, 5940738676, 22964779660, 88983512783, 345532572678, 1344372335524, 5239988770268, 20457802016011, 79992676367108, 313224032098244, 1228088671826973, ... We recognize the beginning of the sequence namely 1 mino, 2 dominoes, 6 triominoes, 19 tetraminoes and so on.

We don't know if it possible to find a closed formula (i.e. a mathematical formula depending on n that counts the number of distinct polyominoes with n unit squares) and this is a research problem for combinatorists.

We have defined polyominoes and now we would like to tile the whole plane by translation of a single polyomino.

Let P be a polyomino. A *tiling by translation* of P is a covering of the whole plane $\mathbb{R}^2$ by translated images of P such that there is no hole in the tiling and no overlapping. A polyomino that tiles the plane by translation is called a *tile* (Fig. 10).

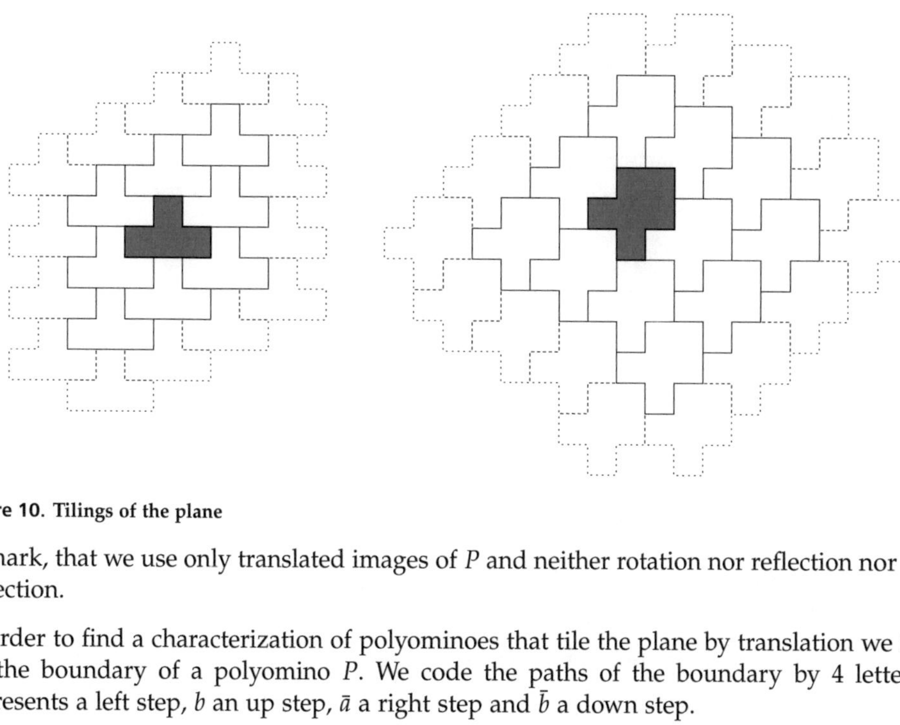

Figure 10. Tilings of the plane

Remark, that we use only translated images of P and neither rotation nor reflection nor glide reflection.

In order to find a characterization of polyominoes that tile the plane by translation we focus on the boundary of a polyomino P. We code the paths of the boundary by 4 letters: a represents a left step, b an up step, $\bar{a}$ a right step and $\bar{b}$ a down step.

Now, we take a starting point o on the boundary and we turn clockwise considering the sequence of steps in order to make a path constituted by unit segments from o to the first return on o. Let $\mathbf{b}(P)$ be a *boundary word of* P that is the path from o to o in clockwise that codes the boundary of the polyomino P in the following way : starting from an origin on the boundary o, the boundary word $\mathbf{b}(P)$ is the concatenation of labels of boundary unit segments read clockwise. Remark that as we turn clockwise we can't have in the step $a\bar{a}$, $\bar{a}a$, $b\bar{b}$ or $\bar{b}b$ otherwise the considered geometric figure is not a union of unit squares and thus the coded object could not be a polyomino.

For example, the boundary word of the mino is $ab\bar{a}\bar{b}$ and all cyclic permutations of letters $b\bar{a}\bar{b}a$, $\bar{a}\bar{b}ab$ and $\bar{b}ab\bar{a}$ leads to different boundary words depending where the starting point is taken. Nevertheless each cyclic permutation of letters define the same polyomino (Fig. 11).

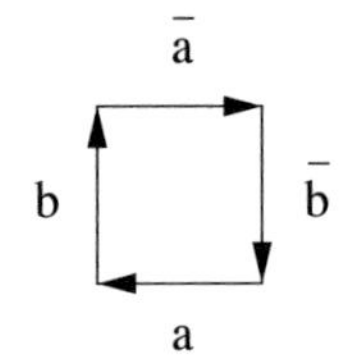

Figure 11. Coding of the boundary of the mino

Thus for a boundary word $w = w_1 w_2 \cdots w_n$ where all the w_i are letters that is $w_i \in \Sigma = \{a, b, \bar{a}, \bar{b}\}$ we define the conjugate of a boundary word w if $w = uv$ then a conjugate is defined by vu. We read on Fig. 11 the following boundary word of a mino $\bar{a}\bar{b}ab$ and we construct the 3 other possible boundary words (by changing the origin of the reading) by the notion of conjugate : we construct $b\bar{a}\bar{b}a$ by taking $u = \bar{a}\bar{b}a$ and $v = b$, we construct $ab\bar{a}\bar{b}$ by taking $u = \bar{a}\bar{b}$ and $v = ab$ and we construct $\bar{b}ab\bar{a}$ by taking $u = \bar{a}$ and $v = \bar{b}ab$.

This notion of boundary word is efficient because each polyomino can be characterized by it's boundary word.

For example the L-triomino could be defined directly by it's boundary word $b\bar{a}\bar{b}\bar{a}\bar{b}aab$ (Fig. 12). Remark that each conjugate of the word $b\bar{a}\bar{b}\bar{a}\bar{b}aab$ leads to the same L-triomino shape.

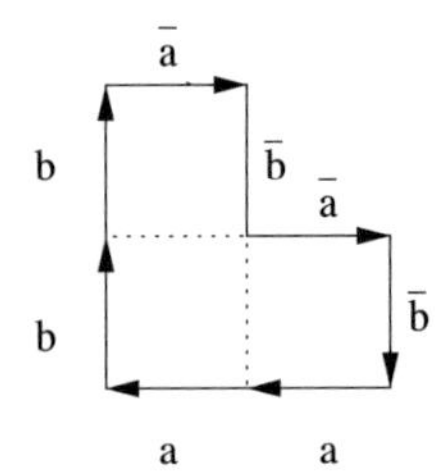

Figure 12. Coding of the boundary of the L-triomino

We know define a formal operation on words over the alphabet $\Sigma = \{a, b, \bar{a}, \bar{b}\}$. This operation will be crucial to define interfaces between tiles and for us abstract interfaces between chains.

We define the $\bar{u}$ operator on the word over the alphabet Σ by

if a word w of length n is written $w = w_1 w_2 \cdots w_n$ where all the w_i are letters that is $w_i \in \Sigma = \{a, b, \bar{a}, \bar{b}\}$

$(i) \overline{w} = \overline{w_n}\, \overline{w_{n-1}} \cdots \overline{w_1}$

$(ii) \bar{\bar{a}} = a$ and $\bar{\bar{b}} = b$.

The point (i) means that to use the bar operator on a word w, we bar each letter of the word w read in reverse order. The point (ii) means that 2 times the operator bar on a letter is equal to this letter.

In fact, if w is a part of a boundary word read clockwise then $\overline{w}$ is the same path read anti clockwise ! In Fig. 13 we see an example of a path clockwise associated with $w = b\bar{a}\bar{b}\bar{a}\bar{b}$ and the same path anti clockwise that is associated with $\overline{w} = \overline{b\bar{a}\bar{b}\bar{a}\bar{b}}$ by using the point (i) we reverse the word and take the bar of each letter and thus find $\overline{w} = \bar{\bar{b}}\bar{\bar{a}}\bar{\bar{b}}\bar{\bar{a}}\bar{b}$ and by simplification according to the point (ii) we finally have $\overline{w} = baba\bar{b}$.

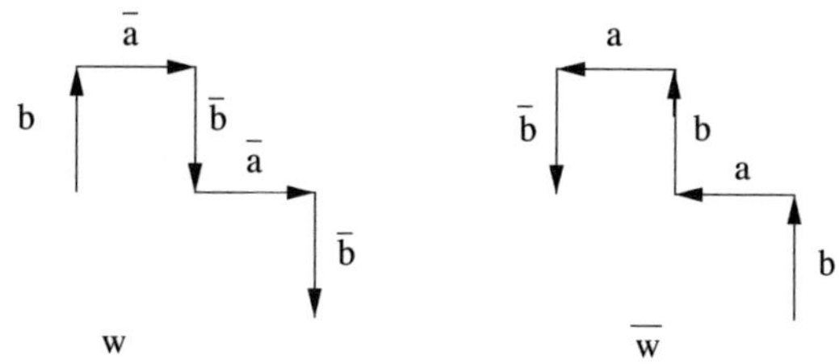

Figure 13. Coding of a finite path clockwise and anti clockwise

In a tiling of the plane each tile are surrounded by a certain number of adjacent copies of P. The spirit of the Beauquier-Nivat's Theorem [1] is that by investigating carefully a boundary word of P we could recover the possible surrounding of P.

Theorem 3.1 (Beauquier, Nivat). *A polyomino P tiles the plane by translation if and only if the boundary word* $\mathbf{b}(P)$ *is equal up to a cyclic permutation of the symbols to the factorization* $X \cdot Y \cdot Z \cdot \overline{X} \cdot \overline{Y} \cdot \overline{Z}$ *where one of the variables in the factorization may be empty.*

If the boundary word is equal to $X \cdot Y \cdot Z \cdot \overline{X} \cdot \overline{Y} \cdot \overline{Z}$ such a polyomino is called *pseudo hexagon,* if it is equal to $X \cdot Y \cdot \overline{X} \cdot \overline{Y}$ such a polyomino is called *pseudo square.*

Examples of factorizations and associated tilings of the plane.

A mino has a unique factorization on pseudo square, indeed the boundary word is $\mathbf{b}(P) = ab\bar{a}\bar{b}$ thus $X = a$ and $Y = b$ and $\overline{X} = \bar{a}$ and $\overline{Y} = \bar{b}$. In fact a mino tiles in a unique way and has the property that each tile is surrounded by 4 adjacent tiles (Fig. 14).

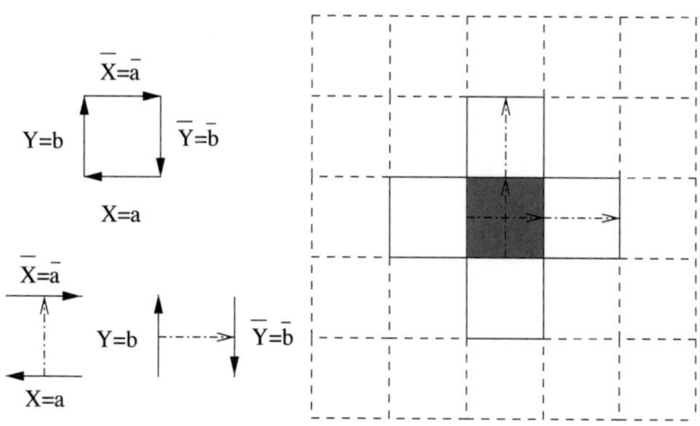

Figure 14. Tiling by a mino and the 4 adjacent tiles of the grey mino

A horizontal domino has exactly 2 factorizations $\mathbf{b}(P) = aab\bar{a}\bar{a}\bar{b}$ one in pseudo square by taking $X = aa$ and $Y = b$ and $\overline{X} = \bar{a}\bar{a}$ and $\overline{Y} = \bar{b}$ and given a tiling like a "non robust brickwall". In this case each tile is surrounded by 4 adjacent tiles (Fig. 15).

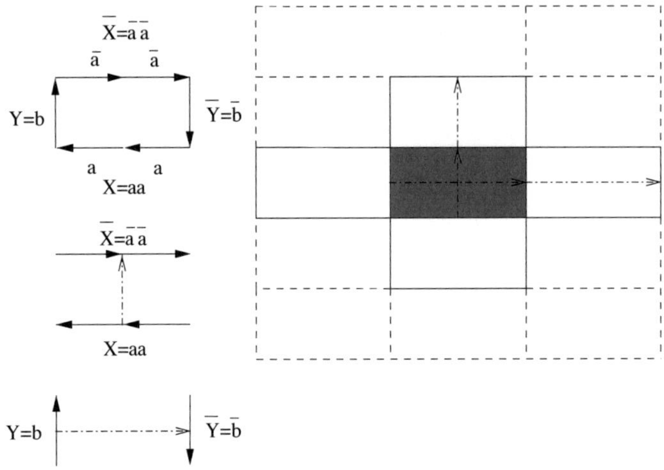

Figure 15. Tiling of the plane by a domino like a pseudo square and the 4 adjacent tiles of the grey domino

One factorization in pseudo hexagon by taking $X = a$, $Y = a$, $Z = b$ and $\overline{X} = \bar{a}$, $\overline{Y} = \bar{a}$, $\overline{Z} = \bar{b}$ which give a tiling like a "robust brickwall". In this case each tile is surrounded by 6 adjacent tiles (Fig. 16).

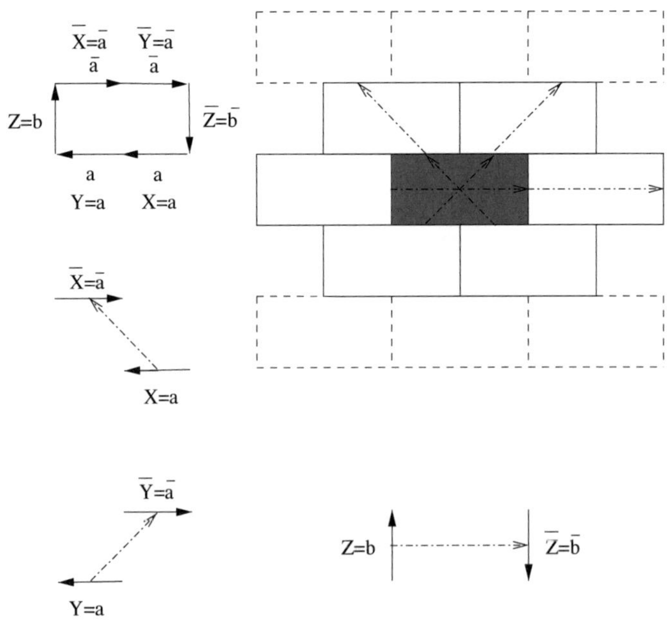

Figure 16. Tiling of the plane by a domino like a pseudo hexagon and the 6 adjacent tiles of the grey domino

Remark that the difference between the 2 tilings by a horizontal domino is the number of adjacent copies of polyominoes that surround the domino at the origin. Thus a domino tiles either like a pseudo square if the number of adjacent tiles of each tile is 4 or like a pseudo hexagon if the number of adjacent tiles of each tile is 6.

http://dx.doi.org/10.5772/58577

For the thin cross with 9 unit squares with boundary word equals to $aab\bar{a}\bar{a}bb\bar{a}\bar{b}\bar{b}\bar{a}\bar{a}\bar{b}aa\bar{b}\bar{b}abb$ it is impossible to find a factorization and this proves that the thin cross doesn't tile the plane anymore (the square with the symbol "?" cannot be covered) (Fig. 17).

Figure 17. **A thin cross that doesn't tile the plane**

The triomino L with boundary word equals to $b\bar{a}\bar{b}\bar{a}\bar{b}aab$ has only one factorization in pseudo hexagon, thus it tiles the plane in a unique way (Fig. 18).

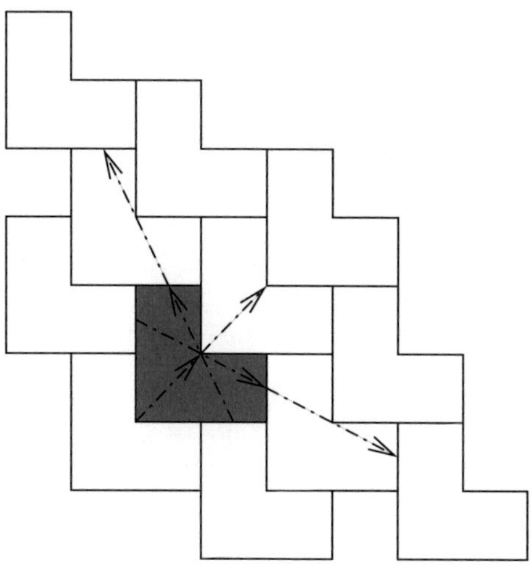

Figure 18. **The L-triomino tiles the plane like a pseudo hexagon**

While the little thin cross with 5 unit squares with boundary words $ab\bar{a}b\bar{a}\bar{b}\bar{a}\bar{b}a\bar{b}ab$ has 2 factorizations in pseudo squares and thus 2 distinct tiling ways according to the 2 factorizations with $X = aba, Y = b\bar{a}b$ and with $X = bab, Y = \bar{a}b\bar{a}$ (Fig. 19).

To summarize, for each polymino we are able to decide if it tiles the plane by translation by considering the factorization of it's boundary word. Now, we investigate regular tilings of the plane.

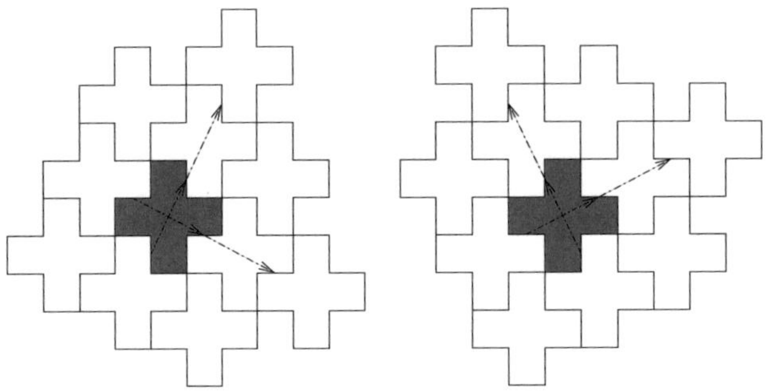

Figure 19. **The 2 tilings like a pseudo square by a thin cross with 5 unit squares**

4. Regular tilings by translation

The theorem of Beauquier-Nivat gives a correspondence between the way of tiling and the factorization that is the number of copies of the polyomino that surrounds a given polyomino. For the mino, we have 4 copies to cover the whole boundary (because a unit square is a pseudo square that tiles like a square). For the triomino L we have 6 copies to cover the whole boundary (because a triomino L is a pseudo hexagon that tiles like a hexagon). In fact we extend the local correspondences between the coding of part of the boundary word to a geometrical translations namely $\vec{v}_1$ translates X to $\overline{X}$ and $\vec{v}_2$ translates Y to $\overline{Y}$ (and possibly $\vec{v}_3$ translates Z to $\overline{Z}$). Remark that we have 2 or 3 translations according to the pseudo square or pseudo hexagon, respectively, and by translation to the whole tiling each polyomino has the same local surrounding.

It is interesting to introduce the notion of regular tiling. In a regular tiling the surrounding by adjacent tiles is the same for each polyomino. The mino and the triomino L tile only in a regular way by extending the local vectors to the whole tiling.

A *regular tiling* is a tiling by translation of a polyomino P such that each tile in the tiling has the same surrounding by translated copies of the tile P according to a given factorization of its boundary word. Each factorization leads to a regular tiling of the plane by translation by extending the local translation to the whole plane in the following way.

If P is a pseudo square, the factorization $\mathbf{b}(P) = X \cdot Y \cdot \overline{X} \cdot \overline{Y}$ defines 4 sides of the tile where the sides in correspondence are identified by the pairings $(X, \overline{X})$ and $(Y, \overline{Y})$. The translations $\vec{v}_1$, $\vec{v}_2$ corresponding to these pairings allow us to tile the whole plane in a regular way by using integral combinations of the translations $\vec{v}_1$ and $\vec{v}_2$ to P in order to generate the whole tiling (Fig. 14 or Fig. 15).

In the case of a pseudo hexagon the construction with 6 sides is similar. If P is a pseudo hexagon, the factorization $\mathbf{b}(P) = X \cdot Y \cdot Z \cdot \overline{X} \cdot \overline{Y} \cdot \overline{Z}$ defines 6 sides of the tile where the sides in correspondence are identified by the pairings $(X, \overline{X})$, $(Y, \overline{Y})$ and $(Z, \overline{Z})$. The translations $\vec{v}_1$, $\vec{v}_2$ and $\vec{v}_3$ corresponding to these pairings allow us to tile the whole plane in a regular way by using integral combination of the translations $\vec{v}_1$ and $\vec{v}_2$ to P in order to generate the whole tiling (Fig. 16 or Fig. 18) . Remark that the third vector of translation $\vec{v}_3$ is an integral combination of $\vec{v}_1$ and $\vec{v}_2$ ($\vec{v}_3 = \vec{v}_1 - \vec{v}_2$ in Fig. 16).

Observe that 2 distinct factorizations of the boundary word of P give 2 distinct regular tilings of the plane (Fig. 19 in the case of 4 adjacent tiles for each tile or Fig. 20 in the case of 6

http://dx.doi.org/10.5772/58577

adjacent tiles for each tile; be careful that while a regular tiling is always given by integral combinations of 2 vectors $\vec{v}_1$ and $\vec{v}_2$, the boundary of each tile in the pseudo hexagon case use 6 sides in correspondence 2-by-2 with the translations $\vec{v}_1$, $\vec{v}_2$ and $\vec{v}_3$). Are you able to recover the 6 adjacent polyominoes of the grey polyomino of Fig. 20 ? Hint: in Fig. 20 one adjacency is only by a single vertical step (for the left regular tiling) or horizontal (for the right regular tiling) ...

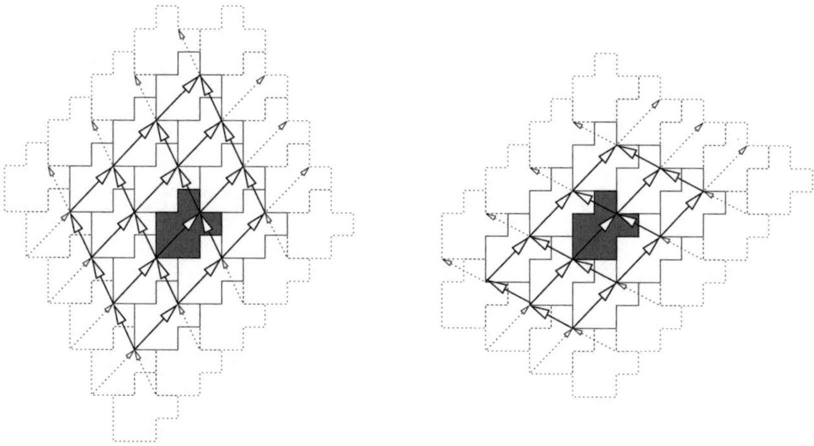

Figure 20. Two regular tilings of the plane by the same polyomino

5. From tilings of the plane to tilings of a fiber

5.1. Regular tiling case

From now on, we are dealing with the form of the oligomerization of proteins either like a cylinder or like a fiber. For us a cylinder has finite height (Fig. 1A) and a fiber is a possibly non finite height object (Fig. 25).

In our model we would like to make the construction of a fiber (that is a non finite height cylinder) by transformation on regular tilings. As the tiling is regular, the whole tiling is invariant by translation of the 2 non null and non collinear vectors $\vec{v}_1$ and $\vec{v}_2$ defined at the previous paragraph. $\vec{v}_1$ and $\vec{v}_2$ are the vectors of translations that respectively send X to $\overline{X}$ and send Y to $\overline{Y}$. In mathematics the parallelogram constructed by $\vec{v}_1$ and $\vec{v}_2$ at a given integer point is called the fundamental domain that is the minimal part of the tiling used to construct the whole tiling by integral combinations of $\vec{v}_1$ and $\vec{v}_2$.

Let $\vec{e}_1$ be a horizontal unit vector. We choose a direction parallel to $\vec{e}_1$ namely $m.\vec{v}_1 + n\vec{v}_2 = k.\vec{e}_1$ with m and n given integers. Then we construct the little circle (that is by usual definition the horizontal circle of the cylinder) whose perimeter is equal to k by superposing the tiles of the original tiling and the tiles translated by $k.\vec{e}_1$. Indeed as the tiling is invariant by translation of integral combination of $\vec{v}_1$ and $\vec{v}_2$ we are able recover the same shape by using the translation $m.\vec{v}_1 + n\vec{v}_2$ (which is by construction equal to $k.\vec{e}_1$). Fig. 21 shows an example of a cylinder constructed with a regular tiling by a mino using the identification of the two bold vertical borders for the translation 8 $\vec{v}_1$ (here for a mino $\vec{v}_1$=$\vec{e}_1$). In the associated cylinder the two vertical borders are collapsed in a single border and this transformation is general as soon as we identify in the tiling two infinite borders in correspondence by a translation of vector $m.\vec{v}_1 + n\vec{v}_2$.

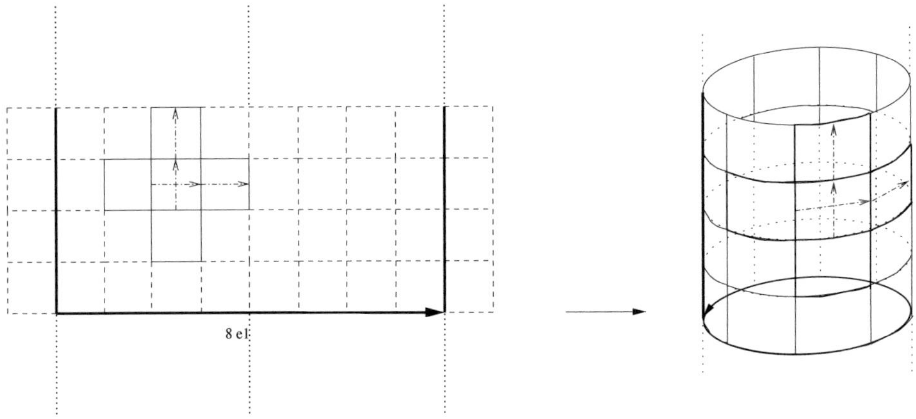

Figure 21. **From tiling to cylinder by using the translation of** 8 **times the vector** $\vec{e}_1$**.**

In the next example (Fig. 22), the vectors $\vec{v}_1$ and $\vec{v}_2$ send respectively $X = b\bar{a}b$ to $\overline{X} = \bar{b}a\bar{b}$ and $Y = aba$ to $\overline{Y} = \bar{a}\bar{b}\bar{a}$. In order to find a horizontal direction we take 2 times the vector $\vec{v}_1$ and 1 time the vector $\vec{v}_2$. Remark that the distance between the point O and the point O translated by $2\vec{v}_1 + \vec{v}_2$ is exactly 5. If we want a little circle with perimeter 10 we consider the translation $2 * (2\vec{v}_1 + \vec{v}_2)$ that is 4 times $\vec{v}_1$ and 2 times $\vec{v}_1$. Now we make a fiber by rolling up the plane by identification the 2 sides in bold (this is exactly the border of the tiling given by a staircase in correspondence by the translation with vector $4\vec{v}_1 + 2\vec{v}_2$) (Fig. 22). In order to have a more general construction and in particular to construct a twist in the fiber, we could use a rational coordinate vector $\vec{r}_1$ instead of $\vec{e}_1$ and find a direction parallel to $\vec{r}_1$ given by the equation $m\vec{v}_1 + n\vec{v}_2 = k\vec{r}_1$.

Conversely, if a polyomino tiles a fiber by translation we recover the whole tiling by finding the little circle and the invariance of the fiber by translation by $k.\vec{e}_1$ and considering in $\mathbb{R}^2$ the strip at the origin and adding all the strips on $\mathbb{R}^2$ by translation $\ell.k.\vec{e}_1$ with $\ell \in \mathbb{Z}$.

To summarize, in order to construct a model of fiber using a regular tiling by a single polyomino, the polyomino must tile the plane in a regular way and thus must have at least a factorization as a pseudo square or as a pseudo hexagon. In the fiber, if the tiling is regular then the number of adjacent copies that surround the border is either 4 or 6. This implies that the number of distinct interfaces is 2 or 3.

Of course the model is very flat and in real fibers the molecules are in a 3 dimensional space nevertheless this model allows us to make hypothesis about the number of interface sites and we try to find either 2 or 3 interface sites in the fiber.

We have explained that to build a fiber with a non finite height cylinder shape, it is necessary to have 2 vectors, namely 2 directions of polymerization growth. We have also seen that each tile of fiber will have 4 or 6 adjacent chains formed via 2 or 3 interfaces, respectively. There are examples of such fibers tiled with a single protein: the famous tobacco mosaic virus which was the first virus to be discovered has 4 adjacent chains to each chain (see Fig. 23 for

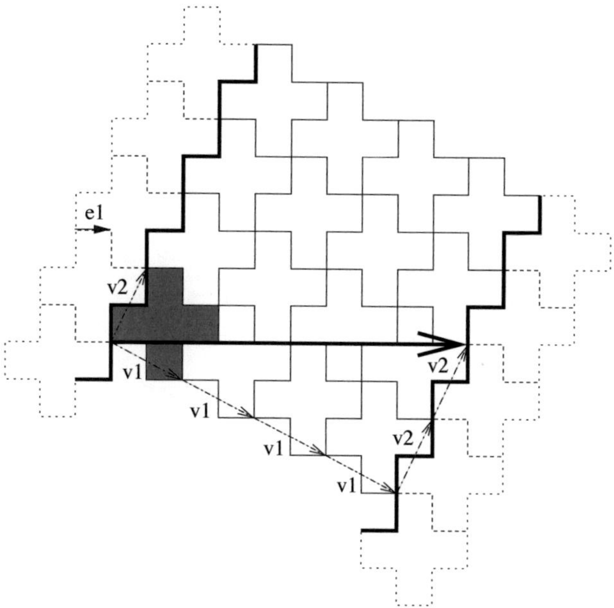

Figure 22. Two boundaries in correspondence by the translation $4\vec{v}_1 + 2\vec{v}_2)$

the abstract view and Fig. 24 for a part of a real fiber) and the polymerization in fiber with name 3J1R has 6 adjacent chains to each chain (see Fig. 25 for the abstract view and Fig.26 for a part of real fiber).

5.2. Tiling fibers by n-mers

This chapter is an investigation of the combinatorial possibilities on the construction of the fiber. Either a fiber is composed by a regular tiling by a single chain (Fig. 23 for the abstract view and Fig. 24 for a part of a real fiber with pseudo square tiling and Fig. 25 for the abstract view and Fig.26 for a part of real fiber with pseudo hexagon tiling) or the regular tiling of the fiber is composed by a n-mer (in mathematics we called this object a meta tile). Thus globally the fiber is a tiling by a meta-tile and this meta-tile could be a single chain or a n-mer.

In real life we find fibers tile either by a single chain or by a n-mer. In Fig. 27 we see a real fiber with name 3J2U constructed with a tetramer of three chains namely A, B and 2 times the chain C. The tetramer is used to tile the fiber of pseudo hexagon type like in Fig. 25 by substituting each pseudo hexagon by the tetramer $\begin{matrix} A & B \\ C & C \end{matrix}$. You will notice in the next section about p53, the final construction of the fiber uses tiling of dimers (see Fig. 33D).

Another example is the construction using oligonucleotides of DNA inspired by the seminal work of Winfree [37]. These technics lead to nano tubes composed by dimers surrounded by either 4 adjacent dimers [23] or 6 adjacent dimers [39].

Conway has a criterium for tiling the plane in the case of a tile and a rotation of the tile by 180 degrees. In his article, he forms a meta-tile using the tile T and it's rotation $\mathrm{Rot}_{180}(T)$ and this meta-tile is able to tile the plane by translation. Guy Cousineau wrote an article to

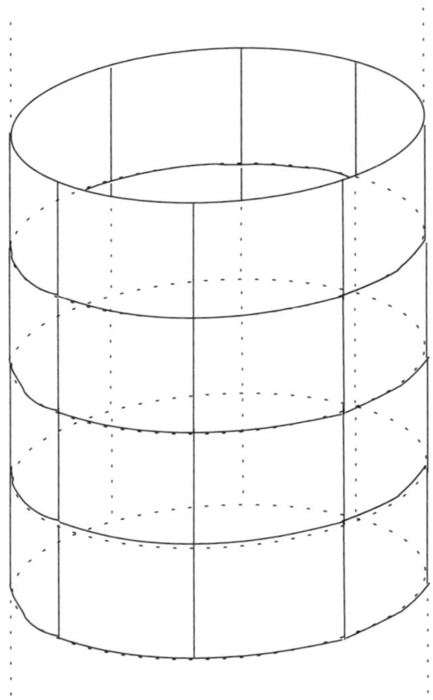

Figure 23. Fiber with a pseudo square shape. Each tile is surrounded by 4 tiles.

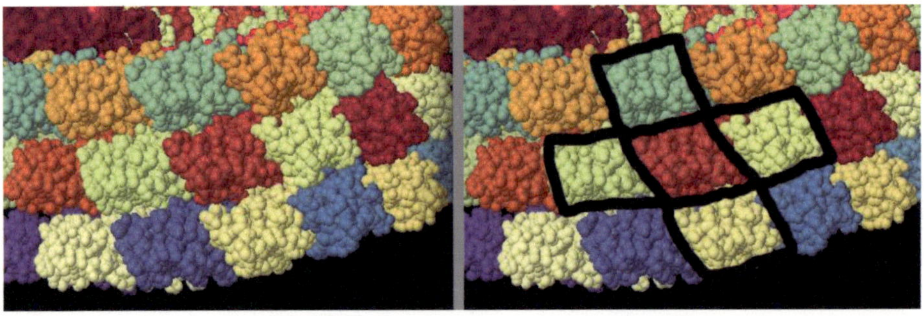

Figure 24. Tobacco Mosaic Virus with 2TMV PDB code: an example of tiling of a fiber with 4 adjacent chains.

explain the combinatorics of tilings with axial symmetry and translation (see the chapter of Cousineau in this book). In real life, fibers could be constructed for example with dimers. In general the dimer in the fiber is either invariant by a rotation of 180 degrees or invariant by an axial symmetry and then the dimer is used to tile by translation a fiber (see Fig. 28). Thus our model allows us to first make an interface for forming the dimer. And eventually to make a fiber by translation of the dimer according to the tiling constraints that is by having exactly 2 or 3 interfaces for the tiling of the fiber by the dimer. To recap we have constructed either 3 (one for the dimer and 2 for a tiling like a pseudo square) or 4 interfaces (one for the dimer and 3 for a tiling like a pseudo hexagon).

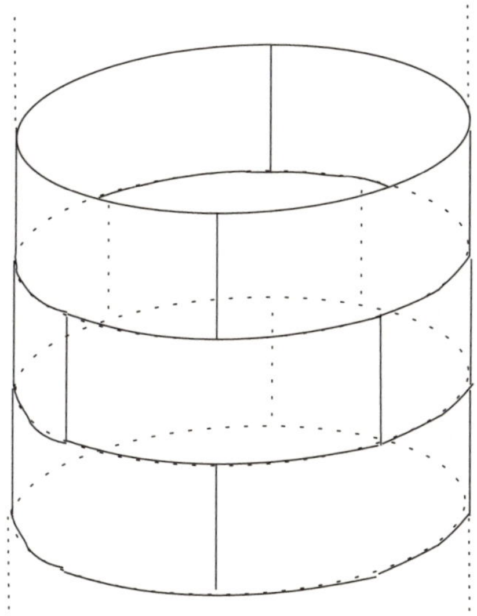

Figure 25. Fiber with a pseudo hexagon shape. Each tile is surrounded by 6 tiles. Remark that pseudo hexagon shapes appear in particular when there is a tilt on the fiber.

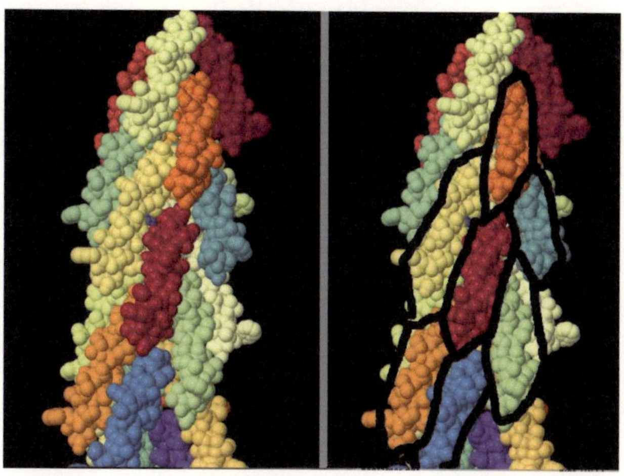

Figure 26. Fiber with 3J1R PDB code: an example of tiling of a fiber with 6 adjacent chains.

For higher stoichiometry, we construct a n-mer invariant by rotation of $360/n$ degrees and we copy this meta-tile in order to tile the whole fiber (see Fig. 29 for an abstract view of the replacement by a tetramer).

According to the previous tiling constraints, as the number of sites is 2 or 3 it is more easy to make a fiber with a monomer, dimer, trimer, tetramer and hexamer because we combine the shape of the n-mers with the form of the interfaces in order to build the whole fiber. We think that the better values are 1, 2, 3, 4, 6 because these values are also in accordance with the crystallographic constraints (the laws of crystallography [31] allow only periodical tilings with rotations of order 1,2,3,4 and 6, that is by rotations of $360, 180, 120, 90$ or 60

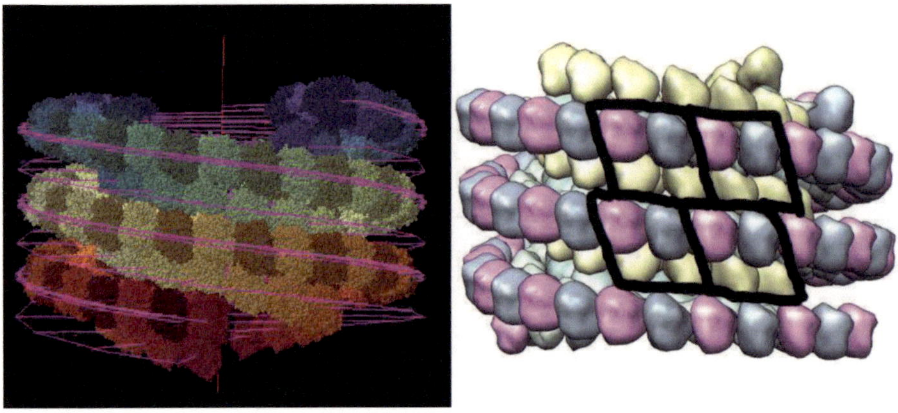

Figure 27. Fiber with 3J2U PBD code containing a tetramer inside each pseudo hexagon

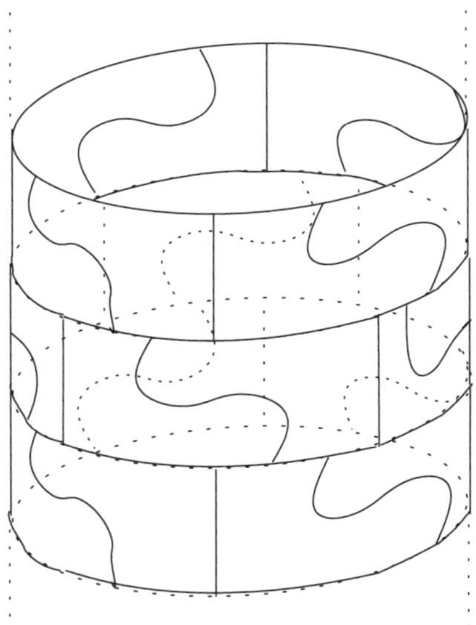

Figure 28. Fiber with a dimer inside each pseudo hexagon

degrees). Of course accordingly a transition from pentamer or heptamer ($n = 5$ or 7) to fiber would be expected to be more difficult at least with the mechanism described in this chapter. In this direction quasicrystals come from aperiodical tilings of the whole plane and non periodical structures could appear in viruses [35], nevertheless non periodical ways to form a fiber by translation of a single tile is described in the next section. Indeed, to form Penrose like tilings we must have two tiles with aperiodic arrangements [5, 31] and it is still an open problem to find an aperiodic tiling of the whole plane with a single tile [15] (we expect by the theorem of Beauquier-Nivat that the boundary of such tile (if it exists) must be fractal...). In our fiber model the rules seem to be restrictive because such symmetry fits directly without deformation into a square or hexagonal tile. Of course in real life, the n-mers could be symmetrical ([17, 22, 24]) but also non-symmetrical ([4, 32]). We think that for tiling constraints on the border of each n-mers the less expansive solution in order to construct a

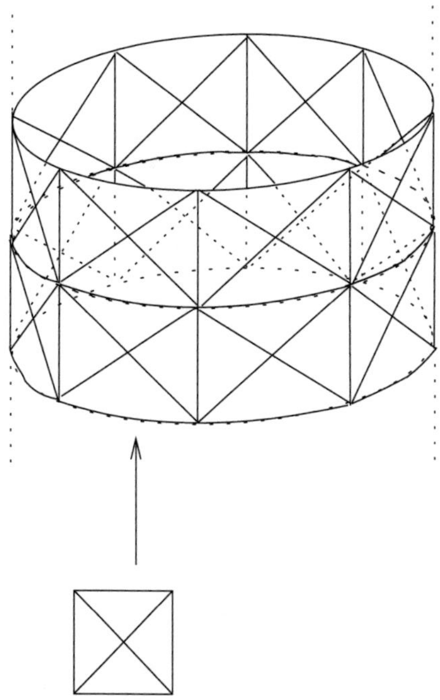

Figure 29. Replacement of each pseudo square by a tetramer.

fiber is given by symmetrical n-mers. On the other hand, dimers of pentamer or heptamers are able to fit with some deformations a hexagonal tile ie not all monomers are identical (Fig. 31). Such a dimer of pentamer has been observed in nature e.g. synthetic tachylectin (PDB 3KIF, Fig. 32) (see [38]).

Likewise the pore forming toxin aerolysin can form a dimer of heptamer after a single amino acid mutation (see [33]). It remains to be established whether the dimeric versions of the pentamer and heptamer are more prompt to form fibers than their single oligomeric counterpart. Of course the combinatoric is more and more complicated for $n > 7$ and to construct the fiber the position of the interfaces on the pseudo square or on the pseudo hexagon is split in different parts for $n > 7$ and this construction is very constrained and maybe this phenomenon is difficult to see in real life. Nevertheless such fibers could be constructed in synthetic biology.

We give some examples of replacements of a pseudo square by a dimer, by a tetramer and by an octamer and of a pseudo hexagon by a dimer, by a trimer and by a hexamer (see Fig. 30).

5.3. From fold plasticity to fibers: the P53 case

Notice that the main difference between tiling a finite height cylinder and tiling a fiber (an infinite height cylinder) is the number of directions of the polymerization growth, one in the former case and two in the latter case. Thus a transition from a cylinder to a fiber implies a transition from a single interface to 2 (4 adjacent chains) or 3 interfaces (6 adjacent chains). For this to occur one simple possibility is to start from an oligomer which has 2 regions of interfaces and loosen the constraints on one so it opens to bind in another direction. Common

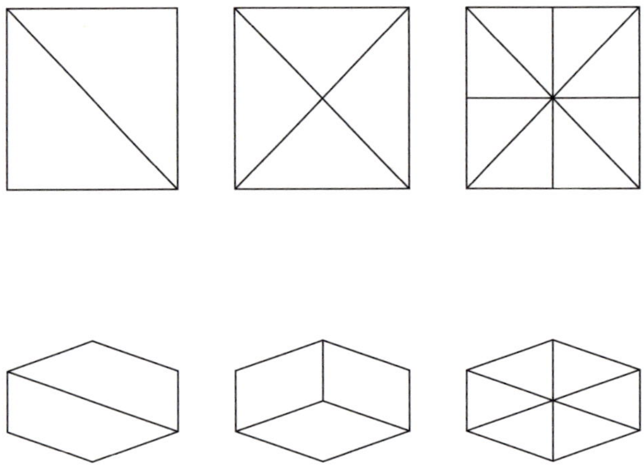

Figure 30. Replacement of a pseudo square and of a pseudo hexagon.

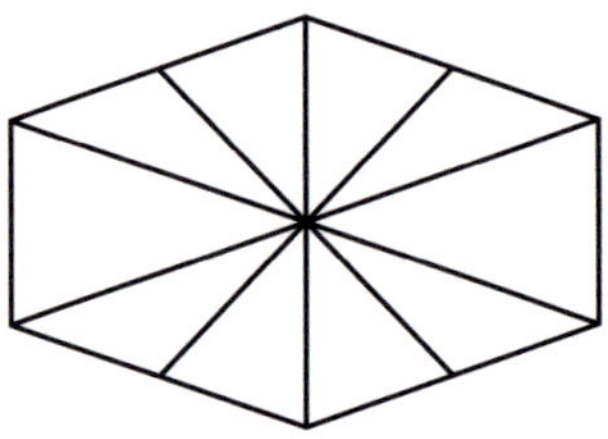

Figure 31. Replacement of pseudo square by 2 non regular pentamers.

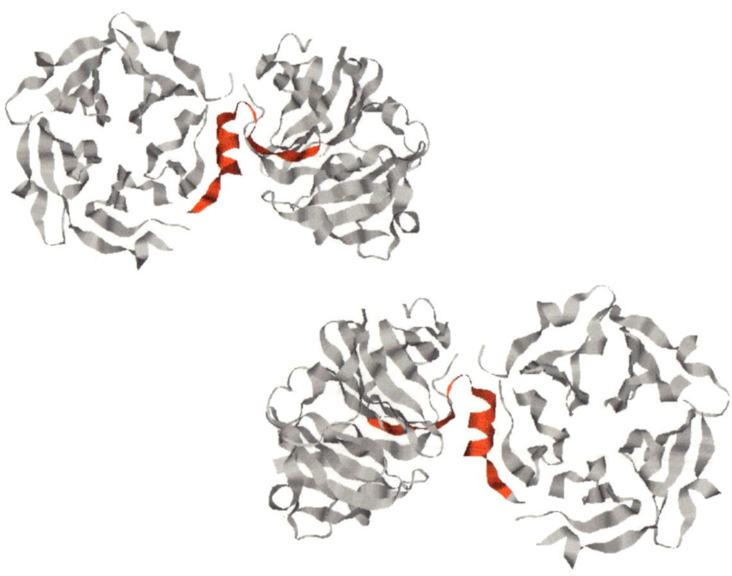

Figure 32. Double asymmetrical pentamer.

biological examples of such transition are domain swapping ([9, 25]) where the presence of a flexible linker frees one interface for fiber formation.

Another possibility is to start from a dimer with 2 regions of interfaces and again loosen one of the region but this time via a mutation. If the modified residue controlled the interactions between the 2 regions, its modification may loosen one region opening it to form a second interface. Since the initial molecule is a dimer, the mutation will free 2 new interfaces enabling the growth in 2 directions. The formation of the fiber will depend on the symmetry of the dimer. To illustrate this possibility, let's consider the p53 tetramer (see Fig. 33). There are several familial point mutations such as G334V or R337 that are responsible for the transition of the p53 into fibers or into a non-native oligomeric state, leading to impairment of the protein function and cancer development [27]. p53 tetramer has a D_2 symmetry and is a dimer of a dimer. Each monomer is made of a β-strand, a small helix and a long α-helix (Fig. 33B). Contacts exist between the residue R337 of the small helix and the residues E349 and D352 of the long helix of the adjacent chain within a dimer of p53 (Fig. 33A). There are contacts between the residues L350 and K351 of one monomer of a dimer and the same residues on another monomer of another dimer (Fig. 33A). The mutation of R337 might inhibit the formation of the small helix and replace it by a more flexible linker which would be too mobile to allow the contacts with the residues E349 and D352 and might also subsequently loosen the contacts between L350 and K351 (Fig. 33B and C). As a consequence the relative position of the β-strand and the long α-helix would be altered allowing the formation of 2 new interfaces (Fig. 33D).

6. Non periodical fiber case and tiling of the whole space case

6.1. From 1-periodical tiling to fibers

In fact, all the preceding constructions are based on regular tilings of the plane. We are able to relax this constraint and to investigate non regular tilings (Fig. 34). Non regular tilings of the plane are constructed by perturbation of regular tilings. We push some strips in the direction $\vec{v}_1$ in order to break the periodicity in direction $\vec{v}_2$. In the seminal article of Beauquier-Nivat [1] an important proposition is proved that a tiling (regular or not) by translation of a polyomino is always 1-periodic that is invariant by a translation by a non null vector. This proposition is crucial for us because it allows to take a non regular tiling by a polyomino P and to map it on a fiber.

We take a given tiling T by translation of a given polyomino P and as the tiling T is 1-periodic it is invariant by vector $\vec{v}_1$. Thus we have the property that the translation by $\vec{v}_1$ of the tiling T noted $\text{Trans}_{\vec{v}_1}(T)$ is equal to the tiling T (here to be more formal and to define the equality between tilings we must introduce equivalent classes of tilings up to translation operator but it is too formal for this chapter written for biologists). By composition of the translations, we have that the translation of k times the vector $\vec{v}_1$ of the tiling T is also equal to the tiling T. That is $\text{Trans}_{k.\vec{v}_1}(T) = T$. Thus globally the tiling T is invariant by the translations $k.\vec{v}_1$ for $k \in \mathbb{N}$.

This global property of invariance of the tiling T implies a strong property on the boundary word of P. In fact in the original article of Beauquier-Nivat the following proposition: if the tiling T is 1-periodic then the boundary word is of the form $XW\overline{X}V$ with $W = U^r$ and $V = \overline{U}^r$. Thus locally in the boundary word we are able to read the periodicity property.

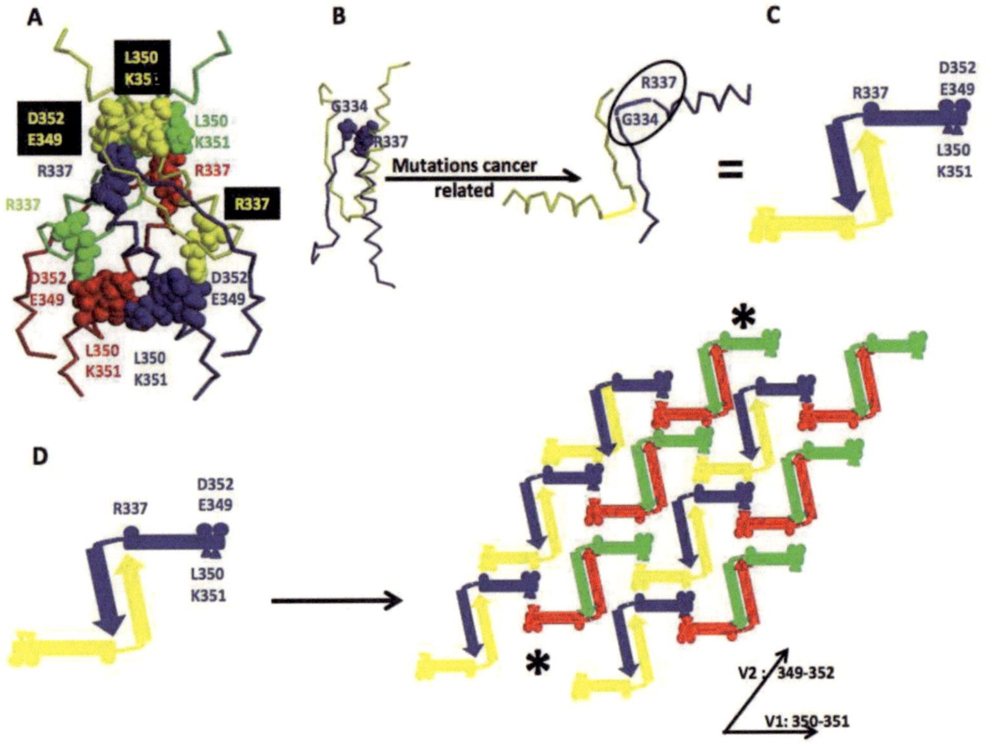

Figure 33. Transition from oligomer to fiber, the example of the p53 case. A. The p53 tetrameric domain is made of 2 dimers. Each monomer is made of a β-strand followed by a small helix ended by a long α-helix parallel to the β-strand (1SAK, PDB code). The backbone of each monomer is indicated by a different color. The spacefill amino acids are the residues in contact with R337, which is a residue sensitive to mutation related to some cancer development. **B. Backbone representation of the p53 dimer.** The model proposes that upon mutation of R337, connections between the 2 monomers within a dimer are loosen, enabling a free movement of the long α-helix relative to the β-strand. C. Schematic of the open-p53 dimer. D. As the structural change takes place in the four monomers, it offers enough new interfaces for growing in 2 directions $\vec{V}_1$ and $\vec{V}_2$.

For example for a bar of 5 unit squares with boundary word $abbbbb\bar{a}\bar{b}\bar{b}\bar{b}\bar{b}\bar{b}$, we have many factorizations one in pseudo square by taking $X = a$ and $Y = bbbbb$ and $\overline{X} = \bar{a}$ and $\overline{Y} = \bar{b}\bar{b}\bar{b}\bar{b}\bar{b}$ and 4 four factorisations in pseudo hexagon by taking $X = a$ and $Y = b$ and $Z = bbbb$ and $\overline{X} = \bar{a}$ and $\overline{Y} = \bar{b}$ and $\overline{Z} = \bar{b}\bar{b}\bar{b}\bar{b}$ or $X = a$ and $Y = bb$ and $Z = bbb$ and $\overline{X} = \bar{a}$ and $\overline{Y} = \bar{b}\bar{b}$ and $\overline{Z} = \bar{b}\bar{b}\bar{b}$ or ... or $X = a$ and $Y = bbbb$ and $Z = b$ and $\overline{X} = \bar{a}$ and $\overline{Y} = \bar{b}\bar{b}\bar{b}\bar{b}$ and $\overline{Z} = \bar{b}$. In this example the periodic part on the boundary word is $U^5 = bbbbb$ with $U = b$ and $\overline{U}^5 = \bar{b}\bar{b}\bar{b}\bar{b}\bar{b}$. Remark that in this case $U^5 = bbbbb$ appears in the factorization in pseudo square.

Another example without factorization in pseudo square is a polyomino constructed by union of 3 triomino L and forming a polyomino with 9 unit squares with boundary word $\bar{b}aababab\bar{a}b\bar{a}\bar{b}\bar{a}\bar{b}\bar{a}\bar{b}$ here the factorizations are

$X = \bar{b}a$ $Y = a$ $Z = babab$ $\overline{X} = \bar{a}b$ $\overline{Y} = \bar{a}$ $\overline{Z} = \bar{b}\bar{a}\bar{b}\bar{a}\bar{b}$ or $X = \bar{b}a$ $Y = aba$ $Z = bab$ $\overline{X} = \bar{a}b$ $\overline{Y} = \bar{a}\bar{b}\bar{a}$ $\overline{Z} = \bar{b}\bar{a}\bar{b}$ or $X = \bar{b}a$ $Y = ababa$ $Z = b$ $\overline{X} = \bar{a}b$ $\overline{Y} = \bar{a}\bar{b}\bar{a}\bar{b}\bar{a}$ $\overline{Z} = \bar{b}$. In this case $U^3 = ababab$ with $r = 3$ and $U = ab$ and $\overline{U}^3 = \bar{b}\bar{a}\bar{b}\bar{a}\bar{b}\bar{a}$ with $r = 3$ and $\overline{U} = \bar{b}\bar{a}$. Remark that in this case U^3 does not appear in the boundary word factorization and is splited in 2 parts, one is Y and

http://dx.doi.org/10.5772/58577

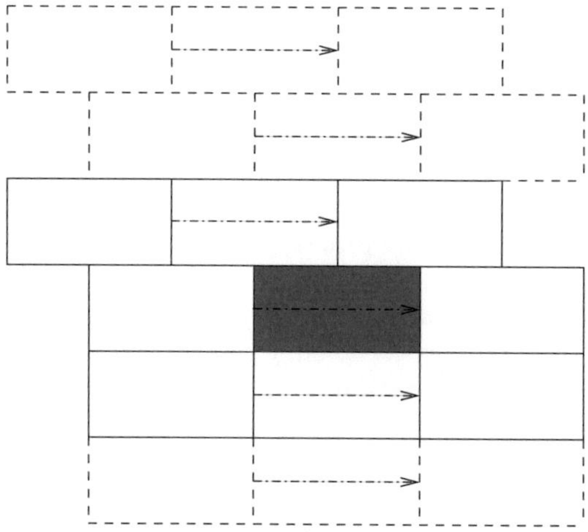

Figure 34. Non regular tiling with dominoes. The grey domino is surrounded by 5 dominoes thus this tiling is certainly not regular. Remark that the horizontal strips are invariant by the horizontal vector $\vec{v}_1$ and also by the vector $3\vec{v}_1$. Thus we are able to map the left border to the right border in order to form a non finite height cylinder.

the second is Z because $W = U^3 = YZ$. Remark in the picture the periodic part in a staircase shape (Fig. 35).

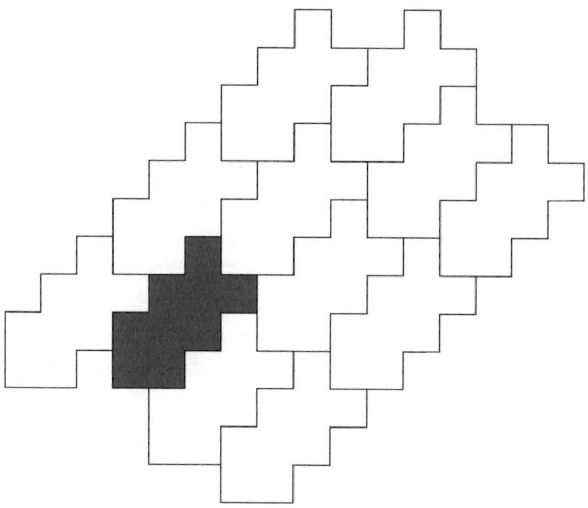

Figure 35. Non regular tiling and stair case shape.

This condition on the border is a necessary condition to find a possible periodicity. In particular, if we want to make the tiling 1-periodic we must push the strip along this direction of periodicity.

In order to construct a fiber we take the tiling and map it on a fiber by superposing tiles according to the direction of periodicity given by $\vec{v}_1$. Now we choose a non null value for k, we are able to construct a fiber because the whole tiling is invariant by the translation $k\vec{v}_1$ and we superpose the tiles of the original tiling and the tiles translated by $k\vec{v}_1$. Thus we make a fiber that contains exactly k polyominoes along the direction $\vec{v}_1$. An important implication of this construction is the possibility of shifting the strips using the stair case shape. This means that the fiber has one degree of liberty and can tilt along the periodic direction.

6.2. From fibers to tilings of the space

To end this section we give some hints about a possible transition from fiber to tiling of the whole 3 dimensional space. In fact the protein chains are in a 3 dimensional space and we could investigate regular tiling of a 3 dimensional space by a translation of a single polycube (generalization in 3D of the polyomino) that is a union of unit cubes face to face. Such kind of regular tilings leads to tile the whole space by using 3 vectors of translation and in [14] Gambini and Vuillon show that the surrounding of a tile could have n adjacent tiles with a fixed integer $n \geq 6$. For tiling the plane or fibers the number of interfaces is finite and equals to 2 or 3 and in contrast for tiling the space by a polycube the number of interfaces could be greater than each integer n with $n \geq 3$. This means that the number of interfaces could not be finite a priori (see [14]). One interesting question is the following, could we observe in real life a transition by fold plasticity from a fiber to a tiling of the whole 3 dimensional space ?

7. Conclusion

In this chapter, we focus on mathematical aspects of fibers. The main goal consists on the explanation of basic rules in order to construct cylinders or fibers based on regular tilings by translation of a single chain or of a juxtaposition of chains forming a n-mer. We notice that the interfaces must be 2 or 3 according to the tiling on pseudo squares or pseudo hexagons. This is the basic construction and leads to a fiber with a pseudo square tile or with a pseudo hexagon tile. Now, in order to add more combinatorial complexity, we are able to replace each tile on the fiber by a n-mer constructed by adding at least another interface. This result inherits the strong combinatorial properties of tilings by polyominoes explored by Beauquier and Nivat [1, 13] and we add the power of combinatorial construction of the n-mers. We could now use this construction to explore the structure of real fibers or to construct synthetic fibers. The steps of investigation will be the following: first we look globally at the fiber by searching the smallest object that tiles the fiber (Fig. 23 for the abstract view with pseudo square tiling and Fig. 24 for a part of a real fiber and Fig. 25 for the abstract view with pseudo hexagon tiling and Fig.26 for a part of real fiber). If this object (called meta-tile) is a single chain then we stop the investigation because the structure is found. Otherwise the meta-tile that tiles the fiber is composed by a n-mer and we investigate carefully the structure of this n-mer (in Fig. 25 the tiling meta-tile is a 4-mer with form $\begin{matrix} A & B \\ C & C \end{matrix}$ and in the p53 fiber, the object is a 2-mer (see Fig. 33D). Thus each fiber is composed if we zoom out by a tiling by translation by a meta-tile which is a pseudo square or is a pseudo hexagon and if we zoom in the meta-tile is a n-mer. Of course if $n = 1$ then the tiling is by a single chain and if $n \geq 2$ the tiling of the fiber is done by a meta-tile which is a n-mer. Remark that for $n = 1$ as we tile by a single shape the number of interfaces is 2 for the pseudo square case or 3 for the pseudo hexagon case. If $n \geq 2$ then we must add at least 1 interface to form the n-mer and thus globally the number of interfaces is at least 3 for the pseudo square case or 4 for the pseudo hexagon case. Our model covers all cases because by zooming in the meta-tile inherits the intrinsic combinatorial complexity of the n-mer construction and by zooming

http://dx.doi.org/10.5772/58577

out we recover the 2 ways of tilings either by a pseudo square or a pseudo hexagon. Now we have a complete theory of fiber construction in biology and new robust bio-mathematical tools in order to reinvestigate our favorite fibers.

A future exploration can select from the entire PDB, the structures with the global symmetry most appropriate for fiber formation. This set can be screened further for isolating oligomers with only 2 interfaces or only 3 interfaces. These oligomers contain the necessary elements for fiber formation. We can then sort the dataset into two categories, cases known to form pathological fibers and the others. Comparing the characteristics of the interfaces of the cases whose fate is to become pathological fibers and of the "non pathological" cases would help distinguishing the parameters providing the plasticity needed for the transition to fiber from the parameters providing resistance to fiber formation.

Acknowledgements

We would like to thank Claudia Billat who reads carefully a previous version of this article.

Author details

C. Lesieur[1] and L. Vuillon[2*]

*Address all corespondence to: Laurent.Vuillon@univ-savoie.fr
1 Université Joseph Fourier, AGIM, Grenoble, France
2 Laboratoire de mathématiques, Université de Savoie, France

References

[1] Beauquier D. and Nivat M., On translating one polyomino to tile the plane, *Disc. Comput. Geom.*, 6 (1991) 575-592.

[2] Bellotti V., Chiti F. (2008) Amyloidogenesis in its biological environment: challenging a fundamental issue in protein misfolding diseases. Current opinion in structural biology 18: 771-779.

[3] Bernstein, F. C., Koetzle, T. F., Williams, G. J., Meyer, E. F., Jr., Brice, M. D., Rodgers, J. R., Kennard, O., Shimanouchi, T., and Tasumi, M. (1977) The Protein Data Bank. A computer-based archival file for macromolecular structures. Eur J Biochem 80, 319-324

[4] Brown, Jerry H. (2006) Breaking symmetry in protein dimers: Designs and functions. Protein Science,15:1,1–13.

[5] de Bruijn, N. G. (1981) Algebraic theory of Penrose's nonperiodic tilings of the plane. I, II. Nederl. Akad. Wetensch. Indag. Math., 1:43, 39–66.

[6] Cheng P.-N., Pham J.D. and Nowick J.S. (2013). The Supramolecular Chemistry of β-Sheets. Journal of the American Chemical Society.

[7] Chiang P.K., Lam M.A. and Luo Y. (September 2008). "The many faces of amyloid beta in Alzheimer's disease". Current molecular medicine 8 (6): 580–4. doi:10.2174/156652408785747951. PMID 18781964.

[8] Claverie, P. and Hofnung, M. and Monod, J. (1968) Sur certaines implications de l'hypothese d'équivalence stricte entre les protomeres des protéines oligomériques. CR Séanc. Acad. Sci 266, 1616–1618.

[9] Eisenberg D. and Jucker M. (2012). The Amyloid State of Proteins in Human Diseases. Cell 148: 1188-1203.

[10] Ferreira S.T., Vieira M.N. and De Felice F.G. (2007). Soluble protein oligomers as emerging toxins in Alzheimer's and other amyloid diseases. IUBMB life 59 (4–5): 332–45. doi:10.1080/15216540701283882. PMID 17505973.

[11] Gebauer, D. and Völkel, A. and Cölfen, H. (2008). Stable prenucleation calcium carbonate clusters. Science 322 5909: 1819–1822.

[12] Gebauer, D. and Cölfen, H.(2011). Prenucleation clusters and non-classical nucleation.Nano Today 6 6: 564–584.

[13] Gambini I. and Vuillon L., An algorithm for deciding if a polyomino tiles the plane, *Theoretical Informatics and Applications*, vol. 41, 2 (2007), 147–155.

[14] Gambini I. and Vuillon L., How many faces can polycubes of lattice tilings by translation of $\mathbb{R}^3$ have?, *the electronic journal of combinatorics*, Vol 18, (2011), P199.

[15] Gambini I. and Vuillon L., Non lattice periodic tilings of $\mathbb{R}^3$ by single polycubes, *Theoretical Computer Science*, 432, (2012), 52–57.

[16] Golomb S. W., Checker boards and polyominoes, *Amer. Math. Monthly*, vol. 61, 10 (1954) 675–682.

[17] Goodsell, D.S. and Olson, A.J. 2000. Structural symmetry and protein function. Annu. Rev. Biophys. Biomol. Struct. 29: 105–153.

[18] Grunbaum B. and Shephard G.C., Tilings with congruent tiles, *Bull. Amer. Maths. Soc.*, vol 3, 3 (1980) 951–974.

[19] Haataja L, Gurlo T, Huang CJ, Butler PC (May 2008). "Islet amyloid in type 2 diabetes, and the toxic oligomer hypothesis". Endocrine Reviews 29 (3): 303–16. doi:10.1210/er.2007-0037. PMC 2528855. PMID 18314421.

[20] Höppener J.W., Ahrèn B. and Lips C.J. (August 2000). "Islet amyloid and type 2 diabetes mellitus". The New England Journal of Medicine 343 (6): 411–9. doi:10.1056/NEJM200008103430607. PMID 10933741.

[21] Irvine G.B., El-Agnaf O.M., Shankar G.M. and Walsh D.M. (2008). "Protein aggregation in the brain: the molecular basis for Alzheimer's and Parkinson's diseases". Molecular

medicine (Cambridge, Mass.) 14 (7–8): 451–64. doi:10.2119/2007-00100.Irvine. PMC 2274891. PMID 18368143.

[22] Janin J., Bahadur R.P. and Chakrabarti P. Protein-protein interaction and quaternary structure. Q Rev Biophys 2008;41(2):133–80.

[23] LaBean T. and Park S. H. (2006). Self-assembled DNA Nanotubes. Nanotechnologies for the Life Sciences.

[24] Levy E.D. and Teichmann S., Structural, Evolutionary, and Assembly Principles of Protein Oligomerization. In Jesús Giraldo and Francisco Ciruela, editors: Progress in Molecular Biology and Translational Science, Vol. 117, Burlington: Academic Press, 2013, pp. 25–51.

[25] Liu, Y., Gotte, G., Libonati, M. and Eisenberg, D. (2001) A domain-swapped RNase A dimer with implications for amyloid formation. Nat Struct Biol 8, 211–214.

[26] Lomas D.A. and Carrell R.W. (2002) Serpinopathies and the conformational dementias. Nature Reviews Genetics 3: 759–768.

[27] Lwin, T. Z., Durant, J. J. and Bashford, D. (2007) A fluid salt-bridging cluster and the stabilization of p53. J Mol Biol 373, 1334–1347.

[28] Monod J., Wyman J., and Changeux J.-P. (1965). On the nature of allosteric transitions: a plausible model. J. Mol. Biol. 12: 88–118.

[29] Ochieng J. and Chaudhuri G. (2010) Cystatin superfamily. J Health Care Poor Underserved 21: 51-70.

[30] Protein Data Bank : http://www.rcsb.org/pdb/home/home.do

[31] Senechal, M. (1995) Quasicrystals and geometry, Cambridge University Press.

[32] Swapna L.S., Srikeerthana K. and Srinivasan N. (2012) Extent of Structural Asymmetry in Homodimeric Proteins: Prevalence and Relevance. PLoS ONE 7(5): e36688. doi:10.1371/journal.pone.0036688

[33] Tsitrin Y, Morton CJ, el-Bez C, Paumard P, Velluz MC, et al. (2002) Conversion of a transmembrane to a water-soluble protein complex by a single point mutation. Nat Struct Biol 9: 729-733.

[34] Tuncbag, N., Kar, G., Keskin, O., Gursoy, A., and Nussinov, R. (2009) A survey of available tools and web servers for analysis of protein–protein interactions and interfaces. Briefings in Bioinformatics 10, 217.

[35] Twarock, R., (2006) Mathematical virology: a novel approach to the structure and assembly of viruses. Philosophical Transactions of the Royal Society A: Mathematical, Physical and Engineering Sciences 364, 1849, 3357–3373.

[36] de Vries, S. J., and Bonvin, A. M. (2008) How proteins get in touch: interface prediction in the study of biomolecular complexes. Curr Protein Pept Sci 9, 394-406.

[37] Winfree, E., Liu, F., Wenzler, L. A., and Seeman, N.C., Design and self-assembly of two-dimensional DNA crystals., Nature 1998, 394, 539-544.

[38] Yadid I, Kirshenbaum N, Sharon M, Dym O, Tawfik DS (2010) Metamorphic proteins mediate evolutionary transitions of structure. Proc Natl Acad Sci U S A 107: 7287-7292.

[39] Yin, P., Hariadi, R. F., Sahu, S., Choi, H. M., Park, S. H., LaBean, T. H., and Reif, J. H. (2008). Programming DNA tube circumferences. Science, 321(5890), 824-826.

Chapter 14

Characterization of Some Periodic Tiles by Contour Words

Guy Cousineau

Additional information is available at the end of the chapter

http://dx.doi.org/10.5772/58631

1. Introduction

Beauquier and Nivat [1] showed that aperiodicity cannot appear in a tiling by a polyomino using translations only : any such tiling is at least half-periodic. This is not the case when we associate a tile with one of its reflected (or rotated) images, which is a natural thing to do since, in real world, the same molecule can appear in various rotational positions or have an isomer.

Figure 1 shows a half-periodic tiling which is quite different from those with only one tile and an aperiodic one in which the reflected image appears only once.

Figure 1. a half-periodic tiling and an aperiodic one

Moreover, the local structure of such tilings offers more possibilities : a tile can be surrounded by an arbitrary numbers of other tiles as can be easily inferred from the example in figure 2.

Therefore, a general study of the tiles involved in such tilings appears to be difficult. Moreover, it is worth noting that there is a lack of notations and methods to study the decompositions of tiles and their possible surroundings.

In this paper, we provide simple word techniques to characterize some polygons that tile the plane together with one of their reflective image. We have mainly concentrated on tilings with symmetry **pg** [5]. Within periodic tilings involving glide reflections, this is the most interesting group to consider since it is a subgroup of all symmetry groups involving glide

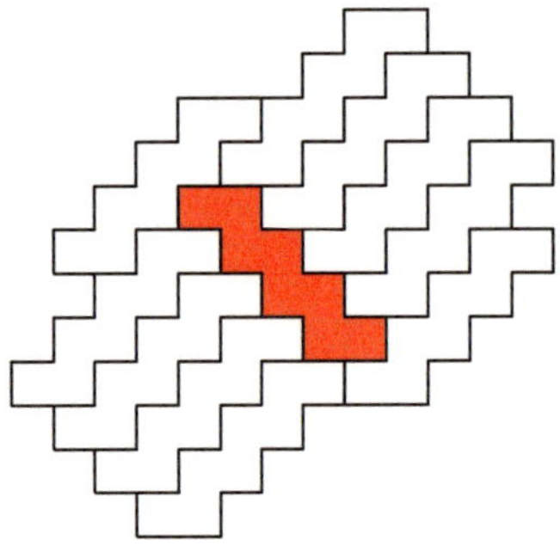

Figure 2. an unusual surrounding

reflections. However, we hope that these techniques will also be useful in the future for investigating more general problems in the spirit of [1].

2. Polygonal tiles

Studies similar to ours are usually placed within the framework of polyominoes. The symmetry group **pg** can indeed be studied in this framework since it involves only one direction of reflexion which can be made for instance horizontal by rotating the axes. However, this is unnecessarily restrictive and we shall use polygons instead. The restriction to polyominoes simplifies the description of tiles by contour words which can be expressed with a simple four letters alphabet (up,down,left,right) whereas the contour word associated with a polygon requires an infinite alphabet (the complex numbers). However, the use of polygons does not make the proofs more complex and allows for more general results.

To be precise, we shall use a notion of "polygonal tile" which in a slight extension of the notion of polygon. A polygonal tile is a sequence of points (vertices) that forms a simple closed line. The edges of a polygonal tile are the segments that join two consecutive vertices. Here, two consecutive edges can have the same direction and this distinguishes polygonal lines from what are usually called polygons. A polygon corresponds to an uncountable infinity of different polygonal lines.

When we consider possible tilings with a given polygonal line, we are only interested in tilings in which vertices of the tilings, i.e. points shared by three tiles or more, correspond to vertices of these tiles and when two tiles are adjacent, the vertices of one tile fit to vertices of the other tile. This restriction is analogous to the restriction to transformations that preserve the "grid" (points with integer coordinates) when one deals with polyominoes. It will enable us to obtain factorizations of the contour words.

2.1. Definitions and notations

We shall represent polygonal tiles by words on the alphabet $\mathbf{C}$ of non null complex numbers. Given a word u in $\mathbf{C}^*$, we denote $|u|$ its length, $\Sigma(u)$ the complex sum of its letters, $R(u)$ and $I(u)$ the real and imaginary parts of $\Sigma(u)$.

http://dx.doi.org/10.5772/58631

The contour word of a polygonal tile is completely defined when a starting point is chosen among its vertices. Otherwise, it is only defined modulo a circular shift of its letters The symbol $\equiv$ will denote equality modulo circular shift and the symbol $=$ will denote identity of words.

Given a word u, we shall denote $(-u)$ the word u in which each letter is replaced by its opposite, $\overline{u}$ the word u in which each letter is replaced by its conjugate, and $\widetilde{u}$ the word u where the order of letters has been reversed.

These three operations commute and are involutive. Therefore, the images of a word u to be considered are exactly

$$u \quad -u \quad \overline{u} \quad \widetilde{u} \quad -\overline{u} \quad -\widetilde{u} \quad \widetilde{\overline{u}} \quad -\widetilde{\overline{u}}$$

2.2. Polygonal tiles, isometries and orientation

The effect of isometries on contour words are the following:

- The translations leave the contour words invariants.
- The rotations by angles θ transform a contour word u into the word obtained by multiplying each letter of u by the complex number with modulus 1 and argument θ .

  In particular, half-turns change u into $(-u)$.
- The reflections according to a horizontal axis change u into $\overline{u}$.
  The other reflections are obtained by applying a rotation to $\overline{u}$.
  In particular, the reflexions according to a vertical axis change u into $-\overline{u}$.
- The glide reflections have the same effect as the associated reflections.

Let us also mention that a change of orientation change u into $-\widetilde{u}$. As a consequence, a contour factor v is centro-symmetric if and only if $v = \widetilde{v}$ (v is a palindrome).

2.3. Periodic tiles

All periodic tilings symmetry groups have as a subgroup **p1**, the group generated by two independent translations. The tiles that can produce tilings with symmetry **p1** are called in [1] pseudo-hexagons and their contour in our notations have the form

$$uvw(-\widetilde{u})(-\widetilde{v})(-\widetilde{w}) \tag{1}$$

We shall use this characterization as a basis to deduce characterizations for other symmetry groups. We shall start with symmetry group **p2** for which we shall give a proof that the Conway criterion [4, 8] is a characterization. Then, we shall proceed to symmetry group **pg**.

Beauquier and Nivat have shown that the form (1) does not only characterize the tiles that tile the plane with symmetry **p1** but more generally the tiles that tile the plane by translations only. Similarly, it might well be the case that the Conway criterion characterizes the tiles that tile the plane by translations only together with their image by half-turn and that the criterion we give in section 4.2 characterizes the tiles that tile the plane by translations only together with their image by reflection. However we have not been able to obtain such a result. This remains, to our knowledge, a conjecture.

2.4. Translation adjacency

A polygonal tile can be made adjacent to a translated image of itself only if its contour has both a factor x and a factor $(-\widetilde{x})$. These two factors are necessarily disjoint because otherwise we have $x = yz$ and $(-\widetilde{x}) = zu$, which implies $(-\widetilde{z})(-\widetilde{y}) = zu$ and finally $(-\widetilde{z}) = z$. Such a word cannot be a factor of a contour word since this implies $\Sigma(z) = 0$.

Therefore, a tile that can be made adjacent to a translated image of itself has the shape $xy(-\widetilde{x})z$ and the translation associated with the superposition of the factors x and $(-\widetilde{x})$, which we shall denote by T_x, is defined by the complex number $\Sigma(y) = -\Sigma(z)$. Such a decomposition $xy(-\widetilde{x})z$ of a contour word will be called a T-decomposition. A T-decomposition will be said exact when x is maximal for translation T_x i.e. when the two following conditions are satisfied:

- y cannot be factorized into $y = y_1 y'(-\widetilde{y_1})$ with $|y_1| \neq 0$
- z cannot be factorized into $z = z_1 z'(-\widetilde{z_1})$ with $|z_1| \neq 0$

A T-decomposition $xy(-\widetilde{x})z$ is therefore exact if and only if the contact of the corresponding tile with its image by translation T_x is exactly $(-\widetilde{x})$.

2.5. Half-turn adjacency

A tile that can be made adjacent to its image by a half-turn must have both a factor x and a factor $\widetilde{x}$. These factors could be disjoint, overlapping or identical. We shall show that the only case we have to consider is the case where x is a palindromic factor equal to $\widetilde{x}$. Such a palindromic factor will be said exact if it is not the center of a bigger palindrome.

To clarify the discussion, let us denote by ht the corresponding half-turn, by A the starting point of factor x, by B its end point, and by A′ and B′ the images of A and B by halft-turn ht.

The case where x and $\widetilde{x}$ are identical correspond to $A' = B$ and $B' = A$. It is not possible to have $A' = A$ and $B' = B$ because a half-turn has only one fixpoint.

The case where the factors x and $\widetilde{x}$ are disjoint correspond to a situation where the order between the four considered points are either A, B, A', B' or A, B, B', A'. However the former case is impossible because, since half-turns are involutive, we have $ht(A') = A$ and $ht(B') = B$. Therefore the image of AB' is $A'B$ a factor strictly contained in AB' which is impossible since half-turns are isometries and preserve the lengthes. We are left with the latter case where the order is A, B, B', A'. Let us denote by w the factor BB'. Since the image by ht of BB' is $B'B$, if ht is a tiling transformation, then w must be a palindrome and $AA' = xw\widetilde{x}$ is a bigger palindrome that includes both x and $\widetilde{x}$ and is its own image under ht.

Now, in the last case where factors x and $\widetilde{x}$ are overlapping, we can eliminate the case where AB is included in $A'B'$ or $A'B'$ included in AB using the length argument and we are left

http://dx.doi.org/10.5772/58631

with the cases A, A', B, B' and A, B', B, A'. Here again, the former can be eliminated by the length argument. In the latter, the factor x is decomposed into $x = uw$ where w corresponds to BB' and must therefore be a palindrome and again x and $\widetilde{x}$ are part of a bigger palindromic factor $uw\widetilde{u}$ which is its own image under ht.

2.6. Reflection adjacency

We are mainly interested here in symmetry **pg** which has only one direction of (glide) reflexion. Also, if we study tilings by a tile and one of its images by reflection, we are also in a situation where one reflection direction is privileged. Therefore, it will be convenient to consider that this unique reflection direction is horizontal. This will enable us to express reflections or glide reflections by applying the complex conjugation operation to its contour word.

A tile that can be made adjacent to its image by reflection or glide reflection must therefore possess both a factor x and a factor $\overline{x}$. These factors can be identical, overlapping or disjoint. When they are identical, x contains only real numbers: it is made of a sequence of horizontal segments. This cannot occur in symmetry **pg** because glide reflexions have no fixpoints.

When the two factors x and $\overline{x}$ overlap, let us denote by u their common factor. We have $x = wu$ and $\overline{x} = uw'$ and thus $uw' = \overline{wu}$. If we are in symmetry **pg**, the transformation that transforms x into $\overline{x}$ is a glide reflection gr which can be decomposed into an horizontal reflection r and an horizontal translation t. The effect of the glide reflection gr on factor x is completely described by the complex number $\Sigma(w)$ which has an imaginary part which correspond to the effect of the reflection r and a real part which corresponds to the effect of the translation t. The complex number $\Sigma(w\overline{w})$, which is a real number, corresponds to applying twice the horizontal translation t.

Let us assume $|\overline{w}| < |u|$. This implies $u = \overline{w}u'$ and $x = w\overline{w}u'$. This implies that x overlaps with its image obtained by applying twice the glide reflection gr or equivalently twice the translation t. But this is impossible in symmetry **pg** because this translation belongs to the symmetry group (as the composition of the glide reflection with itself) and in a periodic tiling, an element of a a tile contour cannot overlap with its image by translation. The only possibility is therefore $|\overline{w}| \geq |\overline{u}|$. In that case, x factorizes into $x = \overline{u}vu$ and $\overline{x}$ is equal to $u\overline{v}\overline{u}$. The tile contour has then a factor $\overline{u}vu\overline{v}\overline{u}$.

Finally, the two factors x and $\overline{x}$ can be disjoint. The tile contour has shape $xy\overline{x}z$ where y may be empty. We can note that the overlapping case is a special case of this last one, where $\overline{u}$ and u play the parts of x and $\overline{x}$.

3. Tiles for symmetry p2

3.1. Presentation of symmetry group p2

The symmetry group **p2** can be generated by three half-turns or by one half-turn and two translations. The latter is more convenient since it leads to a presentation which allows for a characterization of normal forms by a canonical rewriting system.

If we denote the generators by T (the half-turn) and X, Y (the translations), the equations are:

$$XY = YX\ ,\ T^2 = I\ ,\ TXT = X^{-1}\ ,\ TYT = Y^{-1}$$

We give below the associated canonical rewriting system [6] together with an example:

$$\begin{array}{rcl} T^{-1} & \to & T \\ TX^{-1} & \to & XT \\ T^2 & \to & 1 \\ X^{-1}Y^{-1} & \to & Y^{-1}X^{-1} \\ TX & \to & X^{-1}T \end{array} \qquad \begin{array}{rcl} TY^{-1} & \to & YT \\ X^{-1}Y & \to & YX^{-1} \\ XY & \to & YX \\ XY^{-1} & \to & Y^{-1}X \\ TY & \to & Y^{-1}T \end{array}$$

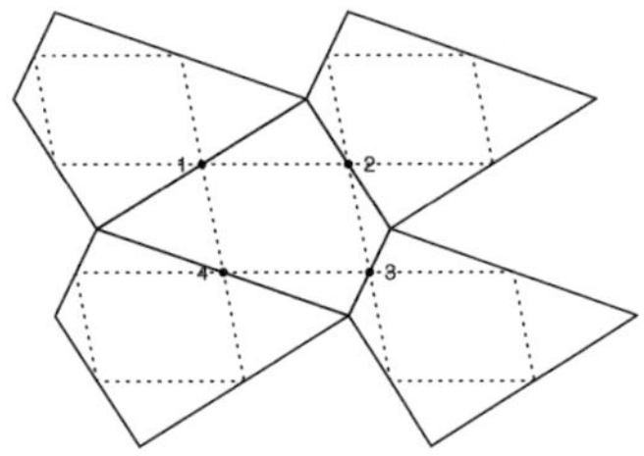

The normal forms are of two sorts: Y^pX^q (translations) and Y^pX^qT (half-turns)

It thus appears that a tile is adequate for a (isohedral) tiling with symmetry **p2** if and only if this tile joined to its image by the half-turn T forms an adequate tile for symmetry group **p1**. This observation explains in a simple way why the Conway criterion is a necessary condition in section 3.3.

3.2. The Conway criterion

A polygonal tile satisfies the Conway criterion if and only if it has the shape $uvw(-\widetilde{u})xy$ where v, w, x and y are palindromes. It is easy to see that if we choose any of its palindromic factors and join it with its half-turn image according to this factor, we obtain a pseudo-hexagon.

Take for instance factor y. We obtain

$$uvw(-\widetilde{u})x(-u)(-v)(-w)\widetilde{u}(-x)$$

or, with a left shift

$$vw(-\widetilde{u})x(-u)(-v)(-w)\widetilde{u}(-x)u$$

Taking $z = (-\widetilde{u})x(-u)$, we have $-z = \widetilde{u}(-x)u$ and the polygon contour appears as:

$$vwz(-v)(-w)(-z)$$

where v, w et z are palindromes.

This is a specific form of pseudo-hexagon which is invariant by half-turn.

3.3. Conway criterion characterizes p2

Reciprocally, let us state the condition for a tile to produce a pseudo-hexagon when joined with its image by half-turn.

First, as mentioned in section 2.5, the tile's contour must have a palindromic factor and therefore have shape xy where y is a palindrome. When the tile is joined to its image by the half-turn, the resulting tile has contour $x(-x)$ and this contour must be that of a pseudo-hexagon. Therefore, there must exist words u, v and w such that:

$$x(-x) \equiv uvw(-\widetilde{u})(-\widetilde{v})(-\widetilde{w})$$

The word $x(-x)$ must thus be identical to a shift of $uvw(-\widetilde{u})(-\widetilde{v})(-\widetilde{w})$.

If $x(-x) = uvw(-\widetilde{u})(-\widetilde{v})(-\widetilde{w})$, then $x = uvw$ and $-x = (-\widetilde{u})(-\widetilde{v})(-\widetilde{w})$, which is equivalent to $x = (\widetilde{u})(\widetilde{v})(\widetilde{w})$. We thus have $uvw = (\widetilde{u})(\widetilde{v})(\widetilde{w})$ which implies $u = \widetilde{u}$, $v = \widetilde{v}$ and $w = \widetilde{w}$. The words u, v et w are palindromes and the primitive tile xy is equal to $uvwy$ where u, v, w and y are palindromes. We are in a special case where the tile is made of four centro-symmetric factors.

If $x(-x) \equiv uvw(-\widetilde{u})(-\widetilde{v})(-\widetilde{w})$ but $x(-x) \neq uvw(-\widetilde{u})(-\widetilde{v})(-\widetilde{w})$, by shifting if necessary the names u, v, w, $\widetilde{u}$, $\widetilde{v}$, $\widetilde{w}$, we can assume without lost of generality that u can be decomposed into $u = u_1u'u_2$ in such a way that $|u_1| = |u_2|$ and either

$$x(-x) = u'u_2vw(-\widetilde{u_2})(-\widetilde{u}')(-\widetilde{u_1})(-\widetilde{v})(-\widetilde{w})u_1$$

or

$$x(-x) = u_2vw(-\widetilde{u_2})(-\widetilde{u}')(-\widetilde{u_1})(-\widetilde{v})(-\widetilde{w})u_1u'$$

The second case differs from the first by the orientation only. Therefore, it is sufficient to deal with the first case.

We have $x = u'u_2vw(-\widetilde{u_2})$ and $-x = (-\widetilde{u}')(-\widetilde{u_1})(-\widetilde{v})(-\widetilde{w})u_1$, which is equivalent to $x = \widetilde{u}'\widetilde{u_1}\widetilde{v}\widetilde{w}(-u_1)$, and thus $u'u_2vw(-\widetilde{u_2}) = \widetilde{u}'\widetilde{u_1}\widetilde{v}\widetilde{w}(-u_1)$.

This implies $u' = \widetilde{u}'$, $u_2 = \widetilde{u_1}$, $v = \widetilde{v}$ et $w = \widetilde{w}$.

The primitive tile's contour can thus be written $u'u_2vw(-\widetilde{u_2})y$ or by shifting it to the left $u_2vw(-\widetilde{u_2})yu'$ where v, w, y and u' are palindromes which corresponds exactly to the Conway criterion.

4. Tiles for symmetry pg

4.1. Presentation of symmetry group pg

The symmetry group **pg** can be generated with two glide reflexions with parallel mirrors and equal associated translations. The presentation has just one equation ${G_1}^2 = {G_2}^2$ where G_1 and G_2 are the two glide reflections.

It can also be generated by a single glide reflection G and a translation X perpendicular to the mirror of G. The presentation still has just one equation which is $XGX = G$. This second presentation is more interesting. We give below the associated canonical rewriting system [6] together with an example:

$$\begin{aligned} GX &\rightarrow X^{-1}G \\ GX^{-1} &\rightarrow XG \\ G^{-1}X &\rightarrow X^{-1}G^{-1} \\ G^{-1}X^{-1} &\rightarrow XG^{-1} \end{aligned}$$

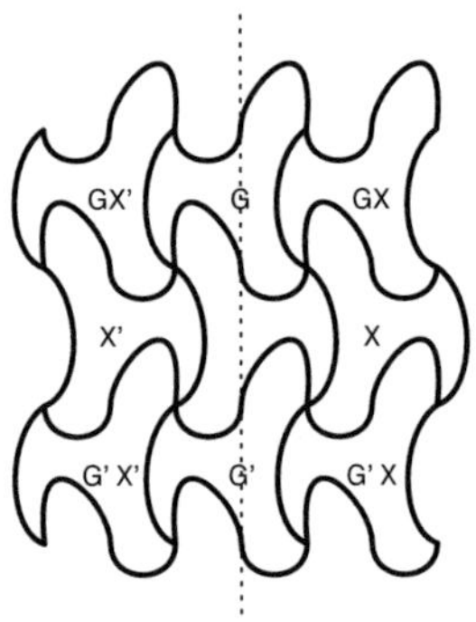

The normal forms are X^pG^{2q} (translations) and X^pG^{2q+1} (glide reflections). Taking $Y = G^2$ (Y is a translation parallel to the mirror of G), the normal forms can be described as X^pY^q (translations) and X^pY^qG (glide reflections).

This permits to see that a tile is adequate for symmetry **pg** if and only if this tile joined to its image by the glide reflexion G forms an adequate tile for symmetry group **p1**.

4.2. Criterion for pg tiles

We are going to show that a tile is adequate for symmetry **pg** if and only if it has one of the shapes

$$u\, v\, w\, \overline{v}\, \overline{u}\, (-\widetilde{w}) \tag{2}$$

$$u\, \overline{u}\, w\, v\, \overline{v}\, (-\widetilde{w}) \tag{3}$$

In both cases, one of the factors may be empty.

Let us first show that we obtain a pseudo-hexagon when we make such a tile adjacent to its reflective image using factor u or v. We shall give the proof for the first form: the second one is similar.

Since the factors u and v play similar parts, we shall use the reflection according to factor u. The reflected tile has contour $\overline{u}\, \overline{v}\, \overline{w}\, v\, u\, (-\widetilde{\overline{w}})$ in the reverse orientation or if we give the right orientation

$$(-\widetilde{\overline{u}})\, \overline{w}\, (-\widetilde{u})\, (-\widetilde{v})\, (-\widetilde{\overline{w}})\, (-\widetilde{\overline{v}})$$

Gluing together the primitive tile and its reflected image using factors $\overline{u}$ and $(-\widetilde{\overline{u}})$, we obtain a polygon with contour

$$u\,v\,w\,\overline{v}\,\overline{w}\,(-\widetilde{u})\,(-\widetilde{v})\,(-\widetilde{\overline{w}})\,(-\widetilde{\overline{v}})\,(-\widetilde{w})$$

Taking $x = w\,\overline{v}\,\overline{w}$ which implies $-\widetilde{x} = (-\widetilde{\overline{w}})\,(-\widetilde{\overline{v}})\,(-\widetilde{w})$, this contour can be written

$$u\,v\,x\,(-\widetilde{u})\,(-\widetilde{v})\,(-\widetilde{x})$$

which is a pseudo-hexagon. The translation X corresponds to the complex number $\Sigma(uv)$ and the translation Y to the complex number $\Sigma(vx) = \Sigma(vw\,\overline{v}\,\overline{w})$

4.3. The criterion characterizes pg tiles

Let us assume that some tile with contour m_1 glued to its reflected image produce a pseudo-hexagon. The conditions for a tile to fit its reflected image have been stated in section 2.6.

In all cases suitable for **pg**, we have

$$m_1 \;=\; x\,y\,\overline{x}\,z \tag{4}$$

and the obtained polygon has contour

$$m_2 \;=\; x\,y\,(-\widetilde{\overline{z}})\,(-\widetilde{x})\,(-\widetilde{\overline{y}})\,z \tag{5}$$

We can also use one more information that comes from the structure of group **pg**. When we compose any glide reflection belonging to the symmetry group **pg** with itself, we obtain a translation that belongs to the symmetry group. Therefore, if in the contour m_1, the factor $\overline{x}$ is the image of the factor x in some glide reflection , in the contour m_2, the factor $(-\widetilde{x})$ is image of the factor x in the corresponding translation. Note that it is written $(-\widetilde{x})$ and not x because it is read in the reverse order in the contour of m_2. However, even if x is an exact factor for the glide reflection, it is not necessarily the case that it is an exact factor for the corresponding translation: we can only assume that x is part of the exact factor for the translation. This distinction will lead to the two possible forms for **pg** tiles.

If we assume that x is an exact factor for the translation then x appears in the pseudo-hexagon decomposition of m_2. Therefore, m_2 can be written as

$$m_2 = x\,u\,v\,(-\widetilde{x})\,(-\widetilde{u})\,(-\widetilde{v}) \tag{6}$$

and therefore, we must have

$$u\,v \;=\; y\,(-\widetilde{\overline{z}}) \tag{7}$$

and

$$(-\widetilde{u})\,(-\widetilde{v}) \;=\; (-\widetilde{\overline{y}})\,z \tag{8}$$

and in the special case where z is empty (the case where y is empty is similar), we get

$$u\,v \;=\; y \tag{9}$$

and

$$(-\widetilde{u})\,(-\widetilde{v}) \;=\; (-\widetilde{\overline{y}}) \tag{10}$$

We shall first deal with this special case. The equation (10) can be rephrased into

$$y \;=\; \overline{v}\,\overline{u} \tag{11}$$

and we have

$$u\,v \;=\; \overline{v}\,\overline{u} \tag{12}$$

This equation has two types of solutions (see lemma 5.1):

- The general solution:
  $u \;=\; (w\overline{w})^p w$ et $v \;=\; \overline{w}(w\overline{w})^q$
  At this point, we have to use the fact that one of the translations which associate $(-\widetilde{u})$ to u or $(-\widetilde{v})$ to v is a vertical translation. This fact implies $R(u) = R(v) = R(x)$ and therefore $p = q$ and $u \;=\; \overline{v}$.
  The tile has shape $m_1 = x\overline{x}(-\widetilde{u})(-\widetilde{\overline{u}})$ or, stating $y \;=\; (-\widetilde{u})$, $m_1 = x\,\overline{x}\,y\,\overline{y}$ (see figure 3).

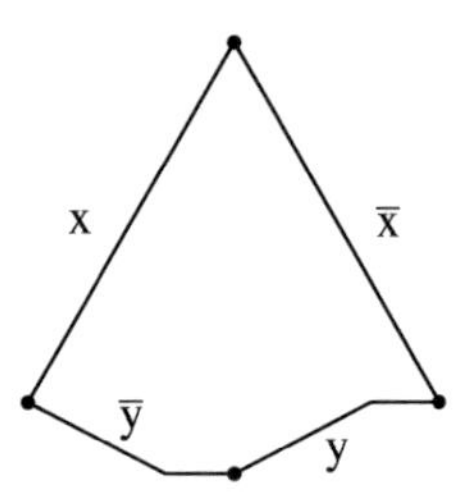

Figure 3. Special case of the first form

- The solution on reals:
  $u = a^p$ et $v = a^q$.
  For the same reason, we must have $p = q$. This is only a subcase of the previous one.

Now we go back to equations (7) and (8). Here again, we shall use the fact that one of the translations corresponding to u and v is a vertical one. Let us assume that the vertical translation is associated with u. Let us also assume that $R(x) \geq 0$ which implies $R(\overline{x}) = R(x) \geq 0$. This is just to ensure that all the vertical projections as they are defined below do exist. The case $R(x) \leq 0$ is similar with x and z having inverted roles.

Let us denote by $1, 2, 3, 4, 5, 6$ the vertices that bound the factors x, y, $(-\widetilde{\overline{z}})$, $(-\widetilde{x})$, $(-\widetilde{\overline{y}})$ and z. Let us also denote $2', 3'$ the images of $2, 3$ in the vertical translation and $5', 6'$ the images of 5 et 6 in its inverse (see figure 4).

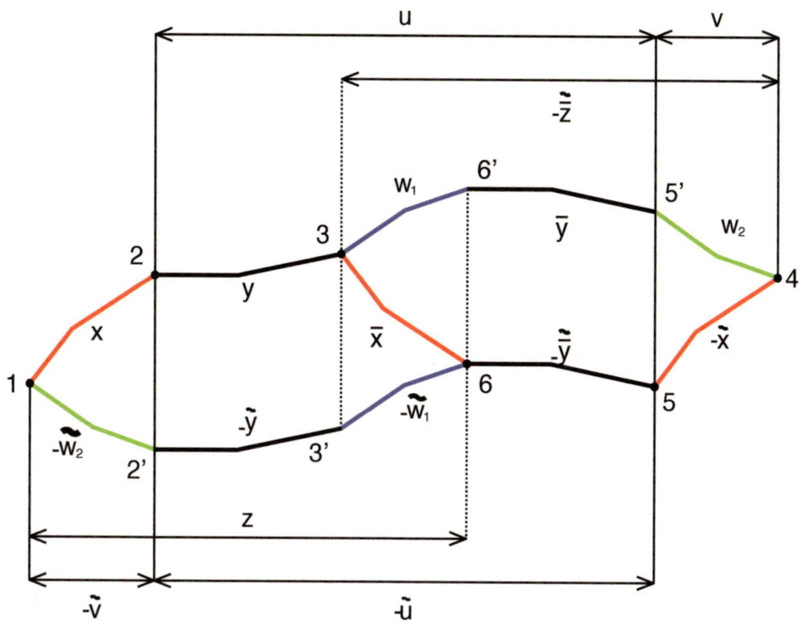

Figure 4. General case 1

The factor u has extremities 2 and $5'$ and the factor v has extremities $5'$ et 4. Let us denote by w_1 the factor with extremities 3 and $6'$ and w_2 the factor with extremities $5'$ and 4. The factor with extremities $3'$ and $2'$ is equal to $(-\widetilde{y})$ and therefore we have

$$(-\widetilde{\overline{z}}) \;=\; w_1\, \overline{y}\, w_2 \tag{13}$$

We also have

$$z \;=\; (-\widetilde{w_1})\,(-\widetilde{y})\,(-\widetilde{w_2}) \tag{14}$$

since the factors $[3,6']$, $[6',5]'$ et $[5',4]$ are images by translation of factors $[3',6]$, $[6,5]$ et $[1,2']$. This last equality can also be written

$$(-\widetilde{\overline{z}}) = \overline{w_2}\, \overline{y}\, \overline{w_1} \tag{15}$$

and we have

$$w_1\, \overline{y}\, w_2 = \overline{w_2}\, \overline{y}\, \overline{w_1} \tag{16}$$

which is an instance of equation

$$x\, y\, z = \overline{z}\, y\, \overline{x}$$

Fortunately, we shall not have to solve this equation in full generality because we can use the fact that $R(w_1) = R(w_2)$ i.e. in the above equation $R(x) = R(z)$. We shall show that in this case $x = \overline{z}$. If this is not the case then either $x = \overline{z}u$ for some u or $z = \overline{x}v$ for some v. These two cases being symmetrical, we shall deal with the first one only. Since x and z have equal real projections, $R(u) = 0$ (u is a pure imaginary number). On the other hand, we have $\overline{z}uyz = \overline{z}yz\overline{u}$ which implies $I(\overline{z}uyz) = I(\overline{z}yz\overline{u})$ which is equivalent to $I(u) = I(\overline{u})$. But this is possible only if $I(u) = 0$ which, since we already know that $R(u) = 0$, implies $\Sigma(u) = 0$ i.e. u would be a closed (looping) factor which cannot exist in a contour word except for the entire contour word. So, we do have $x = \overline{z}$.

So, this shows that in our tile, $w_1 = \overline{w_2}$ and the contour word can finally be written, setting $w = (-\widetilde{w_2})$,

$$w\, x\, y\, \overline{x}\, \overline{w}\, (-\widetilde{y})$$

which is of the required shape (form (2).

Notice that if we do not use the fact that one of the translations is vertical, then the real projections of w_1 and w_2 need not be equal. In that case, we have to consider also solutions such as:

- $w_1 = (w\overline{w})^p w\,,\ (-\widetilde{y}) = (w\overline{w})^n\,,\ w_2 = \overline{w}(w\overline{w})^q$
- $w_1 = (w\overline{w})^p\,,\ (-\widetilde{y}) = (w\overline{w})^n w\,,\ w_2 = (\overline{w}w)^q$

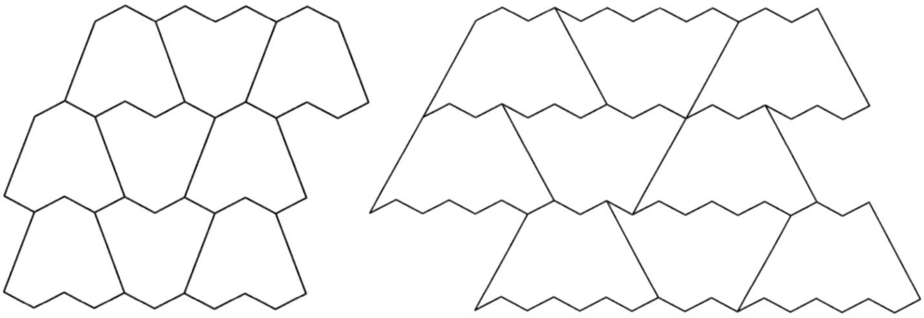

Figure 5. Two tilings with non-**pg** symmetry

These solutions with $p \neq q$ correspond to tiles that produce tilings with other symmetries than **pg** or no symmetry at all as those on figure 5 which are only half-periodic or on figure 1 where aperiodicity appears.

Now, we go back to the case where the factor x of m_2 in equation 5 is not an exact factor of the horizontal translation. This means that x is included inside some translation factor axb where ab is non-empty. We assume that the glide reflection maps x to $\overline{x}$ and x is an exact factor for this glide reflection. If we apply again the glide reflection to $\overline{x}$, we obtain the image of x in the horizontal translation which has to belong to the part of m_2 which is not in m_1 because, otherwise, m_1 would overlap with its image in the horizontal translation.

Let us show that if a and b are both non-empty, they cannot both belong to the contour of m_1. If a is non-empty, its image in the glide reflection does not belong to the contour of m_1 because otherwise, x would not be an exact factor. But then, for the same reason, the image of a under the horizontal translation cannot belong to the contour of m_2 outside m_1, so it has to belong to m_1. The situation is the same for b. Consequently, if we consider the images of x, a, b are such that the fist one should be outside m_1 and the two others inside m_1. But then we would have to assume that the image of x in the horizontal translation is all the part of m_2 that is outside m_1 and m_1 would have to be equal to $x\overline{x}$ and a and b would both be empty.

So, one of the factors a or b, if they are non-empty have to be outside the contour of m_1 and therefore belong to the part of m_2 that is outside m_1. Let us say, it is b. This implies that, in the contour of m_1, the two factors x and $\overline{x}$ are adjacent and b has to be the image of a in the glide reflection as shown on figure 6.

$$m_1 = a\,x\,\overline{x}\,(-\widetilde{a})z \tag{17}$$

and

$$m_2 = a\,x\,(-\widetilde{\overline{a}})\,(-\widetilde{\overline{z}})\,\overline{a}\,(-\widetilde{x})\,(-\widetilde{a})\,z \tag{18}$$

We know that the factor $a\ x\ (-\widetilde{\overline{a}})$ is exact for the horizontal translation. So, we can make explicit the fact that m_2 is a pseudo-hexagon by factorizing $(-\widetilde{\overline{z}})$ into $(-\widetilde{\overline{z}}) = uv$ with $z =$

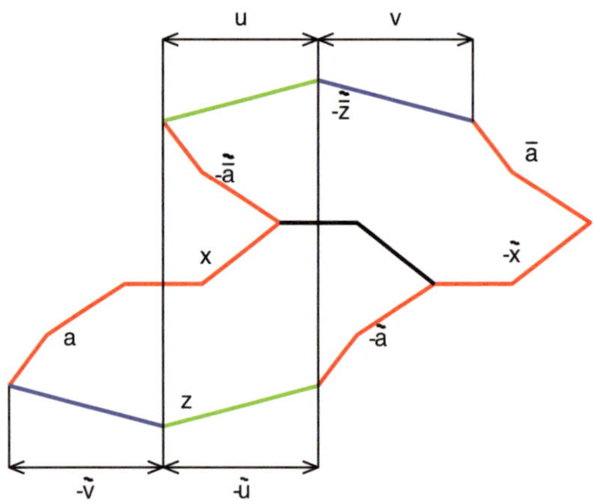

Figure 6. General case 2

$(-\widetilde{u})(-\widetilde{v})$. The two definitions of z lead us to the equation

$$u\,v \;=\; \overline{v}\,\overline{u} \tag{19}$$

and we use the fact that one of the translations associated to factors u and v is vertical to deduce that $R(u) = R(v)$. Therefore $u \;=\; \overline{v}$ and we can conclude that

$$m_1 \;=\; a\,x\,\overline{x}\,(-\widetilde{a})(-\widetilde{u})(-\widetilde{\overline{u}}) \tag{20}$$

which corresponds to the form 3.

This completes our proof. It is clear from the examples of figure 5 that the main result presented here is only a small step toward the characterization of polygons that tile the plane together with one of their reflected images because our proofs are highly dependent on the specificities of symmetry **pg**.. A plausible conjecture is that a polygon has this property if and only if, suitably oriented, its contour has a factorization of form 2 or 3. But the contour word techniques that we have used here will certainly require further development in order to tackle with such a conjecture.

5. Technical lemmas

Solutions of the equation $x \equiv \overline{x}$

This section studies the solutions of the equation $x \equiv \overline{x}$. This equation where the symbol $\equiv$ denotes equality modulo circular shift corresponds to two equations using equality on words.

1. $x \;=\; \overline{x}$
   In that case, each letter of x must be its own conjugate. The word x is made uniquely of reals.

2. $x = u\,v$ for two non empty words u et v and $u\,v = \overline{v}\,\overline{u}$.
   This case is treated by lemma 5.1.

To study such equations, the alphabet of complex numbers divides naturally between reals that are their own conjugate and non reals which have a conjugate distinct from themselves. The lemmas presented below are very similar to those presented in [7] for ordinary equations on words but are slightly more complex due to the partition of the alphabet. The equations will have two kinds of solutions: solutions on R^* which are the solutions of the considered equations where we drop out the conjugation operations and the general solutions.

Lemme 5.1. *The solutions of equation* $u\,v = \overline{v}\,\overline{u}$ *are*

$$u = w^m \;,\; v = w^n$$

where $w \in R^*$ *and*

$$u = (w\overline{w})^m w \;,\; v = \overline{w}(w\overline{w})^n$$

where $w \in C^*$.

To prove this lemma, we shall have to prove also the following one:

Lemme 5.2. *The solutions of equation* $u\,v = \overline{v}\,u$ *are*

$$u = w^m \;,\; v = w^n$$

where $w \in R^*$ *and*

$$u = (w\overline{w})^m w \;,\; v = (\overline{w}w)^n$$

The proofs of these two lemmas can be done simultaneously by induction on $|uv|$. We shall assume $|uv| \geq 2$. The base case will correspond to $|uv| = 2$ and $|uv| = 3$.

- If $|uv| = 2$, u and v are letters and $u = \overline{v}$ in the three cases. This leads to solutions $u = \overline{u} = v = \overline{v}$ on reals and $u = \overline{v}$ on non real complex numbers for lemma 5.1. For lemma 5.2, there is no solution on non real complex numbers.
- If $|uv| = 3$, we have either $u = ab$ et $v = c$ or $u = a,\ v = bc$ where a, b,c are letters. For lemma 5.1 , the two hypotheses lead to $a = \overline{a} = b = \overline{b} = c = \overline{c}$ and we have only the solutions $u = a, v = aa$ et $u = aa, v = a$ where a is a real letter.
  For lemma 5.2 , the hypothesis $u = ab$ et $v = c$ lead also to $a = \overline{a} = b = \overline{b} = c = \overline{c}$ and we have only the two real solutions. On the contrary, the hypothesis $u = a,\ v = bc$ leads to $a = \overline{b} = c$ and we have the solution $u = a,\ v = \overline{a}a$.
- In the inductive case , let us assume that the two lemmas are true when $|uv| \leq n$ and consider two words u et v
  such that $|uv| = n + 1$.
  - Lemme 5.1: $u\,v = \overline{v}\,\overline{u}$. We compare the lengthes of u et v.

  * If $|u| = |v|$, then $u = \overline{v}$.
  * If $|u| > |v|$, take $u = \overline{v}u_1$. We then also have $\overline{u} = u_1 v$ hence $u = \overline{u_1 v}$ and thus $\overline{u_1 v} = \overline{v}u_1$ or $u_1 v = v\overline{u_1}$. By induction hypothesis, the solutions of this equation are $u_1 = w^m$, $v = w^n$ where $w \in R$ which gives $u = w^{m+n}, v = w^n$ and $u_1 = (w\overline{w})^m$, $v = w\,(\overline{w}w)^n$. In that case $u = \overline{v}u_1 = (\overline{w}w)^{m+n}\overline{w}$ and $v = w\,(\overline{w}w)^n$.
  * If $|u| < |v|$, taking $\overline{v} = uv_1$, we also have $v = v_1\overline{u}$ hence $uv_1 = \overline{v_1}u$. By induction hypothesis, the solutions of this equation are $u = w^m$, $v_1 = w^n$ where $w \in R$ which gives $u = w^m, v = w^{m+n}$ and $u = (w\overline{w})^m w$, $v_1 = (\overline{w}w)^n$, which gives $v = (\overline{w}w)^{m+n}\overline{w}$.
- Lemme 5.2: $u\,v = \overline{v}\,u$. We compare again the lengthes.
  * If $|u| = |v|$, then $u = \overline{v} = v$. (real solutions)
  * If $|u| > |v|$, take $u = \overline{v}u_1$. We also have $u = u_1 v$ hence $u_1 v = \overline{v}u_1$. By induction hypothesis, the solutions of this equation are $u_1 = w^m$, $v = w^n$ where $w \in R$ which gives $u = w^{m+n}, v = w^n$ and $u_1 = (w\overline{w})^m w$, $v = (\overline{w}w)^n$, which gives $u = (w\overline{w})^{m+n} w$ and $v = (\overline{w}w)^n$.
  * If $|u| < |v|$, take $\overline{v} = uv_1$. Here again we have $v = v_1 u$, which implies $uv_1 = \overline{v_1 u}$. By induction hypothesis, the solutions of this equation are $u = w^m$, $v_1 = w^n$ where $w \in R$ which gives $u = w^m, v = w^{m+n}$ and $u = (w\overline{w})^m w$, $v_1 = \overline{w}(w\overline{w})^n$, which gives $v = (\overline{w}w)^{m+n+1}$.

The author thanks Luc Boasson, Maurice Nivat and Laurent Vuillon for their help and support in preparing this paper and also the referee whose remarks and questions were extremely useful.

Author details

Guy Cousineau

Fédération de Recherche, CNRS-Université de Savoie
Modélisation, Simulation, Interactions Fondamentales, France

References

[1] D. Beauquier and M Nivat. On translating one polyomino to tile the plane. *Discrete and Computational Geometry*, 6:575–592, 1991.

[2] H.S.M. Coxeter. *Introduction to geometry.* John Wiley and sons, 1980.

[3] H.S.M. Coxeter and W.O.J. Mauser. *Generators and relations for discrete groups.* Springer Verlag, 1972.

[4] M. Gardner. More about tiling the plane: the possibilities of polyominoes, polyamonds and polyhexes. *Scientific American*, 53:112–115, 1975.

[5] B. Grünbaum and G.C. Shephard. *Tilings and patterns.* Wiley and sons, 1987.

http://dx.doi.org/10.5772/58631

[6] Ph. Lechenadec. *Canonical forms in finitely presented algebras*. Pitman, 1986.

[7] M. Lothaire. *Combinatorics on words*. Cambridge University Press, 1997.

[8] D. Schattschneider. Will it tile? try the conway criterion! *Mathematics Magazine*, 53:224–233, 1980.

[9] H.A.G. Vijshoff and J. Van Leeuwen. Arbitrary versus periodic storage schemes and tesselations of the plane using one type of polyomino. *Inform. Control*, 62:1–25, 1984.